Principles of Optimal Design

Second Edition

Principles of Optimal Design puts the concept of optimal design on a rigorous foundation and demonstrates the intimate relationship between the mathematical model that describes a design and the solution methods that optimize it. Since the first edition was published, computers have become ever more powerful, design engineers are tackling more complex systems, and the term "optimization" is now routinely used to denote a design process with increased speed and quality. This second edition takes account of these developments and brings the original text thoroughly up to date. The book now includes a discussion of trust region and convex approximation algorithms. A new chapter focuses on how to construct optimal design models. Three new case studies illustrate the creation of optimization models. The final chapter on optimization practice has been expanded to include computation of derivatives, interpretation of algorithmic results, and selection of algorithms and software. Both students and practicing engineers will find this book a valuable resource for design project work.

Panos Papalambros is the Donald C. Graham Professor of Engineering at the University of Michigan, Ann Arbor.

Douglass J. Wilde is Professor of Design, Emeritus, at Stanford University.

Principles of Optimal Design

Modeling and Computation

SECOND EDITION

PANOS Y. PAPALAMBROS
University of Michigan

DOUGLASS J. WILDE
Stanford University

PUBLISHED BY THE PRESS SYNDICATE OF THE UNIVERSITY OF CAMBRIDGE
The Pitt Building, Trumpington Street, Cambridge, United Kingdom

CAMBRIDGE UNIVERSITY PRESS
The Edinburgh Building, Cambridge CB2 2RU, UK http://www.cup.cam.ac.uk
40 West 20th Street, New York, NY 10011-4211, USA http://www.cup.org
10 Stamford Road, Oakleigh, Melbourne 3166, Australia
Ruiz de Alarcón 13, 28014 Madrid, Spain

First published 2000

Typefaces Times Roman 10.75/13.5 pt. and Univers *System* LaTeX 2_ε [TB]

A catalog record for this book is available from the British Library.

Library of Congress Cataloging in Publication Data

Papalambros, Panos Y.

Principles of optimal design : modeling and computation / Panos Y. Papalambros, Douglass J. Wilde. – 2nd ed.

p. cm.

Includes bibliographical references.

ISBN 0-521-62215-8

1. Mathematical optimization. 2. Mathematical models. I. Wilde, Douglass J. II. Title.

QA402.5.P374 2000
519.3 – dc21

99-047982

ISBN 0 521 62215 8 hardback
ISBN 0 521 62727 3 paperback

Transferred to digital printing 2003

To our families

And thus both here and in that journey of a thousand years,
whereof I have told you, we shall fare well.

Plato (*The Republic*, *Book X*)

Contents

Preface to the Second Edition

A dozen years have passed since this book was first published, and computers are becoming ever more powerful, design engineers are tackling ever more complex systems, and the term "optimization" is routinely used to denote a desire for ever increasing speed and quality of the design process. This book was born out of our own desire to put the concept of "optimal design" on a firm, rigorous foundation and to demonstrate the intimate relationship between the mathematical model that describes a design and the solution methods that optimize it.

A basic premise of the first edition was that a good model can make optimization almost trivial, whereas a bad one can make correct optimization difficult or impossible. This is even more true today. New software tools for computer aided engineering (CAE) provide capabilities for intricate analysis of many difficult performance aspects of a system. These analysis models, often referred to also as simulations, can be coupled with numerical optimization software to generate better designs iteratively. Both the CAE and the optimization software tools have dramatically increased in sophistication, and design engineers are called to design highly complex problems, with few, if any, hardware prototypes.

The success of such attempts depends strongly on how well the design problem has been formulated for an optimization study, and on how familiar the designer is with the workings and pitfalls of iterative optimization techniques. Raw computing power is unlikely to ease this burden of knowledge. No matter how powerful computers are or will be, we will always pose relatively mundane optimal design problems that will exceed computing ability. Hence, the basic premise of this book remains a "modern" one: There is need for a more than casual understanding of the interactions between modeling and solution strategies in optimal design.

This book grew out of graduate engineering design courses developed and taught at Michigan and Stanford for more than two decades. Definition of new concepts and rigorous proof of principles are followed by immediate application to simple examples. In our courses a term design project has been an integral part of the experience, and so the book attempts to support that goal, namely, to offer an integrated procedure of design optimization where global analysis and local interative methods complement each other in a natural way.

A continuous challenge for the second edition has been to keep a reasonable length without ignoring the many new developments in optimization theory and practice. A decision was made to limit the type of algorithms presented to those based on gradient information and to introduce them with a condensed but rigorous version of classical differential optimization theory. Thus the link between models and solutions could be thoroughly shown. In the second edition we have added a discussion of trust region and convex approximation algorithms that remain popular for certain classes of design problems.

On the modeling side we have added a new chapter that focuses exclusively on how to construct optimal design models. We have expanded the discussion on data-driven models to include neural nets and kriging, and we added three complete modeling case studies that illustrate the creation of optimization models. The theory of boundedness and monotonicity analysis has been updated to reflect improvements offered by several researchers since the first edition.

Although we left out a discussion of nongradient and stochastic methods, such as genetic algorithms and simulated annealing, we did include a new discussion on problems with discrete variables. This is presented in a natural way by exploring how the principles of monotonicity analysis are affected by the presence of discreteness. This material is based on the dissertation of Len Pomrehn.

The final chapter on optimization practice has been expanded to include computation of derivatives, interpretation of algorithmic results, and selection of algorithms and software. This chapter, along with the revisions of the previous ones, has been motivated by an effort to make the book more useful for design project work, whether in the classroom or in the workplace.

The book contains much more material than what could be used to spend three lecture hours a week for one semester. Any course that requires an optimal design project should include Chapters 1, 2, and 8. Placing more emphasis on global modeling would include material from Chapters 3 and 6, while placing more emphasis on iterative methods would include material from Chapters 4, 5, and 7. Linear programming is included in the chapter on boundary optima, as a special case of boundary-tracking, active set strategy algorithms, thus avoiding the overhead of the specialized terminology traditionally associated with the subject.

Some instructors may wish to have their students actually code a simple optimization algorithm. We have typically chosen to let students use existing optimization codes and concentrate on the mathematical model, while studying the theory behind the algorithms. Such decisions depend often on the availability and content of other optimization courses at a given institution, which may augment the course offered using this book as a text. Increased student familiarity with high-level, general purpose, computational tools and symbolic mathematics will continue to affect instructional strategies.

Specialized design optimization topics, such as structural optimization and optimal control, are beyond the scope of this book. However, the ideas developed here are useful in understanding the specialized approaches needed for the solution of these problems.

The book was also designed with self-study in mind. A design engineer would require a brush-up of introductory calculus and linear algebra before making good use of this book. Then starting with the first two chapters and the checklist in Chapter 8, one can model a problem and proceed toward numerical solution using commercial optimization software. After getting (or *not* getting) some initial results, one can go to Chapter 8 and start reading about what may go wrong. Understanding the material in Chapter 8 would require selective backtracking to the main chapters on modeling (Chapters 3 and 6) and on the foundations of gradient-based algorithms (Chapters 4, 5, and 7). In a way, this book aims at making "black box" optimization codes less "black" and giving a stronger sense of control to the design engineers who use them.

The book's engineering flavor should not discourage its study by operations analysts, economists, and other optimization theorists. Monotonicity and boundedness analysis in particular have many potential applications for operations problems, not just to the design examples developed here for engineers. We offer our approach to design as a paradigm for studying and solving any decision problem.

Many colleagues and students have reviewed or studied parts of the manuscript and offered valuable comments. We are particularly grateful to all of the Michigan students who found various errors in the first edition and to those who used the manuscript of the second edition as class notes and provided substantial input. We especially acknowledge the comments of the following individuals: Suresh Ananthasuresh, Timothy Athan, Jaime Camelio, Ryan Fellini, Panayiotis Georgiopoulos, Ignacio Grossmann, David Hoeltzel, Tomoki Ichikawa, Tao Jiang, Roy Johanson, John D. Jones, Hyung Min Kim, Justin King, Ramprasad Krishnamachari, Horng-Huei Kuo, Zhifang Li, Arnold Lumsdaine, Christopher Milkie, Farrokh Mistree, Nestor Michelena, Sigurd Nelson, Shinji Nishiwaki, Matt Parkinson, Leonard Pomrehn, Julie Reyer, Mark Reuber, Michael Sasena, Klaus Schittkowski, Vincent Skwarek, Nathaniel Stott, and Irfan Ullah. Special thanks are due to Zhifang Li for verifying many numerical examples and for proofreading the final text.

The material on neural networks and automatic differentiation is based on guest lectures prepared for the Michigan course by Sigurd Nelson. The material on trust regions is also a contribution by Sigurd Nelson based on his dissertation. Len Pomrehn contributed the second part of Chapter 6 dealing with discrete variables, abstracting some of his dissertation's research results. The original first edition manuscript was expertly reworked by Nancy Foster of Ann Arbor.

The second edition undertaking would not have been completed without the unfailing faith of our editor, Florence Padgett, to whom we are indebted. Finally, special appreciation goes to our families for their endurance through yet another long endeavor, whose significance it was often hard to elaborate.

P. Y. P
D. J. W
January 2000

Notation

Integrating different approaches with different traditions brings typical notation difficulties. While one wishes for a uniform and consistent notation throughout, tradition and practice force us to use the same symbol with different meanings, or different symbols with the same meanings, depending on the subject treated. This is particularly important in an introductory book that encourages excursions to other specialized texts. In this book we tried to use the notation that appears most common for the subject matter in each chapter–particularly for those chapters that lead to further study from other texts. Recognizing this additional burden on comprehension, we list below symbols that are typically used in more than one section. The meanings given are the most commonly used in the text but are not exclusive. The engineering examples throughout may employ many of these symbols in the specialized way of the particular discipline of the example. These symbols are not included in the list below; they are given in the section containing the relevant examples. All symbols are defined the first time they occur.

A general notation practice used in this text for mathematical theory and examples is as follows. Lowercase bold letters indicate vectors, uppercase bold letters (usually Latin) indicate matrices, while uppercase script letters represent sets. Lowercase italic letters from the beginning of the alphabet (e.g., a, b, c) often are used for parameters, while from the end of the alphabet (e.g., u, v, x, y, z) frequently indicate variables. Lowercase italic letters from the middle of the alphabet (e.g., i, j, k, l, m, n, p, q) are typically used as indices, subscripts, or superscripts. Lowercase Greek letters from the beginning of the alphabet (e.g., α, β, γ) are often used as exponents. In engineering examples, when convenient, uppercase italic (but not bold) letters represent parameters, and lowercase stand for design variables.

List of Symbols

$\mathbf{A}$	coefficient matrix of linear constraints
$\mathcal{A}$	working set (in active set strategies)

$\mathbf{b}$	right-hand side coefficient vector of linear constraints
$\mathbf{B}$	(1) quasi-Newton approximation to the inverse of the Hessian; (2) "bordered" Hessian of the Lagrangian
$B(x)$	barrier function (in penalty transformations)
$\mathbf{d}$	decision variables
$\mathbf{D}$	(1) diagonal matrix; (2) inverse of coefficient matrix $\mathbf{A}$ (in linear programming)
$\mathcal{D}_i$	feasible domain of all inequality constraints except the ith
$\det(\mathbf{A})$	determinant of $\mathbf{A}$
$\mathbf{e}$	(1) unit vector; (2) error vector
$f(\mathbf{x})$	objective function to be minimized wrt $\mathbf{x}$
$f(x^+)$	function increasing wrt x
$f(x^-)$	function decreasing wrt x
$f^n(x)$	nth derivative of $f(x)$
$\partial f/\partial x_i$	first partial derivative of $f(\mathbf{x})$ wrt x_i
$\partial^2 f/\partial \mathbf{x}^2$, $f_{\mathbf{xx}}$, $\nabla^2 f$	Hessian matrix of $f(\mathbf{x})$; its element $\partial^2 f/\partial x_i \partial x_j$ is ith row and jth column (other symbol: $\mathbf{H}$)
$\partial f/\partial \mathbf{x}$, $f_{\mathbf{x}}$, ∇f	gradient vector of $f(x)$ – a *row* vector (other symbol: $\mathbf{g}^T$)
$\partial \mathbf{f}/\partial \mathbf{x}$, $\nabla \mathbf{f}$	Jacobian matrix of $\mathbf{f}$ wrt $\mathbf{x}$; it is $m \times n$, if $\mathbf{f}$ is an m-vector and $\mathbf{x}$ is an n-vector (other symbol: $\mathbf{J}$)
$\mathcal{F}$	feasible set (other symbol: $\mathcal{X}$)
g_j, $g_j(\mathbf{x})$	jth inequality constraint function usually written in negative null form
$\mathbf{g}(\mathbf{x})$	(1) vector of inequality constraint functions; (2) the transpose of the gradient of the objective function: $\mathbf{g} = \nabla f^T$, a *column* vector
g	greatest lower bound of $f(x)$
$\partial \mathbf{g}/\partial \mathbf{x}$, $\nabla \mathbf{g}$	Jacobian matrix of inequality constraints $\mathbf{g}(\mathbf{x})$
$\partial^2 \mathbf{g}/\partial \mathbf{x}^2$	column vector of Hessians of $\mathbf{g}(\mathbf{x})$; see $\partial^2 \mathbf{y}/\partial \mathbf{x}^2$
h	step size in finite differencing
h_j, $h_j(\mathbf{x})$	jth equality constraint function
$\mathbf{h}(\mathbf{x})$	vector of equality constraint functions
$\partial \mathbf{h}/\partial \mathbf{x}$, $\nabla \mathbf{h}$	Jacobian of equality constraints $\mathbf{h}(\mathbf{x})$
$\partial^2 \mathbf{h}/\partial \mathbf{x}^2$, $\mathbf{h}_{\mathbf{xx}}$	column vector of Hessians of $\mathbf{h}(\mathbf{x})$; see $\partial^2 \mathbf{y}/\partial \mathbf{x}^2$
$\mathbf{H}$	Hessian matrix of the objective function f
$\mathbf{I}$	identity matrix
$\mathbf{J}$	Jacobian matrix
k	(subscript only) denotes values at kth iteration
$\mathcal{K}_i$	constraint set defined by ith constraint
l	lower bound of $f(x)$
$l(x)$	lower bounding function
L	Lagrangian function

$L_{\mathbf{xx}}$	Hessian of the Lagrangian wrt $\mathbf{x}$
$\mathbf{L}$	lower triangular matrix
$\mathbf{LDL}^T$	Cholesky factorization of a matrix
$\mathcal{L}_i$	index set of conditionally critical constraints bounding x_i from below
$\mathbf{M}, \mathbf{M_k}$	a "metric" matrix, i.e., a symmetric positive definite replacement of the Hessian in local iterations
n	number of design variables
$N(0, \sigma^2)$	normal distribution with standard deviation σ
$\mathcal{N}(x)$	normal subspace (hyperplane) of constraint surface defined by equalities and/or inequalities
$\mathcal{N}$	set of nonnegative real numbers including infinity
$\mathbf{P}$	projection matrix
$P(x)$	penalty function (in penalty transformation)
$\mathcal{P}$	set of positive finite real numbers
$q(\mathbf{x})$	quadratic function of $\mathbf{x}$
$r, \mathbf{r}$	controlling parameters in penalty transformations
R	rank of Jacobian of tight constraints in a case
$\mathcal{R}^n$	n-dimensional Euclidean (real) space
$\mathbf{s}$	(1) state or solution variables; (2) search direction vectors (s_k at kth iteration)
$\mathcal{T}(\mathbf{x})$	tangent subspace (hyperplane) of the constraint surface defined by equalities and/or inequalities
$T(\mathbf{x}, \mathbf{r})$	penalty transformation
$T(\mathbf{x}, \lambda, \mathbf{r})$	augmented Lagrangian function (a penalty transformation)
$\mathcal{U}_i$	index set of conditionally critical constraints bounding x_i from above
$x(x_i)$	(ith) design variable
x_L	lower bound on x
x_U	upper bound on x
$\mathbf{x}$	vector of design variables, a point in $\mathcal{R}^n$; $\mathbf{x} = (x_1, x_2, \ldots, x_n)^T$
$\mathbf{x}_0, \mathbf{x}_1, \ldots$	vectors corresponding to points 0, 1, ... – not to be confused with the components $x_0, x_1, \ldots$
$x_i^{(j)}$	ith component of vector $\mathbf{x}_j$ – not used very often
$x_{i,k}$	ith component of vector $\mathbf{x}_k$(k is iteration number)
∂x_i	ith element of $\partial\mathbf{x}$, equals $x_i - x_i^{(0)}$
$\partial\mathbf{x}$	perturbation vector about point $\mathbf{x}_0$, equals $\mathbf{x} - \mathbf{x}_0$; subscript 0 is dropped for simplicity
$\partial\mathbf{x}_k$	perturbation vector about $\mathbf{x}_k$, equals $\mathbf{x}_{k+1} - \mathbf{x}_k$
$\underline{x}(\overline{x})$	argument of the infinum (supremum) of the problem over $\mathcal{P}$

$\underline{x}_i$	argument of the partial minimum (i.e., the minimizer) of the objective wrt x_i
$\mathbf{X}_i$	an $n-1$ vector made from $\mathbf{x} = (x_1, \ldots, x_n)^T$ with all components fixed except x_i; we write $\mathbf{x} = (x_i; \mathbf{X}_i)$
$\underline{\mathbf{x}}$	minimizer to a relaxed problem
X	a subset of $\mathcal{R}^n$ to which $\mathbf{x}$ belongs; the feasible domain; the set constraint
$\underline{X}$	set of $\underline{\mathbf{x}}$
$\underline{X}_i$	set of minimizers to a problem with the ith constraint relaxed
$\mathcal{X}_*$	set of all minimizers in a problem
$\partial^2\mathbf{y}/\partial\mathbf{x}^2$	a vector of Hessians $\partial^2 y_i/\partial\mathbf{x}^2$, $i = 1, \ldots, m$, of a vector function $\mathbf{y} = (y_1, \ldots, y_m)^T$; it equals $(\partial^2 y_1/\partial\mathbf{x}^2, \partial^2 y_2/\partial\mathbf{x}^2, \ldots, \partial^2 y_m/\partial\mathbf{x}^2)$
$z(\mathbf{d})$	reduced objective function, equals f as a function of $\mathbf{d}$ only
$\partial z/\partial\mathbf{d}$	reduced gradient of f
$\partial^2 z/\partial\mathbf{d}^2$	reduced Hessian of f
$(\partial z/\partial\mathbf{h})_*$	sensitivity coefficient wrt equality constraints at the optimum
$\alpha(\alpha_k)$	(kth iteration) step length in line search
δ	a small positive quantity
ε	a small positive quantity – often used in termination criteria
$\lambda_{\min}, \lambda_{\max}$	smallest and largest eigenvalues of the Hessian of f at x_*
$\boldsymbol{\lambda}$	Lagrange multiplier vector associated with equality constraints
μ_k	parameter in modification of $\mathbf{H}_k$ in $\mathbf{M}_k$
$\boldsymbol{\mu}$	Lagrange multiplier vector associated with inequality constraints
$o(x)$	order *higher* than x; it implies terms negligible compared to x
φ	line search function, including merit function in sequential quadratic programming; trust region function
ω_i	weights

Special Symbols

$\leq, \geq$	inequality (active or inactive)
$=$	equality (active or inactive)
$<, >$	inactive inequality
$\geqq, \leqq$	active or critical inequality
$\not\geqq, \not\leqq$	uncritical inequality constraint
$\equiv$	active equality constraint
$\equiv<, \equiv>$	active directed equality
$\|\cdot\|$	norm; a Euclidean one is assumed unless otherwise stated
∂x	perturbation in the quantity x, i.e., a small (differential) change in x

∇f	gradient of f (a row vector)
$\nabla^2 f$	Hessian of f (a symmetric matrix)
$\sum_{i=1}^{n} x_i$	sum over $i; i = 1, 2, \ldots, n (= x_1 + x_2 + \cdots x_n)$
$\prod_{i=1}^{n} x_i$	product over $i; i = 1, 2, \ldots, n (= x_1 x_2 \ldots x_n)$
$\arg\min f(x)$	the value of x (argument) that minimizes f
†	(subscript only) denotes values of quantities at stationary points
*	(subscript only) denotes values of quantities at minimizing point(s)
T	(superscript only) transpose of a vector or matrix
$\triangleq$	definition
$\subset, \subseteq$	subset of
$\in$	belongs

1

Optimization Models

For the goal is not the last, but the best.
Aristotle (Second Book of Physics) (384–322 B.C.)

Designing is a complex human process that has resisted comprehensive description and understanding. All artifacts surrounding us are the results of designing. Creating these artifacts involves making a great many decisions, which suggests that designing can be viewed as a *decision-making process*. In the decision-making paradigm of the design process we examine the intended artifact in order to identify possible alternatives and select the most suitable one. An abstract description of the artifact using mathematical expressions of relevant natural laws, experience, and geometry is the *mathematical model* of the artifact. This mathematical model may contain many alternative designs, and so criteria for comparing these alternatives must be introduced in the model. Within the limitations of such a model, the best, or *optimum*, design can be identified with the aid of mathematical methods.

In this first chapter we define the design optimization problem and describe most of the properties and issues that occupy the rest of the book. We outline the limitations of our approach and caution that an "optimum" design should be perceived as such only within the scope of the mathematical model describing it and the inevitable subjective judgment of the modeler.

1.1 Mathematical Modeling

Although this book is concerned with *design*, almost all the concepts and results described can be generalized by replacing the word *design* by the word *system*. We will then start with discussing mathematical models for general systems.

The System Concept

A system may be defined as a collection of entities that perform a specified set of tasks. For example, an automobile is a system that transports passengers. It follows that a system performs a *function*, or process, which results in an *output*. It is implicit that a system operates under causality, that is, the specified set of tasks is performed because of some stimulation, or *input*. A *block diagram*, Figure 1.1, is

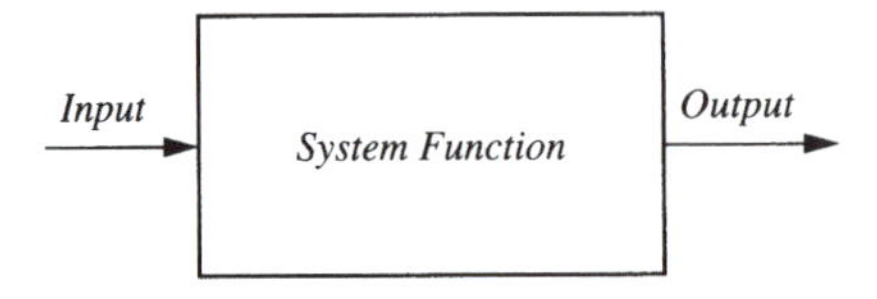

Figure 1.1. Block diagram representation.

a simple representation of these system elements. Causality generally implies that a dynamic behavior is possible. Thus, inputs to a system are entities identified to have an observable effect on the behavior of the system, while outputs are entities measuring the response of the system.

Although inputs are clearly part of the system characterization, what exactly constitutes an input or output depends on the *viewpoint* from which one observes the system. For example, an automobile can be viewed differently by an automaker's manager, a union member, or a consumer, as in Figure 1.2. A real system remains the same no matter which way you look at it. However, as we will see soon, the definition of a system is undertaken for the purpose of analysis and understanding; therefore the goals of this undertaking will influence the way a system is viewed. This may appear a trivial point, but very often it is a major block in communication between individuals coming from different backgrounds or disciplines, or simply having different goals.

Hierarchical Levels

To study an object effectively, we always try to isolate it from its environment. For example, if we want to apply elasticity theory on a part to determine stresses and deflections, we start by creating the *free-body diagram* of the part, where the points of interaction with the environment are substituted by equivalent forces and moments. Similarly, in a thermal process, if we want to apply the laws of mass and energy

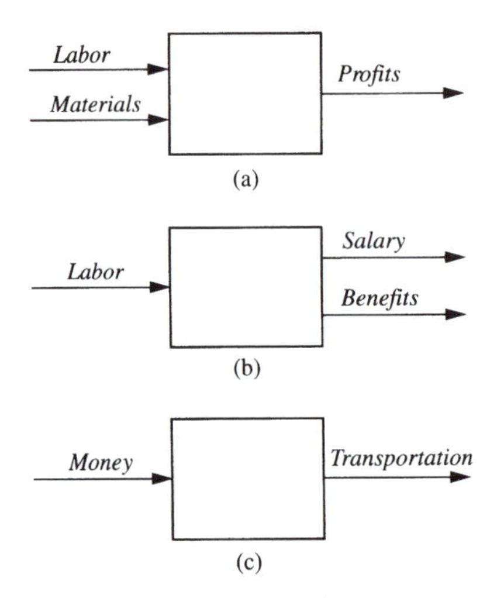

Figure 1.2. Viewpoints of system: automobile. (a) Manufacturer manager; (b) union member; (c) consumer.

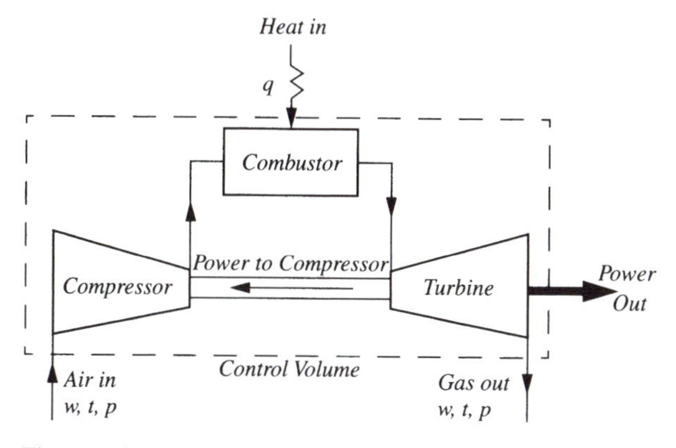

Figure 1.3. A gas-turbine system.

conservation to determine flow rates and temperatures, we start by specifying the *control volume*. Both the control volume and the free-body diagram are descriptions of the *system boundary*. Anything that "crosses" this boundary is a link between the system and its environment and will represent an input or an output characterizing the system.

As an example, consider the nonregenerative gas-turbine cycle in Figure 1.3. Drawing a control volume, we see that the links with the environment are the intake of the compressor, the exhaust of the turbine, the fuel intake at the combustor, and the power output at the turbine shaft. Thus, the air input (mass flow rate, temperature, pressure) and the heat flow rate can be taken as the inputs to the system, while the gas exit (mass flow rate, temperature, pressure) and the power takeoff are the outputs of the system. A simple block diagram would serve. Yet it is clear that the box in the figure indeed contains the components: compressor, combustor, turbine, all of which are themselves complicated machines. We see that the original system is made up of components that are systems with their own functions and input/output characterization. Furthermore, we can think of the gas-turbine plant as actually a component of a combined gas- and steam-turbine plant for liquefied petroleum. The original system has now become a component of a larger system.

The above example illustrates an important aspect of a system study: *Every system is analyzed at a particular level of complexity that corresponds to the interests of the individual who studies the system.* Thus, we can identify *hierarchical levels* in the system definition. Each system is broken down into subsystems that can be further broken down, with the various subsystems or components being interconnected. A boundary around any subsystem will "cut across" the links with its environment and determine the input/output characterization. These observations are very important for an appropriate identification of the system that will form the basis for constructing a mathematical model.

We may then choose to represent a system as a single unit at one level or as a collection of subsystems (for example, components and subcomponents) that must be coordinated at an overall "system level." This is an important modeling decision when the size of the system becomes large.

Mathematical Models

A real system, placed in its real environment, represents a very complex situation. The scientist or the engineer who wishes to study a real system must make many concessions to reality to perform some analysis on the system. It is safe to say that in practice *we never analyze a real system but only an abstraction of it.* This is perhaps the most fundamental idea in engineering science and it leads to the concept of a model:

> *A model is an abstract description of the real world giving an approximate representation of more complex functions of physical systems.*

The above definition is very general and applies to many different types of models. In engineering we often identify two broad categories of models: *physical* and *symbolic*. In a physical model the system representation is a tangible, material one. For example, a scale model or a laboratory prototype of a machine would be a physical model. In a symbolic model the system representation is achieved by means of all the tools that humans have developed for abstraction–drawings, verbalization, logic, and mathematics. For example, a machine blueprint is a *pictorial* symbolic model. Words in language are models and not the things themselves, so that when they are connected with logical statements they form more complex *verbal* symbolic models. Indeed, the artificial computer languages are an extension of these ideas.

The symbolic model of interest here is the one using a *mathematical* description of reality. There are many ways that such models are defined, but following our previous general definition of a model we can state that:

> *A mathematical model is a model that represents a system by mathematical relations.*

The simplest way to illustrate this idea is to look back at the block diagram representation of a system shown in Figure 1.1. Suppose that the output of the system is represented by a quantity y, the input by a quantity x, and the system function by a *mathematical function* f, which calculates a value of y for each value of x. Then we can write

$$y = f(x). \tag{1.1}$$

This equation is the mathematical model of the system represented in Figure 1.1. From now on, when we refer to a model we imply a mathematical one.

The creation of modern science follows essentially the same path as the creation of mathematical models representing our world. Since by definition a model is only an approximate description of reality, we anticipate that there is a varying degree of success in model construction and/or usefulness. A model that is successful and is supported by accumulated empirical evidence often becomes a *law* of science.

Virtual reality models are increasingly faithful representations of physical systems that use computations based on mathematical models, as opposed to realistic-looking effects in older computer games.

Elements of Models

Let us consider the gas-turbine example of Figure 1.3. The input air for the compressor may come directly from the atmosphere, and so its temperature and pressure will be in principle beyond the power of the designer (unless the design is changed or the plant is moved to another location). The same is true for the output pressure from the turbine, since it exhausts in the atmosphere. The unit may be specified to produce a certain amount of net power. The designer takes these as given and tries to determine required flow rates for air and fuel, intermediate temperatures and pressures, and feedback power to the compressor. To model the system, the laws of thermodynamics and various physical properties must be employed. Let us generalize the situation and identify the following model elements for all systems:

System Variables. These are quantities that specify different states of a system by assuming different values (possibly within acceptable ranges). In the example above, some variables can be the airflow rate in the compressor, the pressure out of the compressor, and the heat transfer rate into the combustor.

System Parameters. These are quantities that are given *one* specific value in any particular model statement. They are fixed by the application of the model rather than by the underlying phenomenon. In the example, atmospheric pressure and temperature and required net power output will be parameters.

System Constants. These are quantities fixed by the underlying phenomenon rather than by the particular model statement. Typically, they are natural constants, for example, a gas constant, and the designer cannot possibly influence them.

Mathematical Relations. These are equalities and inequalities that relate the system variables, parameters, and constants. The relations include some type of functional representation such as Equation (1.1). Stating these relations is the most difficult part of modeling and often such a relation is referred to as *the* model. These relations attempt to describe the function of the system within the conditions imposed by its environment.

The clear distinction between variables and parameters is very important at the modeling stage. The choice of what quantities will be classified as variables or parameters is a *subjective* decision dictated by choices in hierarchical level, boundary isolation, and intended use of the model of the system. This issue is addressed on several occasions throughout the book.

As a final note, it should be emphasized that the mathematical representation $y = f(x)$ of the system function is more symbolic than real. The actual "function" may be a system of equations, algebraic or differential, or a computer-based procedure or subroutine.

Analysis and Design Models

Models are developed to increase our understanding of how a system works. A *design* is also a system, typically defined by its geometric configuration, the materials used, and the task it performs. To model a design mathematically we must be able to define it completely by assigning values to each quantity involved, with these values satisfying mathematical relations representing the performance of a task.

In the traditional approach to design it has been customary to distinguish between *design analysis* and *design synthesis*. Modeling for design can be thought of in a similar way. In the model description we have the same elements as in general system models: design variables, parameters, and constants. To determine how these quantities relate to each other for proper performance of function of the design, we must first conduct *analysis*. Examples can be free-body diagram analysis, stress analysis, vibration analysis, thermal analysis, and so on. Each of these analyses represents a descriptive model of the design. If we want to predict the overall performance of the design, we must construct a model that incorporates the results of the analyses. Yet its goals are different, since it is a predictive model. Thus, in a design modeling study we must distinguish between *analysis models* and *design models*. Analysis models are developed based on the principles of engineering science, whereas design models are constructed from the analysis models for specific prediction tasks and are problem dependent.

As an illustration, consider the straight beam formula for calculating bending stresses:

$$\sigma = My/I, \tag{1.2}$$

where σ is the normal stress at a distance y from the neutral axis at a given cross section, M is the bending moment at that cross section, and I is the moment of inertia of the cross section. Note that Equation (1.2) is valid only if several simplifying assumptions are satisfied. Let us apply this equation to the trunk of a tree subjected to a wind force F at a height h above the ground (Alexander 1971), as in Figure 1.4(a). If the tree has a circular trunk of radius r, the moment of inertia is $I = \pi r^4/4$ and the maximum bending stress is at $y = r$:

$$\sigma_{\max} = 4Fh/\pi r^3. \tag{1.3}$$

If we take the tree as given (i.e., $\sigma_{\max}$, h, r are parameters), then Equation (1.3) solved for F can tell us the maximum wind force the tree can withstand before it breaks. Thus Equation (1.3) serves as an analysis model. However, a horticulturist may view this as a design problem and try to protect the tree from high winds by appropriately trimming the foliage to decrease F and h. Note that the force F would depend both on the wind velocity and the configuration of the foliage. Now Equation (1.3) is a design model with h and (partially) F as variables. Yet another situation exists in Figure 1.4(b) where the cantilever beam must be designed to carry the load F. Here the load is a parameter; the length h is possibly a parameter but the radius r would be normally considered as the design variable. The analysis model yields yet another design model.

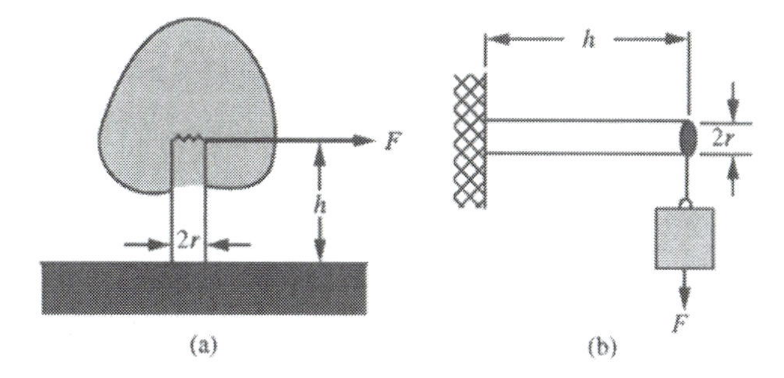

Figure 1.4. (a) Wind force acting on a tree trunk. (b) Cantilever beam carrying a load.

The analysis and design models may not be related in as simple a manner as above. If the analysis model is represented by a differential equation, the constants in this equation are usually design variables. For example, a gear motor function may be modeled by the equation of motion

$$J(d^2\theta/dt^2) + b(d\theta/dt) = -f_g r, \tag{1.4}$$

where J is the moment of inertia of the armature and pinion, b is the damping coefficient, f_g is the tangential gear force, r is the gear radius, θ is the angle of rotation, and t is time. Here J, b, and $f_g r$ are constants *for the differential equation*.

However, the design problem may be to determine proper values for gear and shaft sizes, or the natural frequency of the system, which would require making J, b, and r design variables. An explicit relation among these variables would require solving the differential equation each time with different (numerical) values for its constants. If the equation cannot be solved explicitly, the design model would be represented by a computer subroutine that solves the equation iteratively.

Before we conclude this discussion we must stress that there is no single design model, but different models are constructed for different needs. The analysis models are much more restricted in that sense, and, once certain assumptions have been made, the analysis model is usually unique. The importance of the influence of a given viewpoint on the design model is seen by another simple example. Let us examine a simple round shaft supported by two bearings and carrying a gear or pulley, as in Figure 1.5. If we neglect the change of diameters at the steps, we can say that the design of the shaft requires a choice of the diameter d and a material with associated properties such as density, yield strength, ultimate strength, modulus of elasticity, and fatigue endurance limit. Because the housing is already specified, the length between the supporting bearings, l, cannot be changed. Furthermore, suppose that we have in stock only one kind of steel in the diameter range we expect.

Faced with this situation, the diameter d will be the only design variable we can use; the material properties and the length l would be considered as design parameters. This is what the viewpoint of the shaft designer would be. However, suppose that after some discussion with the housing designer, it is decided that changes in the housing dimensions might be possible. Then l could be made a variable. The project manager,

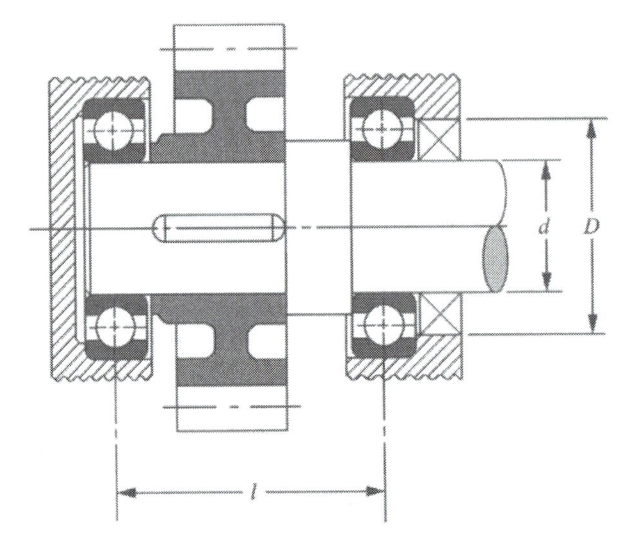

Figure 1.5. Sketch of a shaft design.

who might order any materials *and* change the housing dimensions, would view d, l, and material properties all as design variables. In each of the three cases, the model will be different and of course this would also affect the results obtained from it.

Decision Making

We pointed out already that design models are predictive in nature. This comes rather obviously from our desire to study how a design performs and how we can influence its performance. The implication then is that a design can be modified to generate different alternatives, and the purpose of a study would be to select "the most desirable" alternative. Once we have more than one alternative, a need arises for making a decision and choosing one of them. Rational choice requires a *criterion* by which we evaluate the different alternatives and place them in some form of ranking. This criterion is a new element in our discussion on design models, but in fact it is always implicitly used any time a design is selected.

A criterion for evaluating alternatives and choosing the "best" one cannot be unique. Its choice will be influenced by many factors such as the design application, timing, point of view, and judgment of the designer, as well as the individual's position in the hierarchy of the organization. To illustrate this, let us return to the shaft design example. One possible criterion is lightweight construction so that weight can be used to generate a ranking, the "best" design being the one with minimum weight. Another criterion could be rigidity, so that the design selected would have maximum rigidity for, say, best meshing of the attached gears. For the shop manager the ease of manufacturing would be more important so that the criterion then would be the sum of material and manufacturing costs. For the project or plant manager, a minimum cost design would be again the criterion but now the shaft cost would not be examined alone, but in conjunction with the costs of the other parts that the shaft has to function with. A corporate officer might add possible liability costs and so on.

A criterion may change with time. An example is the U.S. automobile design where best performance measures shifted from maximum power and comfort to maximum fuel economy and more recently to a rather unclear combination of criteria for maximum quality and competitiveness. One may argue that the ultimate criterion is always cost. But it is not always practical to use cost as a criterion because it can be very difficult to quantify. Thus, the criterion quantity shares the same property as the other elements of a model: It is an approximation to reality and is useful within the limitations of the model assumptions.

A design model that includes an evaluation criterion is a *decision-making model.* More often this is called an *optimization model*, where the "best" design selected is called the optimal design and the criterion used is called the *objective* of the model. We will study some optimization models later, but now we want to discuss briefly the ways design optimization models can be used in practice.

The motivation for using design optimization models is the selection of a good design representing a compromise of many different requirements with little or no aid from prototype hardware. Clearly, if this attempt is successful, substantial cost and design cycle time savings will be realized. Such optimization studies may provide the competitive edge in product design.

In the case of *product development*, a new original design may be represented by its model. Before any hardware are produced, design alternatives can be generated by manipulating the values of the design variables. Also, changes in design parameters can show the effect of external factor changes on a particular design. The objective criterion will help select the best of all generated alternatives. Consequently, a preliminary design is developed. How good it is depends on the model used. Many details must be left out because of modeling difficulties. But with accumulated experience, reliable elaborate models can be constructed and *design costs* will be drastically reduced. Moreover, the construction, validation, and implementation of a design model in the computer may take very much less time than prototype construction, and, when a prototype is eventually constructed, it will be much closer to the desired production configuration. Thus, design cycle time may be also drastically reduced.

In the case of *product enhancement*, an existing design can be described by a model. We may not be interested in drastic design changes that might result from a full-scale optimization study but in relatively small design changes that might improve the performance of the product. In such circumstances, the model can be used to predict the effect of the changes. As before, design cost and cycle time will be reduced. Sometimes this type of model use is called a *sensitivity study*, to be distinguished from a complete *optimization study*.

An optimization study usually requires several iterations performed in the computer. For large, complicated systems such iterations may be expensive or take too much time. Also, it is possible that a mathematical optimum could be difficult to locate precisely. In these situations, a complete optimization study is not performed.

Instead, several iterations are made until a sufficient improvement in the design has been obtained. This approach is often employed by the aerospace industry in the design of airborne structures. A design optimization model will use structural (typically finite element) and fluid dynamics analysis models to evaluate structural and aeroelastic performance. Every design iteration will need new analyses for the values of the design variables at the current iteration. The whole process becomes very demanding when the level of design detail increases and the number of variables becomes a few hundred. Thus, the usual practice is to stop the iterations when a competitive weight reduction is achieved.

1.2 Design Optimization

The Optimal Design Concept

The concept of design was born the first time an individual created an object to serve human needs. Today design is still the ultimate expression of the art and science of engineering. From the early days of engineering, the goal has been *to improve the design so as to achieve the best way of satisfying the original need, within the available means.*

The design process can be described in many ways, but we can see immediately that there are certain elements in the process that any description must contain: a *recognition of need*, an *act of creation*, and a *selection of alternatives*. Traditionally, the selection of the "best" alternative is the phase of *design optimization*. In a traditional description of the design phases, recognition of the original need is followed by a technical statement of the problem (problem definition), the creation of one or more physical configurations (synthesis), the study of the configuration's performance using engineering science (analysis), and the selection of "best" alternative (optimization). The process concludes with testing of the prototype against the original need.

Such sequential description, though perhaps useful for educational purposes, cannot describe reality adequately since the question of how a "best" design is selected within the available means is pervasive, influencing all phases where decisions are made.

So what is design optimization?

We defined it loosely as the selection of the "best" design within the available means. This may be intuitively satisfying; however, both to avoid ambiguity and to have an operationally useful definition we ought to make our understanding rigorous and, ideally, quantifiable. We may recognize that a rigorous definition of "design optimization" can be reached if we answer the questions:

1. How do we describe different designs?
2. What is our criterion for "best" design?
3. What are the "available means"?

The first question was addressed in the previous discussion on design models, where a design was described as a system defined by design variables, parameters, and constants. The second question was also addressed in the previous section in the discussion on decision-making models where the idea of "best" design was introduced and the criterion for an optimal design was called an *objective*. The objective function is sometimes called a "cost" function since minimum cost often is taken to characterize the "best" design. In general, the criterion for selection of the optimal design is a function of the design variables in the model.

We are left with the last question on the "available means." Living, working, and designing in a finite world obviously imposes limitations on what we may achieve. Brushing aside philosophical arguments, we recognize that any design decision will be subjected to limitations imposed by the natural laws, availability of material properties, and geometric compatibility. On a more practical level, the usual engineering specifications imposed by the clients or the codes must be observed. Thus, by "available means" we signify a set of requirements that must be satisfied by any acceptable design. Once again we may observe that these design requirements may not be uniquely defined but are under the same limitations as the choice of problem objective and variables. In addition, the choices of design requirements that must be satisfied are very intimately related to the choice of objective function and design variables.

As an example, consider again the shaft design in Figure 1.5. If we choose minimum weight as objective and diameter d as the design variable, then possible specifications are the use of a particular material, the fixed length l, and the transmitted loads and revolutions. The design requirements we may impose are that the maximum stress should not exceed the material strength and perhaps that the maximum deflection should not surpass a limit imposed by the need for proper meshing of mounted gears. Depending on the kind of bearings used, a design requirement for the slope of the shaft deflection curve at the supporting ends may be necessary. Alternatively, we might choose to maximize rigidity, seeking to minimize the maximum deflection as an objective. Now the design requirements might change to include a limitation in the space D available for mounting, or even the maximum weight that we can tolerate in a "lightweight" construction. We resolve this issue by agreeing that *the design requirements to be used are relative to the overall problem definition and might be changed with the problem formulation*. The design requirements pertaining to the current problem definition we will call *design constraints*. We should note that design constraints include all relations among the design variables that must be satisfied for proper functioning of the design.

So what *is* design optimization?

Informally, but rigorously, we can say that design optimization involves:

1. The selection of a set of variables to describe the design alternatives.
2. The selection of an objective (criterion), expressed in terms of the design variables, which we seek to minimize or maximize.

3. The determination of a set of constraints, expressed in terms of the design variables, which must be satisfied by any acceptable design.
4. The determination of a set of values for the design variables, which minimize (or maximize) the objective, while satisfying all the constraints.

By now, one should be convinced that this definition of optimization suggests a philosophical and tactical approach during the design process. It is not a phase in the process but rather a pervasive viewpoint.

Philosophically, optimization formalizes what humans (and designers) have always done. Operationally, it can be used in design, in any situation where analysis is used, and is therefore subjected to the same limitations.

Formal Optimization Models

Our discussion on the informal definition of design optimization suggests that first we must formulate the problem and then solve it. There may be some iteration between formulation and solution, but, in any case, any quantitative treatment must start with a mathematical representation. To do this formally, we assemble all the design variables $x_1, x_2, \ldots, x_n$ into a *vector* $\mathbf{x} = (x_1, x_2, \ldots, x_n)^T$ belonging to a subset $\mathcal{X}$ of the n-dimensional real space $\Re^n$, that is, $\mathbf{x} \in \mathcal{X} \subseteq \Re^n$. The choice of $\Re^n$ is made because the vast majority of the design problems we are concerned with here have real variables. The set $\mathcal{X}$ could represent certain *ranges* of real values or certain *types*, such as integer or standard values, which are very often used in design specifications.

Having previously insisted that the objective and constraints must be quantifiably expressed in terms of the design variables, we can now assert that the objective is a *function* of the design variables, that is, $f(\mathbf{x})$, and that the constraints are represented by *functional relations* among the design variables such as

$$h(\mathbf{x}) = 0 \quad \text{and} \quad g(\mathbf{x}) \leq 0. \tag{1.5}$$

Thus we talk about *equality* and *inequality* constraints given in the form of equal to zero and less than or equal to zero. For example, in our previous shaft design, suppose we used a hollow shaft with outer diameter d_o, inner diameter d_i, and thickness t. These quantities could be viewed as design variables satisfying the equality constraint

$$d_o = d_i + 2t, \tag{1.6}$$

which can be rewritten as

$$d_o - d_i - 2t = 0 \tag{1.7}$$

so that the constraint function is

$$h(d_o, d_i, t) = d_o - d_i - 2t. \tag{1.8}$$

We could also have an inequality constraint specifying that the maximum stress does not exceed the strength of the material, for example,

$$\sigma_{\max} \leq S, \tag{1.9}$$

where S is some properly defined strength (i.e., maximum allowable stress). However, $\sigma_{\max}$ should be expressed in terms of d_o, d_i, and t. If we neglect the effect of bending for simplicity, we can write

$$\sigma_{\max} = \tau_{\max} = M_t(d_o/2)/J, \tag{1.10}$$

where M_t is the torsional moment and J is the polar moment of inertia,

$$J = (\pi/32)\left(d_o^4 - d_i^4\right). \tag{1.11}$$

At this point we may view (1.10) and (1.11) as additional equality constraints with $\sigma_{\max}$ and J being additional design variables. Note that M_t would be a design parameter. Thus, we can rewrite them as follows:

$$\begin{aligned} &\sigma_{\max} - S \leq 0, \\ &\sigma_{\max} - M_t(d_o/2J) = 0, \\ &J - (\pi/32)\left(d_o^4 - d_i^4\right) = 0, \end{aligned} \tag{1.12}$$

so that we have one inequality and two equality constraints corresponding to (1.9). We could also eliminate $\sigma_{\max}$ and J and get

$$16M_t d_o/\pi\left(d_o^4 - d_i^4\right) - S \leq 0, \tag{1.13}$$

that is, just one inequality constraint. This implies that $\sigma_{\max}$ and J were considered *intermediate variables* that with the formulation (1.13) will disappear from the model statement. The above operation from (1.12) to (1.13) is a *model transformation* and it must be always performed judiciously so that the problem resulting from the transformation is *equivalent* to the original one and usually easier to solve. A strict definition of equivalence is difficult. Normally, we simply mean that the solution set of the transformed model is the same as that of the original model.

On the issue of transformation we may observe that the functional constraint representation (1.5) is not necessarily unique. For example, the renderings (1.7) and (1.13) of Equations (1.6) and (1.9), respectively, could have been written as

$$(d_o - d_i)/2t - 1 = 0, \tag{1.14}$$

$$16M_t d_o - S\pi d_o^4 + S\pi d_i^4 \leq 0. \tag{1.15}$$

The functions at the left side of (1.7) and (1.14) as well as (1.13) and (1.15) are *not the same*. For example, the function h in (1.8) varies linearly with t, which is not the case in (1.14). Of course, both functions were arrived at through transformations of the original (1.6). If we are careful, we should arrive at equivalent forms; yet very often careless transformations may confuse the analysis by introducing extraneous information not really there, or by hiding additional information. This is particularly dangerous when expressions are given for processing into a computer. To stress the point further, examine another form of Equation (1.6), namely,

$$(d_o - 2t)/d_i - 1 = 0, \tag{1.16}$$

and suppose that a solution could be obtained for a solid shaft, $d_i = 0$. Using (1.16), this would result in an error in the computer. Measures can be taken to avoid such situations, but we must be careful when performing model transformations.

As a final note, the form (1.5) is not the only one that can be used. Other forms, such as

$$h(\mathbf{x}) = 0, \quad g(\mathbf{x}) \geq 0 \tag{1.17}$$

or

$$h(\mathbf{x}) = 1, \quad g(\mathbf{x}) \leq 1 \tag{1.18}$$

can also be employed equally well. Forms (1.5) and (1.17) are called *negative null form* and *positive null form*, respectively, while (1.18) is the *negative unity form*.

We can now write the formal statement of the optimization problem in the negative null form as

$$\begin{aligned} &\text{minimize } f(\mathbf{x}) \\ &\text{subject to } h_1(\mathbf{x}) = 0, \quad g_1(\mathbf{x}) \leq 0, \\ &\qquad\qquad h_2(\mathbf{x}) = 0, \quad g_2(\mathbf{x}) \leq 0, \\ &\qquad\qquad\quad \vdots \qquad\qquad\quad \vdots \\ &\qquad\qquad h_{m_1}(\mathbf{x}) = 0, \quad g_{m_2}(\mathbf{x}) \leq 0, \\ &\text{and } \mathbf{x} \in \mathcal{X} \subseteq \Re^n. \end{aligned} \tag{1.19}$$

We can introduce the vector-valued functions $\mathbf{h} = (h_1, h_2, \ldots, h_{m_1})^T$ and $\mathbf{g} = (g_1, g_2, \ldots, g_{m_2})^T$ to obtain the compact expression

$$\begin{aligned} &\text{minimize } f(\mathbf{x}) \\ &\text{subject to } \mathbf{h}(\mathbf{x}) = \mathbf{0}, \\ &\qquad\qquad \mathbf{g}(\mathbf{x}) \leq \mathbf{0}, \\ &\qquad\qquad \mathbf{x} \in \mathcal{X} \subseteq \Re^n. \end{aligned} \tag{1.20}$$

We call **h**, **g** the *set* or *system of (functional) constraints* and $\mathcal{X}$ the *set constraint.* Statement (1.19) or (1.20) is a formal *mathematical model* of the optimization problem.

Multicriteria Models

Frequently the natural development of the design model will indicate more than one objective function. For the shaft example, we would really desire minimum weight *and* maximum stiffness. These objectives may be competing, for example, decreasing the weight will decrease stiffness and vice versa, so that some trade-off is required. If we keep more than one function as objectives, the optimization model will have a vector objective rather than a scalar one. The mathematical tools necessary to formulate and solve such *multiobjective* or *multicriteria* problems are quite extensive and represent a special branch of optimization theory.

For a vector objective **c**, the minimization formulation of the multicriteria optimization problem is

$$\begin{aligned} &\text{minimize } \mathbf{c}(\mathbf{x}) \\ &\text{subject to } \mathbf{h}(\mathbf{x}) = \mathbf{0}, \\ &\qquad\qquad \mathbf{g}(\mathbf{x}) \leq \mathbf{0}, \end{aligned} \tag{1.21}$$

where **c** is the vector of I real-valued criteria c_i. The feasible values for $\mathbf{c}(\mathbf{x})$ constitute the *attainable set* $\mathcal{A}$. Several methods exist for converting the multicriteria formulation into a *scalar substitute problem* that has a scalar objective and can be solved with the usual single objective optimization methods. The scalar objective has the form $f(\mathbf{c}, \mathbf{M})$, where **M** is a vector of *preference parameters* (weights or other factors) that can be adjusted to tune the scalarization to the designer's subjective preferences.

The simplest scalar substitute objective is obtained by assigning *subjective weights* to each objective and summing up all objectives multiplied by their corresponding weight. Thus, for min $c_1(\mathbf{x})$ and max $c_2(\mathbf{x})$ we may formulate the problem

$$\min f(\mathbf{x}) = w_1 c_1(\mathbf{x}) + w_2[c_2(\mathbf{x})]^{-1}. \tag{1.22}$$

A generalization of this function is $f = \sum_i f_1(w_i) f_2(c_i, \mathbf{m}_i)$, where the scalars w_i and vectors $\mathbf{m}_i$ are preference parameters. Clearly this approach includes quite subjective information and can be misleading concerning the nature of the optimum design. To avoid this, the designer must be careful in tracing the effect of subjective preferences on the decisions suggested by the optimal solution obtained after solving the substitute problem. Design preferences are rarely known precisely a priori, so preference values are adjusted gradually and trade-offs become more evident with repeated solutions of the substitute problem with different preference parameter values.

A common preference is to reduce at least one criterion without increasing any of the others. Under this assumption the set of solutions for consideration can be reduced to a subset of the attainable set, termed the *Pareto set*, which consists of *Pareto optimal* points. A point $\mathbf{c}_0$ in the attainable set $\mathcal{A}$ is Pareto optimal if and only if there is not another $\mathbf{c} \in \mathcal{A}$ such that $c_i \leq c_{0i}$ for all i and $c_i < c_{0i}$ for at least one i (Edgeworth 1881, Pareto 1971). So in multicriteria minimization a point in the design space is a Pareto (optimal) point if no feasible point exists that would reduce one criterion without increasing the value of one or more of the other criteria. A typical representation of the attainable and Pareto sets for a problem with two criteria is shown in Figure 1.6.

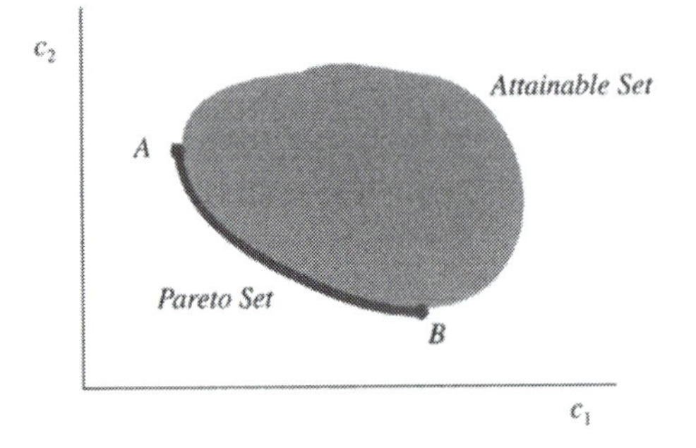

Figure 1.6. Attainable set and Pareto set (line segment AB) for a bicriterion problem.

Each solution of the weighted scalar substitute problem is Pareto optimal. Repeated solutions with different weights will gradually discover the Pareto set. The designer can then select the optimal solution that meets subjective trade-off preferences. The popular linearly weighted scalar substitute function has the limitation that it cannot find Pareto optimal points that lie upon a nonconvex boundary (Section 4.4) of the attainable set (Vincent and Grantham 1981, Osyczka 1984, Koski 1985). A generalized weighted criteria scalar substitute problem is then preferable (Athan 1994, Athan and Papalambros 1996).

Another approach suitable for design problems is to correlate the objective functions with *value functions*, which can then be combined into an overall value function that will serve as a single objective. Essentially, the procedure assigns costs to each objective, converting everything to minimum cost. This idea leads to more general formulations in *utility theory* that are more realistic but also more complicated.

Goal programming (Ignizio 1976) involves an initial prioritization of objective criteria and constraints by the designer. Goals are selected for each criterion and constraint and "slack" variables are introduced to measure deviations from these goals at different design solutions. Goal values are approached in their order of priority and deviations from both above and below the goals are minimized. The result is a compromise decision (Mistree et al. 1993). The concept of Pareto optimality is not relevant to this approach.

Game theory (Borel 1921, von Neumann and Morgenstern 1947, Vincent and Grantham 1981) has also been used in multicriteria optimization formulations (Rao and Hati 1979, Vincent 1983). If there is a natural hierarchy to the design criteria, Stackelberg game models can be used to represent a concurrent design process (Pakala 1994). Some game theoretic strategies will result in points that are not Pareto points, because they make different assumptions about preference structure. For example, a rivalry strategy giving highest priority to preventing a competitor's success would likely result in a non-Pareto point.

The simplest approach, recommended here at least as a first step, is to select from the set of objective functions *one* that can be considered the most important criterion for the particular design application. The other objectives are then treated as constraints by restricting the functions within acceptable limits. One can explore the implied trade-offs of the original multiple objectives by examining the change of the optimum design as a result of changes in the imposed acceptable limits, in a form of sensitivity analysis or parametric study, as explained in later chapters.

Nature of Model Functions

From a modeling viewpoint, functions f, $\mathbf{h}$, and $\mathbf{g}$ can be expressed in different forms. They may be given as explicit algebraic expressions of the design vector $\mathbf{x}$, so that $\mathbf{h}(\mathbf{x}) = \mathbf{0}$ and $\mathbf{g}(\mathbf{x}) \leq \mathbf{0}$ are explicit sets of algebraic equalities and inequalities. The relations are usually derived directly from basic equations and laws of engineering science. However, because basic engineering principles are often incapable of describing the problem completely, we use empirical or experimental data. Explicit relations can be derived through curve fitting of equations into measured data. Another modeling possibility discussed earlier is that the system $\mathbf{h}(\mathbf{x}) = \mathbf{0}, \mathbf{g}(\mathbf{x}) \leq \mathbf{0}$ may not have equations at all but may be the formal statement of a complex procedure involving internal calculations and often realized only as a computer program. In such cases, the term *simulation model* is often used. Typical cases are numerical solutions of coupled differential equations frequently using finite elements. Even then, it is worthwhile to try and derive explicit algebraic equations by repeated computer runs and subsequent curve fitting as discussed in Chapter 2. A model based on explicit algebraic equations generally provides much more insight into the nature of the optimum design.

In practice, mathematical models are mixtures of all the above types. The design analyst must decide how to proceed and one of the goals in this book is to provide assistance for such decisions.

The nature of functions f, $\mathbf{h}$, and $\mathbf{g}$ can also be different from a mathematical viewpoint. If the functions represent algebraic or equivalent relations, then model (1.20) represents a *mathematical programming problem*. These are *finite-dimensional* problems since $\mathbf{x}$ has a finite dimension. If differential or integral operators are explicitly involved and/or the variables $x_i = x_i(t)$, $t \in \Re$, are defined in an *infinite-dimensional* space, then we have the type of problem studied in the *calculus of variations* or *control theory*. These are valid design problems and their study involves suitable extension of our discussions here for finite dimensions, to infinite dimensions. This book is limited to the study of finite-dimensional problems.

Within mathematical programming, when the functions f, h_i, g_j are all linear, then the model is a *linear programming* (LP) one. Otherwise, the model represents a *nonlinear programming* (NLP) problem. As we will see in Chapters 4 and 5, we usually make the assumption that all model functions are continuous and also possess continuous derivatives at least up to first order. This allows the development and application of very efficient solution methods. *Discrete programming* refers to models where all variables take only discrete values, sometimes only integer values, or even just zero or one. These problems are studied in the field of *operations research* under terms such as *integer programming* or *combinatorial optimization*. A common class of design problems comprises *mixed-discrete* models, namely, those that contain both continuous and discrete variables. Solution of such problems is generally very difficult and occasionally intractable. In most of this book we deal with nonlinear programming models with continuous, differentiable functions. Design problems are hardly ever linear and usually represent a mathematical challenge to the traditional methods of nonlinear programming.

The Question of Design Configuration

Any designer knows that the most important and most creative part in the evolution of a design is the synthesis of the configuration. This involves decisions on the general arrangement of parts, how they may fit together, geometric forms, types of motion or force transmission, and so on. This open-ended characteristic of the design process is unique and has always been identified with the human creative potential. The designer creates a new configuration through a spontaneous synthesis of previous knowledge and intuition. This requires both special skill and experience for a truly good (perhaps "best") design.

There can be many configurations meeting essentially the same design goals and one might desire to pose an optimization problem seeking the optimum configuration. To compare configurations we must have a mathematical model that allows us to move from one configuration to another in our search for the optimum. In many design problems each configuration has its own set of design variables and functions. Therefore, combining configurations in a single model where an optimization study will be applied is generally very difficult.

An exciting capability for optimal configuration design has been developed for the optimal layout of structural components. Given a design domain in a two- or three-dimensional space, and boundary conditions describing loads and supports, the problem is to find the best structure (e.g., the lightest or stiffest) that will carry the loads without failure. This configuration (or layout or topology) problem is solved very elegantly by discretizing the design space into cells, usually corresponding to finite elements, and choosing as design variables the material densities in each cell. We now have a common set of design variables to describe all configurations and the problem can be solved in a variety of ways, for example, with a homogenization method (Bendsøe and Kikuchi 1988) or genetic algorithms (Chapman et al. 1994, Schmit and Cagan 1998).

The process is illustrated in Figure 1.7 for the design of a bracket using homogenization to generate the initial topology (Chirehdast et al. 1994). The design domain and associated boundary conditions are shown in Figure 1.7(a). A gray scale image is generated by the optimization process where the degree of "grayness" corresponds to the density levels. Densities are normalized between zero (no material in the cell) and one (cell full of material). The optimal material distribution for a stiff lightweight design derived using homogenization is given in Figure 1.7(b). This image typically needs interpretation and some post-processing to derive a realizable design. This can be achieved by applying image processing techniques such as threshholding,

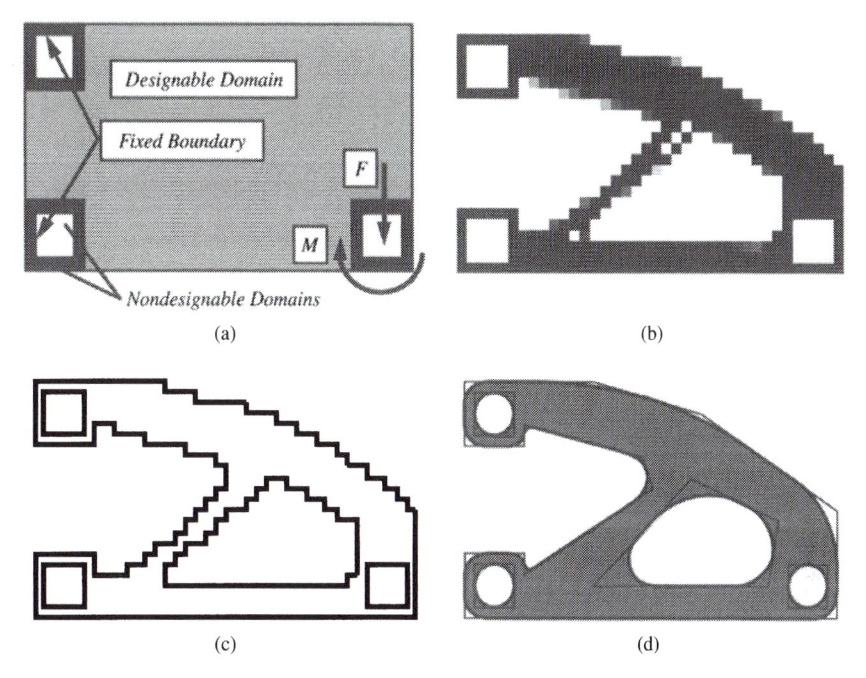

Figure 1.7. Optimal topology design.

smoothing, and edge extraction (Figure 1.7(c)). Practical manufacturing rules can also be applied automatically to derive a part that can be made by a particular process, for example, by casting (Figure 1.7(d)). This method has been successfully used in the automotive industry to design highly efficient structural components with complicated geometry.

Other efforts at obtaining optimal configuration design involve the assignment of design variables with integer zero or one values to each possible design feature depending on whether the feature is included in the design or not. Such models are quite difficult to construct and also tend to result in intractable combinatorial problems. Artificial intelligence methods showed much promise in the 1980s but have produced few operationally significant results. Genetic algorithms seem to be the most promising approach at the present time.

The simplest approach for dealing with optimal configurations, recommended here at least as a first attempt, is to rely on the experience and intuition of the designer to configure different design solutions in an essentially qualitative way. A mathematical model for *each* configuration can be produced to optimize each configuration separately. The resulting optima can then be compared in a quantitative way. The process is iterative and the insights gained by attempting to optimize one configuration should help in generating more and better alternatives.

In our future discussions we will be making the tacit assumption that the models refer to single configurations arrived at through some previous synthesis.

Systems and Components

Recall our discussion in Section 1.1 about hierarchical levels in systems study. Understanding the hierarchy in a system definition has important implications for optimization modeling. When we first define the problem, we must examine at what level we are operating. We should ask questions such as:

Does the problem contain identifiable components?

How are the components linked?

Can we identify component variables and system variables?

Does the system interact with other systems at the same level? At higher levels? At lower levels?

Such questions will clarify the nature of the model, the classification of variables, parameters and constants, and the appropriate definition of objective and constraint functions.

To illustrate the point, consider again the simple shaft example of Section 1.1. A partial system breakdown (one of the many we may devise) is shown in Figure 1.8. Note that if we "optimize" the shaft, what is optimum for the shaft may not be optimum for the transmission. The connections with bearings and gears indicate that if decisions have been made about them, specific constraints may be imposed on the shaft design. Furthermore, suppose that the shaft material is to be chosen. Several design variables representing all the material properties appearing in the mathematical model may be needed, for example, percentages of alloy content in the steel and heat treatment quantities (temperature, time, depth), which moves us to an even lower level in the hierarchy.

Choosing the appropriate analysis level depends on our goals and is often dictated by model complexity and the mathematical size of the problem. The best strategy is to start always with the *simplest meaningful model*, namely, one containing interesting

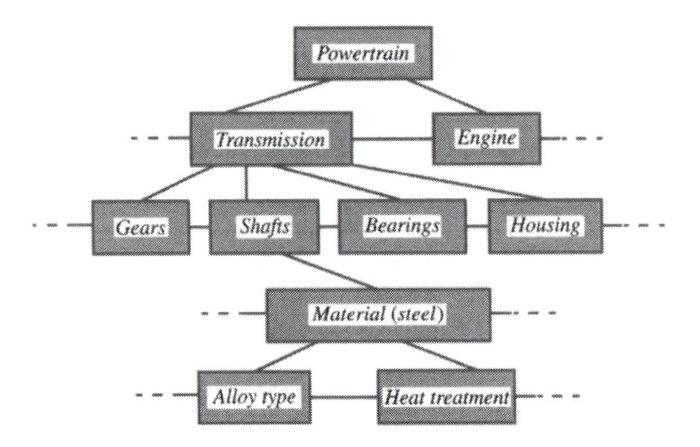

Figure 1.8. Partial representation of a possible system hierarchy for the shaft example.

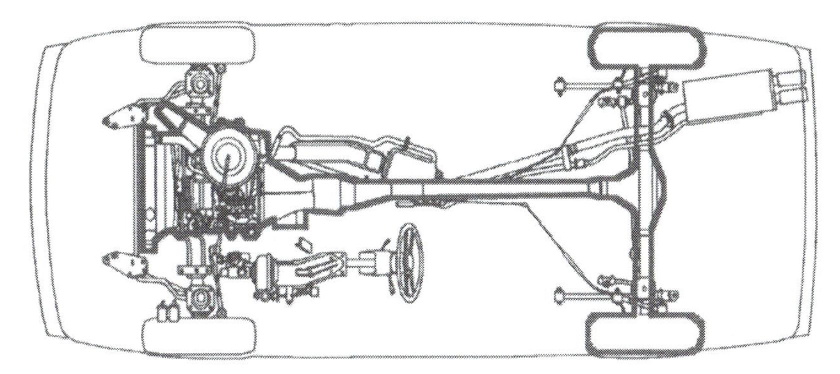

Figure 1.9. Automobile powertrain system.

trade-offs that can be explored by an optimization study. The result will always be *suboptimal*, valid for the subsystem at the level of which we have stopped.

Hierarchical System Decomposition

Another way of looking at the hierarchy shown in Figure 1.8 is to think of the powertrain as a collection of components. We say that the powertrain is *decomposed* into a set of components. In the automobile industry the powertrain of Figure 1.9 is usually decomposed into components as shown in Figure 1.10. This *component* or *object decomposition* appears to be a natural one and design organizations in industry can be constructed in this hierarchical decomposed form to perform a distributed, compartmentalized design activity. To achieve overall system design, component design activities must be properly coordinated. Ideally, the components should be designed in parallel so that we have *concurrent design* of the system.

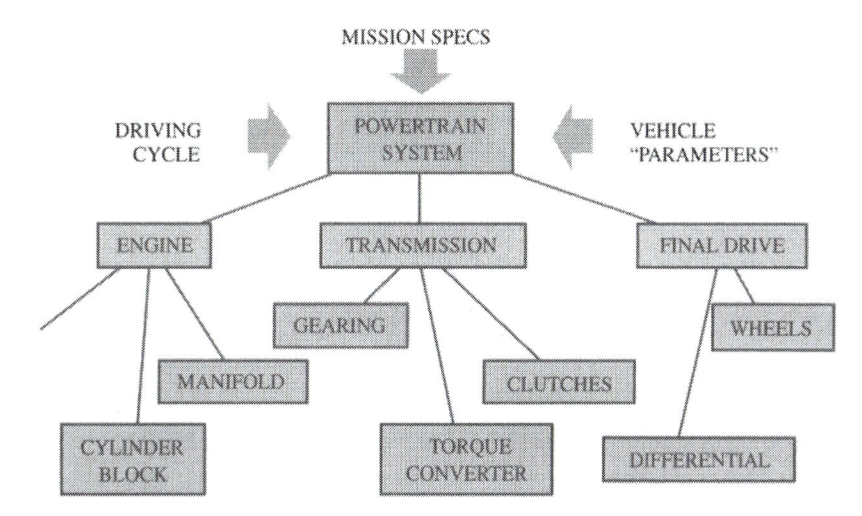

Figure 1.10. Component decomposition of powertrain system.

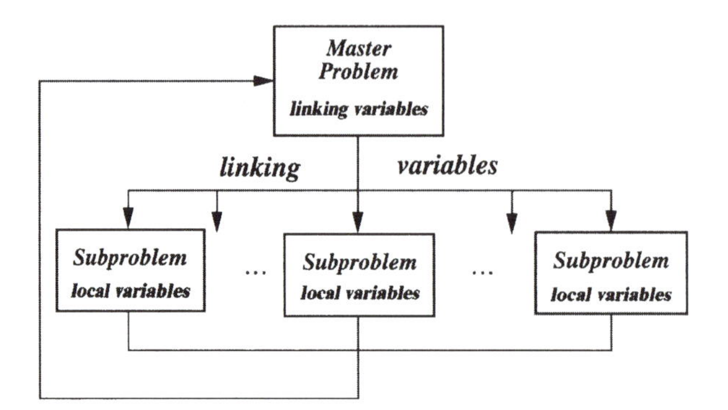

Figure 1.11. Hierarchical coordination.

We may choose to treat the system as a single entity and build a mathematical optimization model with a single objective and a set of constraints. When the size of the problem becomes large, such an approach will encounter difficulties in outputting a reliable solution that we can properly interpret and understand. A desirable alternative then is to model the problem in a decomposed form. A set of independent *subproblems* is coordinated by a *master problem*. The design variables are classified as *local variables* associated with each subproblem and *linking variables* associated with the master problem. The schematic of Figure 1.11 illustrates the idea for a two-level decomposition. Special *problem structures* and *coordination methods* are required to make such an approach successful.

Looking at the powertrain system one can argue that the problem can be decomposed in a different way by looking at what disciplines are required to completely analyze the problem and build a mathematical model. Such an *aspect decomposition* is shown in Figure 1.12. In a mathematical optimization model each aspect or discipline will contribute an analysis model that can be used to generate objective and constraint functions. In a business organization this decomposition corresponds to a functional structure, while object decomposition corresponds to a line structure.

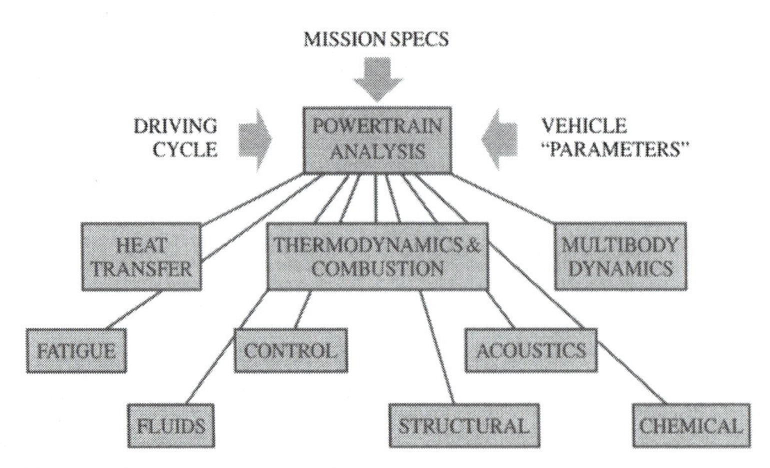

Figure 1.12. Aspect decomposition of powertrain system.

We see then that a system decomposition is not unique. Partitioning a large model into an appropriate set of coordinated submodels is itself an optimization problem. It is not an accident that most industrial organizations today have adopted both an object and an aspect decomposition in what is called a *matrix organization*. The increasing availability of better analysis models allows optimization methods to offer an excellent tool for rigorous design of large complicated systems.

1.3 Feasibility and Boundedness

So far we have been discussing how to represent a design problem as a mathematical optimization model. Any real problem that is properly posed will usually have several acceptable solutions, so that one of them may be selected as optimal. With the precise mathematical definition (1.20) comes the question of when does such a model possess a mathematical solution? This *existence* question is an important theoretical topic in optimization theory and a difficult one. Apart from certain special cases, its practical utility for the type of problem we are concerned with here is still rather minor. Therefore, it is important to accept the fact that many of the arguments we make in solution procedures of practical problems involve a mixture of mathematical rigor and engineering understanding. In other words, having posed a problem with a model such as (1.20) does not complete our contribution from the engineering side in a way that we can hand it over to a mathematician or computer analyst. The problem complexity often defies available mathematical tools, so that only with continuing use of additional engineering judgment can we hope to arrive at a solution that we actually believe.

We say that a problem is *well posed* to imply the assumption of existence of solution for model (1.20). Though a mathematical proof of solution existence may be difficult, many mathematical properties are associated with the model and its solution that can be used to test the engineering statement of the problem. It is not uncommon to have problems not well posed because the model has not been formulated properly. Then mathematical analysis can help clarify engineering thinking and so the process of interplay between physical understanding and associated abstract mathematical qualities of the model is complete.

Let us now examine some issues in the formulation of well-posed problems.

Feasible Domain

The system of functional constraints and the set constraint in (1.20) isolate a region inside the n-dimensional real space. Any point inside that region represents an acceptable design. The set of values for the design variables $\mathbf{x}$ satisfying all constraints is called the *feasible space*, or the *feasible domain of the model*. From all the acceptable designs represented by points in the feasible domain, one must be selected as the optimum, that which minimizes $f(\mathbf{x})$. Clearly, no optimum will exist if the feasible space is *empty*, that is, when no acceptable design exists. This can happen if the constraints are overrestrictive.

To illustrate this, let us look at the design of a simple tensile bar. Constraints are the maximum stress and maximum area limitations. Thus, we have

$$\begin{aligned} &\text{minimize } f(a) = a \\ &\text{subject to } P/a \leq S_{yt}/N, \\ &\qquad\qquad\quad a \leq A_{\max}, \end{aligned} \tag{1.23}$$

where the only design variable is the cross-sectional area a. The four parameters are tensile force P, yield strength in tension S_{yt}, maximum allowable area $A_{\max}$, and a safety factor N. The objective is simply taken to be proportional to the area.

Rearranging and combining the constraints, we have

$$PN/S_{yt} \leq a \leq A_{\max}. \tag{1.24}$$

Any acceptable a must be selected from the interval $[PN/S_{yt}, A_{\max}]$, which is the feasible domain of the model. Now suppose that $A_{\max} = 0.08\ \text{in}^2$, $P = 3{,}000$ lbs with $N = 1$, and the material is hot-rolled steel G10100 with $S_{yt} = 26$ kpsi. Then $PN/S_{yt} = 0.115\ \text{in}^2$. There is no a satisfying (1.24) and the model has *no feasible solution.* We may get a nonempty feasible domain if we use cold-drawn steel G10100 with $S_{yt} = 44$ kpsi, giving $PN/S_{yt} = 0.068\ \text{in}^2$. So now any a in the range [0.068, 0.08] will be feasible. Clearly, to minimize $f(a)$ we must select the smallest acceptable a; therefore $a_* = 0.068$ in (the asterisk signifies optimum value). The design problem would have no solution if we did not have access to a higher strength material.

Boundedness

The values PN/S_{yt} and $A_{\max}$ are examples of lower and upper acceptability limits for the variable a. We call them *lower* and *upper bounds* for a. In this example, the optimum was on the *boundary* of the feasible space. The existence of proper bounds is very important in optimization. Before we discuss this issue, we should make the point that absence of proper bounds may be cause of serious trouble. We can illustrate this with the "seltzer pill" example (Russell 1970).

The problem is to design a disc-shaped seltzer pill so as to minimize the time required for it to dissolve in a glass of water. Assuming the standard dosage is one unit volume, the problem is to maximize the surface area of a cylinder with unit volume. Letting the cylinder have radius r and height h, the problem is modeled as

$$\begin{aligned} &\text{maximize } f(h, r) = 2\pi r^2 + 2\pi r h \\ &\text{subject to } \pi r^2 h = 1. \end{aligned} \tag{1.25}$$

(Note that a maximization problem can be transformed, if we wish, into the equivalent minimization by taking the negative of the objective function.) The variables r, h belong to the two-dimensional strictly positive real space. To solve the problem, we can use the equality constraint to express h in terms of r and substitute in the objective.

Then (1.25) is reduced to

$$\begin{aligned} &\text{maximize } f(r) = 2\pi r^2 + 2/r \\ &\text{subject to } 0 < r < \infty \quad \text{(set constraint).} \end{aligned} \tag{1.26}$$

We see that $f \to +\infty$, for $r \to +\infty$ or $r \to 0$. Both cases correspond to a pill of zero volume, the first one representing a cylinder becoming an infinite two-dimensional plane and the second an infinite one-dimensional line. The problem has no solution because no acceptable value of r can yield a maximum for the surface area. We say that *the problem is not well bounded*. In a mathematical sense the set constraint $0 < r < \infty$ is an *open* set and excludes the values of zero and infinity for r. Although zero is a *lower bound* for r, which is then *bounded from below*, this bound is not valid, that is, it cannot be achieved by the variable. The term *well bounded* is used to make this distinction. These ideas are studied extensively in Chapter 3.

We must stress that this boundedness is relative to the problem statement. For example, if in (1.26) the objective was to *minimize*, then the problem would be bounded because there is an acceptable solution for $r > 0$. For the problem (1.26) we say that the variable r is *unbounded above* and *not well bounded from below*. We will see that, for practical situations, this is often the same as r being *unbounded below*.

Activity

In the simple tensile bar design, we say that the optimal value of the design variable a_* was found by reducing a to its smallest possible value $a_* = PN/S_{yt}$, that is, its lower bound. This bound was imposed on a via the stress constraint in (1.23). Setting the optimal a equal to its lower bound is equivalent to insisting that, *at the optimum*, the inequality $P/a \leq S_{yt}/N$ must be satisfied with *strict equality*. In such cases, we say that the inequality constraint is *active*. *An active constraint is one which, if removed, would alter the location of the optimum*. For an inequality constraint this often means that *it must be satisfied with strict equality at the optimum*. In the tensile bar problem, the stress constraint is active, but the $a \leq A_{\max}$ constraint is *inactive*. Its presence is necessary to define the feasible domain, but, provided that this domain is not empty, it plays no role in the location of the optimum.

The concept of *constraint activity* is very important in design optimization and is one of the common threads in this book. An active constraint represents a design requirement that has direct influence on the location of the optimum. Active inequality constraints most often correspond to critical *failure modes*. This information is very important for the designer. In fact, traditional design procedures were really primitive optimization attempts where certain failure modes were considered critical a priori, relative to some often hidden (or nonanalytically expressed) objective criterion. Essentially, the problem was solved by assembling enough active constraints to make a system of n equations in the n unknown design variables. In formal optimization, this situation may also arise but only as a result of rigorous mathematical arguments.

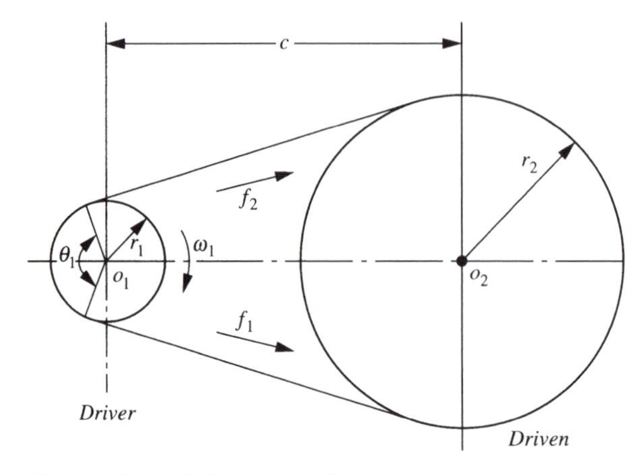

Figure 1.13. A belt-drive configuration.

We will use a simple example to illustrate these ideas. The problem is to select a flat belt for a belt drive employing two pulleys and operating at its maximum capacity as in Figure 1.13. The specifications are as follows. We will use cast iron pulleys and a leather belt with tensile strength $\sigma_{1\,\max} = 250$ psi and specific weight 0.035 lb/in. This combination gives a friction coefficient of about 0.3. The load carrying capacity should be 50 hp with input 3,000 rpm and output 750 rpm.

The design question essentially involves the selection of the pulley radii r_1 and r_2 and a proper cross section a for the belt. The center distance c must be determined also, as well as belt forces f_1 and f_2 at the tight and slack sides, respectively.

One possible design procedure is the following:

Speed ratio requirement:

$$r_1/r_2 = N_2/N_1 = 750/3000 = 0.25. \tag{a}$$

Power transmission requirement:

$$p = t_1\omega_1 = (f_1 - f_2)r_1(2\pi N_1) \geq 50\,\text{hp}. \tag{b}$$

Tensile strength requirement:

$$f_1/a \leq \sigma_{1\,\max} = 250\,\text{psi}. \tag{c}$$

Balance of belt forces requirement:

$$f_1/f_2 = e^{F\theta_1}, \tag{d}$$

where

$$\theta_1 = \pi - 2\arcsin[(r_2 - r_1)/c]. \tag{e}$$

In the above expressions, N_1 and N_2 are the input and output rpm, p is the transmitted horsepower, t_1 is the torque about shaft O_1, F is the coefficient of friction, and θ_1 is the contact angle at pulley 1. Here θ_1 is an "intermediate" variable and Equation (e) could be eliminated immediately along with θ_1. As already discussed, we generally do these eliminations carefully so that no information is lost in the process. Note that for model simplicity we assumed that the equivalent centrifugal force in the force balance of Equation (d) is negligible.

Note that (b) and (c) are written as inequalities, although in a traditional design procedure they would be treated as equalities. Let us keep them here as inequalities. Then we have five relations (a)–(e) and seven unknowns: r_1, r_2, f_1, f_2, c, a, and θ_1. Some more "engineering assumptions" must be brought in. Suppose, for example, that to accommodate the appropriate size of pulley shaft we *select* $r_1 = 3$ in. Then from (a), $r_2 = 12$ in. Following a rule of thumb, we select the center distance as

$$c = \max\{3r_1 + r_2, 2r_2\}, \tag{f}$$

thus getting $c = 24$ in. With this information, after some unit conversions and rearrangements, (b), (c), and (d) become, respectively,

$$f_1 - f_2 \geq 350\,\text{lb}_\text{f}, \tag{b$'$}$$

$$f_1/a \leq 250\,\text{lb}_\text{f}/\text{in}^2, \tag{c$'$}$$

$$f_1/f_2 = 2.038. \tag{d$'$}$$

We can now solve (d$'$) for f_2 and substitute in (b$'$) and (c$'$). After some rearrangement we get

$$f_1 \geq 687(\text{lb}_\text{f}), \tag{b$''$}$$

$$f_1 \leq 250a(\text{lb}_\text{f}/\text{in}^2). \tag{c$''$}$$

We can represent this graphically in Figure 1.14.

The "best" design is selected as the one giving the smallest cross-sectional area a. It is located at the intersection of (b$''$), (c$''$) written as equalities. Thus, the two requirements on stress and power are critical for the design and would represent *active*

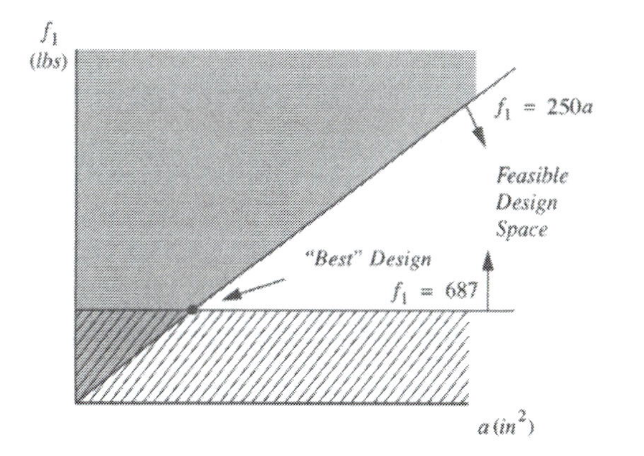

Figure 1.14. Traditional solution for belt-drive design example.

constraints for an optimization problem. The solution is found by setting $f_1 = 687$ lbf, giving $a = 2.75\ \text{in}^2$. This may be a rather wide belt. How could we decrease the size while remaining within the specifications? The answer traditionally would be by trial and error.

Let us see how we would proceed in this problem using the formal optimization representation. The previous information can be assembled in the following model:

$$\begin{aligned}
\text{minimize} \quad & f(a) = Ka \\
\text{subject to} \quad & R_1\colon r_2 - 4r_1 = 0, \\
& R_2\colon 1{,}050 - (f_1 - f_2)r_1 \leq 0, \\
& R_3\colon f_1/a - 250 \leq 0, \\
& R_4\colon f_1/f_2 - e^{0.3\theta_1} = 0, \\
& R_5\colon \theta_1 - \{\pi - 2\arcsin[(r_2 - r_1)/c]\} = 0, \\
& R_6\colon \max\{3r_1 + r_2, 2r_2\} - c \leq 0.
\end{aligned} \tag{1.27}$$

The model (1.27) was set so that it resembles the formulation (1.19) in standard form. This of course is not necessary for our present purposes. We numbered the constraints R_1 to R_6 for easy reference to the restrictions on the design. The objective is loosely defined as proportional to a, with the "cost" parameter K being positive. Also, R_6 is written to represent a *lower bound* on the center distance. The design variables for this problem statement are r_1, r_2, f_1, f_2, θ_1, c, and a – a total of seven. The three equalities suggest that only four variables can be chosen independently. Moreover, if the three inequality constraints turn out to be active, then only one variable could be chosen independently. We say that the original problem has four *degrees of freedom* and then *for every identified active constraint, the degrees of freedom are reduced by one*. The smallest possible number of degrees of freedom for this model is one.

Now we can use equality R_1 to eliminate one variable, r_2, and reduce the problem. We note that since $r_2 = 4r_1$, then $2r_2 > 3r_1 + r_2$ always, so that R_6 can be simply rewritten as $c \geq 8r_1$. We can also use R_4 to eliminate f_2 from R_2. After some rearrangement we can write the problem as follows:

$$\begin{aligned}
\text{minimize} \quad & f(a) = Ka \\
\text{subject to} \quad & R_2\colon f_1 \geq (1{,}050/r_1)(1 - e^{-0.3\theta_1})^{-1}, \\
& R_3\colon a \geq f_1/250, \\
& R_5\colon \theta_1 = \pi - 2\arcsin(3r_1/c), \\
& R_6\colon c \geq 8r_1.
\end{aligned} \tag{1.28}$$

Clearly, the optimal value of a is the smallest acceptable, that is, the lower bound provided by R_3. Thus, R_3 is an *active* constraint, meaning that, at the optimum,

$$a_* = f_{1*}/250. \tag{1.29}$$

This relation can be used now as an equality to eliminate a (we drop the asterisks for convenience). The reduced problem is

$$\begin{aligned} \text{minimize} \quad & f(f_1) = K' f_1 \\ \text{subject to} \quad & R_2\colon f_1 \geq (1{,}050/r_1)\left(1 - e^{-0.3\theta_1}\right)^{-1}, \\ & R_5\colon \theta_1 = \pi - 2\arcsin(3r_1/c), \\ & R_6\colon c \geq 8r_1. \end{aligned} \tag{1.30}$$

Again, the optimal value for f_1 is found by setting it to its lower bound, which requires R_2 to be active. Elimination of f_1 gives the further reduced problem

$$\begin{aligned} \text{minimize} \quad & f(\theta_1, r_1) = K''(1{,}050/r_1)\left(1 - e^{-0.3\theta_1}\right)^{-1} \\ \text{subject to} \quad & R_5\colon \theta_1 = \pi - 2\arcsin(3r_1/c), \\ & R_6\colon c \geq 8r_1. \end{aligned} \tag{1.31}$$

Now a little reflection will show that f will decrease by making θ_1 as large as possible. Moreover, we see from R_5 that θ_1 increases with c so we can minimize the objective if we make c as large as possible. Since R_6 provides a lower bound and is the only constraint left in the model, the problem as posed is *unbounded*. We say that variable c is *unbounded from above* and that R_6 is an *inactive* constraint.

Note that previously in the "traditional" approach we had essentially taken R_6 as active. The same conclusion would have been derived from the more formal optimization process above if in the model (1.27) we had assumed that the rule of thumb provides an *upper bound* on the center distance. The problem, as stated, will not have a solution unless we determine an upper bound for c, which of course would then represent an active constraint. Let us set arbitrarily

$$c \leq 10r_1. \tag{1.32}$$

Then $c_* = 10r_{1*}$ and R_5 gives

$$\theta_{1*} = \pi - 2\arcsin\left(\tfrac{3}{10}\right) = 145^\circ. \tag{1.33}$$

This fixes the objective to a constant, and so c and r_1 can take any values that satisfy (1.32). Clearly, when we convert a design equation to an inequality, the proper *direction* of the inequality must be chosen carefully.

In the discussion preceding the above example, we stated that very often active inequality constraints correspond to the critical failure modes for the design. This idea has been used intuitively in the early work of optimal design for structures. The principle of *simultaneous mode design* in structural design states that "a given form will be optimum, if all failure modes that can possibly intersect occur simultaneously under the action of the load environment." If, for example, a structure could fail because of overstressing and buckling, it should be designed so that yielding and buckling occur at the same time. Original application of this idea together with a minimum weight objective was used to solve such problems formulated as having only equality constraints. An example of such formulations in structural applications is the *fully stressed design*, where the optimization model is set up by minimizing the weight of a system subject to equality constraints that set the stresses in the system components equal to, say, the yield strength.

We should recognize here that the above approach entails a rather a priori decision on the activity of these constraints. This may or may not be the case when more than just stress constraints are present. Constraint activity may also change, if the objective function changes. Therefore, in our analysis we should always try to *prove* that a constraint is active, or inactive, by rigorous arguments based on the model at hand. This proof, being of great importance to the designer, will be the focus of many sections throughout the book, but it will be rigorously addressed beginning in Chapter 3.

1.4 Topography of the Design Space

One can capture a vivid image of the design optimization problem by visualizing the situation with two design variables. The design space could be some part of the earth's surface that would represent the objective function. Mountain peaks would be maxima, and valley bottoms would be minima. An equality constraint would be a road one must stay on. An inequality constraint could be a fence with a "No Trespassing" sign. In fact, some optimization jargon comes from topography. Much can be gained by this visualization and we will use it in this section to describe features of the design space. One should keep in mind, however, that certain unexpected complexities may arise in problems with dimensions higher than two, which may not be immediately evident from the three-dimensional image.

Interior and Boundary Optima

In one-dimensional problems a graphical representation is easy. A problem such as

$$\begin{aligned} &\text{minimize } y = f(x), \quad x \in \mathcal{X} \subseteq R, \\ &\text{subject to } g_1(x) \leq 0, \\ &\qquad\qquad g_2(x) \leq 0 \end{aligned} \tag{1.34}$$

can be represented by a two-dimensional y–x picture, as in Figure 1.15. If the functions behave as shown in the figure, the problem is restated simply as

$$\begin{aligned} &\text{minimize } f(x) \\ &\text{subject to } x_L \leq x \leq x_U. \end{aligned} \tag{1.35}$$

The function $f(x)$ has a unique minimum x_* lying well within the range $[x_L, x_U]$. We say that x_* is an *interior minimum*. We may also call it an *unconstrained minimum*, in the sense that the constraints do not influence its location, that is, g_1 and g_2 are both inactive. It is possible though that problem (1.35) may result in all three situations shown in Figure 1.16. Therefore, if $\hat{x}_*$ is the minimum of the unconstrained function $f(x)$, the solution to problem (1.35) is generally given by selecting x_* such that it is

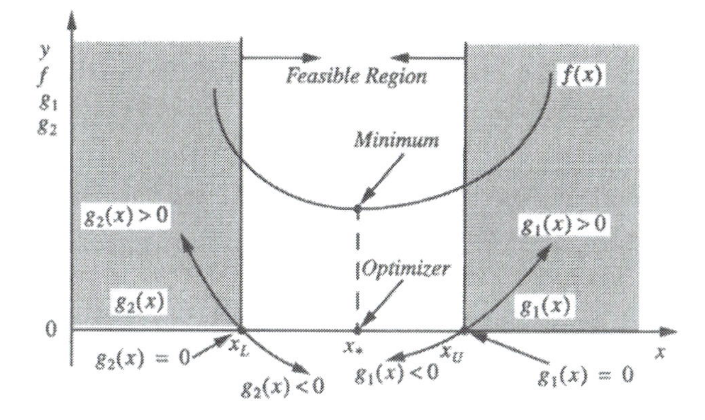

Figure 1.15. One-dimensional representation.

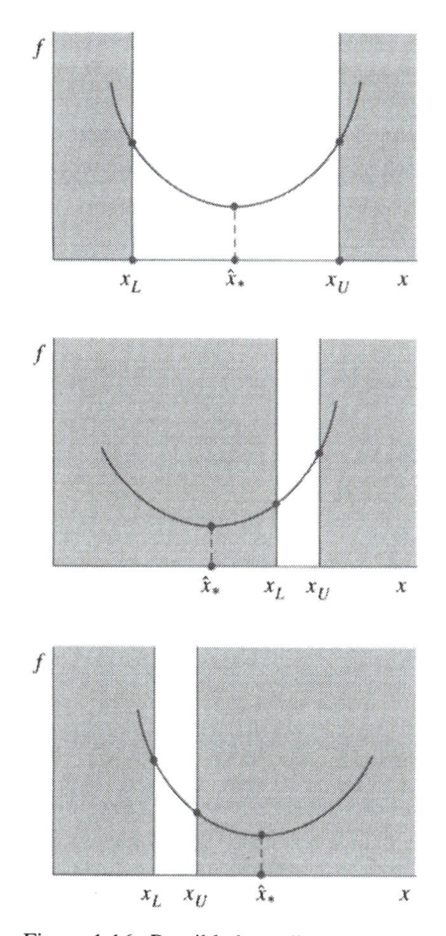

Figure 1.16. Possible bounding of minimum.

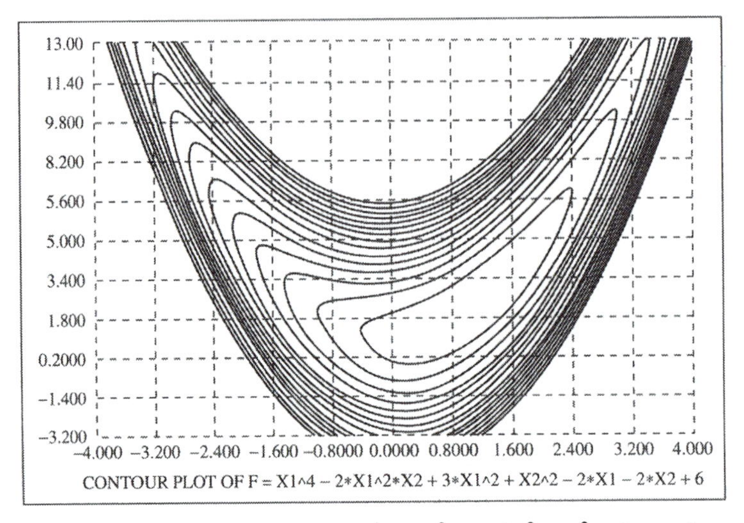

Figure 1.17. Contour plot for $f = x_1^4 - 2x_1^2x_2 + 3x_1^2 + x_2^2 - 2x_1 - 2x_2 + 6$.

the middle element of the set $(x_L, \hat{x}_*, x_U)$ ranked according to increasing order of magnitude, with $\hat{x}_*$ being the unconstrained optimum. In cases (b) and (c) where $x_* = x_L$ and $x_* = x_U$, respectively, the optima are *boundary optima* because they occur at the boundary of the feasible region.

In two-dimensional problems the situation becomes more complicated. A function $f(x_1, x_2)$ is represented by a *surface*, and so the feasible domain would be defined by the intersection of surfaces. It is obviously difficult to draw such pictures; thus a representation using the *orthogonal projection* common in engineering design drawings may be more helpful. Particularly useful is the top view giving contours of the surface on the (x_1, x_2) plane. It is the same picture as we can see on geographical maps depicting elevation contours of mountains and sea beds. Such a picture is shown in Figure 1.17 for the function

$$f(x_1, x_2) = x_1^4 - 2x_1^2x_2 + 3x_1^2 + x_2^2 - 2x_1 - 2x_2 + 6. \tag{1.36}$$

This function has a minimum at the point $(1, 2)^T$. Each contour plots the function f at specific values.

The objective function contours are plotted by setting the function equal to specific values. The constraint functions are plotted by setting them equal to zero and then choosing the feasible side of the surface they represent. The situation in the presence of constraints is shown in Figure 1.18, where the three-dimensional intersection of the various surfaces is projected on the (x_1, x_2) plane. The problem represented has the following mathematical form:

$$\begin{aligned}
&\text{minimize } f = (x_1 - 1)^2 + (x_2 - 1)^2 \\
&\text{subject to } g_1 := (x_1 - 3)^2 + (x_2 - 1)^2 - 1 \le 0, \\
&\qquad g_2 := 2x_1 - x_2 - 5 \le 0, \\
&\qquad x_1 \ge 0, x_2 \ge 0.
\end{aligned} \tag{1.37}$$

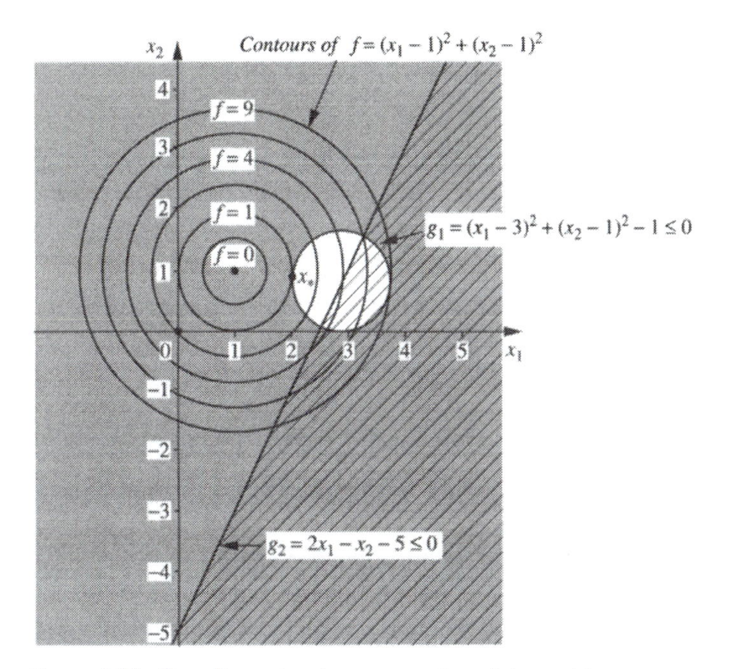

Figure 1.18. Two-dimensional representation of the problem.

Local and Global Optima

An unconstrained function may have more than one optimum, as shown in Figure 1.19. Clearly, points 2 and 5 are minima and 3 and 6 are maxima. Since their optimality is relative to the values of f in the local vicinity of these points, we will call them all *local optima*. Note, however, that point 1 is also a *local maximum*, albeit on the boundary, and similarly point 7 is a *local minimum*. From all the local minima, one gives the smallest value for the objective, which is the *global minimum*. Similarly, we may have a global maximum. There is also the horizontal inflection point x_4, which is neither a maximum nor a minimum although it shares with them the local quality of "flatness," which corresponds to having zero value for the derivative of f. Such points are all called *stationary*, and they can be maxima, minima, or horizontal inflection points.

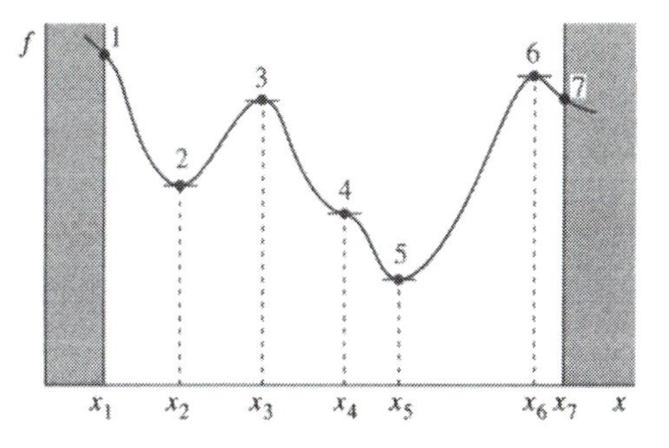

Figure 1.19. Stationary points in one dimension, showing local and global optima.

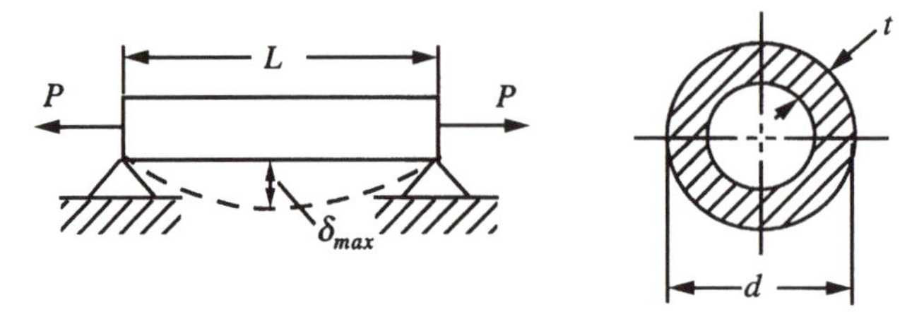

Figure 1.20. Hollow cylinder in tension (after Spunt 1971).

The question of finding global solutions to optimization problems is an important one but is as yet unanswered by the general optimization theory in a practical way. Situations where practical design results may be achieved are described in Wilde (1978). In design problems we may try to bypass the difficulty by arguing that in most cases we should know enough about the problem to make a good intuitive judgment. Yet this attitude would limit us in the use of optimization, since we would be only looking for refinements of existing well-known designs and would exclude the search for perhaps new and innovative solutions.

An observation related to local and global optima is that the optimum (local or global) may not be *unique*. To illustrate this point we will examine a simple example.

The problem is to design a member in tension, as shown in Figure 1.20 (Spunt 1971). The configuration chosen is a hollow cylinder because one of the design requirements is good transverse stiffness demanding the maximum transverse deflection, due to the member's own weight, to be limited to 0.1 percent of its length. The transverse stiffness is evaluated for simple supports at the end points, in a horizontal configuration and under zero axial load. Failure in yielding must also be considered. After choosing a minimum weight objective w, the problem is first written as

$$\begin{aligned}
\text{minimize } w &= \rho\pi dtL \\
\text{subject to } \sigma &\le S_{yt}, \\
\delta_{\max} &\le 0.001L, \\
\sigma &= P/\pi dt, \\
\delta_{\max} &= 5qL^4/384EI, \\
q &= \rho\pi dt, \\
I &\approx \pi d^3 t/8, \\
t &\ge 0.05.
\end{aligned} \tag{1.38}$$

The meaning of the symbols is as follows for a specified steel AISI 1025: $\rho =$ density, 0.283 lb/in^3; $S_{yt} =$ yield strength, 36 kpsi; $E =$ Young's modulus, 30×10^6 psi; $L =$ length of member, 100 in; $P =$ axial load, 10,000 lb; $d =$ outside diameter of member, in; $t =$ wall thickness, in; $\delta_{\max} =$ maximum transverse deflection, in; $q =$ unit length load, lb/in; $I =$ section moment of inertia, in^4. The last inequality constraint says that for manufacturing reasons the thickness must have a minimum value of 0.05 in.

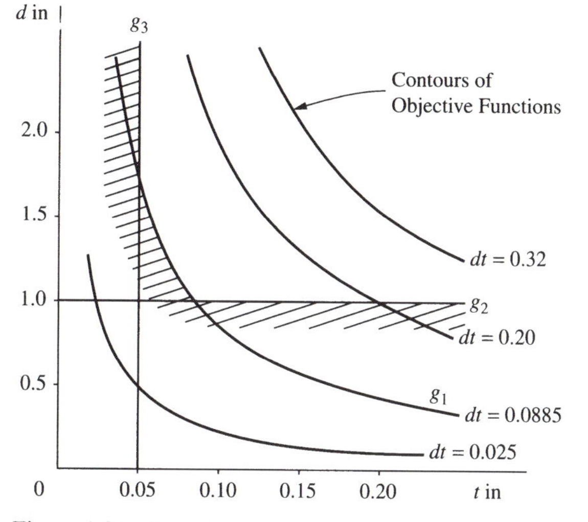

Figure 1.21. Graphical solution for the hollow cylinder example.

Assembling all the above relations and substituting numerical values for the parameters and rearranging, we get

$$\begin{aligned}
&\text{minimize} \quad w = 88.9\,dt \\
&\text{subject to} \quad g_1\colon dt \geq 0.0885, \\
&\qquad\qquad\quad g_2\colon d \geq 0.994, \\
&\qquad\qquad\quad g_3\colon t \geq 0.05.
\end{aligned} \tag{1.39}$$

The solution to this problem is given in graphical form using contour plots on the (d, t) plane in Figure 1.21. Note that the constraint $dt \geq 0.0885$ is always active, giving $w_* = 7.85$ lb over a whole range of values for d and t. There is an infinity of solutions, one of them being the simultaneous failure mode design given by $dt = 0.0885$ and $d = 0.994$. This infinity of solutions can be also detected analytically by observing that the first constraint imposes a stricter bound on dt than the combination of the other two.

Constraint Interaction

Changing parameter values in a given model may be responsible not only for shifting an optimum from the interior to the boundary and vice versa but also for shifting the global solution from one local optimum to another. Intuitively thinking, the best design for one application does not also have to be the best for another (described by another set of parametric values). In design optimization, a thorough study should aim at identifying all the local optima and determining the global one for different ranges of parameters. This may turn out to be a difficult task from the mathematical viewpoint, or expensive from the computational viewpoint.

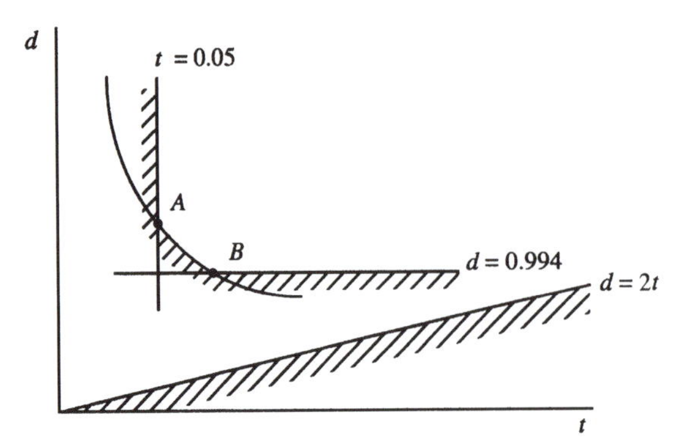

Figure 1.22. A redundant constraint.

If we look carefully at the previous example of the hollow cylinder, we will discover that there is a physical constraint not included in the model, namely, that

$$d \geq 2t \tag{1.40}$$

if the problem is to make any sense. If we resketch the solution and include the new, forgotten constraint (1.40), we get the situation in Figure 1.22. We see that the forgotten constraint actually lies in the infeasible region of the design space of the original model (1.39). It does not modify further the feasible domain at least in the region of interest, and its demand is "covered" by the other constraints. We say that in this case, the constraint is *redundant*. Its presence does not add anything to the problem (although this is not strictly true globally). We should not confuse a redundant constraint with an inactive one. *Redundancy pertains to feasibility while activity pertains to optimality.*

Now it was luck that (1.40) turned out to be redundant. The relative positions of the constraint lines in Figure 1.22 depend on the chosen values of the parameters. If we let

$$\begin{aligned} &C_0 = \rho\pi L, \quad C_1 = P/\pi S_{yt}, \\ &C_2 = (104\rho L^3/E)^{1/2}, \quad C_3 = 0.05, \end{aligned} \tag{1.41}$$

then the revised problem is written as

$$\begin{aligned} &\text{minimize } w = C_0 dt \\ &\text{subject to } g_1{:}\, dt \geq C_1, \quad g_3{:}\, t \geq C_3, \\ &\qquad\qquad\quad g_2{:}\, d \geq C_2, \quad g_4{:}\, d \geq 2t, \end{aligned} \tag{1.42}$$

and the positive values of C_1, C_2, and C_3 will determine the relative positions of the curves in Figure 1.22. What makes g_4 redundant is the presence of $d \geq 0.994$. We say that $d \geq 0.994$ *dominates* $d \geq 2t$. Similarly, in Figure 1.23(a), g_1 is redundant and is dominated by the combination of g_2 and g_3. However, g_4 is dominated

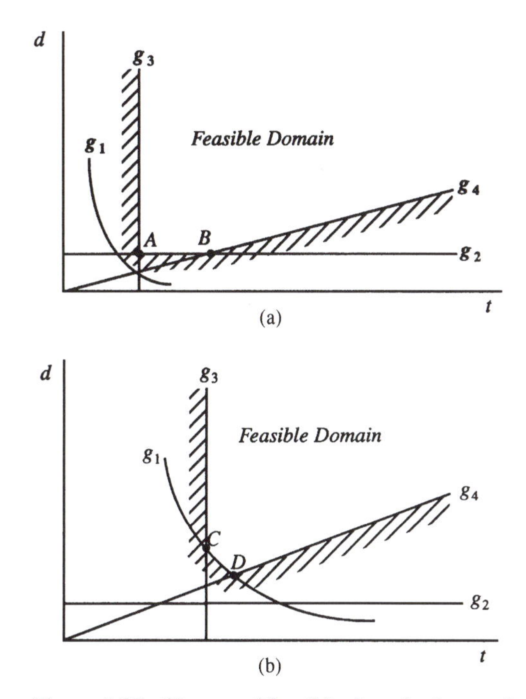

Figure 1.23. Change of feasible domain due to changes in parameter values.

by g_2 only up to point B, after which it takes over and dominates g_2. This situation is exaggerated in Figure 1.23(b) where g_4 dominates g_2 throughout. Note that for (a) the optimum is at point A, whereas at (b) it is in the entire range CD as earlier.

The concept of constraint dominance was used here in conjunction with redundancy, that is, the dominant constraint makes the dominated one redundant. However, the same concept of dominance can be used in conjunction with constraint activity – the dominant constraint may be active, making the dominated one inactive. Determining dominance, or regions of dominance among constraints, can enhance our capabilities of locating the optimum and interpreting it properly. In design terms, a dominant constraint will represent a requirement that is primarily responsible for feasibility and/or criticality at the optimum. A more rigorous study of these issues is undertaken in Chapter 3.

Another hidden constraint is actually included in the above model. The assumption of "thin" walls is necessary for theoretical validity of the entire model. What thin means depends on what accuracy we may take as acceptable, but a condition $d \geq 10t$ may be reasonable. This then should be another constraint in the model, which would clearly dominate $d \geq 2t$ and make it redundant. Moreover, if $d \geq 10t$ proves to be active, one should become immediately suspicious of the meaning of such an optimum, since its location would depend on a modeling limitation – not a physical one.

1.5 Modeling and Computation

The model, once complete, is a mathematical problem statement. Consequently, mathematical methods can be used to identify and compute the solution to the problem. This is obviously why the abstract model of the design is created in the first place. However, it would be a tactical error to address the solving task in a purely mathematical way divorced from the physical significance of the model and its elements. Optimization problems are generally difficult to solve mathematically. For example, defective models having no solution may not be easily recognized as such by a typical solution procedure, and much effort may be spent searching for something that does not exist. However, "best" designs need not be mathematically proven optima.

One main thesis in this book is that modeling and solving optimization problems must be viewed in a unified way. Understanding of the underlying physical significance of the model can allow rigorous simplification of the mathematical solution task. Conversely, rigorous study of the mathematical properties of the model can show modeling deficiencies and incomplete consideration of the design problem. Chapter 2 offers some guidance and examples for the initial construction of design optimization models. Chapter 3 on model boundedness presents rigorous ways for checking whether a model is properly constructed and for identifying obvious optima that lie on the boundary. The foundations of the method of *monotonicity analysis* are set there. The principles of monotonicity analysis are utilized later in Chapter 6 to perform rigorous model reduction before any numerical computation.

The typical solution methods in mathematical programming are iterative numerical searches. Starting at an initial point, a direction of improvement is identified and a step is taken in that direction. If the objective is to minimize, the new point should give a lower value for the objective function, while care is taken not to violate any constraints. The goal is to create a sequence of points that converges to the minimum, whether that is interior or on the boundary. What guides such a search is knowledge about mathematical properties of the minimizing point. In Chapters 4 and 5, these properties and some basic iterative strategies are developed for interior and boundary optima, respectively. The exposition there should give enough of a taste of the complexities involved so that the model simplification and reduction techniques of Chapter 6 would be properly motivated. The full power of modern iterative methods comes a bit later in Chapter 7.

Numerical iterative computation is driven by *local knowledge*, that is, knowledge pertaining to a single point in the design space and the neighborhood of this point. Model analysis and reduction methods are driven by *global knowledge*, that is, mathematical properties of the entire model and knowledge pertaining to *all* points of the design space. Global knowledge is superior and results in better techniques but it is difficult to acquire or the desirable properties may simply be nonexistent. Local knowledge is easy to get but its range is limited. A major goal of this book is to show how to use a combination of both types of knowledge to solve design problems.

1.6 Design Projects

A very effective way to study the ideas and methods in this book in a university course environment is through student term projects. Several exercises are aimed at such project work, and the checklists in Chapter 8 should prove useful. Experienced designers will also find this material helpful in their early attempts at applying the ideas of the book in their work.

A term project should be undertaken as early as possible. After studying the first two chapters, a student should be able to formulate the optimization problem and develop an initial mathematical model. As progress is achieved through the book, the various methods and ideas can be applied to modify, simplify, and eventually solve the optimization problem.

Typical project milestones are: a *project proposal* that contains the description on the design optimization trade-offs and initial mathematical model; an *interim report* that outlines efforts to analyze and simplify the model based primarily on the material in Chapters 3 through 6; and a *final report* that contains the final model statement, model reduction, numerical solutions, parametric studies, and interpretation of results based on material throughout the book.

Project topics may be assigned or chosen by the students. Both approaches have merit and in a typical class a mixture is usually required. For the students who do not have a project idea of their own, topics may be selected from journal articles published in the engineering literature. However, students should be strongly encouraged to take full responsibility in accepting a published model or problem. This usually forces a more than perfunctory study of the problem at hand and a familiarization with the model, its sources, and limitations, which is necessary for an eventually successful optimization study.

Modern mathematical software that combine modeling and symbolic and numerical computation capabilities are dramatically increasing the scope and ease of formulating and solving optimal design problems. This book offers many opportunities for the inspired reader to implement or test the ideas and methods presented using such software. These efforts are strongly recommended.

1.7 Summary

Design optimization models are representations of a decision-making process that is restricted to what can be modeled mathematically. A proper identification of the system and its elements that represent the design is crucial for successful formulation and solution of mathematical optimization models.

The important ideas of feasibility, boundedness, and constraint activity were introduced and were identified as areas where substantial effort will be spent in subsequent chapters. The complexity of the design optimization problem from the mathematical viewpoint became evident with examination of the topography of the design space. One may face not only multiple optima, including some on the boundary, but also changing values of the parameters that may drastically change the shape of the design space and shift the location of optima.

Small changes in the values of design parameters are studied by *sensitivity analysis*, which is a by-product of the optimization solution. Large changes in values are more useful but also more difficult to study. In general, a *parametric study* will involve solving the problem several times with different sets of parameter values, so that the optimal solutions can be found for a range of parameters. *Continuation* or *homotopy* methods can be used to perform these studies rigorously and efficiently. This is important because, as we may recall, the classification of variables and parameters is subjective: An important parameter may be switched to a variable status and the model resolved. Both sensitivity analysis and parametric study are sometimes referred to as *post-optimal analysis*. This final step is always desirable for a complete optimization study because it provides information about the nature of the optimum design, beyond a set of mere numerical values.

Design variables may be frequently required to take discrete values. Some configuration design problems can be successfully formulated as combinatorial optimization problems where the variables are binary (i.e., can take only 0 or 1 as values). Some design requirements may be formulated as "branch" functions that may be discontinuous or nondifferentiable. Occasionally such problems can be reformulated into an equivalent problem with continuous variables and (also differentiable) functions, but this is not always possible. It this book we address only the optimization of continuous and differentiable models. Algorithms for discrete or combinatorial models have substantially different mathematical structure, including concepts from probability theory and statistics.

Two themes in this book were highlighted and should be kept in mind while reading the following chapters. The first is that modeling and computation affect each other in a pervasive manner and should be viewed in an integrated way. The second is that local iterative methods and global analysis methods are complementary and both should be utilized for a successful design optimization study.

An underlying requirement of any optimization study is a good *quantitative* understanding of the behavior of the artifact or process to be optimized. A recurring criticism voiced against design optimization is that it is meaningless in view of the complexity of the real design situation. Accepting the wisdom of such caution, one will still argue that an optimization model does not require any more acceptance than traditional quantitative engineering analysis. Moreover, the continuing push for reducing design costs and cycle time using computer-based models makes the use of optimization tools inevitable.

Notes

Complete citations are given in the references at the end of the book.

Mathematical modeling is an extensive subject with a very large bibliography spanning all fields where quantification of phenomena has been attempted. Some general-purpose references are the texts by Bender (1978), Dym and Ivey

(1980), Gajda and Biles (1979), and Williams (1978). A particularly fascinating book is *Thinking with Models* by Saaty and Alexander (1981). Engineers may find rather mind opening the modeling of natural objects rather than artifacts as presented by Wainwright, Biggs, Currey, and Gosline (1982), or earlier by Alexander (1971).

Strong modeling emphasis is often given in introductory texts on dynamic systems. Examples are the texts by Beachley and Harrison (1978) and at a higher level by Wellstead (1979) and Fowler (1997), which contain rich methodologies for general systems. An extensive discussion of different types of mathematical models, including optimization and dynamic systems, can be found in *Patterns of Problem Solving* by Rubinstein (1975). A comprehensive collection of modeling examples from physics to music and sports, including optimization, can be found in the volume edited by Klamkin (1987).

An introduction to modern visualization techniques and their use in optimization studies is given in an appealing book by Jones (1996). Mathematical modeling in the context we use the term here, when implemented on a computer, is frequently referred to as *simulation*. See, for example, the books by Shannon (1975) and by Law and Kelton (1991). Complex simulation models can be approximated sometimes by simpler ones called *surrogate* models or *metamodels*; see, for example, Friedman (1996).

Design books in the various disciplines are good sources of models, although modeling usually must be extracted from the text because traditional design methods do not utilize an explicit modeling approach. There are several texts with a modeling bent – often including optimization: Carmichael (1982), Dimarogonas (1989), Jelen (1970), Johnson (1961, 1980), Leitman (1962), Miele (1965), Rudd, Powers, and Siirola (1973), Siddall (1972), Spunt (1971), Stark and Nichols (1972), Stoecker (1971, 1989), Ullman (1992), and Walton (1991).

Modeling objective functions and preferences in a multicriteria context has been studied extensively and is an altogether separate branch of decision analysis. Useful introductory references are the books by Saaty (1980), Roubens and Vincke (1985), Eschenauer, Koski, and Osyczka (1990), Statnikov and Matusov (1995), Hazelrigg (1996), Roy (1996), and Lootsma (1997).

Global optimization methods, both deterministic and stochastic, have received renewed attention in recent years, often in conjunction with heuristics or the handling of discrete variables. Useful references in this area are: Nemhauser and Wolsey (1988), Ratschek and Rokne (1988), Horst and Tuy (1990), Horst, Pardalos, and Thoai (1995), Floudas (1995), Pintér (1996), Ansari and Hou (1997) – which includes chapters on heuristic search, neural nets, simulated annealing, and genetic algorithms – Kouvelis and Yu (1997), and Glover and Laguna (1997). Specifically for genetic algorithms, useful introductions are the texts by Goldberg (1989) and Davis (1991).

Texts on design optimization often contain examples that demonstrate mathematical modeling. This is as good a place as any to list references on design optimization that may be consulted: Aris (1964), Arora (1989), Avriel and Golany (1996), Avriel,

Rijckaert, and Wilde (1973), Belegundu and Chandrupatla (1999), Bracken and McCormick (1967), Fox (1971), Haftka and Kamat (1985), Haug and Arora (1979), Himmelblau (1972), Johnson (1971), Kamat (1993), Kirsch (1981, 1993), Lev (1981), Mickle and Sze (1972), Morris (1982), Reklaitis, Ravindran, and Ragsdell (1983), Siddall (1982), Vanderplaats (1984), Wilde (1978), Zener (1971). An industrial viewpoint is given in Dixon (1976).

The technical journal literature is replete with examples of models and optimization. Examples are transaction journals of U.S. engineering societies such as AIAA, AIChE, ASCE, ASLE, and ASME. In ASME, optimization-related articles are often contributions of members of the Design Automation Committee and they appear in particular in the *Journal of Engineering for Industry* (up to 1974), *Journal of Mechanical Design* (1974–82 and 1990–present) and *Journal of Mechanisms, Transmissions and Automation in Design* (1982–90). Other journals are *Design Optimization, Engineering Optimization, Computer Aided Design, Journal of Optimization Theory and Applications, and Structural Optimization* (which includes multidisciplinary applications). Several computer-oriented journals often include articles with mathematical models, for example, *Artificial Intelligence in Design and Manufacturing* and *Integrated Computer-Aided Engineering*.

Exercises

1.1 Consult an elementary text in a selected discipline (e.g., thermodynamics or structures) and find an example of a system. Identify all its elements, its hierarchical level, and boundary. Does the text include a mathematical model? Define carefully the analysis model, including a list of assumptions. How could you use it as a design model?

1.2 Consult a design text in a selected discipline and find a worked-out example of an elementary design such as the belt drive example of Section 1.3. In that same spirit, describe the traditional design approach and then formulate an optimization model statement. Derive any results, if you can, and compare the two approaches. Particularly examine how many degrees of freedom you gained in the optimization model statement and how many constraints you think are active.

1.3 In the belt-drive model (1.31) verify that the objective function decreases as the center distance c increases.

1.4 Find or write a computer program that plots one-dimensional function plots and two-dimensional contour plots. Experiment using these tools to draw various functions. Do not try only polynomials, but include generalized polynomials (e.g., $f = c_1x_1^{\alpha}x_2^{\beta} + c_2x_1^{\gamma}x_2^{\delta}$ where $c_1, c_2, \alpha, \beta, \gamma, \delta$ are real) and transcendental functions (e.g., trigonometric and logarithmic).

1.5 Use the program of Exercise 1.4 to develop model representations such as those in Figures 1.17 and 1.18. For example, sketch and interpret figures for the following:

(a) $f(\mathbf{x}) = (x_2 - x_1)^4 + 8x_1x_2 - x_1 + x_2 + 3$

in the interval $-2 \leq x_i \leq 2$,

(b) $f(\mathbf{x})$ as above and the constraint

$g(x) = x_1^4 - 2x_2x_1^2 + x_2^2 + x_1^2 - 2x_1 \geq 0.$

1.6 Derive with rigorous analytical arguments the complete solution to the model given in (1.39).

1.7 Find values for the parameters $C_i (i = 0, 1, 2, 3)$ in problem (1.42) so that the feasible domain is represented by the schematics (a) and (b) in Figure 1.23, where g_4 is $d \geq 10t$ (not $d \geq 2t$).

1.8 Using the program of Exercise 1.4 plot the function

$$f(x) = -x_2 - 2x_1x_2 + x_1^2 + x_2^2 - 3x_1^2x_2 - 2x_1^3 + 2x_1^4$$

and examine its behavior around point $(1, 1)^T$. Can you identify a minimum for this function?

2

Model Construction

It seems that we reach perfection not when we have nothing more to add, but when we have nothing more to subtract.
Antoine de Saint-Exupéry (Terre des Hommes) (1900–1944)

Building the mathematical model is at least half the work toward realizing an optimum design. The importance of a good model cannot be overemphasized. But what constitutes a "good" model? The ideas presented in the first chapter indicate an important characteristic of a good optimal design model: *The model must represent reality in the simplest meaningful manner.* An optimization model is "meaningful" if it captures trade-offs that provide rigorous insights to whoever will make decisions in a particular context. One should start with the simplest such model and add complexity (more functions, variables, parameters) only as the need for studying more complicated or extensive trade-offs arises. Such a need is generated by a previous successful (and simpler) optimization study, new analysis models, or changing design requirements. Clearly the process is subjective and benefits from experience and intuition.

Sometimes an optimization study is undertaken *after* a sophisticated analysis or simulation model has already been constructed and validated. Optimization ideas are then brought in to convert an analysis capability to a design capability. Under these circumstances one should still start with the simplest model possible. One way to reduce complexity is to use *metamodels*: simpler analysis models extracted from the more sophisticated ones using a variety of data-handling techniques. The early optimization studies are then conducted using these metamodels.

In this chapter we provide some examples of constructing simple design optimization models, primarily to help the reader through an initial modeling effort. Before we get to these examples we discuss some basic techniques on creating metamodels, from the more classical forms of curve fitting to the newer ideas of neural networks and kriging.

2.1 Modeling Data

Design models often need to incorporate experimental data usually given in the form of tables or graphs. In this section we describe how data can be put into useful equation form. This is a very desirable modeling practice, whether for further analysis or for computation.

Graphical and Tabular Data

Data collected from experiments or other empirical sources are usually represented in tabular or graphical form. This is another way of depicting a function. In a mathematical model it is advantageous to convert such function representation into a more convenient algebraic form. This is the general idea behind *curve fitting*.

Polynomial representations are the most obvious way to curve fit. A function $y = f(x)$ is represented by an nth degree polynomial

$$y = a_0 + a_1 x + a_2 x^2 + \cdots + a_n x^n \tag{2.1}$$

and the data are used to calculate the constants a_i. If we have $n + 1$ data points, the a_is are uniquely defined. If we have more than $n + 1$, some "best-fit" procedure must be used, as we will discuss soon.

A useful approximation tool is the *Taylor expansion*. If $f(x)$ has derivatives of every order it can be represented *exactly* by an *infinite* series of terms centered about a given point x_0, that is,

$$f(x) = \sum_{n=0}^{\infty} \frac{f^{(n)}(x_0)}{n!}(x - x_0)^n, \tag{2.2}$$

where $f^{(n)}(x_0)$ is the nth derivative at x_0 and $n! = 1 \cdot 2 \cdot 3 \cdot \ldots \cdot n$. Since an infinite series is not convenient, we replace (2.2) by the first N terms and a remainder R_N, which we assume to be a small quantity. Then we have

$$f(x) = \sum_{n=0}^{N} \frac{f^{(n)}(x_0)}{n!}(x - x_0)^n, \tag{2.3}$$

which is now an *approximation* of the function. The magnitude of the remainder will depend, in part, on how many terms we have included in (2.3).

Note the resemblance of the Taylor approximation (2.3) to the polynomial expression (2.1). The difference between a Taylor series representation and a general polynomial is that the Taylor series utilizes derivative information that is *localized* (Figure 2.1), whereas the polynomial fitting uses information from different points (Figure 2.2). Notice, however, that this distinction starts disappearing if the derivatives are calculated *numerically* with *finite differences*. This situation is typical of all numerical schemes involving derivatives and is discussed further in Section 8.2.

Returning to the curve-fitting question, suppose that we choose a linear fit for curve a (Figure 2.2). We read off the curve two pairs of values (x_0, y_0), (x_1, y_1) and form the system

$$\begin{aligned} y_0 &= a_0 + a_1 x_0, \\ y_1 &= a_0 + a_1 x_1, \end{aligned} \tag{2.4}$$

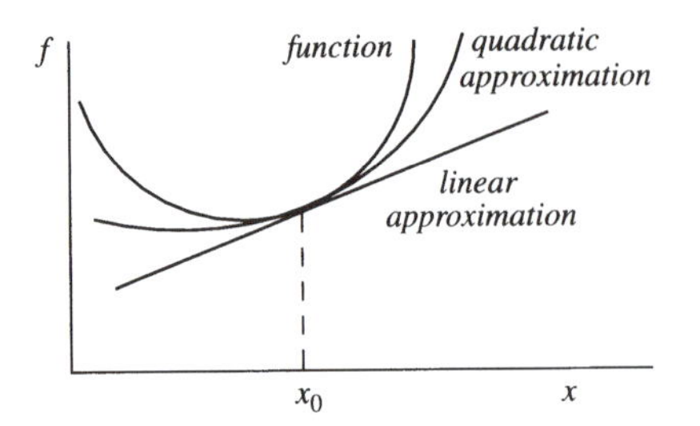

Figure 2.1. Linear and quadratic approximations at a point x_0.

which we solve for the unknowns a_0, a_1. For a quadratic fit, curve b in Figure 2.2, we use an additional pair (x_2, y_2) and have the system

$$\begin{aligned} y_0 &= a_0 + a_1x_0 + a_2x_0^2, \\ y_1 &= a_0 + a_1x_1 + a_2x_1^2, \\ y_2 &= a_0 + a_1x_2 + a_2x_2^2. \end{aligned} \tag{2.5}$$

This type of curve fitting requires solving a system of linear equations to determine the a_is. A method such as Gaussian elimination could be utilized for solution.

The pairs (x_i, y_i) can be read from a table instead of a curve. One advantage of having a curve is that we can decide where to pick the sampling points (x_i, y_i). For example, *equally spaced* points along the x axis simplify the solution of the linear system by having coefficients calculated symbolically in advance.

Functional forms, other than polynomials, can be used for curve fitting. One example is *generalized polynomials* such as

$$y = a + bx^m, \tag{2.6}$$

where a, b, and m are real numbers. Determining these constants is easy if we take the logarithms of (2.6), that is,

$$\log(y - a) = \log b + m \log x. \tag{2.7}$$

This relation, being linear in the logarithms, allows us to use a log–log plot to read off the values of b and m, for an estimated a. The estimate on a can be updated until a satisfactory fit is achieved (see Exercise 2.2).

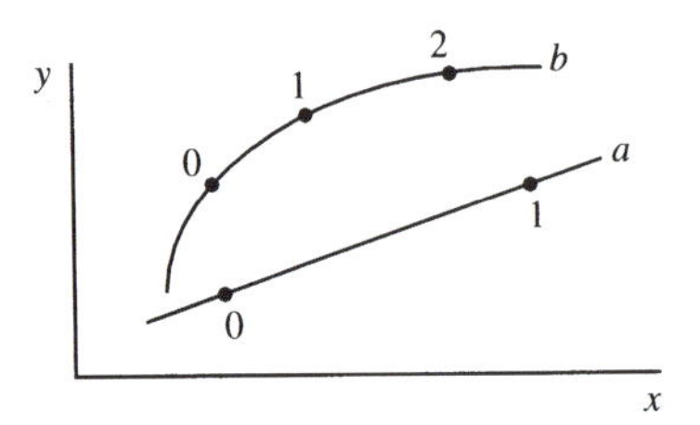

Figure 2.2. Polynomial curve fitting. Two points describing a linear equation a; three points describing a quadratic equation b.

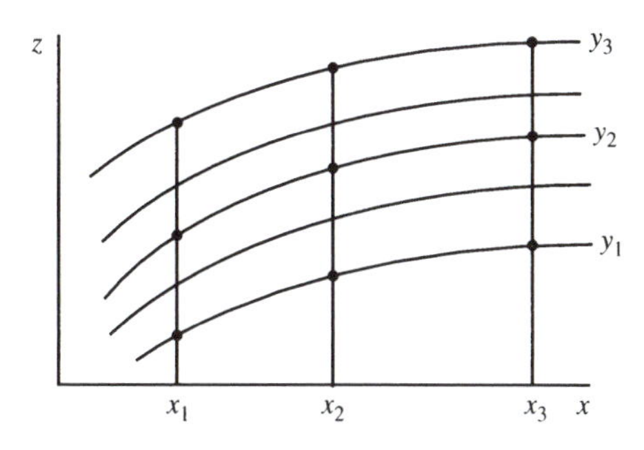

Figure 2.3. Family of curves.

We should stress here that in optimization models it is usually preferable to search for curve-fitting functions that result in as few terms as possible. High-degree polynomials are difficult to visualize and manipulate algebraically. Even for numerical computation, the use of simple fitting functions is a useful approach.

Families of Curves

Engineering data are often represented by a family of "parametric" curves, as in Figure 2.3. The quantity z is plotted against quantity x and different curves are drawn for different values of the "parameter" y. Examples in machine design are performance curves of pumps and motors, fatigue–strength diagrams for materials, geometry factors for gear layouts, and torque–speed–slip curves for fluid couplings. These families of curves are simply graphical representations of a function of two variables:

$$z = f(x, y). \tag{2.8}$$

Sometimes the function f is known from analytical considerations, and the graphical representation is just a convenient way to convey the behavior of the system. However, quite often the curves represent experimental data and the function f is not explicitly known. In this case, a two-dimensional curve-fitting procedure is required. We can do this by extending the ideas of the one-dimensional case.

Consider the situation in Figure 2.3 and assume that we decide to represent z as a function of x, for given y, by a quadratic:

$$z = a_0 + a_1 x + a_2 x^2. \tag{2.9}$$

At least some of the a_is will depend on y. Let us assume again that each a_i is a quadratic function of y:

$$a_i = A_{0,i} + A_{1,i} y + A_{2,i} y^2, \quad i = 0, 1, 2. \tag{2.10}$$

Thus, the total functional expression is given by

$$\begin{aligned} z = {} & (A_{00} + A_{10} y + A_{20} y^2) + (A_{01} + A_{11} y + A_{21} y^2) x \\ & + (A_{02} + A_{12} y + A_{22} y^2) x^2. \end{aligned} \tag{2.11}$$

To evaluate the nine coefficients A_{ij} we need nine data points $[(x, y), z]$, selected, for example, as shown in Figure 2.3. The result will be nine equations in nine unknowns that are solved to find the A_{ij}s.

This procedure can be used for more than two variables and is particularly useful if the functions are relatively simple. Obviously, any combination of functions may be used for the representation of the various coefficients. For example, one may decide that a function $z = f(x, y)$ is given by

$$z = a_1 x + a_2 x^2, \tag{2.12}$$

where

$$a_1 = A_0 + A_1 y, \qquad a_2 = B y^m, \tag{2.13}$$

so that the resulting curve-fitting equation is

$$z = (A_0 + A_1 y)x + B y^m x^2. \tag{2.14}$$

Now four data points $[(x, y), z]$ are needed for evaluation of the constants.

Numerically Generated Data

Any algorithm, whether it is based on an analysis or on a simulation model, can be viewed as a function, calculating an output for a given input. If we have many inputs and outputs, we can represent the algorithm by a system such as

$$\begin{aligned}
y_1 &= f_1(x_1, x_2, \ldots, x_n),\\
y_2 &= f_2(x_1, x_2, \ldots, x_n),\\
&\vdots\\
y_m &= f_m(x_1, x_2, \ldots, x_n).
\end{aligned} \tag{2.15}$$

The functions $f_1, f_2, \ldots, f_m$ may represent a complicated set of computations that has no explicit form. In some cases we can use the outputs $y_1, y_2, \ldots, y_m$ directly in some other part of the model. In other cases, however, it is desirable to represent these functions in a more explicit form.

Knowledge of what the functions $f_1, \ldots, f_m$ look like generally helps our intuitive understanding of the model represented by the algorithm. But there is often another practical reason for attempting to get the functions (2.15) in an explicit form: Algorithmic calculation of $\mathbf{y}$ from $\mathbf{x}$ may be computationally slow or expensive. In an iterative subsequent calculation that uses $\mathbf{y}$ values, this may become an impractical burden. In such cases a good approach may be to use the algorithm as a source of "computational experiments" that can supply us with the data points, just as if we had performed a physical experiment. Then curve-fitting or other model reduction techniques can be used based on these data points to derive new simpler functions that represent the functions (2.15) explicitly with some acceptable accuracy.

These new models are frequently referred to as *reduced order* models, *surrogate* models, or *metamodels*, depending on the techniques used to obtain them. Any subsequent use of the algorithm is replaced by the surrogate models containing the explicit functions. This may reduce drastically the computational load in a large design model that incorporates many analysis models. When a final design is reached, the original algorithms can be used to obtain more precise estimates. The development of optimization algorithms that can combine models of varying complexity while maintaining original convergence properties is an active area of current research.

2.2 Best Fit Curves and Least Squares

The curve-fitting methods we have examined so far assume that the number of data points available (or used) is exactly equal to the number of data points needed. For example, an nth degree polynomial is determined by exactly $n+1$ data points. When more points are available, this creates a difficulty: We would like to use the additional information, presumably for a better fit, but it would be exceptional to have the additional points fall on any one polynomial passing through any selected $n+1$ points.

We can easily find situations where the polynomial that has the *least total deviation* from the data points does not pass through any of the points. We would like to define the *best fit* polynomial through some formal procedure. One approach is to calculate the absolute values of the deviations from the data points and find the polynomial that minimizes the sum of these values. Another approach, which is generally considered preferable for our present purpose, is to find the polynomial that minimizes the sum of the squares of the deviations, a method known as a *least-squares fit*.

Consider the linear curve fitting first, with the equation

$$y = a_0 + a_1 x \tag{2.16}$$

to be fitted through m data points, that is, $(x_1, y_1), (x_2, y_2), \dots, (x_m, y_m)$. The deviation of a point (x_i, y_i) from that calculated from Equation (2.16) is $(a_0 + a_1 x_i - y_i)$. The least-squares problem is then to find the values of a_0 and a_1 so as to

$$\text{minimize} \sum_{i=1}^{m} (a_0 + a_1 x_i - y_i)^2. \tag{2.17}$$

The minimum is found by setting the partial derivatives with respect to a_0 and a_1 equal to zero, as explained in Chapter 4. This yields two equations, called the *normal* equations:

$$\begin{aligned} &\sum_i 2(a_0 + a_1 x_i - y_i) = 0, \\ &\sum_i 2(a_0 + a_1 x_i - y_i)x_i = 0. \end{aligned} \tag{2.18}$$

After rearranging and solving for a_0 and a_1, we get

$$a_0 = \frac{\sum x_i^2 \sum y_i - \sum x_i \sum x_i y_i}{m \sum x_i^2 - \left(\sum x_i\right)^2},$$
$$a_1 = \frac{m \sum x_i y_i - \sum x_i \sum y_i}{m \sum x_i^2 - \left(\sum x_i\right)^2}, \tag{2.19}$$

where all summations are over $i = 1, \ldots, m$. These equations give the *least-squares best linear fit*.

Using the techniques of Chapter 4, one can easily find that the normal equations for a least-squares fit of an nth degree polynomial using m data points are

$$\begin{pmatrix} m & \sum x_i & \sum x_i^2 & \ldots & \sum x_i^m \\ \sum x_i & \sum x_i^2 & \sum x_i^3 & \ldots & \sum x_i^{m+1} \\ \vdots & \vdots & \vdots & \vdots & \vdots \\ \sum x_i^m & \sum x_i^{m+1} & \sum x_i^{m+2} & \ldots & \sum x_i^{2m} \end{pmatrix} \begin{pmatrix} a_0 \\ a_1 \\ \vdots \\ a_n \end{pmatrix} = \begin{pmatrix} \sum y_i \\ \sum x_i y_i \\ \vdots \\ \sum x_i^m y_i \end{pmatrix}, \tag{2.20}$$

which must be solved simultaneously to obtain $a_0, a_1, \ldots, a_n$. Numerical procedures for solving this linear system efficiently exploit the structure of the coefficient matrix, particularly for large problems.

The least-squares technique can be applied to any function we may select for representing the data, not just polynomials. But using the least-squares method does not mean that the curve fitting is actually good. Two misuses of the method are shown in Figure 2.4. In the first case the method gives a correct answer that is meaningless since the data are scattered. In the second case, the degree of the fitting curve is too low, and the method gives the best possible *bad* solution. Visual inspection and preliminary checks can be very useful before one rushes to complicated analysis.

Polynomial curve fits have normal equations that are linear with respect to the parameters a_i, such as (2.20) and represent a *linear regression* problem. Curve fits such as $y = bx^m$ that can be linearized by using logarithms are called *intrinsically linear*.

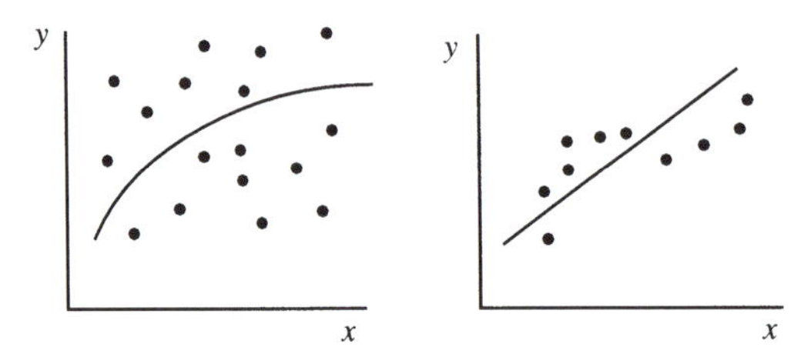

Figure 2.4. Least-squares curves used incorrectly.

However, the equation

$$y = [a_1/(a_1 - a_2)][\exp(-a_1x) - (-a_2x)] \qquad (2.21)$$

is *intrinsically nonlinear* because it cannot be linearized by transformations. Then we speak of *nonlinear regression*. Least-squares methods for nonlinear regression are significantly more difficult to implement, particularly for large problems. They comprise a special class of nonlinear programming methods.

The main difficulties we face when we work with least squares can be summarized as follows: (i) A linear or lower degree model is used for a curve that is actually of a higher degree (underfitting); (ii) a function form is too "wiggly" between data points and we try to follow it too closely (overfitting); (iii) when we increase the polynomial degree to achieve fidelity we also introduce ill-conditioning, due to the different order of magnitude between low and high degree terms; and (iv) as the model dimensionality increases it is difficult to decide how to include cross terms without overburdening the model.

Many models arising from engineering data are intrinsically nonlinear and considerable effort may be needed to construct a good model. However, in models used for design optimization, the accuracy of the model need not be great initially. As we will see in the discussion of monotonicity properties in Chapter 3, it is usually a *trend* of behavior that leads to constraint activity and eventual determination of the optimal design. Once the nature of the optimized design is understood, more detailed calculations with more accurate data and functions can be performed to finalize the values of the variables.

2.3 Neural Networks

Visual inspection of a given set of plotted data may suggest an obvious nonlinear relationship, but the exact relationship may not be apparent. A least-squares model determines the coefficients within an algebraic expression, but one must have some idea of the desired form of this algebraic expression before seeking the coefficient values.

An automated approach that addresses this need is the use of *neural networks* (*neural nets*, for short) to fit the data. The ideas behind neural nets are the same as nonlinear least squares: searching for a set of coefficients within an algebraic function that minimizes the error of the function evaluation when compared with data. However, the terminology and algebra are a little different.

In the same way that power terms a_mx^m can be viewed as building blocks for least-square fits, the building blocks of neural nets are the *neurons* or *nodes*. As shown in Figure 2.5(a), the output y of a neuron depends on its inputs $x_1, x_2, \ldots, x_n$, which correspond to stimuli in the biological analogy. In principle, each neuron behaves in a simple manner by providing an output between 0 and 1, depending on a weighted sum of its inputs. If this weighted sum is greater than a certain *bias*, which may be

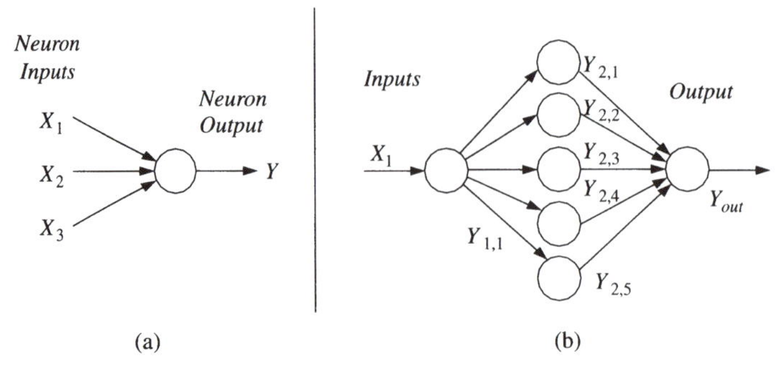

Figure 2.5. Schematic diagram of (a) single neuron, (b) neural net with $M = 5$ nodes in the hidden layer.

different for each neuron, the output will be "on" (greater than $1/2$); otherwise the output will be "off" (less than $1/2$).

One popular method of modeling an individual neuron uses the *logsig sigmoid function*. For a single input the function is defined in (2.22a) and plotted in Figure 2.6, and the extension for multiple inputs is defined in (2.22b):

$$y = \log \text{sig}(x) = \frac{1}{1 + \exp(b - wx)}, \tag{2.22a}$$

$$y = \log \text{sig}(x_1, x_2, \ldots, x_n) = \frac{1}{1 + \exp\left(b - \sum_{i=1}^{n} w_i x_i\right)}. \tag{2.22b}$$

Each neuron has its own weight w and bias b that define how the neuron behaves for a given input. The bias shifts the curve to the right or to the left and the weight determines the slope of the curve at the midpoint (Figure 2.6). For multiple inputs x_i

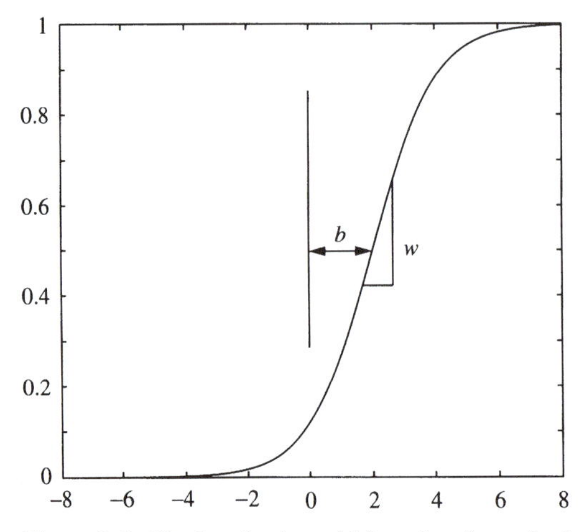

Figure 2.6. The logsig sigmoid function for a single input neuron, Equation (2.22a).

there is a weight w_i per input. The neuron combines the weighted inputs linearly, adds a bias, and gives an output.

To model complex functions neurons are strung together so that the inputs to some neurons are the outputs from other neurons, all linked in a neural net, as shown in Figure 2.5(b). The neural net combines several sigmoid functions that are shifted, stretched, added, or multiplied, in order to fit a given set of data. The output of each neuron depends only on its own inputs, and so the final output of the neural net in Figure 2.5(b) is given by the set of equations

$$\begin{aligned}
(1 + \exp(b_{1,1} - w_{1,1}x))^{-1} &= y_{1,1}, \\
(1 + \exp(b_{2,1} - w_{2,1}y_{1,1}))^{-1} &= y_{2,1}, \\
&\vdots \\
(1 + \exp(b_{2,5} - w_{2,5}y_{1,1}))^{-1} &= y_{2,5}, \\
b_{3,1} - \sum_{i=1}^{5} w_{3,i}y_{2,i} &= y_{\text{out}}.
\end{aligned} \tag{2.23}$$

There is one equation for each neuron, and each quantity $y_{2,i}$ is an intermediate value passed on to the next neuron. The middle layer of neurons that produces these intermediate values is often called the *hidden layer*, and so the neural net in Figure 2.5(b) can be described as *a neural net with a single hidden layer of five nodes*. Additionally, the sigmoid function has a value between 0 and 1; thus the last neuron often uses its weights and bias to scale the output to a more meaningful range.

Weights and biases are analogous to the coefficients in nonlinear least squares in that they determine what the response y_{out} is for any given stimuli x. Therefore weights and biases must be determined for a particular set of data before the neural net will become useful. The process of finding these weights and biases is known as *training the neural net*. Like all curve-fitting techniques, it is necessary to have some data $\{(x_1, y_1), (x_2, y_2), \ldots, (x_m, y_m)\}$ from the function to be approximated. The neural net needs to be trained to respond with the output y_k when the input is x_k. The extent to which the neural net reproduces the data is usually measured by the sum of the squares of the error, just as in nonlinear least squares:

$$\varepsilon = \sum_{k=1}^{m} (y_k - y_{\text{out}}(x_k))^2. \tag{2.24}$$

Numerical techniques such as those described in Chapter 4 are used to find the weights and biases such that ε is minimized. Neural net users should be warned that weights and biases are by no means unique, owing to their highly nonlinear nature. That is, two neural nets that have completely different weights and biases may model a particular set of data equally well. Judging when a neural net model is sufficiently accurate is somewhat of an art. One should always try to check a neural net's responses visually against actual data before using the net model approximation.

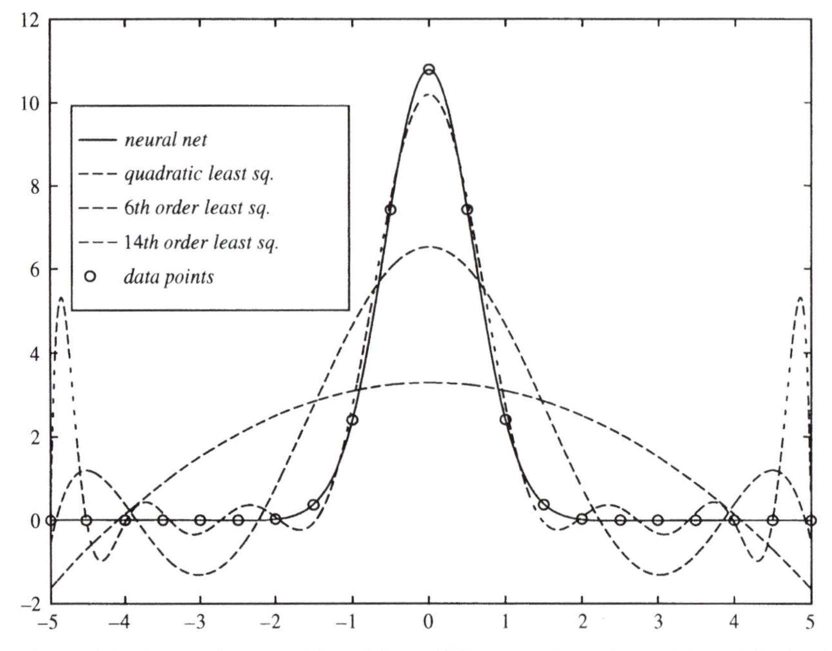

Figure 2.7. A neural net model and three different-order polynomial models for the same data.

Nonetheless, neural nets can be a quick and effective method of modeling nonlinear data, especially when no algebraic form is readily available. For example, Figure 2.7 compares the response of a neural net with four hidden nodes to least-squares polynomials of different orders.

2.4 Kriging

In the metamodels presented so far there resides an underlying assumption that the true function $y(x)$ is approximated by a function $f(x)$ with some error, namely, $y(x) = f(x) + \varepsilon$, and that the error ε assumes an independent, identically distributed normal distribution $N(0, \sigma^2)$ with standard deviation σ. However, one may assert that the errors in the predicted values are *not* independent; rather, they are a systematic function of x. This is based on the expectation that if the predicted value $f(x_i)$ is far from the true value $y(x_i)$, then a point $x_i + \delta$ very close to x_i will have a predicted value $f(x_i + \delta)$ that is also far from the true value $y(x_i + \delta)$.

The kriging metamodel, named after the South African geologist D. G. Krige who first developed the method, is an interpolation model comprised of two parts: a polynomial $f(x)$ and a functional departure from that polynomial, $Z(x)$:

$$y(x) = f(x) + Z(x). \tag{2.25}$$

Here Z is a stochastic Gaussian process that represents uncertainty about the mean of $y(x)$ with an expected value $E(Z(x)) = 0$ and a covariance for two points x and w

given by $\text{cov}(Z(x), Z(w)) = \sigma^2 R(w, x)$, where σ^2 is called the *process variance* and $R(w, x)$ is the *spatial correlation function*. The choice of $R(w, x)$ quantifies how quickly and smoothly the function moves from point x to point w and determines how the metamodel fits the data. A common expression used in kriging models is

$$R(w_i, x_i) = e^{-\theta |w_i - x_i|^2}, \tag{2.26}$$

where the subscript refers to the data points $i = 1, \ldots, m$ and θ is a positive model parameter. A spatial correlation function is designed to go to zero as $|w - x|$ becomes large. The implication is that the influence of a sampled data point on the point to be predicted becomes weaker as the two points get farther away from each other. The magnitude of θ dictates how quickly that influence deteriorates.

When the function to be modeled depends on n variables, the most common practice is to multiply the correlation functions for each dimension using the product correlation rule

$$R(w, x) = \prod_{j=1}^{n} R_j(w_j - x_j), \tag{2.27}$$

where $j = 1, \ldots, n$ indicates the dimension. A different spatial correlation function may be used for each R_j, including different choices for θ in Equation (2.26).

A polynomial choice in Equation (2.25) is of the form

$$f(x) = \beta_0 + \beta_1 x + \beta_2 x^2.$$

If only a constant term β is used for the polynomial (i.e., $f(x) = \beta$) the predicted values are found by

$$\mathbf{y}(x) = \beta + \mathbf{r}^T(x)\mathbf{R}^{-1}(\mathbf{y} - \mu\mathbf{I}) \tag{2.28}$$

where $\mathbf{y}$ is the $m \times 1$ vector of observed responses, $\mathbf{I}$ is an $m \times 1$ "identity vector" (a vector of unit elements), $\mathbf{R}$ is the $m \times m$ matrix of the correlations $R(x_i, x_j)$ among the design points, and $\mathbf{r}$ is the $m \times 1$ vector of the correlations $R(x, x_i)$ between the point of interest x and the sampled design points.

The only parameters that appear in the kriging metamodel are the coefficients β_i associated with the polynomial part $f(x)$ and the coefficients θ_i associated with the stochastic process $Z(x)$. Interestingly, when a function is smooth, the degree of the polynomial $f(x)$ does not affect significantly the resulting metamodel fit. Unlike other metamodels, there is no need to determine a specific functional form for the kriging metamodel, because $Z(x)$ captures the most significant behavior of the function. This is an important advantage of kriging models. Usually a simple constant term β is sufficient for a good $f(x)$ and occasionally a linear polynomial is used.

The generalized least-squares estimate for β can be calculated from the expression

$$\beta = (\mathbf{I}^T\mathbf{R}^{-1}\mathbf{I})^{-1}\mathbf{I}^T\mathbf{R}^{-1}\mathbf{y}, \tag{2.29}$$

where $\mathbf{R}$ is the $m \times m$ symmetric correlation matrix with ones along the diagonal. The estimated variance from the underlying global model (that is, the process variance),

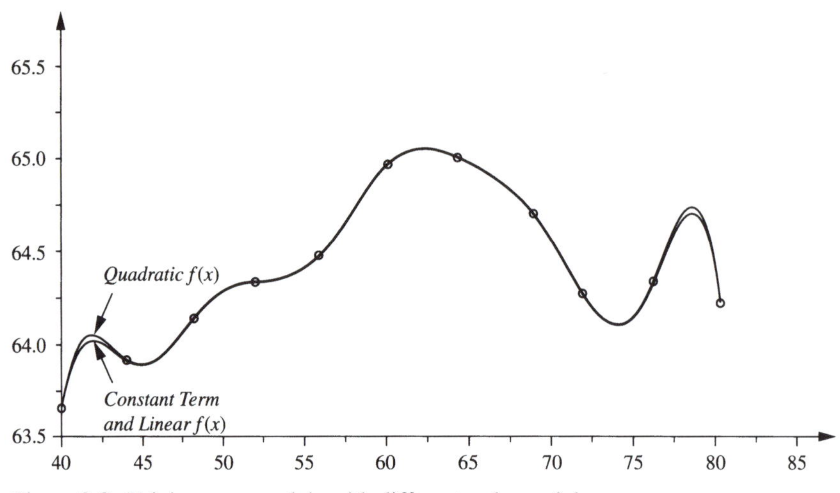

Figure 2.8. Kriging metamodels with different polynomials.

as opposed to the variance in the sampled data, is given by Sacks et al. (1989) as

$$\sigma^2 = [(\mathbf{y} - \mathbf{I}\beta)^T \mathbf{R}^{-1}(\mathbf{y} - \mathbf{I}\beta)]/m. \tag{2.30}$$

Both the variance σ^2 and the covariance matrix $\mathbf{R}$ are functions of the parameters θ_i. The θ_is can be found with a maximum likelihood estimate, namely, by maximizing the function

$$-(1/2)(m \times \ln(\sigma^2) + \ln|R|) \tag{2.31}$$

using the methods of Chapter 4. In standard computing packages for kriging models this calculation would be typically performed in a default way from the given set of data.

To illustrate the above ideas consider a problem in which eleven data points have been sampled as indicated in Figure 2.8. On the same figure we show kriging models derived using a commercial optimization program (LMS Optimus 1998), assuming constant, linear, and quadratic polynomial models for $f(x)$. The parameter θ is computed by the default method of the package. The kriging metamodel appears unaffected by the choice of polynomial model, as predicted by the theory.

The ability of kriging models to capture functional behavior is illustrated in Figure 2.9. The function modeled is

$$\begin{aligned} f(x_1, x_2) = {} & 2 + 0.01\left(x_2 - x_1^2\right)^2 + (1 - x_1)^2 + 2(2 - x_2)^2 \\ & + 7\sin(0.5x_1)\sin(0.7x_1x_2) \end{aligned}$$

and its actual graph is shown in Figure 2.9 in three-dimensional form (a) and projected on the x_1–x_2 plane (b). The kriging approximation is shown in Figures 2.9(c) and (d). The kriging model assumes a constant polynomial with $\beta = 15.54$, $\theta_1 = 25.94$, and $\theta_2 = 7.32$. A sampling grid with 21 points is used, shown as circles on the contour

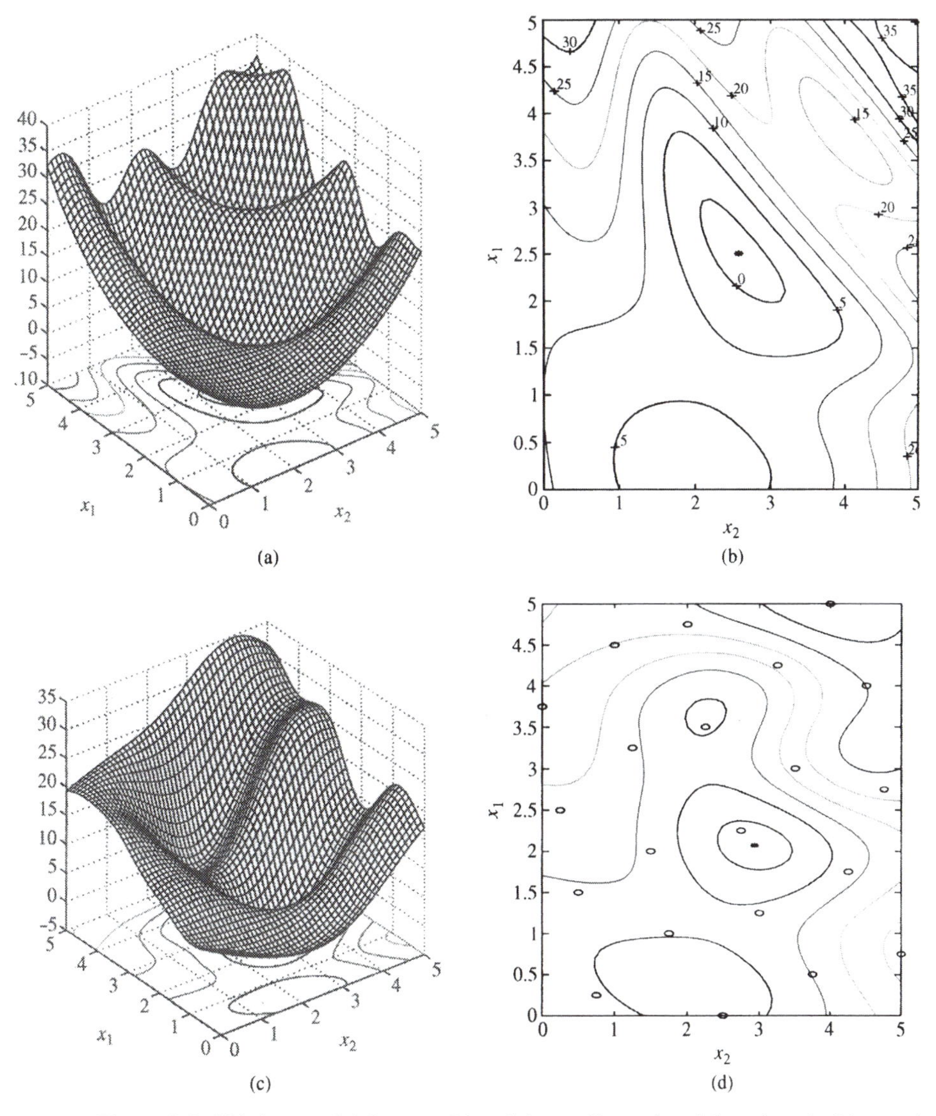

Figure 2.9. Kriging model for a multimodal two-dimensional function; (a, b) actual function, (c, d) kriging metamodel.

plot. The input values are rescaled in the interval [0, 1] to reduce ill-conditioning of the **R** matrix.

2.5 Modeling a Drive Screw Linear Actuator

We now present the first modeling example. The drive screw modeled here is part of a power module assembly that converts rotary to linear motion, so that a given load is linearly oscillated at a specified rate. The device is used in a household appliance (a washing machine). The system assembly consists of an electric motor,

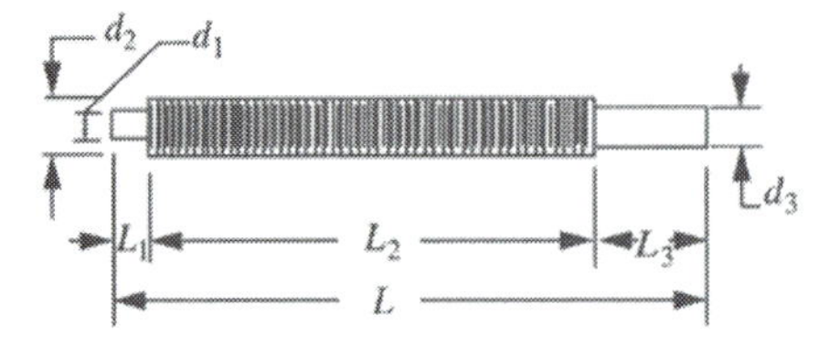

Figure 2.10. Schematic of drive screw design.

drive gear, pinion (driven) gear, drive screw, load-carrying nut, and a chassis providing bearing surfaces and support. The model addresses only the design of the drive screw, schematically shown in Figure 2.10. The drive screw can be made out of metal or plastic. Only a metal screw model is presented below.

Assembling the Model Functions

The objective function is to minimize product cost consisting of material and manufacturing costs. Since machining costs are considered fixed for relatively small changes in the design, only material cost is included in the objective, namely,

$$f = C_m(\pi/4)\big(d_1^2 L_1 + d_2^2 L_2 + d_3^2 L_3\big). \tag{2.32}$$

Here C_m is the material cost ($\$/\text{in}^3$) and d_1, d_2, and d_3 are the diameters of the gear/drive screw interface, the threaded, and the bearing surface segments, respectively. The respective segment lengths are L_1, L_2, and L_3.

There are operational, assembly, and packaging constraints. For strength against bending during assembly we set

$$(Mc_1/I) \leq \sigma_{\text{all}}, \tag{2.33}$$

where $M = F_a L/2$ is the bending moment, F_a is the force required to snap the drive screw into the chassis during assembly, and L is the total length of the part, namely,

$$L = L_1 + L_2 + L_3. \tag{2.34}$$

Furthermore, $c_1 = d_2/2$, $I = \pi(d_2^4/64)$ is the moment of inertia, and σ_{all} is the maximum allowable bending stress for a given material.

During operation, a constraint against fatigue failure in shear must be imposed:

$$KTc_3/J \leq \tau_{\text{all}}. \tag{2.35}$$

Here, K is a stress concentration factor, T is the applied torque, $c_3 = d_1/2$, $J = \pi(d_1^4/32)$ is the polar moment of inertia, and τ_{all} is the maximum allowable shear stress. The torque is computed from the equation

$$T = T_m c_2 N_s/N_m, \tag{2.36}$$

where T_m is the motor torque, $c_2 = 1/16$ (lb/ounce) is a conversion factor, and N_s and N_m are the numbers of teeth on the screw (driven) and motor (drive) gear, respectively.

To meet the specified linear cycle rate of oscillation, a speed constraint is imposed:

$$(c_4 N_m S_m)/(N_s N_T) \leq S, \tag{2.37}$$

where $c_4 = 60^{-1}$ (no. of threads/rev)(min/s) is a conversion factor, S_m is the motor speed (rpm), N_T is the number of threads per inch, and S is the specified linear cycle rate (in/s).

For the screw to operate in a drive mode a no-slip constraint must be satisfied (Juvinall 1983):

$$\left(W\frac{d_2}{2}\right)\left[\frac{\pi f d_2 + N_T^{-1}\cos\alpha_n}{\pi d_2 \cos\alpha_n - N_T^{-1} f}\right] \leq T. \tag{2.38}$$

Here W is the drive screw load, f is the friction coefficient, N_T^{-1} is the number of screw threads, and α_n is the thread angle measured in the normal plane.

There is an upper bound on the number of threads per inch imposed by mass production considerations,

$$N_T \leq 24. \tag{2.39}$$

From gear design considerations, particularly avoidance of tooth interference, limits on the numbers of gear teeth are imposed:

$$N_m \geq 8, \quad N_s \leq 52. \tag{2.40}$$

Packaging considerations impose the following constraints:

$$8.75 \leq L_1 + L_2 + L_3 \leq 10.0, \tag{2.41}$$

$$7.023 \leq L_2 \leq 7.523, \tag{2.42}$$

$$1.1525 \leq L_3 \leq 1.6525, \tag{2.43}$$

$$d_2 \leq 0.625. \tag{2.44}$$

There are also some obvious geometric inequalities on the diameters that can be expressed by the constraints

$$d_1 \leq d_2, \quad d_3 \leq d_2. \tag{2.45}$$

Note, however, that allowing any of these constraints to be active would lead to questionable designs. For example, if $d_1 = d_2$, there will be no shoulder step to facilitate assembly. It would be better to consider the relations (2.45) as a *set* constraint $d_1 < d_2$, $d_3 < d_2$ and not include them explicitly as constraint functions in the model.

Model Assumptions

There is a very important general practice that should be followed in all modeling efforts from the start: *Be aware and keep track of all modeling assumptions*. For example, in the model above several assumptions were invoked: To keep the objective a simple function of the material volume, manufacturing costs are assumed fixed and a high volume production is planned; "Standard Unified Threads" are used; the assembly force for the drive screw is concentrated at the midpoint; and frictional forces are considered only between threads and load nut, with friction assumed negligible everywhere else.

Assumptions are always made in the modeling process, as was discussed in Chapter 1. However, one must always check whether subsequent results from optimization conform to these assumptions, lest they are violated. Violation by the optimal values will indicate that the model used is inappropriate for the optimal design obtained and the optimization results are at least suspect and possibly erroneous. The remedy is usually a more accurate, probably also more complicated, mathematical model of the phenomenon under question. For example, Equation (2.33) is valid only if the length to diameter ratio is more than ten ($L/d_i \geq 10$). If the optimal solution violates this requirement, the simplification of treating the screw as a "beam" may be inadequate and a more elaborate model using elasticity theory may be necessary.

Including such *model validity constraints* explicitly in the optimization model, so that they should not be violated, is not appropriate. Finding a model validity constraint as active at the optimum would be meaningless, since the location of the optimal solution would be then dictated by model deficiencies. Model validity constraints should be considered as an open set constraint $\mathcal{X}$ and they must be inactive at the optimum. The proper inclusion of the beam assumption above in the model is the set constraint $L/d_i > 10$.

Model Parameters

Deciding on model parameters and assigning values is an important final step in the model construction. For example, although in the model above we looked at only a metal screw, a significant trade-off exists on the choice of using stainless steel versus plastic. A steel screw will have higher strength and be smaller in size but will require secondary processing, such as rolling of threads and finishing of bearing surfaces. A plastic screw would be made by injection molding in a one-step process that is cheaper, but more material would be used because plastic has a lower strength. Specialty plastics with high strength would be even more expensive and less moldable. Thus the choice of material must be based on a model that contains more information than the current one. A constant term representing manufacturing costs should be included in the objective. Indeed a more accurate cost objective should include capital investment costs for manufacturing.

Table 2.1. List of Parameters (Material Values for Stainless Steel)

C_m	material cost ($/in^3)
f	friction coefficient (0.35 for steel on plastic)
F_a	force required to snap fit screw into chassis during assembly (6 lb)
K	stress concentration factor (3)
L_1	length of gear/drive screw interface segment (0.405 in)
S	linear cycle rate (0.0583 in/s)
S_m	motor speed (300 rpm)
T_m	motor torque (2 in-ounces)
W	drive screw load (3 lb)
α_n	thread angle in normal plane (60°)
σ_{all}	maximum allowable bending stress (20,000 psi)
τ_{all}	maximum allowable shear stress (22,000 psi)

Nevertheless, substantial insight can be gained from the present model if we include material as a *parameter*; in fact each material is represented by four parameters: C_m, σ_{all}, τ_{all}, and f. It would be much more difficult to treat material as a variable, because then we would have four additional variables with discrete values and implicitly linked, perhaps through a table of material properties. As mentioned in Chapter 1, this would destroy the continuity assumed in nonlinear programming formulations and would introduce substantial mathematical complications.

Negative Null Form

The model is now summarized in the negative null form, with all parameters represented by their numerical values (Table 2.1), except for material parameters. In a model analysis that could follow based on the material in Chapter 3 and beyond, we would prefer to keep these parameters in the model with their symbols, rather than giving them numerical values as far as possible. This way we would attempt to derive results independently of the material used, substantially facilitating a post-optimal parametric study on the choice of material.

In the complete model assembly all "intermediate" variables (those defined through equalities) are eliminated together with the associated equality constraints by direct substitution. This should be done when possible, in order to arrive at a model with only inequality constraints, making constraint activity analysis much easier, as discussed in Chapter 3. Elimination of equality constraints is a *model reduction* step, since the number of design variables is reduced. Note however that model reduction may not be always a *model simplification* step, as the resulting expressions may become more complicated for analysis and/or computation. Some judgment must be exercised here. Occasionally, "directing an equality" (Section 3.6) may be useful in avoiding direct elimination.

In the model below, the intermediate variables M, L, T, I, c_1, c_3, and J together with the corresponding defining equalities have been eliminated. The variables left are d_1, d_2, d_3, L_2, L_3, N_m, N_s, and N_T. We have

$$
\begin{aligned}
&\text{minimize} f = C_m(\pi/4)\big(0.405 d_1^2 + d_2^2 L_2 + d_3^2 L_3\big) \\
&\text{subject to} \\
&g_1 = 38.88 + 96L_2 + 96L_3 - \pi\sigma_{\text{all}}\, d_2^3 \le 0, \qquad (2.46) \\
&g_2 = 6(N_s/N_m) - \pi\tau_{\text{all}} d_1^3 \le 0, \\
&g_3 = 8.345 - L_2 - L_3 \le 0, \\
&g_4 = -9.595 + L_2 + L_3 \le 0, \\
&g_5 = L_2 - 7.523 \le 0, \\
&g_6 = 7.023 - L_2 \le 0, \\
&g_7 = L_3 - 1.6525 \le 0, \\
&g_8 = 1.1525 - L_3 \le 0, \\
&g_9 = d_2 - 0.625 \le 0, \\
&g_{10} = 5(N_m/N_s) - 0.0583 N_T \le 0, \\
&g_{11} = 1.5 d_2 \left[\frac{\pi f d_2 + 0.5 N_T^{-1}}{0.5\pi d_2 - f N_T^{-1}}\right] - 0.125(N_s/N_m) \le 0, \\
&g_{12} = N_T - 24 \le 0, \\
&g_{13} = 8 - N_m \le 0, \\
&g_{14} = N_s - 52 \le 0,
\end{aligned}
$$

and $\mathcal{X} = \{$all variables positive, N_m, N_T, and N_s integer, $d_3 < d_2$, $d_1 < d_2\}$.

As there are no equality constraints, there are eight degrees of freedom corresponding to the eight design variables. Since N_m, N_T, and N_s must take integer values, this problem is a mixed continuous-integer nonlinear programming problem and its solution will require more than standard numerical NLP methods.

2.6 Modeling an Internal Combustion Engine

The second modeling example involves the design of an internal combustion engine from a thermodynamic viewpoint. The goal is to obtain preliminary values for a set of combustion chamber variables that maximize the power output per unit displacement volume while meeting specific fuel economy and packaging constraints. Two chamber geometries are studied: (1) a simple flat head design with one exhaust valve and one or two intake valves and (2) a compound valve head design.

Both models are based on fundamental thermodynamics relations, augmented by some empirical data. These explicit models begin with an expression for the ideal thermal efficiency using the basic air-cycle definition, and then adjust that to account for air–fuel effects, including exhaust gas recirculation and combustion time losses.

Table 2.2. Nomenclature for Engine Model

A_f	air/fuel ratio
b	cylinder bore, mm
BKW	brake power/liter, kW/l
$BMEP$	break mean effective pressure
c_r	compression ratio
C_s	port discharge coefficient
d_E	exhaust valve diameter, mm
d_I	intake valve diameter, mm
EGR	exhaust gas recirculation, %
$FMEP$	friction mean effective pressure, bars
h	compound valve chamber deck height, mm
H	distance dome penetrates head, mm
$IMEP$	indicated mean effective pressure, kPa
$isfc$	indicated specific fuel consumption, g/kW-h
N_c	number of cylinders
N_v	number of valves
Q	lower heating value of fuel, kJ/kg
r	radius of compound valve chamber curvature, mm
s	stroke of piston, mm
S_v	surface to volume ratio, mm^{-1}
V, v	displacement volume, mm^3
v_c	clearance volume, mm^3
v_d	dome volume, mm^3
V_p	mean piston speed, m/min
w	revolutions per minute at peak power, $\times 10^{-3}$
Z_b	RPM factor in volumetric efficiency
Z_n	Mach Index of port and chamber design
g	ratio of specific heats
η_v	volumetric efficiency
η_{vb}	base volumetric efficiency
η_t	thermal efficiency
η_{tad}	adiabatic thermal efficiency
η_{tw}	thermal efficiency at representative part load point
ρ	density of inlet charge, kg/m^3
ϕ	equivalence ratio

The resulting thermal efficiency value is further corrected for heat transfer losses in the engine and for engine speed effects. The symbols used in the models are summarized in Table 2.2.

Flat Head Chamber Design

The geometry for a flat head design is shown in Figure 2.11. The objective is to maximize the brake power per unit engine displacement *BKW/V*:

$$BKW/V = K_0(BMEP)w, \tag{2.47}$$

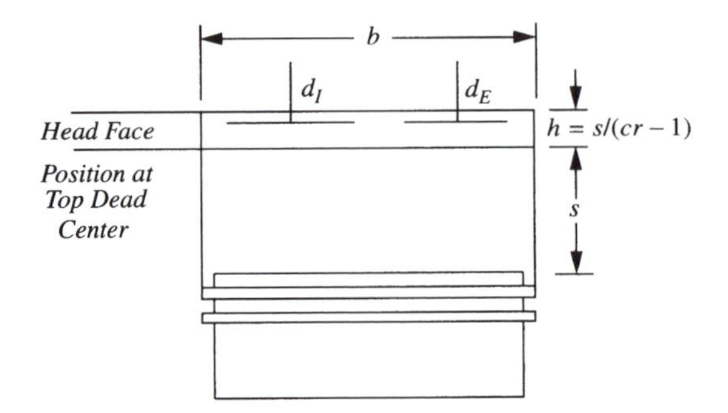

Figure 2.11. Schematic of geometry for flat head design.

where $K_0 = 1/120$ is a unit conversion constant and

$$BMEP = IMEP - FMEP. \tag{2.48}$$

Here *BMEP*, *IMEP*, and *FMEP* are brake, indicated, and friction mean effective pressures, respectively, and w is revolutions per minute at peak power.

The *IMEP* is computed from

$$IMEP = \eta_t \eta_v (\rho Q / A_f), \tag{2.49}$$

where η_t is the thermal efficiency, η_v is the volumetric efficiency, ρ is the density of inlet charge, Q is the lower heating value of fuel, and A_f is the air/fuel ratio. The term $(\rho Q/A_f)$ is the amount of energy available in the fuel–air mixture per unit volume. The volumetric efficiency accounts for flow losses and the product $\eta_v(\rho Q/A_f)$ is the energy per unit volume available in the mass inducted into the combustion chamber. The thermal efficiency accounts for the thermodynamics associated with the Otto cycle.

The volumetric efficiency can be expressed as

$$\eta_v = \eta_{vb}\left(1 + Z_b^2\right)/\left(1 + Z_n^2\right), \tag{2.50}$$

where η_{vb} is the base volumetric efficiency, Z_b is the rpm factor in volumetric efficiency, and Z_n is the Mach Index of the port and chamber design. The base volumetric efficiency for a "best-in-class" engine may be expressed empirically with a curve-fitting formula in terms of w (recall Section 2.2), as

$$\eta_{vb} = \begin{cases} 1.067 - 0.038e^{w-5.25} & \text{for } w \geq 5.25, \\ 0.637 + 0.13w - 0.014w^2 + 0.00066w^3 & \text{for } w \leq 5.25. \end{cases} \tag{2.51}$$

Also empirically (Taylor 1985), for a speed of sound of 353 m/s we set

$$Z_b = 7.72(10^{-2})w, \tag{2.52}$$

$$Z_n = 9.428(10^{-5})ws(b/d_I)^2/C_s, \tag{2.53}$$

where s is the piston stroke, b is the cylinder bore, d_I is the intake valve diameter, and C_s is the port discharge coefficient, a parameter characterizing the flow losses of a particular manifold and port design.

The thermal efficiency is expressed as

$$\eta_t = \eta_{\text{tad}} - S_v(1.500/w)^{0.5}, \tag{2.54}$$

where η_{tad} is the adiabatic thermal efficiency given by

$$\eta_{\text{tad}} = \begin{cases} 0.9\left(1 - c_r^{(1-\gamma)}\right)(1.18 - 0.225\phi) & \text{for } \phi \leq 1, \\ 0.9\left(1 - c_r^{(1-\gamma)}\right)(1.655 - 0.7\phi) & \text{for } \phi > 1. \end{cases} \tag{2.55}$$

Here c_r is the compression ratio, γ is the ratio of specific heats, and ϕ is the equivalence ratio that accommodates air/fuel ratios other than stoichiometric. In the optimal design model stoichiometry will be assumed, so that $\phi = 1$ and the two expressions in Equation (2.55) will give the same result. The thermal efficiency for an ideal Otto cycle is $1 - c_r^{(1-\gamma)}$ and the 0.9 multiplier accounts empirically for the fact that the heat release occurs over finite time, rather than instantaneously as in an ideal cycle. It is assumed valid for displacements in the order of 400 to 600 cc/cylinder and bore to stroke ratios typically between 0.7 and 1.3. Heat transfer is accounted for by the product of the surface-to-volume ratio of the cylinder, S_v, and an rpm correction factor. The surface-to-volume ratio is expressed as

$$S_v = [(0.83)(12s + (c_r - 1)(6b + 4s))]/[bs(3 + (c_r - 1))] \tag{2.56}$$

for the reference speed of 1,500 rpm and 60 psi of *IMEP*.

Finally the *FMEP* is derived with the assumption that the operating point of interest will be near wide open throttle (WOT), the point used for engine power tests, and that engine accessories are ignored. Under these conditions pumping losses are small and the primary factors affecting engine friction are compression ratio and engine speed. The resulting expression is

$$FMEP = (4.826)(c_r - 9.2) + \left(7.97 + 0.253V_p + 9.7(10^{-6})V_p^2\right), \tag{2.57}$$

where V_p is the mean piston speed given by

$$V_p = 2ws. \tag{2.58}$$

Constraints are imposed by packaging and efficiency requirements.

Packaging constraints are as follows. A maximum length of $L_1 = 400$ mm for the engine block constrains the bore based on a rule of thumb that the distance separating the cylinders should be greater than a certain percentage of the bore dimension. Therefore, for an in-line engine

$$K_1 N_c b \leq L_1, \tag{2.59}$$

where the constant $K_1 = 1.2$ for a cylinder separation of at least 20% of the bore, and N_c is the number of cylinders in the block. Similarly, an engine height limit of $L_2 = 200$ mm constrains the stroke

$$K_2 s \leq L_2, \tag{2.60}$$

where $K_2 = 2$. For a flat cylinder head, geometric and structural constraints require the intake and exhaust valve diameters to satisfy the relationship

$$d_I + d_E \leq K_3 b, \tag{2.61}$$

where $K_3 = 0.82$, and the ratio of exhaust valve to inlet valve diameter is restricted as

$$K_4 \leq d_E/d_I \leq K_5, \tag{2.62}$$

where $K_4 = 0.83$ and $K_5 = 0.87$. Finally, the displacement volume is a given *parameter* related to design variables by

$$V = \pi N_c b^2 s/4 \tag{2.63}$$

with 400cc $< V/N_c <$ 600cc for model validity reasons, as mentioned above.

We now examine efficiency-related constraints. To preclude significant flow losses due to compressibility of the fuel/air charge during induction the Mach Index of the port and chamber design must be less than $K_6 = 0.6$ (Taylor 1985):

$$Z_n = 9.428(10^{-5}) ws(b/d_I)^2/C_s \leq K_6. \tag{2.64}$$

The knock-limited compression ratio for 98 octane fuel can be represented by (Heywood 1980)

$$c_r \leq 13.2 - 0.045b. \tag{2.65}$$

The rated rpm at which maximum power occurs should not exceed the limits of the torque converter in conventional automatic transmissions, $K_7 = 6.5$. Therefore,

$$w \leq K_7. \tag{2.66}$$

Fuel economy at part load (1,500 rpm, $A_f = 14.6$) is a representative restriction on overall fuel economy. Therefore, a constraint is imposed on the indicated specific fuel consumption, *isfc*, at this part load:

$$isfc = 3.6(10^6)(\eta_{tw} Q)^{-1} \leq K_8, \tag{2.67}$$

where η_{tw} is the part-load thermal efficiency and $K_8 = 240$ g/kW-h.

In order to assign parameter values, we select specifications for a 1.9L four-cylinder engine. For this typical engine configuration maximizing power is of definite importance. The following values are then used for the parameters:

$$\begin{aligned} &\rho = 1.225\,\text{kg/m}^3,\ V = 1.859(10^6)\,\text{mm}^3,\ Q = 43,958\,\text{kJ/kg}, \\ &A_f = 14.6,\ C_s = 0.44,\ N_c = 4,\ \gamma = 1.33. \end{aligned} \tag{2.68}$$

The ratio of specific heats is computed from the expression

$$\gamma = 1.33 + 0.01(EGR/30) \tag{2.69}$$

with zero recirculation assumed.

Many of the expressions above (e.g., for friction or surface/volume) are valid only within a limited range of bore-to-stroke ratios, namely,

$$0.7 \leq b/s \leq 1.3. \tag{2.70}$$

This is again a *model validity constraint* that must be satisfied and be inactive at the optimum. As before, we will treat Equation (2.70) as a set constraint.

The model is now assembled after elimination of the stroke variable using the equality constraint on displacement volume, Equation (2.63), and is cast into negative null form, the objective being to *minimize* negative specific power (*BKW/V*).

MODEL A

Minimize $f = K_0(FMEP - (\rho Q/A_f)\eta_t\eta_v)w$ (in kW/liter),

where

$$FMEP = (4.826)(c_r - 9.2) + \left(7.97 + 0.253V_p + 9.7(10^{-6})V_p^2\right),$$
$$V_p = (8V/\pi N_c)wb^{-2},$$
$$\eta_t = \eta_{\text{tad}} - S_v(1.5/w)^{0.5},$$
$$\eta_{\text{tad}} = 0.8595\left(1 - c_r^{-0.33}\right),$$
$$S_v = (0.83)[(8 + 4c_r) + 1.5(c_r - 1)(\pi N_c/V)b^3]/[(2 + c_r)b],$$
$$\eta_v = \eta_{vb}[1 + 5.96 \times 10^{-3}w^2]\big/\left[1 + \left[(9.428 \times 10^{-5})(4V/\pi N_c C_s)\left(w/d_I^2\right)\right]^2\right],$$

$$\eta_{vb} = \begin{cases} 1.067 - 0.038e^{w-5.25} & \text{for } w \geq 5.25, \\ 0.637 + 0.13w - 0.014w^2 + 0.00066w^3 & \text{for } w \leq 5.25 \end{cases}$$

subject to (2.71)

$$g_1 = K_1 N_c b - L_1 \leq 0 \quad \text{(min. bore wall thickness)},$$
$$g_2 = (4K_2V/\pi N_c L_2)^{1/2} - b \leq 0 \quad \text{(max. engine height)},$$
$$g_3 = d_I + d_E - K_3 b \leq 0 \quad \text{(valve geometry and structure)},$$
$$g_4 = K_4 d_I - d_E \leq 0 \quad \text{(min. valve diameter ratio)},$$
$$g_5 = d_E - K_5 d_I \leq 0 \quad \text{(max. valve diameter ratio)},$$
$$g_6 = (9.428)(10^{-5})(4V/\pi N_c)\left(w/d_I^2\right) - K_6 C_s \leq 0 \quad \text{(max. Mach Index)},$$
$$g_7 = c_r - 13.2 + 0.045b \leq 0 \quad \text{(knock-limited compression ratio)},$$
$$g_8 = w - K_7 \leq 0 \quad \text{(max. torque converter rpm)},$$
$$g_9 = 3.6(10^6) - K_8 Q\eta_{tw} \leq 0 \quad \text{(min. fuel economy)},$$

where $\eta_{tw} = 0.8595\left(1 - c_r^{-0.33}\right) - S_v$

and $\mathcal{X} = \{\text{all variables positive}, 0.7 < b/s < 1.3\}$.

There are five design variables, b, d_I, d_E, c_r, and w, nine inequality constraints, and no equality constraints. All equalities that appear in the model above are simple definitions of intermediate quantities appearing in the inequalities. Significant

Table 2.3. Engine Design Specification Parameters (Base Case)

Parameter	Value	Specification
K_0	1/120	unit conversion, 4-stroke engine
K_1	1.2	cylinder separation as % of bore
K_2	2	engine height as a multiple of stroke
K_3	0.82	valve spacing as % of bore (flat head)
K_4	0.83	lower bound on valve ratio
K_5	0.89	upper bound on valve ratio
K_6	0.6	upper bound on Mach Index
K_7	6.5	upper bound on rpm
K_8	230.5 g/kW-h	upper bound on *isfc*
K_9	$2.4(10^6)$ mm^3	upper bound on displacement volume
K_{10}	$1.6(10^6)$ mm^3	lower bound on displacement volume
K_{11}	1/64	bore fraction spec. for deck height
K_{12}	0.125	bore fraction spec. for valve distance
L_1	400 mm	upper bound on engine block length
L_2	200 mm	upper bound on engine block height

parameters, for which numerical values were given in Equation (2.68), are maintained in the model with their symbols for easy reference in subsequent parametric post-optimality studies.

Parameter values dictated by current practice or given design specifications are indicated by the $K_i (i = 0, 1, \dots, 12)$ and $L_i (i = 1, 2)$ coefficients and summarized in Table 2.3.

In subsequent model analyses, such as the boundedness analysis described in Chapter 3, it may be convenient to rewrite Model A in a more compact functional form as follows:

MODEL A1

$$\min f(c_r, w, b, d_I) = K_0 w[FMEP(c_r, w, b) - P_0 \eta_t(c_r, b)\eta_v(w, d_I)]$$

subject to (2.72)

$$g_1(b) = b - P_1 \le 0 \quad \text{(min. bore wall thickness)},$$

$$g_2(b) = P_2 - b \le 0 \quad \text{(max. engine height)},$$

$$g_3(b, d_E, d_I) = d_I + d_E - K_3 b \le 0 \quad \text{(valve geometry and structure)},$$

$$g_4(d_E, d_I) = K_4 d_I - d_E \le 0 \quad \text{(min. valve diameter ratio)},$$

$$g_5(d_E, d_I) = d_E - K_5 d_I \le 0 \quad \text{(max. valve diameter ratio)},$$

$$g_6(w, d_I) = P_3 w - d_I^2 \le 0 \quad \text{(max. port and chamber Mach Index)},$$

$$g_7(c_r, b) = c_r - 13.2 + 0.045b \le 0 \quad \text{(knock-limited compression ratio)},$$

$$g_8(w) = w - K_7 \le 0 \quad \text{(max. torque converter rpm)},$$

$$g_9(c_r, b) = P_4 - 0.8595(1 - c_r^{-0.33}) + S_v(c_r, b) \leq 0 \quad \text{(min. fuel economy)},$$

and $\mathcal{X} = \{\text{all variables positive}, 0.7 < b/s < 1.3\}$.

The functions $FMEP(c_r, w, b)$, $\eta_t(c_r, b)$, $\eta_v(w, d_I)$, and $S_v(c_r, b)$ are abbreviations of the defining equalities in (2.71). Several *parametric functions* P_i have been introduced to simplify the presentation of the model. These are defined below with the numerical values corresponding to parameter values for the base case:

$$\begin{aligned} P_0 &= \rho Q/A_f = 3{,}688 \text{ kPa}, \\ P_1 &= L_1/K_1 N_c = 83.33 \text{ mm}, \\ P_2 &= (4K_2 V/\pi N_c L_2)^{1/2} = 76.90 \text{ mm}, \\ P_3 &= (9.428)(10^{-5})(4V/\pi N_c)/(K_6 C_s) = 215.46 \text{ mm}^2 \text{ s}, \\ P_4 &= 3.6(10^6)/K_8 Q = 1.1339(10^6) \text{ kg}^{-1}. \end{aligned} \tag{2.73}$$

The parametric functions above present an interesting observation: Parameter values may be correlated, namely, concurrent changes in some parameters may "cancel" each other and not affect the solution. For example, L_1 and K_1 may both increase in proportion to each other so that the value of P_1 remains the same or changes very little. It may be more interesting to conduct post-optimal parametric studies with respect to the P_is and then examine the effect of the original parameters on the optimal values.

Compound Valve Head Chamber Design

The geometry for the compound valve head design is shown in Figure 2.12. This new geometry will change the model above, adding new design variables and

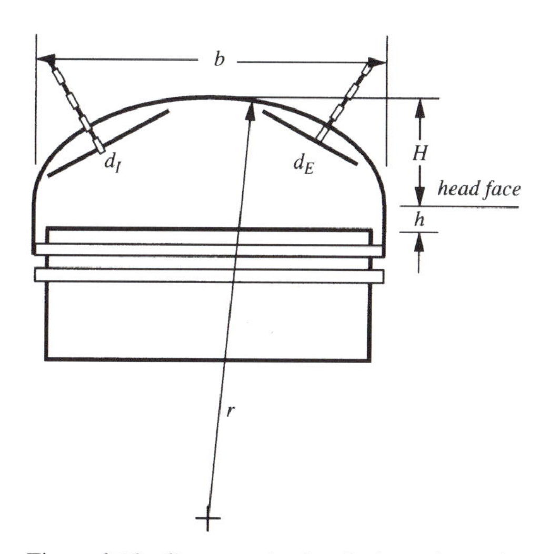

Figure 2.12. Compound valve design schematic.

constraints. The new variables are the displacement volume v (considered a parameter in the flat head model), the deck height h, and the radius of curvature r. A relationship among clearance volume v_c, displacement volume v, and compression ratio is imposed by the definition of the compression ratio:

$$c_r = (v/N_c + v_c)/v_c. \tag{2.74}$$

The clearance volume is the sum of deck volume and dome volume v_d,

$$v_c = \pi hb^2/4 + v_d, \tag{2.75}$$

where

$$\begin{aligned} v_d = {} & (1/3)\pi\left[(r^2 - b^2/4)^{1.5} - \left(r^2 - d_I^2/4\right)^{1.5} - \left(r^2 - d_E^2/4\right)^{1.5}\right] \\ & - \pi r^2[(r^2 - b^2/4)^{0.5} - \left(r^2 - d_I^2/4\right)^{0.5} - \left(r^2 - d_E^2/4\right)^{0.5}] - (2/3)\pi r^3. \end{aligned} \tag{2.76}$$

A typical design specification on the deck height is

$$h = K_{11}b, \tag{2.77}$$

where $K_{11} = 1/64$. A least distance of $K_{12}b$ must separate the two valves (Taylor 1985), where $K_{12} = 0.125$. Geometrically this can be approximated by setting

$$\left(d_I^2 - H^2\right)^{0.5} + \left(d_E^2 - H^2\right)^{0.5} \leq (1 - K_{12})b, \tag{2.78}$$

where H is the distance the dome penetrates the head and is defined as

$$H = r - (r^2 - b^2/4)^{0.5}. \tag{2.79}$$

In negative null form the problem of maximizing power per displacement for the compound valve head geometry becomes

MODEL B

Min $f(c_r, w, b, d_I) = K_0 w[FMEP(c_r, w, b) - P_0\eta_t(c_r, b)\eta_v(w, d_I)]$

subject to (2.80)

$h_1 = c_r - (v/N_c + v_c)/v_c = 0$ (compression ratio definition),

$v_c = \pi hb^2/4 + v_d$,

$$\begin{aligned} v_d = {} & (1/3)\pi\left[(r^2 - b^2/4)^{1.5} - \left(r^2 - d_I^2/4\right)^{1.5} - \left(r^2 - d_E^2/4\right)^{1.5}\right] \\ & - \pi r^2\left[(r^2 - b^2/4)^{0.5} - \left(r^2 - d_I^2/4\right)^{0.5} - \left(r^2 - d_E^2/4\right)^{0.5}\right] - (2/3)\pi r^3, \end{aligned}$$

$h_2 = h - K_{11}b = 0$ (deck height specification),

$$
\begin{aligned}
&g_1(b) = b - P_1 \leq 0 && \text{(min. bore wall thickness)},\\
&g_2(b) = (P_2/V^{0.5})v^{0.5} - b \leq 0 && \text{(max. engine height)},\\
&g_3(b, d_E, d_I) = \left(d_I^2 - H^2\right)^{0.5} + \left(d_E^2 - H^2\right)^{0.5} &&\\
&\qquad\qquad - K_{3c}b \leq 0 && \text{(min. valve distance)},\\
&\text{where } H = r - (r^2 - b^2/4)^{0.5}, &&\\
&g_4(d_E, d_I) = K_4 d_I - d_E \leq 0 && \text{(min. valve diameter ratio)},\\
&g_5(d_E, d_I) = d_E - K_5 d_I \leq 0 && \text{(max. valve diameter ratio)},\\
&g_6(w, d_I) = (P_3/V)vw - d_I^2 \leq 0 && \text{(max. port/chamber Mach Index)},\\
&g_7(c_r, b) = c_r - 13.2 + 0.045b \leq 0 && \text{(knock-limited compression ratio)},\\
&g_8(w) = w - K_7 \leq 0 && \text{(max. torque converter rpm)},\\
&g_9(c_r, b) = P_4 - 0.8595\left(1 - c_r^{-0.33}\right) &&\\
&\qquad\qquad + S_v(c_r, b) \leq 0 && \text{(min. fuel economy)},
\end{aligned}
$$

and $\mathcal{X} = \{\text{all variables positive}, 0.7 < b/s < 1.3, K_{10} < 4v/N_c < K_9\}$.

The empirical parameters have the same values as in the flat head case. In addition we have defined

$$K_{3c} = 1 - K_{12} = 0.875,\ K_9 = 2.4(10^6),\ K_{10} = 1.6(10^6). \tag{2.81}$$

In contrast to the flat head case the displacement volume is now treated as a variable; therefore a model validity set constraint on v is included. Two equality constraints are added to the model that can be directly eliminated in principle, as they are explicitly solvable for at least one of the design variables. Also, constraint g_3 has been rewritten.

This concludes the initial modeling effort for the engine example.

2.7 Design of a Geartrain

Multispeed gearboxes are used to provide a range of output rotational velocities for any input rotational velocity. This range is delivered using combinations of mating gears mounted upon parallel shafts. Determining the kinematic layout of the gearbox is a configuration design problem (Koenigsberger 1965). Input and output speeds are usually provided to the designer as requirements. The important decisions for the layout are the number of shafts, distances between shafts, and the number and placement of gears upon the shafts. Tooth and diameter sizing (related through the gear module) are based upon strength considerations.

The model here is taken from Athan (1994) and is based on previous work by Osyczka (1984). The layout used is shown in Figure 2.13. In Chapter 1 we mentioned that examining different configurations in the same mathematical design model is difficult. However, in the gearset here, a second configuration is considered easily

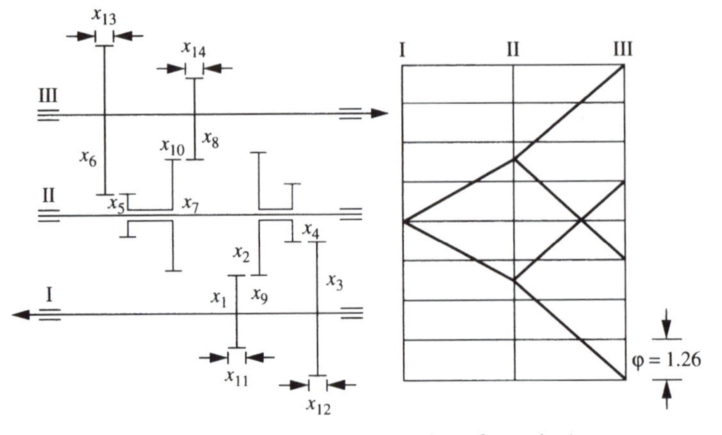

Figure 2.13. Gearbox speed layout (crossed configuration).

under the same model. The configuration in Figure 2.14 is termed "open" because no paths cross in the speed layout. The previous one is then termed "crossed." The difference between the two configurations is only their transmission ratios.

Model Development

As the model is somewhat involved we will only provide a summary development. For brevity, the nomenclature in Table 2.4 indicates the conventional symbols used in gear design and their corresponding symbols used in the mathematical model here. The entire model is summarized in Equation (2.87) below.

The problem is formulated as a multicriteria optimization problem with four objective functions to be minimized. The first objective is the volume of material used for the gears, approximated by the volume of a cylinder with diameter equal

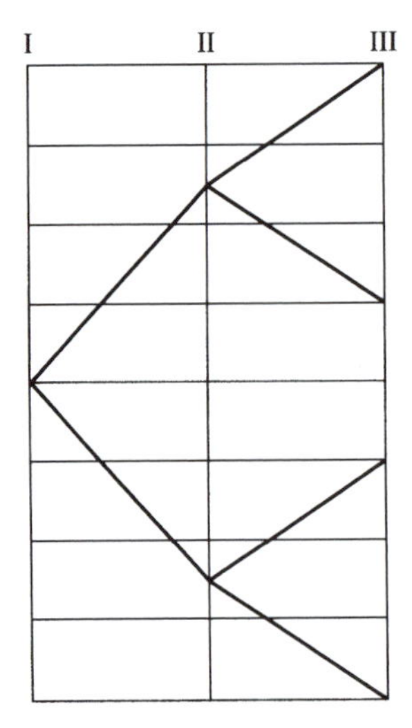

Figure 2.14. Gearbox speed layout (open configuration).

Table 2.4. Nomenclature for Gearbox Design

Symbol	Description
C_p	elastic coefficient $(\text{MPa})^{1/2}$
d_i	pitch circle diameter of the ith gear (mm)
D_i	shaft diameter (mm), $i = 1, 2, 3$
d_p	diameter of pinion (mm)
E_1, E_2	efficiencies of mechanisms on shafts I and II
F	face width (mm)
I	AGMA geometry factor for surface fatigue
J	AGMA geometry factor for bending
K_0	overload factor
K_m	load distribution factor
K_v	AGMA velocity or dynamic factor
m	module (mm)
m_g	speed ratio $(= N_g/N_p)$
m_i	module of the gearset (mm), $i = 1, 2$ $(= x_9$ and $x_{10})$
N_g	number of teeth on the gear
N_i	number of teeth in the ith gear $(= x_1, \ldots, x_8)$
N_p	number of teeth on the pinion
P	power on the input shaft (kW)
S_0	minimum rotational speed of the input shaft (rev/min)
S_i	specified output rotational speeds (rev/min), $i = 1, 2, 3, 4$
S_M	maximum rotational speed of the input shaft (rev/min)
V	pitch-line velocity (m/s)
W	fractional error allowed in the transmitted speeds
W_t	tangential force on the gear tooth (N)
z_i	face width of mating pinion gear (mm), $i = 1, 2, 3, 4$ $(= x_{11}, x_{12}, x_{13}, x_{14})$
ϕ	pressure angle (20°)
σ_P	permissible bending stress (MPa)
σ_H	permissible compressive stress (MPa)

to the addendum circle diameter and with height equal to the tooth width. The second objective is the maximum peripheral velocity, which occurs between gears with numbers of teeth N_7 and N_8. This velocity corresponds to operation at the highest rotational speed of the output shaft, S_4, and is pertinent to dynamic considerations, such as vibration and noise. The third and fourth objectives are the gearbox width and the distance between shafts, respectively. The design variables are the number of teeth on the gears $x_1, \ldots, x_8$, the modules of the gear wheels of the double-shaft assemblies, x_9 and x_{10}, and the tooth widths x_{11}, x_{12}, x_{13}, and x_{14}.

The geometry of the geartrain requires that all gear pairs that span the same distance between shafts must have the same combined length. This gives constraints g_1 and g_2. To ensure that the gear wheels will fit between the shafts, constraints g_3 to g_6 must hold. A pressure angle of 20 degrees is assumed for all gears. To prevent undercutting, a practical constraint requires all gears to have at least 14 teeth, yielding constraints g_7 through g_{14}. Another practical constraint requires the sum of teeth of the mating pinion and gear to be contained within the range [35, 100], yielding constraints g_{15} through g_{22}.

For space considerations, or when the numbers of gear teeth or the tangential velocities on the pitch circle have to be limited, it may be necessary to keep the transmission ratios between the two gears below 2 : 1 and above 1 : 4. This restriction also promotes approximately equal tooth wear between the gear and pinion. The resulting constraints are g_{23} to g_{30}. Limits imposed upon the deviation from the desired output speeds give constraints g_{31} through g_{38}. To ensure that the annular portion of the gear blank is strong enough to transmit the torque, a practical constraint requires that the pitch circle diameter of the gear should be greater than the bore diameter by at least 2.5 times the module, namely, constraints g_{39} through g_{46} are imposed.

Another practical design rule requires the face width to be 6 to 12 times the value of the module, constraints g_{47} to g_{54}. Gearsets having face widths greater than 12 times the module are susceptible to nonuniform distribution of the load across the face of the tooth because of torsional deflection of the gear and shaft, machining inaccuracies, and the difficulty of maintaining accurate and rigid mountings. If the face width is less than 6 times the module, larger gears are needed to carry the load per unit of face width. Large gears require more space in the gear enclosure and make the finished machine bigger and more expensive.

Contact ratio limits are important to gear design. Gears should not be designed having contact ratios less than about 1.4 since inaccuracies in mountings might reduce the contact ratio even more, increasing the possibility of impact between the teeth and also the noise level (Shigley 1977). These limits yield constraints g_{55} to g_{58}.

Stress constraints against failure of a gear tooth in bending are expressed as (Shigley 1977)

$$\sigma_P \geq W_t K_0 K_m / (K_v m_i J F), \tag{2.82}$$

with K_v determined from the Barth equation

$$K_v = 6/(6 + V). \tag{2.83}$$

An analytical expression for J is determined from Carroll and Johnson (1988):

$$J = (1.763476 + 17.36320/N_p + 6.676833/N_g)^{-1}. \tag{2.84}$$

$K_0 = 1.25$ for a uniform source of power and a moderate shock in the driven machinery, and $K_m = 1.25$. Thus bending stress considerations translate into the four constraints g_{59} to g_{62}.

To guard against failure by surface pitting due to contact stresses the general equation

$$(\sigma_H / C_p)^2 \geq W_t K_0 K_m / (K_v F d_p I) \tag{2.85}$$

can be used (Shigley 1977), where

$$I = ((\sin 2\phi)/4)(m_g/(m_g + 1)) \tag{2.86}$$

and $C_p = 191(\text{MPa})^{1/2}$ for a steel gear and pinion. The surface pitting consideration translates to the four constraints g_{63} to g_{66}. The material 826M40 with a Brinell Hardness Number of 341 and a permissible contact stress of 1,197.8 N/mm^2 is specified.

Finally, note that the number of gear teeth must be integers, and that the gear modules must be values from the discrete set [2.5, 3.0, 4.0, 5.0, 6.0, 8.0]. The tooth widths are continuous real numbers.

Some additional parameter values are as follows: $S_0 = 280$ rev/min, $S_M = 1,400$ rev/min, $S_1 = 56$ rev/min, $S_4 = 355$ rev/min, $E_1 = 1.0$, $E_2 = 0.97$, $P = 14.5$ kW, $D_1 = 46$ mm, $D_2 = 60$ mm, $D_3 = 100$ mm, $W = 0.02$, $\sigma_P = 393$ N/mm^2, and $\sigma_H = 1,197.8$ MPa. For the crossed configuration, $S_2 = 112$ rev/min and $S_3 = 180$ rev/min, while for the open configuration, $S_2 = 180$ rev/min and $S_3 = 112$ rev/min. These two parameter values constitute the only difference between the two configurations.

Model Summary

The mathematical model is now stated as follows:

Minimize

$$\begin{aligned}
c_1(\mathbf{x}) = 7.86(10^{-7})\big(&x_9^2 x_{11}((x_1+2)^2 + (x_2+2)^2) \\
&+ x_9^2 x_{12}((x_3+2)^2 + (x_4+2)^2) + x_{10}^2 x_{13}((x_5+2)^2 \\
&+ (x_6+2)^2) + x_{10}^2 x_{14}((x_7+2)^2 + (x_8+2)^2\big) \text{ dcm}^3,
\end{aligned}$$

$$c_2(\mathbf{x}) = 0.0732 x_3 x_7 x_{10}/x_4 \text{ m/s},$$

$$c_3(\mathbf{x}) = 2(x_{11} + x_{12} + x_{13} + x_{14}) \text{ mm},$$

$$c_4(\mathbf{x}) = 0.5(x_9(x_1 + x_2) + x_{10}(x_5 + x_6)) \text{ mm},$$

subject to constraints (2.87)

$$\begin{aligned}
&g_1\colon x_1 + x_2 - x_3 - x_4 = 0, && g_7\colon 14 - x_1 \le 0,\\
&g_2\colon x_5 + x_6 - x_7 - x_8 = 0, && g_8\colon 14 - x_2 \le 0,\\
&g_3\colon 46 + x_{10}x_5 - x_9(x_1 + x_2) \le 0, && g_9\colon 14 - x_3 \le 0,\\
&g_4\colon 46 + x_{10}x_7 - x_9(x_1 + x_2) \le 0, && g_{10}\colon 14 - x_4 \le 0,\\
&g_5\colon 100 + x_9x_2 - x_{10}(x_5 + x_6) \le 0, && g_{11}\colon 14 - x_5 \le 0,\\
&g_6\colon 100 + x_9x_4 - x_{10}(x_5 + x_6) \le 0, && g_{12}\colon 14 - x_6 \le 0,
\end{aligned}$$

$$\begin{aligned}
&g_{13}\colon 14 - x_7 \le 0, && g_{18}\colon x_3 + x_4 - 100 \le 0,\\
&g_{14}\colon 14 - x_8 \le 0, && g_{19}\colon 35 - x_5 - x_6 \le 0,\\
&g_{15}\colon 35 - x_1 - x_2 \le 0, && g_{20}\colon x_5 + x_6 - 100 \le 0,\\
&g_{16}\colon x_1 + x_2 - 100 \le 0, && g_{21}\colon 35 - x_7 - x_8 \le 0,\\
&g_{17}\colon 35 - x_3 - x_4 \le 0, && g_{22}\colon x_7 + x_8 - 100 \le 0,
\end{aligned}$$

$$g_{23}: 0.25 - x_1/x_2 \le 0,$$
$$g_{24}: x_1/x_2 - 2 \le 0,$$
$$g_{25}: 0.25 - x_3/x_4 \le 0,$$
$$g_{26}: x_3/x_4 - 2 \le 0,$$
$$g_{27}: 0.25 - x_5/x_6 \le 0,$$
$$g_{28}: x_5/x_6 - 2 \le 0,$$
$$g_{29}: 0.25 - x_7/x_8 \le 0,$$
$$g_{30}: x_7/x_8 - 2 \le 0,$$
$$g_{31}: S_1(1 - W) - (280x_1x_5)/x_2x_6 \le 0,$$
$$g_{32}: (280x_1x_5)/x_2x_6 - S_1(1 + W) \le 0,$$
$$g_{33}: S_2(1 - W) - (280x_1x_7)/x_2x_8 \le 0,$$
$$g_{34}: (280x_1x_7)/x_2x_8 - S_2(1 + W) \le 0,$$
$$g_{35}: S_3(1 - W) - (280x_3x_5)/x_4x_6 \le 0,$$
$$g_{36}: (280x_3x_5)/x_4x_6 - S_3(1 + W) \le 0,$$
$$g_{37}: S_4(1 - W) - (280x_3x_7)/x_4x_8 \le 0,$$
$$g_{38}: (280x_3x_7)/x_4x_8 - S_4(1 + W) \le 0,$$
$$g_{39}: 46 - x_9(x_1 - 2.5) \le 0,$$
$$g_{40}: 46 - x_9(x_3 - 2.5) \le 0,$$
$$g_{41}: 60 - x_9(x_2 - 2.5) \le 0,$$
$$g_{42}: 60 - x_9(x_4 - 2.5) \le 0,$$
$$g_{43}: 60 - x_{10}(x_5 - 2.5) \le 0,$$
$$g_{44}: 60 - x_{10}(x_7 - 2.5) \le 0,$$
$$g_{45}: 100 - x_{10}(x_6 - 2.5) \le 0,$$
$$g_{46}: 100 - x_{10}(x_8 - 2.5) \le 0,$$
$$g_{47}: 6 - x_{11}/x_9 \le 0,$$
$$g_{48}: x_{11}/x_9 - 12 \le 0,$$
$$g_{49}: 6 - x_{12}/x_9 \le 0,$$
$$g_{50}: x_{12}/x_9 - 12 \le 0,$$
$$g_{51}: 6 - x_{13}/x_{10} \le 0,$$
$$g_{52}: x_{13}/x_{10} - 12 \le 0,$$
$$g_{53}: 6 - x_{14}/x_{10} \le 0,$$
$$g_{54}: x_{14}/x_{10} - 12 \le 0,$$
$$g_{55}: 1.4 - \{[(x_9(x_1+2)/2)^2 - (x_1x_9\cos\phi/2)^2]^{1/2} - [(x_9(x_2+2)/2)^2 - (x_2x_9\cos\phi/2)^2]^{1/2} - (x_9(x_1+x_2)\sin\phi/2)\}/(px_9\cos\phi) \le 0,$$
$$g_{56}: 1.4 - \{[(x_9(x_3+2)/2)^2 - (x_3x_9\cos\phi/2)^2]^{1/2} - [(x_9(x_4+2)/2)^2 - (x_4x_9\cos\phi/2)^2]^{1/2} - (x_9(x_3+x_4)\sin\phi/2)\}/(px_9\cos\phi) \le 0,$$
$$g_{57}: 1.4 - \{[(x_{10}(x_5+2)/2)^2 - (x_5x_{10}\cos\phi/2)^2]^{1/2} - [(x_{10}(x_6+2)/2)^2 - (x_6x_{10}\cos\phi/2)^2]^{1/2} - (x_{10}(x_5+x_6)\sin\phi/2)\}/(px_{10}\cos\phi) \le 0,$$
$$g_{58}: 1.4 - \{[(x_{10}(x_7+2)/2)^2 - (x_7x_{10}\cos\phi/2)^2]^{1/2} - [(x_{10}(x_8+2)/2)^2 - (x_8x_{10}\cos\phi/2)^2]^{1/2} - (x_{10}(x_7+x_8)\sin\phi/2)\}/(px_{10}\cos\phi) \le 0,$$
$$g_{59}: 1.34859(10^6)(36 + 0.028\pi x_1x_9) \times (1.763476 + 17.36320/x_1 + 6.676833/x_2)/\left(\pi x_1x_{11}x_9^2\right) - 393 \le 0,$$
$$g_{60}: 1.34859(10^6)(36 + 0.028\pi x_3x_9) \times (1.763476 + 17.36320/x_4 + 6.676833/x_3)/\left(\pi x_3x_{12}x_9^2\right) - 393 \le 0,$$
$$g_{61}: 1.30813(10^6)(36x_2 + 0.028\pi x_1x_5x_{10}) \times (1.763476 + 17.36320/x_5 + 6.676833/x_6)/\left(\pi x_1x_5x_{13}x_{10}^2\right) - 393 \le 0,$$

$$g_{62}: 1.30813(10^6)(36x_2 + 0.028\pi x_1 x_7 x_{10})$$
$$\times (1.763476 + 17.36320/x_8 + 6.676833/x_7)/(\pi x_1 x_7 x_{14} x_{10}^2) - 393 \le 0,$$
$$g_{63}: 5.39434(10^5)(36 + 0.028\pi x_1 x_9)(x_1 + x_2)/(\pi \sin 2\phi x_{11} x_2 x_1^2 x_9^2)$$
$$- (44.2/191)^2 \le 0,$$
$$g_{64}: 5.39434(10^5)(36 + 0.028\pi x_3 x_9)(x_3 + x_4)/(\pi \sin 2\phi x_{12} x_4 x_3^2 x_9^2)$$
$$- (44.2/191)^2 \le 0,$$
$$g_{65}: 5.23251(10^5)(36x_2 + 0.028\pi x_1 x_5 x_{10})(x_5 + x_6)/(\pi \sin 2\phi x_{13} x_1 x_6 x_5^2 x_{10}^2)$$
$$- (44.2/191)^2 \le 0,$$
$$g_{66}: 5.23251(10^5)(36x_2 + 0.028\pi x_1 x_7 x_{10})(x_7 + x_8)/(\pi \sin 2\phi x_{14} x_1 x_8 x_7^2 x_{10}^2)$$
$$- (44.2/191)^2 \le 0.$$

The restrictions on the values of the variables mentioned above represent the set constraint.

Model Reduction

Some reduction to the apparent complexity of this model is possible by examining whether some constraints dominate others – which can then be deleted. The four contact ratio constraints g_{55}–g_{58} are simplified by cancelling x_9 from the first two and x_{10} from the others, so each becomes a function of just two gear number variables. Each function can be evaluated over the full range of allowable variable values, using a spreadsheet program, the results showing that the constraints that define the ranges for the gear number variables (g_7–g_{14}, g_{16}, g_{18}, g_{20}, and g_{22}) dominate the contact ratio constraints. Therefore they can be eliminated from the model but to avoid confusion the constraints are not renumbered.

The bending stress constraints, g_{59}–g_{62}, use the same variables as the contact stress constraints, g_{63}–g_{66}, and in a similar functional structure. The difference between their corresponding values can be found to be monotonic in all variables. Dominance cannot be proven definitively for the complete range of variable values. Yet the worst case scenario requires an unlikely combination of extreme values for the design variables. One may decide to eliminate the presumably dominated constraints during optimization, with the proviso that the resulting solutions will be checked for feasibility against these eliminated constraints.

Constraints g_1 and g_2 are equality constraints and can be used to eliminate variables x_3 and x_7 in only part of the model (i.e., the equalities are retained in the model). When substitutions are made in constraints g_{17}, g_{18}, g_{21}, and g_{22}, they become exactly the same as constraints g_{15}, g_{16}, g_{19}, and g_{20}, and therefore they can be eliminated.

A change of variables is considered next. Using $R_1 = x_1/x_2$, $R_2 = x_3/x_4$, $R_3 = x_5/x_6$, and $R_4 = x_7/x_8$ the equations are recast to explore possible simplifications.

Constraints g_{31}–g_{38} are rewritten as

$$\begin{aligned} &0.196 \leq R_1 R_3 \leq 0.204, \qquad 0.392 \leq R_2 R_3 \leq 0.408, \\ &0.63 \leq R_1 R_4 \leq 0.656, \qquad 1.2425 \leq R_2 R_4 \leq 1.293. \end{aligned} \tag{2.88}$$

But from g_{23}–g_{30}:

$$\begin{aligned} &0.25 \leq R_1 \leq 2.0, \qquad 0.25 \leq R_3 \leq 2.0, \\ &0.25 \leq R_2 \leq 2.0, \qquad 0.25 \leq R_4 \leq 2.0. \end{aligned} \tag{2.89}$$

Thus

$$\begin{aligned} &0.196/R_{3u} \leq R_1 \leq 0.204/R_{3l}, \quad 0.196/R_{1u} \leq R_3 \leq 0.204/R_{1l}, \\ &0.63/R_{4u} \leq R_1 \leq 0.656/R_{4l}, \quad 0.392/R_{2u} \leq R_3 \leq 0.408/R_{2l}, \\ &0.392/R_{3u} \leq R_2 \leq 0.408/R_{3l}, \quad 0.63/R_{1u} \leq R_4 \leq 0.656/R_{1l}, \\ &1.2425/R_{4u} \leq R_2 \leq 1.293/R_{4l}, \quad 1.2425/R_{2u} \leq R_4 \leq 1.293/R_{2l}, \end{aligned} \tag{2.90}$$

where the subscripts u and l represent upper and lower bounds on the respective variables. Substituting for $R_{3l} = 0.25$, $R_{3u} = 2.0$, $R_{4l} = 0.25$, and $R_{4u} = 2.0$ into Equations (2.90) gives

$$0.315 \leq R_1 \leq 0.816 \text{ and } 0.6212 \leq R_2 \leq 1.632. \tag{2.91}$$

Using these updated values for $R_{1l} = 0.315$, $R_{1u} = 0.816$, $R_{2l} = 0.6212$, and $R_{2u} = 1.632$ in Equations (2.90) again gives

$$0.2402 \leq R_3 \leq 0.6476. \tag{2.92}$$

But from g_{27}, $0.25 \leq R_3$ and so the new limits on R_3 are

$$0.25 \leq R_3 \leq 0.6476. \tag{2.93}$$

Similarly using now g_{30} we get

$$0.772 \leq R_4 \leq 2.0. \tag{2.94}$$

The new bounds for R_3 and R_4 do not change the range of feasible values for R_1 and R_2. In summary then, we have

$$\begin{aligned} &0.315 \leq R_1 \leq 0.816, \qquad 0.25 \leq R_3 \leq 0.6476, \\ &0.6212 \leq R_2 \leq 1.632, \qquad 0.772 \leq R_4 \leq 2.0. \end{aligned} \tag{2.95}$$

This result immediately implies that constraints g_{23}–g_{26}, g_{28}, and g_{29} can never be active and could thus be eliminated from the model. Also combining this information with the fact that the sum of the gear teeth must lie between 35 and 100, new ranges on the minimum and maximum values of x_1 to x_8 can be found. For example,

$$35 \le x_1 + x_2 \le 4.1746x_1 \quad \text{and} \quad 100 \ge x_1 + x_2 \ge 2.225x_1 \tag{2.96}$$

imply

$$14 \le x_1 \le 45. \tag{2.97}$$

Similarly,

$$\begin{aligned} &19 \le x_2 \le 76, \quad 4 \le x_3 \le 62, \quad 14 \le x_4 \le 61, \quad 14 \le x_5 \le 39, \\ &22 \le x_6 \le 80, \quad 16 \le x_7 \le 66, \quad 14 \le x_8 \le 56. \end{aligned} \tag{2.98}$$

Since the lower bounds on x_2, x_6, and x_7 were found to be higher than 14, constraints g_8, g_{12}, and g_{13} are dominated and discarded. Also, $g_{39} + g_{41} = g_{40} + g_{42}$ implies that one of the constraints (e.g., g_{41}) can be discarded. Similarly g_{44} can be discarded as well.

Thus a total of 23 constraints were eliminated during this model reduction phase leaving the model with 41 inequality and 2 equality constraints. The original design variables may be kept or the new variables R_i can be used to replace four of the original ones.

2.8 Modeling Considerations Prior to Computation

In this section we summarize some general ideas and procedures for analyzing the feasible domain prior to any attempts at computation.

Natural and Practical Constraints

Constraints can be classified into two categories. The first category we call *natural constraints*. These may express natural laws and experimental or geometric relations among the design variables. Such constraints describe functional relations of the design variables and can be either equality or inequality constraints. They must be included if a feasible and functioning design is to be achieved.

The second category contains the *practical constraints*, which are usually inequality constraints imposing simple bounds on the design variables. These bounds are either estimates of limits based on current common practice in the particular technology or simple reasonable assumptions about the feasible domain of the design variables. They are particularly necessary for numerical searches where only such a reduction of the feasible domain can make computations manageable.

When modeling a design problem for optimization, we should include at first only the natural constraints and exclude the practical ones. There are two reasons for this. First, the naturally constrained model may have an optimum that, although violating the practical constraints slightly, is significantly better than the optimum of the model that includes the practical constraints. The designer should be left to decide if the constraints can be relaxed slightly to include the optimal design. Second, finding an optimum at an estimated bound means that the optimum is only as good as the estimate of the bound. If the problem is so structured that the optimum lies always on the boundary for different bound values (a situation not uncommon as we will see in Chapters 3 and 6), then the optimum is essentially fixed by the adopted model. Thus, by excluding practical constraints, the designer avoids generating an artificial optimum limited by the model itself.

A problem modeled without practical constraints may become poorly bounded. This means that the problem may not have been posed properly. The designer must ensure that all physical constraints have been included. Alternatively, the objective function must be reexamined for the appropriate expression of the primary decision criterion. If, for example, the objective is cost, it may not include some cost factor that appears unimportant but which, when excluded, makes the optimization model unrealistic. Practical limits established with an (often hidden) objective in the engineer's mind should be uncovered and incorporated in a properly posed optimization model.

Sometimes, practical constraints are unavoidable because the extra modeling effort associated with a more precise and rigorous model is too great or too costly. Thus, although we may often include practical constraints in the design model, we should do this intentionally and always check whether the optimal design requires any of the practical constraints to be active. Inclusion of practical constraints in favor of a simpler objective function may often be the source of monotonicity properties in the problem. In such cases, the *monotonicity analysis* introduced in Chapter 3 will point out the influence of the practical constraints and provide a rational basis for deciding whether to pursue a more complicated model or not.

Theoretical limitations for a valid *analysis model*, on which the design model is based, require special attention. This was already mentioned in Sections 1.4 and 2.5. The majority of natural phenomena studied in engineering are described analytically by theories giving satisfactory results only within certain ranges of values for the variables (or parameters) involved. When such theories are used for designing, it is tacitly assumed that the design will be within the range of applicability of the theory. These assumptions will usually be included in the model as *inequality* constraints. The point here is that such *model validity* constraints must be considered as *inactive in advance*. If the search for the optimum leads to violation or even activity of such a constraint, then the result is of questionable utility. The validity of a theory at the boundary of its range of applicability may be unclear, with another theory possibly valid as well. Moreover, accepting the model validity limitation as a basis for optimal design can be dangerously misleading. In such cases, a closer look at the model is necessary. The designer/analyst may have to divide the design space into regions where different theories apply, examine them separately, and compare the results. Alternatively, a more general but elaborate theory that does not have the previous limitations can be used alone. In both situations, the modeling and analysis effort will increase substantially since analysis simplifications must be abandoned as unacceptable.

Asymptotic Substitution

The analytical expression of the objective or constraints often includes nonlinear terms known to be functions possessing upper and/or lower bounds. Elementary transcendental functions are typical examples. Bounds may also be needed to give the mathematical expression physical significance. For example, when x is a finite positive number, then

$$\begin{aligned} &0 \leq \sin x \leq 1, 0 \leq \cos x \leq 1, \exp x > 1, \\ &0 \leq \arctan x < \pi/2, 0 < \operatorname{arccot} x \leq \pi/2, \\ &1 \leq \cosh x, 1 < \coth x, 0 \leq \tanh x < 1. \end{aligned} \tag{2.99}$$

Usually such bounds are not included explicitly in the optimization models, although some measures may be taken to assure they are not violated. Examination of the model for such bounds may simplify the constraints. Moreover, it may keep a numerical algorithm from going astray.

Of particular interest are functions possessing asymptotes. Let $f_1(x_1) = f_2(x_2)$ be an equality constraint in which the monotonically increasing function $f_1(x_1)$ has the asymptotic value A as x_1 approaches x_A. This implies an additional constraint of the form

$$f_1(x_1) \leq A. \tag{2.100}$$

Assume further that the variable x_1 appears only in this equality constraint. Then the two constraints

$$f_1(x_1) \leq A \quad \text{and} \quad f_1(x_1) = f_2(x_2) \tag{2.101}$$

can be replaced in the model by the single constraint

$$f_2(x_2) \leq A. \tag{2.102}$$

The above simple modification, called *asymptotic substitution*, eliminates the asymptotic expression by deleting the "asymptotic" variable x_1.

If the asymptotic variable appears also in other constraints, it cannot be deleted. The asymptotic substitution can still be performed by breaking down the optimization study into two separate cases:

1. An "asymptotic solution" case, where the constraint containing the function with the asymptote is considered *active* so that (2.102) is equivalent to

$$f_2(x_2) = A. \tag{2.103}$$

2. An "interior solution" case where the constraint is considered *inactive*, so that (2.102) is equivalent to the set constraint

$$f_2(x_2) < A. \tag{2.104}$$

These two cases are generally easier to analyze and optimize separately and the results compared afterwards. More detaiis on case analysis are presented in Chapter 6.

Feasible Domain Reduction

A rigorous way of reducing the feasible domain is achieved by close examination of the constraints. Additional bounds on the design variables may often be implied by constraint interaction. This is sometimes referred to as *constraint propagation*, particularly in artificial intelligence approaches. To establish such bounds, manipulations of the constraints are necessary. The simple theorems of Chapter 3 can prove very useful in this. The analysis, more an art than a rigid procedure, requires only simple algebraic manipulations. The gearbox model provided some examples.

As another example, consider the following constraint set:

$$\begin{aligned} h_1 &= x_4 - \frac{0.4987x_1 \exp(-0.45x_2^{0.585})}{0.245x_1 + 0.0012} = 0, \\ h_2 &= x_4 - 1 - \frac{0.916x_3^2}{0.245x_1 + 0.0012} = 0, \\ h_3 &= x_1 + x_2 - 1 = 0, \end{aligned} \tag{2.105}$$

and $x_1 < 0.06$ (a set constraint), where all variables are strictly positive.

We can eliminate the variables x_2 and x_4 using h_1 and h_3. After some rearrangement, we get the set

$$\begin{aligned} h_2 &= [0.544F(x_1^+) - 0.267]x_1 - 0.0013 - x_3^2 = 0, \\ F(x_1^+) &\triangleq \exp[-0.45(1 - x_1)^{0.585}], \end{aligned} \tag{2.106}$$

the latter being an increasing function of x_1 defined with the symbol $F(x_1^+)$ for simplicity of representation. However, there is some implicit information in (2.105) that can be uncovered easily before further computation. From h_2 in (2.105) we get $x_4 > 1$; hence, h_1 and h_3 imply

$$0.4987x_1F(x_1^+) > 0.245x_1 + 0.0012. \tag{2.107}$$

Rearrangement gives a function form with a lower bound:

$$f(x_1) \triangleq 204x_1[2.044F(x_1^+) - 1] > 1. \tag{2.108}$$

For $x_1 = 0$, $F(0) = 0.6376$ so the factor $[2.044F(x_1^+) - 1]$ is always positive for every feasible value of x_1 within the range $0 < x_1 < 0.06$. But then $f(x_1)$ is increasing with respect to x_1. Since the solution of $f(x_1^+) = 1$ is $x_1 = 0.016$, the inequality $f(x_1^+) > 1$ is equivalent to $x_1 > 0.016$. Thus, the feasible domain for x_1 is really

$$0.016 < x_1 < 0.06. \tag{2.109}$$

With this information, constraint h_2 in (2.106) implies the following inequalities:

$$x_3^2 < [0.544F(0.06) - 0.267]0.06 - 0.0013,$$
$$x_3^2 > [0.544F(0.016) - 0.267]0.016 - 0.0013, \tag{2.110}$$

which can be solved to give the following range for x_3:

$$0.0012 < x_3 < 0.062. \tag{2.111}$$

This set constraint, implicit in model (2.105), can be explicitly used for further monotonicity analysis and for imposing rigorous bounds in local computation, should it be necessary. This would be preferable to often arbitrary practical bounds on the variables.

2.9 Summary

This chapter's discussion on curve fitting, regression analysis, neural networks, and kriging offers alternative ways to organize quantitative data in a form that both conveys the behavior of the system and is suitable for mathematical treatment. Each of these modeling strategies represents an area of expertise in itself. Their treatment in this chapter was introductory, limited to the goal of presenting them as a modeling idea. The modeling examples served to illustrate concepts introduced in Chapter 1.

In a class setting the development presented here would correspond to what might be included in a proposal for a term project: Explain the key design trade-offs and show how they can be quantified in a mathematical optimization problem statement.

Modeling considerations prior to computation are critical to the eventual success of a numerical algorithm. In practice, we may initiate some numerical computation with a model that may not be fully developed and analyzed. Preliminary numerical results (or inability to obtain them) can be used to "debug" the mathematical model, question modeling assumptions, rethink the constraints, or review parameter values. This deliberate interplay between modeling and computation is characteristic of a designer conversant with optimization tools.

Notes

Curve fitting is studied in many texts on numerical methods under the subject of interpolation. A very readable exposition is given by Hornbeck (1975). More in-depth study is provided in the classic texts by Dahlquist and Björck (1974) and by Carnahan, Luther, and Wilkes (1969). Least squares are covered in almost every book on numerical optimization methods. For design modeling, useful ideas can be found in Stoecker (1989) and Johnson (1980). A good reference for a more general approach to regression analysis is Draper and Smith (1981).

A good introductory exposition on neural nets can be found in Smith (1993). For a commercial computing tool see the *Matlab Neural Net Toolbox* (Matlab 1997, Beale and Demuth 1994).

Many scientists were made aware of kriging methods by Cressie (1988, 1990), but it wasn't until four statisticians wrote a paper on the topic (Sacks, Welch, Mitchell, and Wynn 1989) that kriging's appeal became more widely realized. The work was groundbreaking in that it attempted to link the relatively disjoint fields of statistical analysis and deterministic computer experiments. Commercial tools make the derivation of kriging models quite easy (LMS Optimus 1998). The material in Section 2.4 is based on M. Sasena's master's thesis (Sasena 1998).

The drive screw model is from a 1990 student class project by B. Alexander and B. Rycenga at Michigan based on a device at Whirlpool Corporation. The problem is fully explored and solved in Papalambros (1994) using monotonicity analysis. The engine model is due to Terry Wagner and was a starting point for the model in his dissertation (Wagner 1993) as well as for design synthesis tools at Ford Motor Company. The gearbox model is based on Athan (1994), which in turn was based on a design optimization class project conducted by Prashant Kulkarni and Tim Athan at Michigan.

Exercises

2.1 Derive general expressions for the coefficients of linear, quadratic, and cubic approximations, when the sampling points are equally spaced along the x-axis.

2.2 Consider an electric motor series cost model with the data given below (Stoecker 1971):

hp	Cost/$	$/hp
0.50	50	100.00
0.75	60	80.00
1.00	70	70.00
1.50	90	60.00
2.00	110	55.00
3.00	150	50.00
5.00	220	44.00
7.50	305	40.50
15.00	560	37.30

Derive the curve-fitting equation $\$/\text{hp} = 34.5 + 36(hp)^{-0.865}$.

Hint: Draw the curve using the table values and estimate a value for the constant term. For the steep part of the curve, draw its representation on a log–log plot to get values for the coefficient of the second term. Iterate as necessary.

2.3 Helical compression spring design is an often-used example of optimization formulation because of its simplicity. Formulate such a model with spring index and wire diameter as the two design variables. Choose an objective function (e.g., weight) and create as many constraints as you can think of. Typically, these include surging, buckling, stress, clash allowance, geometric limitations, and minimum number of coils. Select parameter values and find the solution graphically.

2.4 Sometimes the rate of flow of viscous substances can be estimated by measuring the rate that vortices are shed from an obstacle in the flow. This is the principle behind a vortex meter. A sensor gives a pulse every time a vortex passes and the volumetric rate of flow can be estimated by measuring the pulse rate. The (fictional) data in the table were taken to calibrate such a meter.

Fictional Data Representing the Pulse Rate of a Vortex Meter as a Function of the Velocity of the Fluid Passing the Meter

Flow Rate V	**Pulse Rate** ρ
1.18	1.28
1.45	1.65
1.83	2.12
2.36	2.72
3.14	3.49
4.26	4.48
5.91	5.75
8.39	7.38
12.1	9.49
18.1	12.1
27.8	15.6
44.0	20.0
72.1	25.7
123.0	33.1
218.8	42.5
407.8	54.5
798.3	70.1
1645.2	90.0
3573.9	115.5
8186.7	148.4

(a) Plot the data on a log–log scale. (b) Fit the data to the equation $\rho = aV^b$. (c) This fit can be improved; specifically, using the relation from (b) employ a neural net as a correction factor, namely, train a small neural net to fit the equation $\rho = \varphi(V)aV^b$ or, more appropriately, find a correction factor that is a function of V:

$$\varphi(V) = \frac{\rho}{aV^b}.$$

(d) Using the same log–log graph from part (a), plot the relations from parts (b) and (c), namely, plot V versus $\varphi(V)aV^b$.

2.5 Consider the case where there is no correlation between any of the data points. (a) If a constant term were used for $f(x)$, what would the kriging model degenerate to? (b) Consider the opposite extreme where there is perfect correlation between data points, say, as in a straight line. What happens to the kriging system?

3

Model Boundedness

The dragon exceeds the proper limits; there will be occasion for repentance.
The Book of Changes (Yi Qing) (c. 1200 B.C.)

In modeling an optimization problem, the easiest and most common mistake is to leave something out. This chapter shows how to reduce such omissions by systematically checking the model before trying to compute with it. Such a check can detect formulation errors, prevent wasteful computations, and avoid wrong answers. As a perhaps unexpected bonus, such a preliminary study may lead to a simpler and more clearly understandable model with fewer variables and constraints than the original one.

The methods of this chapter, informally referred to as boundedness checking, should be regarded as a model reduction and verification process to be carried out routinely before attempting any numerical optimization procedure. At the same time, one should be cautious about the limitations of boundedness arguments because they are based on *necessary conditions*, namely mathematical truths that hold assuming an optimal solution exists. Such existence, derived from sufficient conditions, is not always easy to prove. The complete optimality theory in Chapters 4 and 5 provides important additional tools to those presented in this chapter.

The chapter begins with the fundamental definitions of bounds and optima, allowing a precise definition of a well-bounded model. Since poor model boundedness is often a result of extensive *monotonicities* in the model functions, the boundedness theory presented here has become known as *Monotonicity Analysis*. The concepts of constraint activity, criticality, dominance, and relaxation are presented formally, along with two *monotonicity principles* that allow quick, practical boundedness checking.

3.1 Bounds, Extrema, and Optima

This section develops formally the ideas of minimum and boundedness discussed informally in Section 1.3. The rigor of some definitions may seem uncomfortable at first; yet such rigor is needed to study with sufficient precision certain modeling situations that can prevent later costly errors.

Well-Bounded Functions

Let $f(x)$ be a real-valued function with x in the domain $\mathcal{R}$, the set of real numbers. If there is a finite number l such that $f(x) \geq l$ for all x in $\mathcal{R}$, then it is mathematical practice to call l a *lower bound* for $f(x)$. The *greatest lower bound* (glb) or *infimum* is any number g, itself a lower bound, that is larger than, or equal to, any distinct lower bound; that is, for all x, $g \geq l$ for all $l \leq f(x)$. Note that l or g may be negative or zero. Henceforth the phrase "over $\mathcal{R}$" will be added to remind us of the domain of x; that is, g will be called the "glb over $\mathcal{R}$". The glb over $\mathcal{R}$ may not exist, as when $f(x) = 1/x$, for which of course there is no finite lower bound.

In what follows, the function $f(x)$ may still be defined over the set $\mathcal{R}$ of real numbers or any of its subsets. However, our attention will be focused on two subsets of $\mathcal{R}$: the set $\mathcal{N}$ of nonnegative numbers (including zero and infinity) and the set $\mathcal{P}$ of positive finite numbers. The set $\mathcal{P}$ is given special attention here because most physical problems are defined in this positive finite domain. We therefore have

$$\mathcal{N} = \{x: 0 \leq x \leq \infty\}; \qquad \mathcal{P} = \{x: 0 < x < \infty\}. \tag{3.1}$$

Let g_0 be the glb for $f(x)$ over $\mathcal{N}$ and g^+ be the glb over $\mathcal{P}$. The set inclusion relations $\mathcal{R} \supset \mathcal{N} \supset \mathcal{P}$ imply that $g \leq g_0 \leq g^+$ when these numbers exist. To represent that g^+ is the infimum of $f(x)$ (over $\mathcal{P}$) we write

$$g^+ = \inf_{x \in P} f(x).$$

Suppose there is a nonnegative number $\underline{x}$ such that $f(\underline{x}) = g^+$. Then $\underline{x}$ is called an *argument of the infimum over* $\mathcal{P}$. In case the infimum has more than one argument, we let $\underline{\mathcal{X}}$ represent the nonempty set of all of them: $\underline{\mathcal{X}} = \{x: f(\underline{x}) = g^+\}$. Notice that not all arguments have to belong to $\mathcal{P}$. If all of them do, that is, if all $\underline{x}$ are positive and finite, then $f(x)$ is said to be *well bounded (below) over* $\mathcal{P}$. Otherwise $f(x)$ is said to be *not well bounded (below) over* $\mathcal{P}$.

This definition of well boundedness is slightly more restrictive than previously used (see Notes at end of chapter) by requiring *all* infima to be in $\mathcal{P}$.

Example 3.1 Consider the following functions:

1. $f(x) = x$: no g exists, but $g_0 = g^+ = 0$, so $\underline{x} = 0 \notin \mathcal{P}$ and hence $f(x)$ is not well bounded below over $\mathcal{P}$.
2. $f(x) = x^2 + 1 : g = g_0 = g^+ = 1$. Since the argument $\underline{x} = 0$, $f(x)$ is not well bounded below over $\mathcal{P}$.
3. $f(x) = (x - 1)^2 : g = g_0 = g^+ = 0$, and since $\underline{x} = 1 \in \mathcal{P}$, $f(x)$ is well bounded below over $\mathcal{P}$.
4. $f(x) = -x : g$, g_0, and g^+ do not exist, so no arguments exist, and $f(x)$ is not well bounded below over $\mathcal{P}$.

5. $f(x) = 1/x^2 : g = g_0 = g^+ = 0$. Although $f(x) = 0$ for x equal to positive or negative infinity, only the positive one qualifies as an argument of the infimum. Since $\underline{x} \notin P$, $f(x)$ is not well bounded below over $\mathcal{P}$.
6. $f(x) = 1/x$: no g exists, but $g_0 = g^+ = 0$ for $\underline{x} = \infty$, so $f(x)$ is not well bounded below over $\mathcal{P}$.
7. The infimum itself can be negative, for example, $f(x) = (x-1)^2 - 1 : g = g_0 = g^+ = -1$ where the argument $\underline{x} = 1$; well bounded over $\mathcal{P}$.
8. $f(x) = \exp(-x) : g = g_0 = g^+ = 0$; not well bounded over $\mathcal{P}$ because their arguments are infinite.
9. $f(x) = (x-1)^2(x-2)^2 : g = g_0 = g^+ = 0$. There are two arguments: $\underline{\mathcal{X}} = \{1, 2\}$; well bounded over $\mathcal{P}$.
10. $f(x) = (x^2 - 1)^2 : g = g_0 = g^+ = 0$; $f(-1) = f(1) = g^+$, but there is a negative as well as a positive argument $\underline{x} = \pm 1$; not well bounded over $\mathcal{P}$.
11. $f(x_1, x_2) = 3 + (x_2 - 1)^2$: Here the bivariate function does not depend on x_1; consequently $g = g_0 = g^+ = 3 = f(x_1, 1)$, which gives the same value not only in $\mathcal{P}$ but also when $x_1 = 0$ (and ∞). Hence f is well bounded with respect to x_2, although not with respect to x_1. ■

The word *infimum* subsequently will refer only to g^+. If *all* arguments of g^+ are positive and finite, that is, $\mathcal{P} \supseteq \underline{\mathcal{X}}$, the infimum will be called the *minimum for $f(x)$ over $\mathcal{P}$*, or *minimum* for short. For any other case, in this book $f(x)$ will be said to have no minimum (over $\mathcal{P}$) unless otherwise stated. This assumption simplifies much of the theory and proofs in this chapter and is consistent with model formulations of most engineering design problems. The finite positivity assumption is relaxed in the theory of Chapters 4 and 5, as it is not necessary there.

The argument of a minimum is written x_* when it is unique. The set of all finite positive arguments of a minimum is written $\mathcal{X}_*$.

Notice that minima exist whenever $f(x)$ is well bounded, but not vice versa. A function having an infimum at 0 or $+\infty$ is not considered well bounded, even when minima exist elsewhere. This definition intends to handle the special situation where $f(x) = K$, a constant. In this case $f(0) = f(\infty) = K = g^+$ and $\underline{\mathcal{X}} \supset \mathcal{P}$, violating the definition of well boundedness even though all positive finite x are arguments of the infimum. This refinement has two objectives. In the first place it keeps computer algorithms from generating physically absurd solutions. Secondly it simplifies proof of the *Monotonicity Principles* to follow, especially the second.

A function having a finite infimum whose argument is $+\infty$ is said to be *asymptotically bounded*, as in cases 5, 6, 8, and 11 of Example 3.1. A function whose argument of the infimum is zero is said to be *bounded at zero*, as in cases 1, 2, and 11 of Example 3.1.

Example 3.2 In Example 3.1, only cases 3, 7, and 9 are well bounded, although case 11 has minima for every x_1 in $\mathcal{P}$. In case 9, $\mathcal{X}_* = \{1, 2\}$; in case 10, $x_* = 1$. ■

The analogous concepts involving upper instead of lower bounds are given in the following table:

Bound	**Extremum**	**Arg**	**Optimum**	**Arg**
Lower (lb)	Greatest lb; inf(imum)	$\underline{x}$	Min(imum)	x_*
Upper (ub)	Least ub; sup(remum)	$\overline{x}$	Max(imum)	x^*

Keep in mind that the infima $f(\underline{x})$ and minima $f(x_*)$ are images in the range of a function. They are never in the function's domain of pre-images x containing the arguments. Well boundedness concerns only the domain of x, never the range of $f(x)$. A common source of confusion is to apply the word "minimum" not only to the image $f(x_*)$, but also to the argument x_*, which, strictly speaking, is incorrect. To avoid this confusion, the word "minimizer" will be used henceforth for x_* synonymously with "argument of the minimum."

Nonminimizing Lower Bound

When finding a minimum is difficult, a more convenient lower bound may be good enough even though it is not the true minimum. Consider, for example, the function

$$f(x) = 25{,}100x + 341x^2 + 1.34x^3 + 50{,}000x^{-1}, \tag{3.2}$$

where $x \in \mathcal{P}$. Section 4.3 will prove the well-known result from calculus that df/dx, the first derivative of f with respect to (wrt) x, must vanish at the minimum of f. That is,

$$\left.\frac{df}{dx}\right|_{x=x_*} = 25{,}100 + 682x_* + 4.02x_*^2 - 50{,}000x_*^{-2} = 0. \tag{3.3}$$

Although numerical solution of this fourth-degree equation is not difficult, there is no closed form equation for the minimizer x_*, a deficiency that could inhibit further analysis. If, however, the second and third terms of f and df/dx were deleted, the resulting derivative equation would be solvable in closed form. Let this approximating function be denoted by $l(x)$:

$$l(x) = 25{,}100x + 50{,}000x^{-1}. \tag{3.4}$$

Since, for every positive value of x,

$$f(x) = l(x) + 341x^2 + 1.34x^3 > l(x), \tag{3.5}$$

$l(x)$ is called a *lower bounding function* for $f(x)$.

The value $\hat{x}$ that minimizes $l(x)$, although not $f(x)$, satisfies the condition

$$\left.\frac{dl}{dx}\right|_{x=\hat{x}_*} = 25{,}100 - 50{,}000\hat{x}^{-2} = 0. \tag{3.6}$$

The closed form solution is $\hat{x} = (50{,}000/25{,}100)^{1/2} = 1.41$. The corresponding minimum value of $l(x)$ is $l(\hat{x}) = 70.9(10^3)$, which is a lower bound on the original function $f(x)$:

$$f(x) > l(x) \geq l_*(x) = l(\hat{x}) = 70.9(10^3). \tag{3.7}$$

Notice that $\hat{x}$ does not minimize $f(x)$, whose value at $\hat{x}$ is

$$\begin{aligned} f(\hat{x}) &= l(\hat{x}) + 341\hat{x}^2 + 1.34\hat{x}^3 \\ &= 70.9(10^3) + 679 + 4 = 71.6(10^3). \end{aligned} \tag{3.8}$$

Although neither the minimum f_* nor its argument x_* have been found, the true minimum has in this case been closely *bracketed*, since $71.6(10^3) \geq f_* > 70.9(10^3)$. This interval of uncertainty may in some practical cases be acceptable.

The possible range of x_* can be bounded, at least on one side, by determining the sign of the derivative of f at $\hat{x}$ (not x_*):

$$\left.\frac{df}{dx}\right|_{x=\hat{x}} = \left.\frac{dl}{dx}\right|_{x=\hat{x}} + 682\hat{x} + 4.02\hat{x}^2 > 0. \tag{3.9}$$

Hence, f can be decreased only by decreasing x, and so $x_* < \hat{x} = 1.41$. This rigorous approach for obtaining quick approximate solutions to optimization problems is useful in several situations, including problems with discrete variables discussed in Chapter 6.

Multivariable Extension

Instead of a single variable, let there now be n finite positive variables x_i, where $i = 1, \ldots, n$. The domain $\mathcal{P}^n$ of the positive finite vector $\mathbf{x} = (x_1, \ldots, x_n)^T$ is the Cartesian product:

$$\mathcal{P}^n = \mathcal{P}_1 \times \cdots \times \mathcal{P}_n = \{x_i\colon 0 < x_i < \infty, \quad i = 1, \ldots, n\}, \tag{3.10}$$

where

$$\mathcal{P}_i = \{x_i\colon 0 < x_i < \infty\}. \tag{3.11}$$

The concepts of real-valued function $f(\mathbf{x})$, its lower bounds, greatest lower bound, infimum, and minimum all extend immediately to the vector $\mathbf{x}$, where it is understood that any argument of the infimum is now an n-component vector $\underline{\mathbf{x}}$.

Air Tank Design

Consider the volume of metal in the flat-headed cylindrical air tank shown in Figure 3.1 (Unklesbay, Staats, and Creighton 1972). The metal volume m depends on the inside radius r, the shell thickness s, the shell length l, and the head thickness h according to the geometric formula

$$m = \pi[(r + s)^2 - r^2]l + 2\pi(r + s)^2 h. \tag{3.12}$$

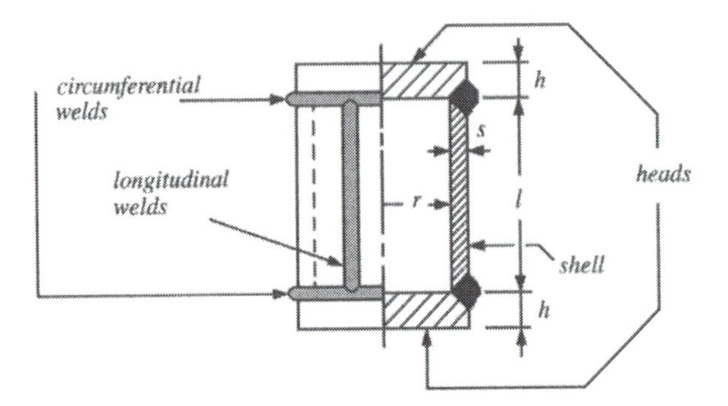

Figure 3.1. Vertical flat-headed air tank.

Let these four positive finite variables be numbered in alphabetical order: $h = x_1$, $l = x_2$, $r = x_3$, and $s = x_4$, and let $f(\mathbf{x})$ be identified with the metal volume m in cubic centimeters. Then

$$\begin{aligned} f(\mathbf{x}) &= \pi\left\{\left[(x_3 + x_4)^2 - x_3^2\right]x_2 + 2(x_3 + x_4)^2 x_1\right\} \\ &= \pi\left[\left(2x_3x_4 + x_4^2\right)x_2 + 2(x_3 + x_4)^2 x_1\right]. \end{aligned} \tag{3.13}$$

Since $f(\mathbf{x})$ is a sum of positive terms, it must itself be positive: $f(\mathbf{x}) > 0$. Thus lower bounds for $f(\mathbf{x})$ include -10, -2.5, and 0, with zero being the greatest lower bound. Hence, $\inf f(\underline{\mathbf{x}}) = 0$, for which the argument is $\underline{\mathbf{x}} = (0, 0, 0, 0)^T$. Here $\mathcal{P}^4 = \mathcal{P}_1 \times \mathcal{P}_2 \times \mathcal{P}_3 \times \mathcal{P}_4$. Since $\underline{\mathbf{x}} \notin \mathcal{P}^4$, f has no positive finite minimizer. Evidently constraints are needed.

3.2 Constrained Optimum

The domain of $\mathbf{x}$ may be restricted further by constraints, for example, equalities, inequalities, discreteness restrictions, and/or logical conditions, defining a *constraint set* $\mathcal{K}$. The set $\mathcal{K}$ is said to be *consistent* if and only if $\mathcal{K} \neq \{\}$.

Example 3.3 Consider the constraint sets

$$\mathcal{K}_1 = \{x: x \geq 4\} \neq \{\};\ \mathcal{K}_2 = \{x: x < 3\} \neq \{\}.$$

Each constraint set is consistent, but $\mathcal{K}_1 \cap \mathcal{K}_2 = \{\}$ is *inconsistent*. Engineering problems should have consistent constraints with at least one positive finite element in the set, that is, $\mathcal{K} \cap \mathcal{P} \neq \{\}$. Note that $\mathcal{K}_3 = \{x: x \leq -2\}$ is consistent but not positive. ■

Constrained bounds, extrema, and optima are defined as before using the *feasible* set $\mathcal{F} = \mathcal{K} \cap \mathcal{P}$ instead of $\mathcal{P}$. Let $f(\mathbf{x})$ be an *objective function* defined on $\mathcal{F}$. Let g^+ be the greatest lower bound (infimum) on $f(\mathbf{x})$, $f(\mathbf{x}) \geq g^+$ for all $\mathbf{x} \in \mathcal{F}$. If there exists $\mathbf{x}_* \in \mathcal{F}$ such that $f(\mathbf{x}_*) = g^+$, then $f(\mathbf{x}_*)$ is the (*constrained*) *minimum* of $f(\mathbf{x})$, and $\mathbf{x}_*$ is the *minimizer* (or *minimizing argument*). We write $\mathbf{x}_* = \arg\min f(\mathbf{x})$ for $\mathbf{x} \in \mathcal{F}$.

By analogy with the concepts of well boundedness for unconstrained functions, if all constrained minimizers are in $\mathcal{F}$, f is said to be *well constrained (below)*. Since optimization models should at least be well constrained, it is good to know the conditions under which this does or does not happen (the main theme in this chapter).

The concept of constraint is also easily extended to the multivariable case. Just as for the objective function $f(\mathbf{x})$, let the constraint functions now depend on the vector $\mathbf{x}$. In the air tank example, consider the following inequality constraints. The volume $\pi r^2 l (= \pi x_2 x_3^2)$ must be at least $2.12(10^7)\text{cm}^3$, so $\pi x_2 x_3^2 \geq 2.12(10^7)$. In negative null form this is

$$\mathcal{K}_1 = \{\mathbf{x}: g_1 = -\pi x_2 x_3^2 + 2.12(10^7) \leq 0\}. \tag{3.14}$$

The ASME code for flat-headed unfired pressure vessels limits the ratio of head thickness to radius $h/r = x_1 x_3^{-1} \geq 130(10^{-3})$, whence

$$\mathcal{K}_2 = \{\mathbf{x}: g_2 = -x_1 x_3^{-1} + 130(10^{-3}) \leq 0\}. \tag{3.15}$$

It also restricts the shell thickness by $s/r = x_3^{-1} x_4 \geq 9.59(10^{-3})$, and so

$$\mathcal{K}_3 = \{\mathbf{x}: g_3 = -x_3^{-1} x_4 + 9.59(10^{-3}) \leq 0\}. \tag{3.16}$$

To allow room to attach nozzles, the shell must be at least 10 cm long,

$$\mathcal{K}_4 = \{\mathbf{x}: g_4 = -x_2 + 10 \leq 0\} \tag{3.17}$$

Finally, space limitations prevent the outside radius $r + s = x_3 + x_4$ from exceeding 150 cm:

$$\mathcal{K}_5 = \{\mathbf{x}: g_5 = x_3 + x_4 - 150 \leq 0\}. \tag{3.18}$$

This preliminary model has five inequality constraints for four variables. As for a single variable, $\mathcal{K}$ is the intersection of all constraints and its intersection with $\mathcal{P}^n$ is the feasible set $\mathcal{F}$. For m constraints,

$$\mathcal{F} = \left[\bigcap_{j=1}^{m} \mathcal{K}_j\right] \cap \mathcal{P}^n. \tag{3.19}$$

In the example, this would be written

$$\mathcal{F} = \mathcal{K}_1 \cap \mathcal{K}_2 \cap \mathcal{K}_3 \cap \mathcal{K}_4 \cap \mathcal{K}_5 \cap \mathcal{P}^4. \tag{3.20}$$

Partial Minimization

If all variables but one are held constant, it may be easy to see which constraints, if any, bound the remaining variable away from zero. To this end, let all variables except x_1 be fixed at values $x_i = X_i (i > 1)$, and define $\mathbf{X}_1 = (X_2, \ldots, X_n)^T$, an $n - 1$ vector. Let $\mathbf{x} = (x_1, \mathbf{X}_1)^T$ with only x_1 being a variable. The functions $g_j(x_1, \mathbf{X}_1)$ for $j = 0, 1, \ldots, m_2$ and $h_j(x_1, \mathbf{X}_1)$ for $j = 1, \ldots, m_1$ all depend only on a single variable $x_1 \in \mathcal{P}_1$. Hence, the infimum and minimum, as well as their arguments, are defined in the usual ways but now these quantities depend on the values $X_2, \ldots, X_n$.

In the air tank design, let $x_1 (= h)$ vary while the other variables are fixed at values $(x_2, x_3, x_4)^T = (X_2, X_3, X_4)^T = \mathbf{X}_1$. Then the *partial minimization problem* with only x_1 variable is to minimize

$$f(x_1, \mathbf{X}_1) = \pi\left[\left(2X_3X_4 + X_4^2\right)X_2\right] + 2\pi(X_3 + X_4)^2 x_1, \tag{3.21}$$

which can be abbreviated $f(x_1, \mathbf{X}_1) = a(\mathbf{X}_1) + b(\mathbf{X}_1)x_1$ where $a(\mathbf{X}_1)$ and $b(\mathbf{X}_1)$ depend only on $\mathbf{X}_1$.

Only constraint $\mathcal{K}_2$ depends on x_1, with the latter restricted only by $x_1 \geq 130(10^{-3})$ $X_3 > 0$. The remaining constraints influence the problem only in restricting the values of x_2, x_3, and x_4 that produce feasible designs. We implicitly assume then that

$$\begin{aligned} \pi X_2 X_3^2 &\geq 2.12(10^7), \\ X_3^{-1} X_4 &\geq 9.59(10^{-3}), \\ X_2 &\geq 10, \\ X_3 + X_4 &\leq 150. \end{aligned} \tag{3.22}$$

For any such feasible positive finite (X_2, X_3, X_4), the function $b(\mathbf{X}_1) > 0$. Hence $f(x_1, \mathbf{X}_1)$ is the minimum for $x_1 = 130(10^{-3})X_3$, no matter what feasible positive finite value X_3 takes. Therefore, $x_1(\mathbf{X}_1) = x_{1_*}(\mathbf{X}_1) = 130(10^{-3})X_3$ is the argument of what is called the *partial minimum* of f wrt x_1. This partial minimum is $a(\mathbf{X}_1) + b(\mathbf{X}_1)[130(10^{-3})X_3]$. Since this partial minimum exists, the objective is well constrained wrt x_1.

Formally, define the *feasible set* for x_1, given $\mathbf{X}_1$, as $\mathcal{F}_1 = \{\mathbf{x}\text{: } \mathbf{x} \in \mathcal{F} \text{ and } x_i = X_i \text{ for } i \neq 1\}$. Let $\mathbf{x}'$ be any element of $\mathcal{F}_1(\mathbf{X}_1)$. Then $f(\mathbf{x}') \geq \inf f(x_1, \mathbf{X}_1)$ and $x_1(\mathbf{X}_1) = \arg\inf f(\mathbf{x}')$. If $x_1(\mathbf{X}_1) \in \mathcal{F}_1$, then $x_1(\mathbf{X}_1) = x_{1*}(\mathbf{X}_1)$, and $f(\mathbf{x}') \geq \min f(x_1, \mathbf{X}_1)$ for $\mathbf{x}' \in \mathcal{F}_1$. The function $\min f(x_1, \mathbf{X}_1)$ for $\mathbf{x}' \in \mathcal{F}_1$ is called the *partial minimum* of f wrt x_1.

In the air tank design, the approach used for x_1 can also be applied to the other three variables. One could use the abstract formalism just presented, in which only the "first" variable is allowed to change, simply by renumbering the variables so that the new one to be studied is called x_1. But in practice this would be unnecessarily formal. Often, the variables do not have to be numbered at all; their original symbols are good enough and in fact may clarify the model by reminding the analyst of their physical significance, at least until numerical processing is necessary.

Thus let us continue the partial minimization study by working with the air tank model in its original form, retaining indices for objective and constraints to facilitate reference:

(0) $\min m = \pi[(2rs + s^2)l + 2(r + s)^2 h]$

subject to

(1) $\pi r^2 l \geq 2.12(10^7)$,

$$
\begin{aligned}
&(2) \quad h/r \geq 130(10^{-3}),\\
&(3) \quad s/r \geq 9.59(10^{-3}),\\
&(4) \quad l \geq 10,\\
&(5) \quad r+s \leq 150.
\end{aligned}
\tag{3.23}
$$

Consider the shell thickness s. Let the other variables be fixed at any feasible values H, L, and R satisfying $\pi R^2 L \geq 2.12(10^7)$. The capitalizations remind us which variables have been temporarily fixed. Then we have

$$
\begin{aligned}
&(0s) \quad \min m(s) = \pi[(2Rs+s^2)L + 2(R+s)^2 H]\\
&\qquad\quad \text{subject to}\\
&(3s) \quad s/R \geq 9.59(10^{-3}),\\
&(5s) \quad R+s \leq 150.
\end{aligned}
\tag{3.24}
$$

Notice that as s increases, so does the objective $m(s)$. Hence, to minimize s we must make s as small as allowed by the constraints. The outside diameter constraint $(5s)$ bounds s from above and so does not prevent s, and hence $m(s)$, from decreasing without bound. However, the shell thickness constraint $(3s)$ bounds s from below. A partial minimum wrt s therefore exists where $s = 9.59(10^{-3})R$, for any feasible R.

Constraint Activity

It is useful to study what happens when a (nonredundant) constraint is hypothetically removed from the model. If this changes the optimum, the constraint is called *active*; otherwise, it is termed *inactive*, provided the optimizing arguments are unaffected. Important model simplifications are possible any time the activity or inactivity of a constraint can be proven before making detailed optimization calculations.

Formally, let $\mathcal{D}_i = \cap_{j \neq i} \mathcal{K}_j$ be the set of solutions to all constraints *except* g_i. Such solutions may or may not be in $\mathcal{K}_i$. Then the set of all feasible points is $\mathcal{F} = \mathcal{D}_i \cap \mathcal{K}_i \cap \mathcal{P}^n$. Let f be well bounded in $\mathcal{F}$, and let $\mathcal{X}_*$ be the set of arguments of $\{\min f, \mathbf{x} \in \mathcal{F}\}$. The minimization problem with g_i deleted, that is, for $\mathbf{x} \in \mathcal{D}_i \cap \mathcal{P}^n$, is called the *relaxed problem*; let $\mathcal{X}_i$ represent the set of its arguments. If $\mathcal{X}_i$ and $\mathcal{X}_*$ are the same ($\mathcal{X}_i = \mathcal{X}_*$), then constraint g_i is said to be *inactive* because its deletion would not affect the solution of the minimization problem.

At the other extreme, if $\mathcal{X}_i$ and $\mathcal{X}_*$ are disjoint ($\mathcal{X}_i \cap \mathcal{X}_* = \{\}$), then g_i is said to be *active*. Its deletion would of course give the wrong answer.

There is also an intermediate case in which some of the relaxed solutions $\mathcal{X}_i$ satisfy g_i while others do not. In this situation, which can occur when the objective does not depend on all the independent variables, $\mathcal{X}_i$ strictly contains $\mathcal{X}_*$, $\mathcal{X}_i \supset \mathcal{X}_*$, since any $\mathbf{x}' \in \mathcal{X}_i$ and also in the subset satisfying g_i belongs to $\mathcal{X}_*$. When this happens, g_i is said to be *semiactive* (see Figure 3.2). This subtle concept is needed to prove the Relaxation Theorem of Section 3.5, as well as in the proof of the Second Monotonicity Principle in Section 3.7.

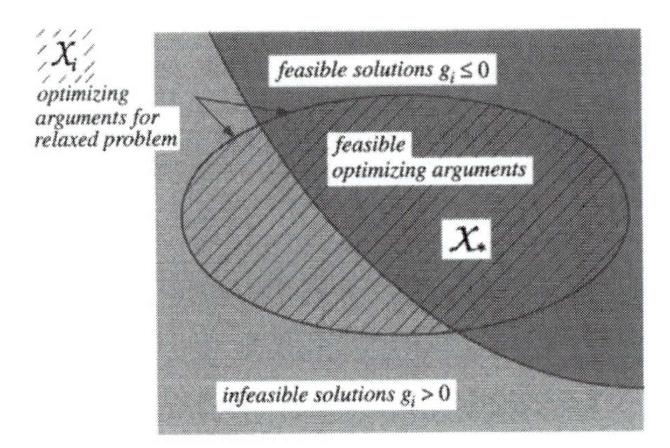

Figure 3.2. A semiactive constraint.

Example 3.4 Consider the model

$$\min f = x_1^2 + (x_2 - 1)^2(x_2 - 3)^2(x_2 - 4)^2$$

subject to

$$g_1 : x_1 - 1 \ge 0, g_2 : x_2 - 2 \ge 0, \quad g_3 : -x_2 + 5 \ge 0.$$

Partial minimization wrt x_1, only in the first term of f, shows that the left side of g_1 vanishes at any minimum, and so $x_{1*} = 1$ for all values of x_2. The other term of f is the nonnegative product $(x_2 - 1)^2(x_2 - 3)^2(x_2 - 4)^2$, which attains its minimum of 0 whenever any of its factors vanish, that is, when $x_2 = 1$, $x_2 = 3$, or $x_2 = 4$. All three solutions satisfy g_3, but case $x_2 = 1$ violates constraint g_2. Thus, the set of arguments for the constrained problem has two elements: $\mathcal{X}_* = \{(1, 3), (1, 4)\}$. The minimum is $f(\mathcal{X}_*) = 1$. If g_1 is deleted, the relaxed solution is $\mathcal{X}_1 = \{(0, 3), (0, 4)\}$ and $f(\mathcal{X}_1) = 0 < f(\mathcal{X}_*) = 1$. Since $\mathcal{X}_* \cap \mathcal{X}_1 = \{\}$, g_1 is active. If g_2 is deleted, the relaxed solution is $\mathcal{X}_2 = \{(1, 1), (1, 3), (1, 4)\}$, which overlaps but does not equal $\mathcal{X}_*$. Hence, g_2 is semiactive, and $f(\mathcal{X}_2) = f(\mathcal{X}_*)$. Deletion of g_3 has no effect ($\mathcal{X}_3 = \mathcal{X}_*$), and so g_3 is inactive. Figure 3.3 illustrates these facts. ■

These definitions concerning activity will ease the description and exploitation of an important phenomenon too often overlooked in the optimization literature. Any constraint proven active in advance reduces the degrees of freedom by one and can possibly be used to eliminate a variable, while any constraint probably inactive can be deleted from the model. Both types of simplification reduce computation, give improved understanding of the model, and at times are absolutely necessary to obtain all correct optimizing arguments. In fact, many numerical optimization procedures would find the preceding example difficult to solve, though it is simple when analyzed properly in advance. In practical numerical implementations, activity information would not be used to change the model. Rather, it should be properly incorporated in the logic of an *active set strategy*. Active set strategies will be discussed in Chapter 7.

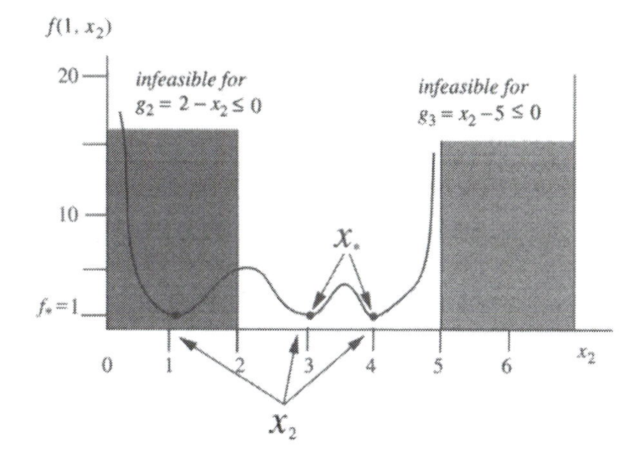

Figure 3.3. Semiactive and inactive constraint.

Semiactive constraints occupy a twilight zone, for they can neither be deleted like the inactive constraints nor used to eliminate a variable like the active constraints. Provisional relaxation is permissible in an active set strategy, provided the relaxed solution is checked against the semiactive constraint as in Example 3.4, in which (1, 1) would be found to be infeasible for g_2: $x_2 \geq 2$. However, it would have been incorrect to use g_2 as a strict equality to eliminate x_2, for the result (1, 2) is clearly not the minimizer.

The following theorem is useful for numerically testing a constraint for activity when a priori analysis is not adequate. It is proven here for future reference.

Activity Theorem Constraint g_i is active if and only if $f(\mathcal{X}_i) < f(\mathcal{X}_*)$. That is, the value of the objective at the minimizers of the relaxed problem is less than its value at the minimizers of the original problem.

Proof Let $\mathbf{x}' \in \mathcal{X}_*$ be any argument of min $f(\mathbf{x})$ for $\mathbf{x} \in \mathcal{F}$. Since $\mathcal{X}_* \subset \mathcal{F} \subset \mathcal{D}_i$, it follows that $\mathbf{x}' \in \mathcal{D}_i$. When g_i is active, $\mathcal{X}_* \cap \mathcal{X}_i = \{\}$, and so $\mathbf{x}' \notin \mathcal{X}_i$. Then by definition of $\mathcal{X}_i$, $f(\mathbf{x}') > f(\mathcal{X}_i)$ because $\mathbf{x}' \in \mathcal{D}_i$ but $\mathbf{x}' \notin \mathcal{X}_i$.

When g_i is semiactive, the intersection $\mathcal{X}_* \cap \mathcal{X}_i$ is nonempty and so let $\mathbf{x}'' \in \mathcal{X}_* \cap \mathcal{X}_i$. This time the definition of $\mathcal{X}_i$ gives $f(\mathbf{x}'') = f(\mathcal{X}_i)$ because now $\mathbf{x}'' \in \mathcal{X}_i \subset \mathcal{D}_i$. When g_i is inactive, $\mathcal{X}_i = \mathcal{X}_*$, and so $f(\mathcal{X}_i) = f(\mathcal{X}_*)$. This completes the proof.

Example 3.5 Consider the model with

$$\min f = (x_1 - 2)^2$$

subject to

$$g_1 = -x_1 + 3 \leq 0.$$

The relaxed minimum is zero, where the minimizing argument is $\mathcal{X}_1 = 2$. Thus $f(\mathcal{X}_1) = 0$. However, this cannot be the minimum for the original problem because it violates the constraint $g_1(\mathcal{X}_1) = -2 + 3 > 0$. The constrained minimizer is $x_* = 3$, where $g_1(x_*) = 0$ and $f_* = f(x_*) = (3 - 2)^2 = 1$. Since $f(\mathcal{X}_1) < f(x_*)$, the constraint is active by the Activity Theorem. The same theorem proves that a second constraint $g_2 = -x_1 + 1 \leq 0$ would be inactive, since $f(\mathcal{X}_2) = f(x_*)$ and $\mathcal{X}_2 = x_*$. ■

Cases

In a constrained optimization problem, the constraints are a mixture of equalities and inequalities. At the optimum, certain of these, called the *active set*, are satisfied as strict equalities. Thus, solving the smaller problem in which all the inactive constraints are deleted and all the semiactive constraints are satisfied would give the optimum for the original problem. This smaller optimization problem, having the same optimum as the original problem, will be called the *optimum case*. It corresponds to the situation where the correct set of active and semiactive constraints has been precisely identified.

It will be useful to define the idea of *case* more precisely to cover more general situations. Let the set of all equality and inequality constraint function indices be denoted by $\mathcal{J} = [1, \ldots, m]$, and let $\mathcal{W}$ be any subset of $\mathcal{J}$: $\mathcal{W} \subseteq \mathcal{J}$. Here $\mathcal{W}$ may be the entire set $\mathcal{J}$ or, at the other extreme, the empty set. The constraints whose indices are in $\mathcal{W}$ form the *working set*, also called the *currently active (and semiactive) set*.

The problem of minimizing the original objective $f(\mathbf{x})$ subject only to the constraints whose indices belong to $\mathcal{W}$ is called the *case* associated with $\mathcal{W}$. Thus, there is a case for every subset $\mathcal{W}$. The number of such cases is of course quite large even for small values of constraints m and variables n, because many combinations are possible. In fortunate circumstances, however, most of these cases can, with little or no computation, be proven either nonoptimal or inconsistent. For instance, cases in which $m > n$ need not be considered, because such a system usually would be inconsistent and have no solution. Methods to be developed starting in Section 3.5 will also disqualify many cases as being nonoptimal. Thus every time a constraint is proven active (or semiactive), all cases not having that constraint active (or semiactive) can be eliminated from further consideration.

The detailed study of how to decompose a model into cases, most of which can be disqualified as nonoptimal or inconsistent, will be deferred to Chapter 6. Meanwhile, we will continue to develop the basic definitions and theorems, applying them to such appropriately simple problems as the air tank. *Case decomposition* during model analysis is the counterpart of an *active set strategy* during numerical computations. The difference is that case decomposition operates with global information to screen out poorly bounded cases, while typical active set strategies use only local information to move from one case to another.

3.3 Underconstrained Models

The number of cases can usually be reduced drastically by identifying and eliminating cases probably not well constrained. This section shows how to recognize and exploit the simple and widespread mathematical property of monotonicity to see if a model has too few constraints to be well bounded. An especially useful kind of constraint activity, known as *criticality*, will be defined and developed.

Monotonicity

The requirement that an optimization model be well constrained often indicates in advance which constraints are active, leading to simplified solution methods and increased insight into the problem. *Monotonicity* of objective or constraint functions can often be exploited to obtain such simplifications and understanding.

A function $f(x)$ is said to *increase* or be *increasing* with respect to the single positive finite variable x in $\mathcal{P}$, if for every $x_2 > x_1$, $f(x_2) > f(x_1)$. Such a function will be written $f(x^+)$. For the continuously differentiable functions usually encountered in engineering, this means the first partial derivative $\partial f/\partial x$ is everywhere strictly positive; the definition with inequalities is meant to include the rarer situations where f is nondifferentiable or even discontinuous in either domain or range.

For an increasing function, $f(x_2^{-1}) > f(x_1^{-1})$ for every $x_2^{-1} > x_1^{-1}$, so that $f(x_1) < f(x_2)$ for every $x_1 > x_2$. Hence, $f(x^{-1})$ is said to *decrease* with respect to x and is written $f(x^-)$. Consequently, properties of increasing functions can be interpreted easily in terms of decreasing functions. Functions that are either increasing or decreasing are called *monotonic*.

For completeness, the theory is extended to functions that have a flat spot. This situation, although rare for constraint functions, occurs quite naturally in an objective function that does not depend on all the independent variables. Flat spots occur when the strict inequality "<" between successive function values is replaced by the inequality "≤." In this circumstance f is said to increase weakly (≤) rather than strictly (<) as before. The word "strictly" will be omitted in situations that are either clearly unambiguous or clearly intend to include both the weak and strict monotonicity.

As is the case throughtout this book, all theorems and results are valid, as stated, for models posed in negative null form.

The Monotonicity Theorem If $f(x)$ and the consistent constraint functions $g_i(x)$ all increase weakly or all decrease weakly with respect to x, the minimization problem domain is not well constrained.

Proof Since the constraints are consistent, then for every index i there exists a positive finite constant $A_i (0 < A_i < \infty)$ such that $g_i(A_i) = 0$. If all functions increase, then $g_i(0) \leq g_i(A_i) = 0$, and so $x = 0$ satisfies all constraints. Moreover $f(0) \leq f(x)$ for all $x > 0$ because $f(x)$ increases weakly. Hence arg inf $f(x) = 0 \notin \mathcal{F}$ satisfies all constraints $g_i(x) \leq 0$, and therefore the minimization problem domain is not well constrained. When all functions decrease weakly, then $g_i(\infty) \leq g_i(A_i) = 0$, and so $x = \infty$ satisfies all constraints, and $f(\infty) \leq f(x)$ for all positive finite x because $f(x)$ decreases weakly. Hence arg inf $f(x) = \infty \notin \mathcal{F}$ satisfies all constraints, and the minimization problem domain is not well constrained.

This perfectly obvious theorem, so easy to understand and prove, is not very useful directly. It has however two important corollaries obtained by logical contrapositive statements – the negation of both hypothesis and conclusion, followed by their interchange. Negating the conclusion gives "the problem domain *is* well constrained," which becomes the hypothesis of the corollaries, to be known as *Monotonicity Principles*. Thus it is always assumed that the model is well constrained until contradicted by a violation of either Monotonicity Principle.

Even more important is the consideration (Hansen, Jaumard, and Lu 1989a) of what is meant by functions that are *not* increasing. The set of nonincreasing functions includes not only the decreasing functions but also the much larger class of all nonmonotonic and even constant functions. This important extension will be used in Section 3.8.

First Monotonicity Principle

The concepts of partial minimization and constraint activity permit immediate extension of the properties of monotonic functions of a single variable to those with many variables. A variable that is monotonic in every (objective and constraint) function in which it appears is called a *monotonic variable*. For all monotonic variables that occur in the objective function, the resulting summary is the *First Monotonicity Principle* (MP1). Its name is capitalized to emphasize that, despite its simplicity, the principle is widely applicable and very useful. It is a contrapositive corollary of the Monotonicity Theorem.

> *First Monotonicity Principle (MP1)* In a well-constrained minimization problem every increasing variable is bounded below by at least one nonincreasing active constraint. ("Flat" spots in the objective can by coincidence generate a semiactive constraint.)

The major value of MP1 is that it can sometimes prove a constraint is active without finding the optimum first. This happens when there is only one constraint that can bound a certain variable positively and finitely. Such a constraint is said to be *critical*.

Criticality

Consider an objective function $f(x_i, \mathbf{X}_i)$ increasing in a variable x_i. Suppose all (inequality) constraints but one also increase in x_i, the remaining constraint $g_j(x_i, \mathbf{X}_i) \leq 0$ being either decreasing or nonmonotonic. Then by MP1, g_j is active and bounds x_i from below. Such a constraint is said to be *critical* for x_i because if it were relaxed, the objective would no longer be well constrained wrt x_i. An inequality constraint that is critical is indicated by adding a second line to the inequality symbol. For example, $g_j \leqq 0$.

To reduce the danger of confusing criticality with activity, regard criticality as a special case of the more general concept of activity. Thus, although all critical constraints are active, not all active constraints are critical. Criticality is constrained activity imposed by monotonicity, which may not be present in other types of active constraint. The advantage of criticality is that it is a particularly easy kind of activity to identify.

In the air tank design example, volume constraint (1) is critical with respect to r, head thickness constraint (2) is critical wrt h, and shell thickness constraint (3) is critical wrt s. Thus the problem, Equation (3.23), may now be written

$$\begin{aligned}
&(0) \quad \min m(h^+, l^+, r^+, s^+) \\
&\qquad \text{subject to} \\
&(1) \quad \pi r^2 l \geqq 2.12(10^7) \quad (\text{wrt} \quad r), \\
&(2) \quad h/r \geqq 130(10^{-3}) \quad (\text{wrt} \quad h), \\
&(3) \quad s/r \geqq 9.59(10^{-3}) \quad (\text{wrt} \quad s), \\
&(4) \quad l \geq 10, \\
&(5) \quad r + s \leq 150.
\end{aligned} \tag{3.25}$$

The parentheses on the right remind us of the variables for which the various constraints are critical.

Notice that even though volume constraint (1) bounds f from below wrt l, so does the minimum length constraint (4). Hence, neither of them is critical for l, although either by itself would be critical if the other were not there. A partial minimum can therefore be found wrt three of the variables (h, r, and s), but at the moment the situation is unclear for the fourth variable l. The next section shows how to resolve this.

Optimizing a Variable Out

Suppose that an objective $f(\mathbf{x}) = f(x_1, \mathbf{X}_1)$ has been minimized partially wrt x_1. The minimizing argument $x_{1*}(\mathbf{X}_1)$ is a function of the remaining $n-1$ variables $\mathbf{X}_1 = (x_2, \ldots, x_n)^T$, so that the objective and all the constraints now depend only on $\mathbf{X}_1$: $g_j(x_{1*}(\mathbf{X}_1), \mathbf{X}_1) = g_j(\mathbf{X}_1)$.

If, as in the air tank design, $x_{1*}(\mathbf{X}_1)$ is obtained explicitly as the closed-form solution of one of the original constraints, say, $g_{j*}(\mathbf{x}) = 0$, written as an equality, then $g_{j*}(x_{1*}(\mathbf{X}_1), \mathbf{X}_1) = 0$ and the constraint is satisfied trivially by any partial minimum wrt x_1. Hence, this constraint does not restrict the remaining variables and should be deleted from the model. This deleted constraint is used later to recover the numerical values of x_{1*} once the optimal values of the remaining variables are known. Thus the variable x_1 disappears from the model, along with a constraint, after such a partial minimization. When this happens, we say that x_1 has been *optimized out* (in this case *minimized out*), by analogy with integrating a variable out of a function. Remember, however, that this can be done only when the argument is the explicit solution of one of the constraints as a strict equality.

Occasionally, a critical constraint restricts some variable implicitly and cannot be deleted in this way. For example, the critical constraint $-x_1^2 + x_2^2 - 3 = 0$ implies that $x_2 > \sqrt{3}$ for all positive x_1 and hence cannot be deleted. In general, critical and active constraints should be treated with the same care as equality constraints, particularly when deletions are contemplated.

In the air tank design, partial minimization wrt head thickness h gave $h_* = 130(10^{-3})r$ where r is yet to be determined. Substitution of h_* for h everywhere in the original problem gives

$$m(h_*, l, r, s) = \pi[(2rs + s^2)l + 2(r + s)^2(130)(10^{-3})r]$$

subject to the four constraints (1), (3), (4), and (5), which do not depend on h. Since h_* was determined as the solution to constraint (2), the latter is trivially satisfied by h_*, that is, $h_*/r = 130(10^{-3}) \geq 130(10^{-3})$. Hence, it is deleted, leaving four constraints in the three remaining variables. The head thickness h has been minimized out.

After a variable has been optimized out, the reduced model should be examined again to see if further partial minimization is easily possible. Such reduction should be continued as long as the partial minimizations can be done by inspection. In this way one ends with a completely optimized model, a determination that the model is not well constrained, or a model worthy of attack by more powerful methods.

The reader can verify in the partially optimized air tank that the shell thickness s must be made as small as possible, forcing it to its only possible lower bound in constraint (3). Thus $s_* = 9.59(10^{-3})r$, whence the reduced problem becomes

$$\begin{aligned}
&(0) \quad \min_{r,l} m(h_*, s_*, r, l) = \pi\{[(2r)(9.59(10^{-3})r) + (9.59)^2(10^{-3})^2r^2]l \\
&\qquad\qquad\qquad\qquad + 2(1.00959r)^2(0.130)r\} \\
&\qquad\qquad\qquad\quad = \pi r^2(0.01927l + 0.2650r) \\
&\qquad \text{subject to} \qquad\qquad\qquad\qquad\qquad\qquad (3.26) \\
&(1) \quad \pi r^2 l \geq 2.12(10^7), \\
&(4) \quad l \geq 10, \\
&(5) \quad r \leq 150/1.00959 = 148.6.
\end{aligned}$$

Now the radius r can be optimized out, since only the first constraint prevents the radius and therefore the metal volume from vanishing. Thus $r_* = (2.12(10^7)/\pi)^{1/2}l^{-1/2} = 2598l^{-1/2}$ and (3.26) is further reduced to

$$\begin{aligned}
&(0) \quad \min_{l} m(h_*, s_*, r_*, l) = \pi[130.1(10^3) + 4.647(10^9)l^{-3/2}] \\
&\qquad \text{subject to} \qquad\qquad\qquad\qquad\qquad\qquad (3.27) \\
&(4) \quad l \geq 10, \\
&(5) \quad 2598l^{-1/2} \leq 148.6 \quad \text{or} \quad l \geq 306.
\end{aligned}$$

This triply reduced problem is *not* well bounded. Since the objective decreases with l, which although bounded below is not bounded above by either remaining constraint, the infimum is $130.1(10^3)\pi$, where the argument $l = \infty$. But because this is not a finite argument, no minimum exists.

Adding Constraints

Monotonicity analysis has thus identified an incompletely modeled problem without attempting fruitless numerical computations. As already discussed in Chapter 2, one way to deal with this situation is to add an appropriate constraint.

Suppose it is found that the widest plate available is 610 cm. This imposes a sixth inequality constraint:

$$(6) \quad l \leq 610. \tag{3.28}$$

Recall from Equation (3.27) that the reduced objective decreases with l and that the two other constraints remaining provide only lower bounds on l. Hence, the new constraint (3.28) is critical in the reduced problem and is therefore active, $l \leqq 610$. Now the problem has been completely solved, for

$$\begin{aligned} l_* &= 610 \text{ cm}, & s_* &= 9.59(10^{-3})r_* = 1.0 \text{ cm}, \\ r_* &= 2598 l_*^{-1/2} = 105 \text{ cm}, & h_* &= 130(10^{-3})r_* = 13.6 \text{ cm}. \end{aligned} \tag{3.29}$$

Care in ensuring that the model was well constrained revealed an oversight, guided its remedy, and produced the minimizing design–without using any iterative optimization technique.

3.4 Recognizing Monotonicity

Things simplify greatly when monotonicity is present. Even fairly complicated functions can be monotonic, although this important property can go undetected without a little analysis. This section shows how to recognize and prove monotonicity, not only in common, relatively simple functions, but in composite functions and even integrals.

Simple and Composite Functions

Recall that a function $f(\mathbf{x})$ is said to be increasing wrt x_1, one of its independent variables, if and only if $\Delta f/\Delta x_1 > 0$ for all $\mathbf{x} > \mathbf{0}$ and for all $\Delta x_1 \neq 0$. This definition includes the case where f is differentiable wrt x_1, so that then $\partial f/\partial x_1 > 0$ for all $\mathbf{x} > \mathbf{0}$. If $f(\mathbf{x})$ increases wrt x_1, then $-f(\mathbf{x})$ is said to decrease wrt x_1. Notice that the same function can increase in one variable while decreasing in another. Finally, $f(\mathbf{x})$ is called *independent* of x_1 if and only if $\Delta f/\Delta x_1 = 0$ for all $\mathbf{x} > \mathbf{0}$ and all $\Delta x_1 \neq 0$.

A set of functions is said collectively to be *monotonic* wrt x_1 if and only if every one of them is either increasing, decreasing, or independent wrt x_1. The term *monotonic* is reserved for sets of functions that are not all independent. Similarly, a set of functions is said to be increasing (decreasing) wrt x_1 if and only if all functions are increasing (decreasing) wrt x_1. If one function is increasing and the other decreasing, both wrt x_1, they are said to have *opposite monotonicity* wrt x_1. Two functions that either both increase or both decrease are said to have the *same monotonicity*, or to be *monotonic in the same sense*.

The functions in most engineering problems are built up from simpler ones, many of which exhibit easily detected monotonicity. Simple rules are derived now for establishing monotonocity of such functions. The functions studied here are assumed differentiable to first order and positive over the range of their arguments. Then f increases (decreases) wrt x_1 if and only if $\partial f/\partial x_1$ is positive (negative).

Let f_1 and f_2 be two positive differentiable functions monotonic wrt x_1 over the positive range of $\mathbf{x}_1$. Then $f_1 + f_2$ is monotonic if both f_1 and f_2 are monotonic in the same sense, a fact easily proven by direct differentiation. The monotonic functions f and $-f$ have opposite monotonicities.

Now consider the product $f_1 f_2$. Since

$$\partial(f_1 f_2)/\partial x_1 = f_1(\partial f_2/\partial x_1) + f_2(\partial f_1/\partial x_1) \tag{3.30}$$

the product $f_1 f_2$ will be monotonic if both f_1 and f_2 are positive and have the same monotonicities.

Let f be raised to the power a. Then $\partial(f^a)/\partial x_1 = af^{a-1}(\partial f/\partial x_1)$. Hence f^a will have the same monotonicity as f whenever a is positive. If $a < 0$, f^a will have opposite monotonicity.

Finally, consider the composite function $f_1(f_2(\mathbf{x}))$. Differentiation by the chain rule gives $\partial f_1(f_2)/\partial x_1 = (\partial f_1/\partial f_2)(\partial f_2/\partial x_1)$. Hence, the composition $f_1(f_2)$ is also monotonic. It increases (decreases) whenever f_1 and f_2 have the same (opposite) monotonicity. For example, in the composite function $\ln[x(1-x^2)^{-1/2}]$, the function x^2 increases for $x > 0$, but $(1 - x^2)$ is decreasing and positive only for $0 < x < 1$. In this restricted range, however, $(1 - x^2)^{-1/2}$ increases, as does $x(1 - x^2)^{-1/2}$ and, since the logarithmic function increases, the composite function does too.

Integrals

Since integrals can be hard to analyze, or expensive to evaluate numerically, it is worthwhile to learn that monotonicities of integrals are often easy to determine. Let $g(x)$ be continuous on $[a, b]$. Then, by the Fundamental Theorem of Integral Calculus, there exists a function $G(x)$ called the *indefinite integral* of $g(x)$ for all $a \le x \le b$ such that $g(x) = dG/dx$. The *definite integral* having $g(x)$ as integrand is

$$\int_a^b g(x)\,dx = G(b) - G(a) = f(a, b). \tag{3.31}$$

Differentiation of f wrt its *limits* a and b gives

$$\frac{\partial f}{\partial a} = \left.\frac{dG}{dx}\right|_{x=a} = -g(a), \qquad \frac{\partial f}{\partial b} = g(b). \tag{3.32}$$

It follows that if the integrand is positive on $[a, b]$, that is, $g(x) > 0$ for all $a \le x \le b$, then $f(a^-, b^+)$. The definite integral of a positive function increases wrt its upper limit and decreases wrt its lower limit (see Figure 3.4). Note that changing the sign of the integrand at either a or b will reverse the monotonicity wrt to a or b, respectively. Next, consider the effect of monotonocity in the integrand. Let $g(x, y^+)$ be

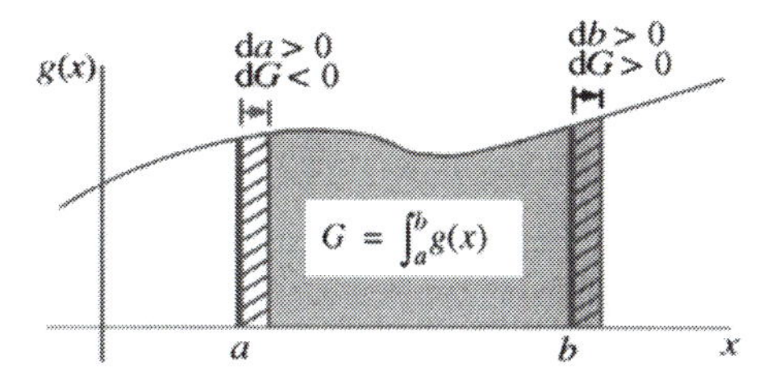

Figure 3.4. Monotonicity of an integral with respect to its limits.

continuous for $x \in [a, b]$ and increasing in y. That is, $g(x, y_2^+) > g(x, y_1^+)$ for all $x \in [a, b]$ if and only if $y_2 > y_1$. Let $G(a, b, y)$ be the definite integral

$$G(a, b, y) = \int_a^b g(x, y)\, dx. \tag{3.33}$$

Theorem $G(a, b, y)$ increases wrt y.

Proof Let $y_2 > y_1$. Then

$$\begin{aligned} G(a, b, y_2) - G(a, b, y_1) &= \int_a^b g(x, y_2)\, dx - \int_a^b g(x, y_1)\, dx \\ &= \int_a^b [g(x, y_2) - g(x, y_1)]\, dx > 0 \end{aligned}$$

since the integrand is positive for all $x \in [a, b]$.

Thus, if a function is monotonic, so is its integral (see Figure 3.5).

3.5 Inequalities

This section develops five concepts of monotonicity analysis applicable to inequality constraints. The first concerns *conditional criticality* in which there are several constraints capable of bounding an objective. *Multiple criticality*, the second concept, posts a warning in cases where the same constraint can be critical for more than one variable. *Dominance*, the third, shows how a conditionally critical constraint

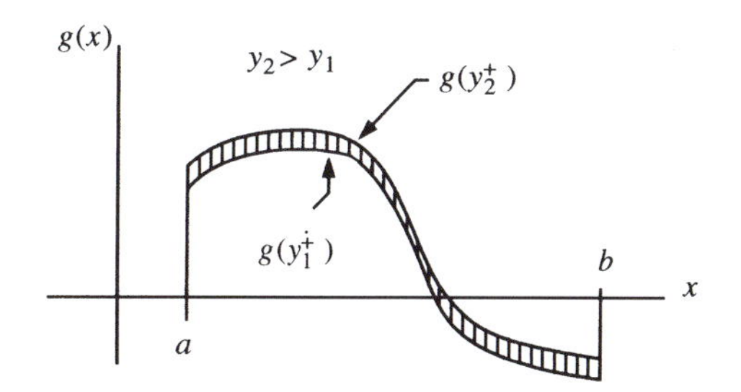

Figure 3.5. Monotonic integrand.

can sometimes be proven inactive. The fourth idea, *relaxation*, develops a tactic for cutting down the number of cases to be searched for the optimum. Finally, the curious concept of *uncriticality* for relaxing constraints or detecting inconsistency is examined.

Let us continue investigating reasonable upper bounds on the length in the air tank problem. Suppose the vessel is to be in a building whose ceiling allows no more than 630 cm for the vessel from end to end. In terms of the original variables, this gives a seventh inequality constraint:

$$l + 2h \leq 630. \tag{3.34}$$

In the reduced model this becomes

$$(7) \quad l + 2(130)(10^{-3})(2598)l^{-1/2} = l + 675.5l^{-1/2} \leq 630. \tag{3.35}$$

This constraint is not monotonic, but since it does not decrease as does the substituted objective function, it could bound the feasible domain of l away from infinity. There are now two constraints: (6) having monotonicity different from that of the reduced objective and (7) being nonmonotonic. The next section introduces concepts for studying this situation abstractly.

Conditional Criticality

To generalize the situation at the current state of the air tank design, let there be a monotonic variable x_i appearing in an objective $f(x_i, \mathbf{X}_i)$ for several constraints $g_j(x_i, \mathbf{X}_i)$, $j = 1, \ldots, m$ having opposite monotonicity wrt x_i to that of the objective. Suppose further, for reasons justified in the next subsection, that *none of these* $m(>1)$ *constraints are critical for any other variable*. Then, by MP1, at least one of the constraints in the set must be active, although it is unclear which. Such a set will be called *conditionally critical* for x_i. Constraints (3.28) and (3.35) form such a conditionally critical set in the air tank design.

Multiple Criticality

Criticality might be regarded as the special case of conditional criticality in which $m = 1$, were it not for the requirement in the definition that a constraint already critical for one variable cannot then be conditionally critical for another. To reconcile these definitions, let us refine the notion of criticality according to the number of variables I bounded by a given critical constraint. If $I = 1$, such a constraint is *uniquely critical*, as are the head and shell thickness constraints (2) and (3) in the original air tank problem, Equation (3.23). But if $I > 1$ as in the volume constraint (1) (critical for both l and r relative to the original objective), the constraint is called *multiply critical*.

Thus, it is only *unique* criticality that is a special case of conditional criticality. Multiple criticality obscures our understanding of the model from the standpoint not only of formal definition but also of seeing if a model is well constrained. The air tank problem in its original formulation of Section 3.2 demonstrated this. All four variables appeared in a critical constraint; yet the problem was not well constrained. The trouble was that the volume constraint was *multiply* critical, bounding both radius and length. Not until this constraint was used to eliminate one variable did it become apparent that the objective was not well bounded with respect to the other.

Multiply critical constraints should be eliminated if possible. *The notion of multiple criticality is a warning to the modeler not to jump to conclusions about well boundedness before all multiply critical constraints have been eliminated.*

Dominance

Sometimes a constraint can be proven inactive, even though it may be conditionally critical. If two constraints g_1 and g_2 are in the relation $g_2(\mathbf{x}) < g_1(\mathbf{x}) \leq 0$ for all feasible $\mathbf{x}$, then the constraint $g_2(\mathbf{x}) \leq 0$ cannot be active without violating $g_1(\mathbf{x}) \leq 0$. When this relation occurs, g_2 is said to be *(globally) dominated* by g_1 and is called *redundant*. For the theorem following, recall from Section 3.2 that the deletion space D_2 is the feasible region after deletion of g_2 from the constraint set.

Dominance Theorem If $g_2(\mathbf{x}) < g_1(\mathbf{x}) \leq 0$, then $\mathcal{F} = \mathcal{D}_2 \cap \mathcal{P}^n$. Deletion of a globally dominated constraint does not affect the feasible space.

Proof Since all solutions of g_1 satisfy g_2 but not vice versa, the constraint spaces satisfy $\mathcal{K}_1 \subseteq \mathcal{K}_2$. By definition of the deletion space $\mathcal{D}_2 = \cap_{j \neq 2} \mathcal{K}_j$, $\mathcal{F} = \mathcal{K}_2 \cap \mathcal{D}_2 \cap \mathcal{P}^n \supseteq \mathcal{K}_1 \cap \mathcal{D}_2 \cap \mathcal{P}^n = \mathcal{D}_2 \cap \mathcal{P}^n$. But also $\mathcal{F} = \mathcal{K}_2 \cap \mathcal{D}_2 \cap \mathcal{P}^n \subseteq \mathcal{D}_2 \cap \mathcal{P}^n$ and so $\mathcal{F} = \mathcal{D}_2 \cap \mathcal{P}^n$.

The Dominance Theorem permits deletion of any globally dominated constraint from an optimization problem. The resulting minimum will automatically satisfy the deleted constraint, which therefore will not be checked. The deletion of a dominated constraint is indicated by enclosing it in square brackets, for example, $[g_2(\mathbf{x}) \leq 0]$.

Example 3.6 Consider the problem $\{\min f = x_1$, subject to $g_1 = -x_1 + 2 \leq 0$, $g_2 = -x_1 + 1 \leq 0\}$. The two constraints are conditionally critical, but $g_2 = -x_1 + 1 \leq -x_1 + 2 = g_1 \leq 0$, and so by the Dominance Theorem g_2 may be deleted, leaving the problem $\{\min x_1$, subject to $-x_1 + 2 \leq 0, [-x + 1 \leq 0]\}$. Deletion of g_2 makes g_1 critical and therefore active, reducing the problem to $\{\min x_1$, subject to $-x_1 + 2 \leq 0\}$, which has the solution $x_{1*} = 2$. ■

Relaxation

Recall from the definition of activity that removing an active constraint from the model will change the location of the optimum. Now consider the situation where a constraint has been left out of the model and an optimum has been identified. Leaving a constraint out we referred to as *constraint relaxation* (Section 3.2). If the relaxed constraint is brought in the model and is found to be violated, one would expect that this constraint must be active. The next theorem confirms that this is indeed the case.

Before tackling the theorem, the reader should review the definitions in the subsection on constraint activity in Section 3.2.

Relaxation Theorem For a consistent, well-constrained problem, if any relaxed argument $\mathbf{x}'$ violates g_1, that is, $g_1(\mathbf{x}') > 0$, then g_1 is active or semiactive.

Proof Since $\mathcal{F}$ is consistent and well constrained, $\mathcal{X}_*$ exists in $\mathcal{F}$. Also $\mathbf{x}' \notin \mathcal{F}$ since $\mathbf{x}' \notin \mathcal{K}_1 \supset \mathcal{F}$, and so $\mathbf{x}' \notin \mathcal{X}_* \subset \mathcal{F}$. By definition, $\mathbf{x}' \in \mathcal{X}_1$, being a relaxed argument. If there exists another relaxed argument $\mathbf{x}'' \in \mathcal{X}_1$ that happens to satisfy g_1 and is therefore in $\mathcal{X}_*$, then since $\mathbf{x}' \notin \mathcal{X}_*$, $\mathcal{X}_1 \supset \mathcal{X}_*$ and g_1 is semiactive (Figure 3.2). If no relaxed argument $\mathbf{x}''$ exists, then $\mathcal{X}_1 \cap \mathcal{X}_* = \{\}$ and g_1 is active.

Corollary If $\mathcal{X}_*$ is unique, that is, $\mathcal{X}_* = \mathbf{x}_*$, then g_1 is active.

For a proof, note that in this case $\mathcal{X}_1 \cap \mathbf{x}_* = \{\}$, since $\mathbf{x}' \notin \mathcal{X}_*$ for every $\mathbf{x}' \in \mathcal{X}_1$. Hence g_1 is active.

This idea is illustrated in the air tank design. The optimal solution found before adding length constraint (3.34) was $l_* = 610$. However, this violates the new constraint (3.35). The relaxation theorem shows that we can conclude that the new constraint is active and that the new optimum is $l_* = 602.5$, $r_* = 105.8$, $h_* = 13.75$, and $s_* = 1.02$, slightly shorter and wider. The overall length is of course the limiting value of 630 cm.

The relaxation theorem suggests a test of activity: Delete the constraint being tested, find the infimum for the relaxed problem, and see if its solution satisfies the deleted constraint. If it does not, the constraint is active. If it does, the constraint is inactive for a unique minimum and at most semiactive for multiple minima. This is a useful tactic if the relaxed problem is much easier to solve or if one suspects the deleted constraint may not be active.

Uncriticality

Now consider a well-constrained problem having an inequality constraint $g_1(x_1) \leq 0$ with g_1 increasing with x_1 as does the objective $f(x_1)$. Such a constraint is said to be (*partially*) *uncritical* wrt x_1 because it would be critical if f were being maximized instead of minimized. Such a constraint can be critical or active wrt the other variables, in which case this partial uncriticality is uninteresting and will be ignored. But if it is uncritical with respect to all variables on which it depends, it is said simply to be *uncritical* and warrants special attention. To indicate uncriticality, draw a vertical line through the inequality sign, for example, $g_1(\mathbf{x}) \nleq 0$.

Uncriticality Theorem Let there be a constraint $g_1(x_1) \leq 0$ that is critical wrt x_1, and suppose there is another constraint $g_2(x_1) \nleq 0$ that is uncritical for x_1 and depends on no other variables. Then, either g_2 is inactive or the constraints are inconsistent.

Proof Without loss of generality, assume f and g_2 increase wrt x_1, while g_1 decreases wrt x_1. Then the constraint space for g_1 is $\mathcal{K}_1 = \{x_1: x_1 \geq A_1 > 0, \mathbf{x} \in \mathcal{F}\}$, where A_1 is the first component of the argument of the minimum. Since $g_2(x_1)$ increases, there is a unique value A_2 such that $g_2(A_2) = 0$, and the constraint space for g_2 is $\mathcal{K}_2 = \{x_1: x_1 \leq A_2, \mathbf{x} \in \mathcal{F}\}$. The feasible region $\mathcal{F}$ is a subset of $\mathcal{K}_1 \cap \mathcal{K}_2 = \{x_1: A_1 \leq x_1 \leq A_2, \mathbf{x} \in \mathcal{F}\}$. So if $A_1 \leq A_2$, f is minimized at A_1 over both $\mathcal{F}$ and $\mathcal{D}_2$, in which case g_2 is inactive. If $A_1 > A_2$, $\mathcal{F} = \{\,\}$, that is, the problem is inconsistent.

In the proof of the theorem it was shown that an uncritical constraint can be satisfied with strict equality at the minimum without being active. This happens when the minimum is the only feasible solution, as when by coincidence $A_1 = A_2$ in the proof.

The theorem suggests that uncritical constraints be deleted to create a relaxed problem. Then the minimum found can be checked against the deleted constraints to see if they are satisfied. If they do, the minimum has been found; otherwise the Uncriticality Theorem assures us that either the violated constraint is active or the constraints are inconsistent. And if there is only one variable involved, as in the theorem just proven, the only possibility is inconsistency. This single-variable case, although it may seem special, occurs quite often in practice.

In the air tank design, minimum shell length constraint (4) and maximum radius constraint (5) in Equation (3.27) are uncritical, written as

$$l \not\geq 10 \quad \text{and} \quad l \not\geq 306.$$

By the Dominance Theorem only the second of these need be retained, since its satisfaction implies that of the first. Both were relaxed in the preceding analysis, which gave $l_* = 602.5$. Since this satisfies all constraints, the solution is indeed optimal. If, however, the height restriction were reduced from 630 cm to 300 cm, then l_* would accordingly be something less than 300, which would violate uncritical maximum radius constraint (5) in Equation (3.27). The Uncriticality Theorem would then indicate that the constraints were inconsistent, in this case because the longest and widest allowable vessel would have a volume less than required by the capacity constraint (1).

3.6 Equality Constraints

We now show how to apply the results derived for inequalities to problems constrained by strict equalities. After discussing activity in equalities, the concept of *directing an active equality* is developed, that is, replacing the equality by an active inequality in such a way that the optimum is not affected. Finally, a theory of *regional monotonicity* is developed to extend *constraint direction* to nonmonotonic situations.

Equality and Activity

The First Monotonicity Principle implies that a critical constraint is satisfied with strict equality at the minimum. That is, if any inequality constraint $g_j \leq 0$ is critical, then $g_j(\mathcal{X}_*) = 0$.

Of course an active equality constraint is trivially satisfied as an equality. However, not all equality constraints are active. Consider, for example, {min $f = 2x_1$, subject to $g_1: x_1 \geq 1, g_2: x_2 = 5$}. Here g_1 is critical, and so $\mathcal{X}_* = (1, 5)$, and $f(\mathcal{X}_*) = 2$. Deletion of the second constraint gives $\mathcal{X}_* = (1, x_2)$ for every positive finite value of x_2. But still the minimum is $f(\mathcal{X}_*) = 2$, and so the second constraint is only semiactive. We indicate when an equality is known to be active by placing a third horizontal line below the equals sign, for example, $h_j(\mathbf{x}) \equiv 0$.

A critical constraint is certainly active, but when there are several constraints *conditionally* critical for some variable, it is not obvious which will be active. All that can be said is that at least one constraint of every conditionally critical set must be active with strict equality if the objective is to be well constrained. More than one member of a conditionally critical constraint set can be active.

Replacing Monotonic Equalities by Inequalities

The theory so far has focused on monotonic inequality constraints. Hence, it remains unclear what to do when monotonic variables occur in a strict equality. As a last resort, the analyst can use such an equality to eliminate a variable, but this will be seen to be dangerous when the activity of the constraint has not been established. Moreover, not all equations are solvable explicitly, and indiscriminate elimination of a variable sometimes destroys useful monotonicities of other variables. Therefore, this section will show how one can often replace an equality with an inequality constraint when it is monotonic.

To motivate the study of this subtle bit of theory, consider the following modification of the air tank problem. Let the volume and total length be represented explicitly by v and t, respectively. Then inequalities (1), Equation (3.25), and (7), Equation (3.35), become

$$\begin{aligned} &(1\text{-}i) \quad v \geq 2.12(10^7), \\ &(7\text{-}i) \quad t \leq 630. \end{aligned} \tag{3.36}$$

The new variables are related to the old ones through equalities:

$$\begin{aligned} &(1\text{-}e) \quad v = \pi r^2 l, \\ &(7\text{-}e) \quad t = l + 2h. \end{aligned} \tag{3.37}$$

This replacement of an inequality by an equality and another inequality in a new variable, being totally artificial in this example, is not at all recommended. It is done here simply to make an example for the following theory on how to "direct" equalities. However, there are situations where introduction of such new variables is an unavoidable modeling tactic. Whenever this happens, equality constraints are inevitable.

For example, an inequality of the form

$$(x_1 + x_2)^{1/2} + x_3 \le 1 \qquad (3.38)$$

can be replaced by the change of variable equation

$$x_4 = x_1 + x_2 \qquad (3.39)$$

and the resulting inequality

$$x_4^{1/2} + x_3 \le 1. \qquad (3.40)$$

This last form can have theoretical and computational advantages over the original inequality but the equality must be included in the model as well.

Directing an Equality

Let $f(\mathbf{x})$ and $h_1(\mathbf{x})$ be monotonic functions of the first variable x_1, and consider minimizing f subject to the equality constraint $h_1(\mathbf{x}) = 0$ as well as other inequality and equality constraints. Discussion of nonmonotonic functions is deferred to the next subsection. The problem is: min $f(\mathbf{x})$ subject to $\mathbf{x} \in \mathcal{F} = \mathcal{K}_1 \cap \mathcal{D}_1 \cap \mathcal{P}^n$, where $\mathcal{K}_1$ is the constraint space of $h_1 = 0$, and $\mathcal{D}_1 = \cap_{j>1} \mathcal{K}_j$. Let f be well constrained with its minimum in $\mathcal{X}_*$.

Consider now a second minimization problem in which the equality constraint is replaced with the inequality constraint $h_1(\mathbf{x}) \le 0$. Let $\mathcal{K}_1' = \{\mathbf{x} : h_1(\mathbf{x}) \le 0\}$. This new *inequality-constrained* problem is to minimize f subject to $\mathbf{x} \in \mathcal{F}' = \mathcal{K}_1' \cap \mathcal{D}_1 \cap \mathcal{P}^n$. Let its minimum, if it exists, be in $\mathcal{X}_*'$. Assume that f is increasing and h_1 is decreasing. Then we have the following:

Monotonic Direction Theorem If h_1 is active, then the inequality-constrained problem is well constrained, and its solution set $\mathcal{X}_*'$ is identical to $\mathcal{X}_*$, the solution set of the equality-constrained problem.

Proof Given a minimum $\mathbf{x}_* = (x_{1*}, x_{2*}, \ldots, x_{n*})^T$ of the equality constrained problem, let $\mathbf{x}' = (x_{1*}', x_{2*}, \ldots, x_{n*})^T$ with $x_1' \ne x_{1*}$. Require $\mathbf{x}'$ to satisfy $h_1(\mathbf{x}') < 0$, that is, to be feasible but not active for the inequality-constrained problem. Then, $x_1' > x_{1*}$ because $h_1(\mathbf{x})$ decreases with x_1. Since $f(x)$ increases with x_1, $f(\mathbf{x}') > f(\mathbf{x}_*)$. Moreover, $h_1(\mathbf{x}_*) = 0$ since h_1 is active, and so $\mathbf{x}_*$ is feasible for the inequality-constrained problem. Hence, $\mathbf{x}_*$ is the minimum for both problems, making $\mathcal{X}_*' = \mathcal{X}_*$ and ensuring that the inequality-constrained problem is well constrained.

It is advantageous to replace (active) equality constraints with inequalities in this way because this can facilitate further monotonicity analysis. This procedure, called *directing the equality* is symbolized by placing an inequality sign next to the original active equality sign. For example, $h_1(\mathbf{x}) \equiv 0$, after direction, becomes $h_1(\mathbf{x}) \equiv <0$. This notation is deliberately different from $h_1(x) \leqq 0$, which would imply that the inequality is given rather than derived from an equality. Even though such a directed equality must be satisfied with strict equality at the minimum, recall that the equality itself might not be active in general.

Example 3.7 The problem $\{\min f = x_1^2 + x_2^2$, subject to $h_1 = x_1 + x_2 - 2 = 0\}$ has its minimum at $\mathbf{x}_* = (1, 1)^T$, and $f_* = 2$. If constraint h_1 is replaced by $h_1 = x_1 + x_2 - 2 \geq 0$ then $h_1(\mathbf{x}_*) = 0$, $h_1 \geq 0$, and $\mathbf{x}_* = (1, 1)^T$ with $f_* = 2$ again. So we can write $h_1(\mathbf{x}) \equiv\, >0$, or $-h_1(\mathbf{x}) \equiv\, <0$. ■

In the latest version (with Equations 3.36 and 3.37) of the air tank design, the constraints involving r are now

$$\begin{aligned} v &= \pi r^2 l, \\ h/r &\geq 130(10^{-3}), \\ s/r &\geq 9.59(10^{-3}), \\ r + s &\leq 150. \end{aligned} \tag{3.41}$$

Since the objective increases wrt r, at least one of these constraints must bound r from below, but none of the inequalities do this. Consequently, the *equality* must be critical for r. It can therefore be written with three lines to symbolize its criticality: $v \mathrel{\dot{\equiv}} \pi r^2 l$. Moreover, the Monotonic Direction Theorem permits its replacement by the lower bounding inequality $v \leq \pi r^2 l$, which is written $v \equiv\, <\pi r^2 l$ to retain the information that the relation was originally a strict equality and is critical.

A different situation occurs with regard to the total length in Equation (3.37), namely

$$(7\text{-}e) \quad t = l + 2h.$$

After using critical constraints to eliminate all variables except l and t, we are left with the remaining constraints

$$\begin{aligned} &(4) & l &\geq 10, \\ &(5) & 2598 l^{-1/2} &\leq 148.6, \\ &(6) & l &\leq 610, \\ &(7'\text{-}e) & l + 675.5 l^{-1/2} &= t. \end{aligned} \tag{3.42}$$

Since the reduced objective of Equation (3.27) decreases with l, we seek constraints bounding l from above. The first two bound l from below instead; so they cannot be critical. Shell length maximum constraint (6) is monotonic and bounds l from above as required. But total length constraint ($7'$-e) is not monotonic in l, which would appear to prevent immediate use of the Monotonic Direction Theorem, although the First Monotonicity Principle still applies. After dealing with this problem in the next section, we will return to the example for further application of the direction theorem.

Regional Monotonicity of Nonmonotonic Constraints

The nonmonotonic function $g(l) = l + 675.5l^{-1/2}$ that is the left member of total length constraint ($7'$-e), Equation (3.42) in the example, strictly decreases wrt positive l to its minimum at $l_\dagger = 48.5$, after which it increases. The first derivative $\partial g/\partial l$ vanishes only at $l_\dagger = 48.5$ and

$$\partial^2 g/\partial l^2 > 0 \tag{3.43}$$

for positive finite l.

Thus, $g(l)$ can be regarded as a *piecewise* monotonic function of l, with the sense of the monotonicity changing at the stationary point $l_\dagger$. Let $g^-(l)$ be the decreasing function $g(l)$ defined only where $\partial g/\partial l < 0$, that is, where $0 < l < l_\dagger = 48.5$, and, similarly, let $g^+(l)$ be the increasing function $g(l)$ defined for $l > l_\dagger = 48.5$. For an upper bound on l, the Monotonic Direction Theorem applied to $g^+(l)$ indicates that the strict equality $g^+(l) = t$ can be replaced by the inequality

$$g^+(l) = < t \tag{3.44}$$

provided one retains the inequality restricting the domain of l, that is,

$$(7+) \quad l > 48.5. \tag{3.45}$$

This last strict inequality cannot provide a bound for l. Its presence is necessary, however, to permit writing $g^+(l)$ in its original form $g(l)$, so that now constraint ($7'$-e) can be written

$$(7'') \quad l + 675.5l^{-1/2} = < t. \tag{3.46}$$

The full set of constraints now is

$$\begin{aligned}
&(4) && l \geq 10,\\
&(5') && l \geq 306,\\
&(6) && l \leq 610,\\
&(7'') && l + 675.5l^{-1/2} = < t,\\
&(7+) && l > 48.5,\\
&(7'\text{-}i) && t \leq 630.
\end{aligned} \tag{3.47}$$

Only (6) and ($7''$) have opposite monotonicity wrt l from the objective, and neither is uniquely critical. They form instead a conditionally critical set in which at least one of them must be active. We have already seen that total length constraint ($7''$) happens to be active for this particular set of parameters. That is, its deletion from the system would allow too long a shell.

3.7 Variables Not in the Objective

This section deals with models that have monotonic variables occurring in the constraints but not in the objective, a situation not covered by the First Monotonicity Principle. The additional information available by working with all the variables can be crucial, particularly for directing equalities to make MP1 applicable to the variables in the objective. These ideas form the Second Monotonicity Principle. Conditional criticality is extendable to this situation.

Hydraulic Cylinder Design

Consider Figure 3.6 showing a hydraulic cylinder, a device for lifting heavy loads as in a car hoist or elevator, or for positioning light ones as in an artificial limb. In the most general design context, it has five design variables: inside diameter i, wall thickness t, material stress s, force f, and pressure p. It is desired to select i, t, and s to minimize the outside diameter $(i + 2t)$ subject to bounds on the wall thickness, $t \geq 0.3$ cm, the force, $f \geq 98$ Newtons, and the pressure, $p \leq 2.45(10^4)$ Pascals. There are two physical relations. The first relates force, pressure, and area $f = (\pi/4)i^2 p$. The second gives the wall stress $s = ip/2t$. The model is summarized as follows:

$$
\begin{aligned}
\text{minimize } g_0: &\quad i + 2t \\
\text{subject to } g_1: &\quad t \geq 0.3, \\
g_2: &\quad f \geq 98, \\
g_3: &\quad p \leq 2.45(10^4), \\
g_4: &\quad s \leq 6(10^5), \\
h_1: &\quad f = (\pi/4)i^2 p, \\
h_2: &\quad s = ip/2t.
\end{aligned}
\tag{3.48}
$$

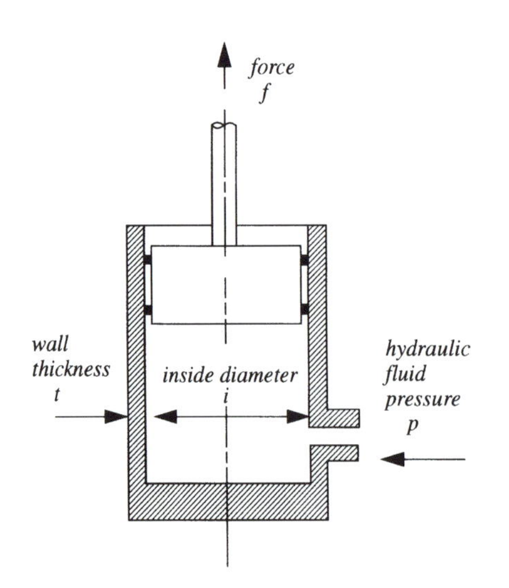

Figure 3.6. Hydraulic cylinder.

Applying the First Monotonicity Principle gives no new information on constraint activity, or even direction, since both objective variables i and t are in several constraints that include undirected equations. Here is where a second monotonicity principle for the nonobjective variables f, p, and s needs to be derived, after which the model analysis will be continued.

A Monotonicity Principle for Nonobjective Variables

Strictly speaking, MP1 could be applied directly to the nonobjective variables (Hansen, Jaumard, and Lu 1989a). The objective function happens to satisfy the definition of a *weakly* increasing function of all nonobjective variables, because it is independent of the nonobjective variables and, consequently, flat throughout $\mathcal{P}$ for each of them. Hence every nonobjective variable must be well constrained below by a nonincreasing constraint function; this prevents solutions at zero. Moreover, the objective also decreases weakly with respect to the nonobjective variables and therefore must be well bounded above by a nondecreasing constraint function to exclude infima at infinity.

Rather than require this double application of MP1, let us express the situation and its resolution as follows.

> *Second Monotonicity Principle* (*MP2*) In a well-constrained minimization problem every nonobjective variable is bounded both below by at least one nonincreasing semiactive constraint and above by at least one non-decreasing semiactive constraint.

There are two new things to notice about MP2. Firstly, semiactivity is the norm, rather than activity as in MP1, because the objective function is by definition flat near the minimum with respect to the nonobjective variables. This makes nonunique minima possible and forces further analysis to prove uniqueness and consequently activity, a matter to be illustrated in a continuation of the hydraulic cylinder example. The second consideration is that although two separate bounding constraints are needed if they are strictly monotonic, a single nonmonotonic constraint could bound the problem both above and below.

In the hydraulic cylinder design the nonobjective force variable f appears in two constraints: the inequality g_2, bounding f from below, and the equation h_1. By MP2, h_1 must constrain f from above; so it must be directed as

$$h_1': \quad f = < (\pi/4) i^2 p.$$

Notice that no triple line has been used, for it has not been proven that the constraint is any more than semiactive. But since f has now been proven to be well constrained in both directions, it can be eliminated by combining h_1' with g_2 into a single inequality

$$(h_1', g_2): \quad i^2 p \geq 124.8.$$

Similarly, the nonobjective variable s is bounded above by inequality g_4 and below semiactively by the properly directed equation h'_2, permitting their combination to eliminate s:

$$(h'_2, g_4): \quad ip/t \leq 1.2(10^6).$$

The third nonobjective variable p appears in three constraints: the upper bound g_3, the new inequality (h'_1, g_2), which as directed bounds p from below, and the new inequality (h'_2, g_4), which again as directed provides another upper bound.

So the Second Monotonicity Principle has led to eliminating two well-constrained nonobjective variables and the two original strict equalities.

It is time now to reapply MP1 to the objective variables. This exposes an interesting criticality, for the increasing internal diameter i can only be constrained below by the new inequality (h'_1, g_2), which therefore must be critical and written with a double line:

$$(h'_1, g_2): \quad i^2 p \geqq 124.8.$$

The final nonobjective variable p can now be eliminated to give

$$p = 124.8(10^6)/i^2 \text{ Pa},$$

which when substituted into g_3 gives

$$(h'_1, g_2, g_3): \quad i \geq 7.14 \text{ cm}$$

and into (h'_2, g_4) yields

$$(h'_2, g_4; h'_1, g_2): \quad it \geq 1.04 \text{ cm}^2.$$

There remain but three inequalities: g_1, (h'_1, g_2, g_3), and $(h'_2, g_4; h'_1, g_2)$, the first and third conditionally critical for t, and the second and third conditionally critical for i, the only two remaining variables, both in the objective function.

For this particular set of parameter values the last inequality turns out to be inactive, giving the solution $i = 7.14$ cm, $t = 0.3$ cm, $f = 98$ N, $p = 2.45(10^4)$ Pa, and $s = 2.92(10^5)$ Pa $< 6(10^5)$ Pa. The four cases resulting from using general parameter values will be analyzed in Section 6.2.

3.8 Nonmonotonic Functions

The Monotonicity Theorem, in the most general form developed in Section 3.3, applies also to nonmonotonic functions. While this extension from previous theory greatly expands the number of problems amenable to Monotonicity Analysis, there is a complication introduced thereby, which can cause error if not taken into account. The complication springs from the fact that whereas monotonic functions can have no more than one root, nonmonotonic functions can have several, even in fairly practical engineering situations.

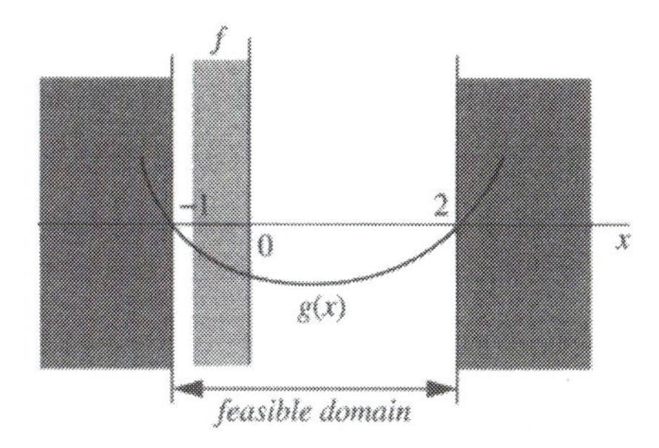

Figure 3.7. Convex constraint: partly negative feasible domain.

This situation in fact occurs in the air tank problem, whose reduced model of Section 3.6 called for minimizing a decreasing function of length l subject conditionally to both an increasing constraint $l - 610 \leq 0$ and a nonmonotonic constraint $l + 675.5l^{-1/2} - 630 \leq 0$. The latter constraint function has two roots, not only the one $l = 602.5$, which happens to be active, but also $l = 1.15$, ignored previously because the constraint $l \geq 1.15$ generated by it is uncritical. Thus only one of the roots could satisfy the MP1 requirement that the constraint be nonincreasing, and it was possible to decide which root to use in advance without any computation. In general this decision could be difficult to resolve correctly.

Example 3.8 Consider the following problem (suggested by Y. L. Hsu), which, although having a simple form to make the algebra clear, illustrates a situation that could certainly arise in an engineering problem:

$$\min x$$
$$\text{subject to } g_1\colon x^2 - x - 2 \leq 0.$$

By MP1, the nonincreasing constraint must be critical if the problem is well constrained. Hence any well-constrained minimum must be at a root of g_1. The constraint function is easily factored as $g_1 = (x + 1)(x - 2)$, so it has two roots, -1 and $+2$. Since the negative root lies outside the positive definite domain, one might be tempted to assume that the minimum is at $x = 2$, the only positive root. This would be grossly incorrect, however, for Figure 3.7 displays this point as the global maximum! It also shows the source of this seeming paradox. The infimum $x = 0$ satisfies the constraint g_1 but is not a minimum because it is not in the positive finite domain $\mathcal{P}$. The problem is therefore not well constrained from below, and the hypothesis of MP1 is not satisfied.

To avoid errors of this sort, pay attention to the *local* monotonicity at any root considered a candidate for the minimum. MP1 explicitly requires the constraint here to be nonincreasing, whereas at the false optimum $x = 2$ the constraint strictly increases. The only possible minimum must therefore be at $x = -1$, where g_1 decreases as required by MP1. Since this point is not in $\mathcal{P}$, no positive finite minimum exists. Figure 3.7 depicts the geometry of the situation.

The fact that the lower root must be the nonincreasing one bounding the objective could have been inferred in advance, since the second derivative of g_1 is the strictly

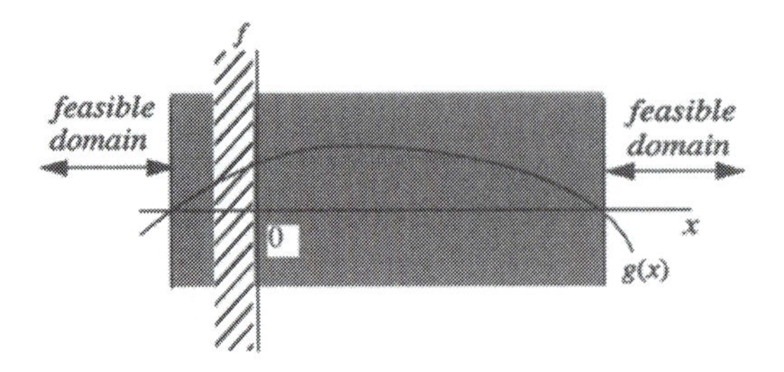

Figure 3.8. Concave constraint: one feasible region negative.

positive number 2. Another indicative characteristic of the constraint, to be developed in Chapter 4, is its convexity, in this case suggested by its being unbounded above as x approaches plus or minus infinity. All these properties require the smaller root to be the bounding one if the bound is in $\mathcal{P}$. ■

In the air tank design the bound existed because the bounding root (the larger one for that decreasing objective) was positive. In Example 3.8 the negativity of the smaller (bounding) root exposed the problem as not well constrained. If (as in Exercise 3.16) the feasible region for the constraint $(x-1)(x-4) = x^2 - 5x + 4$ is shifted two units right so that it is entirely in $\mathcal{P}$, the problem will be well constrained and the smaller root minimizing.

An interesting but nasty situation occurs when the constraint is concave, in the continuous case leading to a negative second derivative. This is equivalent to reversing the original inequality. Then for an increasing function the larger root, not the smaller, is where the constraint is nonincreasing and could be active.

Example 3.9 Consider Example 3.8 but with the sign of the constraint function changed to illustrate the point above:

$$\min x$$
$$\text{subject to } g_1'\colon -x^2 + x + 2 \le 0.$$

The constraint roots are -1 and 2, exactly as in Example 3.8, but this time it is the larger root 2 that is nonincreasing. Since it is in $\mathcal{P}$ the temptation is strong to accept it as the minimizing solution, which Figure 3.8 shows it actually is. But to prove this, one really must verify that the smaller root -1 is *not* in $\mathcal{P}$. This negativity guarantees that the left-hand region of x for which it is an *upper* bound is not in $\mathcal{P}$ where it could cause trouble by being unbounded from below. Thus in this case one must prove that the nonoptimizing root is *not* in $\mathcal{P}$, for if it were, the problem would not be well constrained. Example 3.10 illustrates this latter situation. ■

Example 3.10 Consider the problem

$$\min x$$
$$\text{subject to } g_1'\colon -x^2 + 5x - 4 \le 0.$$

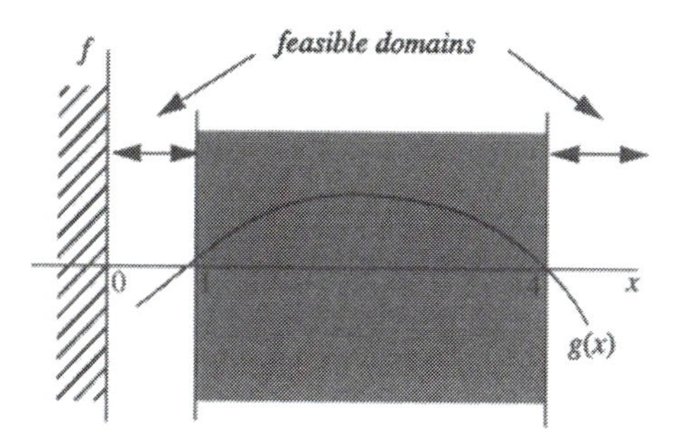

Figure 3.9. Concave constraint: disjoint feasible regions, one partly negative.

This constraint is the reverse of that in Exercise 3.16, and so $g_1' = -g_1$ is concave rather than convex. Its roots are 1 and 4 as in Exercise 3.16, but here the larger root 4 is the lower bound specified by MP1, provided the other root does not generate an unbounded region. But the unbounded region does intersect $\mathcal{P}$; thus the model is not well constrained (see Figure 3.9). ■

What is dangerous about concave constraints is this ability to split the feasible domain into disjoint regions. When this happens, the local search methods of gradient-based numerical optimization can miss the correct answer if they are started in the wrong region. Hence it is necessary, as in Example 3.8, to prove that the unbounded region is entirely outside $\mathcal{P}$, if the constraint is to be active at the nonincreasing larger root.

The complexities generated by constraint functions with more than two roots will not be explored here because they are rather rare situations in practice. For three or more constraint roots the appropriate analysis would be along the lines outlined here.

3.9 Model Preparation Procedure

The following procedure informally organizes systematic application of the many properties, principles, and theorems developed in this chapter. The goal is to refine the original model through repeated monotonicity analysis until it has been proven to be, if not well constrained, at least not obviously unbounded – with as few variables and as little residual monotonicity as possible. The procedure is incomplete in that some steps are ambiguous in their requests for choices of variables or constraints. Moreover, some principles derived in the chapter (relaxation, for example) are not involved in the procedure. At this point, they are left to the designer for opportunistic application.

The intention is to have the analyst go through the loop as long as any new information or reduction is possible. The final result will be a tightly refined model that gives the final design, displays inconsistency, or is suitable for further numerical work. A more complete procedure that includes solution steps may be devised, but only after the ideas in subsequent chapters are explored.

Begin with a model comprised of an objective function with inequality and/or equality constraints in which the distinction between variables and parameters is clear. Proceed to examine the following possible situations:

1. Dominance
   Examine constraints for dominance. If dominance is present, remove dominated constraints.
2. Variable selection
   Choose a design variable for boundedness checking, preferably one appearing in few functions. If there are no new variables, go to step 8.
3. Objective monotonicity
   Check objective for monotonicity wrt the variable selected in step 2.
   a. If monotonic, MP1 is potentially applicable; note whether increasing or decreasing; go to step 4.
   b. If independent, MP2 is potentially applicable; make a note; go to step 4.
   c. Otherwise, return to step 2.
4. Constraint monotonicity
   If the variable appears in an equality constraint, the equality must be first directed, deleted, or substituted by an implicit constraint before inequalities are examined. If MP1 and MP2 wrt this variable do not apply (see a, b below), choose another variable appearing in the equality and return to step 2 using that as the new variable.
   If the variable does not appear in an equality, choose an inequality constraint depending on this variable and check it for monotonicity.
   See if either MP applies to the variable selected.
   a. If neither does, return to step 2.
   b. If one does, use it to see if constraint can bound this variable. If it can, add constraint to conditionally critical set. Otherwise, identify constraint as uncritical wrt the variable.
   c. Choose a new constraint and repeat a, b. If no more constraints exist, go to step 5.
5. Criticality
   Count the number of conditionally critical constraints.
   a. If zero, stop; *model is not well bounded!*
   b. If one, constraint is *critical*; note constraint and variable; go to step 6.
   c. If more than one, note set; return to step 2.
6. Multiple criticality
   a. If constraint is critical for some other variable, it is *multiply critical*; use constraint to eliminate some variable; a reduced problem is now

generated, so return to step 1. If no elimination is possible, go to step 8.

b. Otherwise, constraint is *uniquely critical*; go to step 7.

7. Elimination

First try implicit elimination, since this does not involve algebraic or numeric details, only the current monotonicities. If this does not reduce the model, try explicit elimination unless this destroys needed monotonicity in other functions. Nonmonotonic functions can have multiple roots, so be careful to use the right one as discussed in Section 3.7. A reduced model is now generated, so return to step 1. If no elimination is performed, go to step 8.

8. Uncriticality

Note constraints that are uncritical wrt all variables on which they depend.

a. Relax uncritical constraints; remaining criticalities may have now changed, so return to step 1.

b. If none, go to step 9.

9. Consistency check

a. If numerical solution has been determined, substitute it into all relaxed uncritical constraints. If solution is feasible, it is *optimal.*
If solution is infeasible, *stop*; *model is inconsistent*!

b. If solution is not yet determined, save reduced model for methods in rest of book.

Anyone reaching step 9b with conditionally critical sets having only two members will be pardoned if they succumb to the urge to relax one of them and try again. How to do this intelligently is, in fact, a major topic of Chapter 6.

3.10 Summary

Most books on optimization devote at most a few paragraphs to the fundamentals covered in this chapter – bounds and the impact of constraints upon them – before plunging into the theory on which numerical calculations are based. The more detailed development here justifies itself not only by preventing attempts at solving bad models but also by the potentially significant reduction in model size it permits. Such reduction both increases the designer's understanding of the problem and eases the subsequent computational burden.

Careful development of the process of identifying constraint activity using monotonicity arguments allows prediction of active constraints before any numerical work is initiated. Even when all the active constraints cannot be identified a priori, partial knowledge about constraint activity can be useful in solidifying our faith in solution obtained numerically by methods described in later chapters.

The next two chapters develop the classical theory of differential optimization, starting with the derivation of optimality conditions and then showing how iterative algorithms can be naturally constructed from them. In Chapter 6 we revisit the ideas of the present chapter but with the added knowledge of the differential theory, in order to explore how optimal designs are affected by changes in the design environment defined by problem parameters. We also discuss how the presence of discrete variables may affect the theory developed here for continuous variables.

Notes

In the first edition, this chapter superseded, summarized, expanded, or updated a number of the authors' early works on monotonicity analysis during the decade starting in 1975, and so those works were not cited there or here. The only references of interest then and now are the original paper by Wilde (1975), in which the idea of monotonicity as a means of checking boundedness was introduced, and the thesis by Papalambros (1979), where monotonicity analysis became a generalized systematic methodology.

This second edition integrates the important extension to nonmonotonic functions published by Hansen, Jaumard, and Lu (1989a,b). In the first edition a function was considered well bounded if any of the infima were in $\mathcal{P}$. Requiring all infima to be in $\mathcal{P}$ simplified and shortened the discussion. The Section 3.8 treatment of how to handle the multiple root solutions generated by the nonmonotonic extension is new, partly in response to questions raised by Dr. Hsu Yeh-Liang when he was a teaching assistant for the Stanford course based on the first edition. Hsu also pointed out the overly strong statement of MP2 in the first edition, which has been appropriately weakened in this one.

Several efforts for automating monotonicity analysis along the lines of Section 3.9 have been made. Hsu (now at Yuan-Ze Institute of Technology, Taiwan) has published an optimization book in Chinese containing an English language program automating much of the monotonicity analysis outlined in Section 3.9, but for strictly monotonic functions. An earlier version can be found in Hsu (1993). Rao and Papalambros also had developed the artificial intelligence code PRIMA (Papalambros 1988, Rao and Papalambros 1991) for automating monotonicity analysis, particularly for handling implicit eliminations automatically. Other programs for automatic monotonicity analysis have been developed by Hansen, Jaumard, and Lu; Agogino and her former students Almgren, Michelena, and Choy; by Papalambros and his former students Azarm and Li; and by Zhou and Mayne. Mechanical component design texts have many practical engineering problems amenable to monotonicity analysis. At the Stanford and Michigan courses more than one student project in any given semester has shown to the class the failure of numerical optimization codes in a design study unaided by monotonicity analysis.

Exercises

3.1 Classify functions in Example 3.1 according to existence of upper bounds rather than lower bounds.

3.2 Determine the monotonicity, or lack of it, for variables x and y, together with the range of applicability, for the following functions:

(a) $\exp[x^{-1}(1 - x^2)]^{1/2}$,

(b) $\sin(x^2 + \ln y^{-1})$,

(c) $\exp(-x^2)$,

(d) $\exp(x)/\exp(1/x)$,

(e) $\ln(1/x^2)$,

(f) $\int_0^x \exp(-t)\,dt$,

(g) $\int_a^b \exp(-xt)\,dt$.

3.3 Suppose $\underline{x}$ minimizes $f(x)$ over $\mathcal{P}$. Prove that if b is a positive constant, $\underline{x}$ maximizes $g(x) = a - bf(x)$, where a is an arbitrary real number.

3.4 (From W. Braga, Pontificia Universidade Catolica do Rio de Janeiro, Brazil.) A cubical refrigerated van is to transport fruit between Sao Paulo and Rio. Let n be the number of trips, s the length of a side (cm), a the surface area (cm^2), v the volume (cm^3), and t the insulation thickness (cm). Transportation and labor cost is $21\,n$; material cost is $16(10^{-4})a$; refrigeration cost is $17(10^{-4})an/(t + 1.2)$; and insulation cost is $41(10^{-5})at$ – all in Brazilian currency at a given date. The total volume of fruit to be transported is $34(10^6)$cm^3. Express the problem of designing this van for minimum cost as a constrained optimization problem.

3.5 Consider Braga's fruit-van problem, Exercise 3.4. Where possible, replace equalities by active inequalities, and determine which inequalities are active at the minimum. Is this problem constraint bound? Is it well constrained?

3.6 In the minimization problem

$$\begin{aligned} &\min x_3x_4 + 10x_5 && (0)\\ &\text{subject to } x_1x_4 \le 100, && (1)\\ &x_2 = x_3 + x_4, && (2)\\ &x_3 \ge x_4, && (3)\\ &1/x_1 + x_4 = x_5, && (4) \end{aligned}$$

direct equalities and prove criticality where possible. To the right of each constraint for which you draw conclusions, indicate the order (1, 2, ...) in which the constraint was analyzed, the Monotonicity Principle used (1, 2), and the variable studied ($x_1, \ldots, x_5$). Use inequality signs ($>$, $<$) to show direction of an equality, and use an underline to indicate criticality. Then use critical constraints to eliminate variables, obtaining a nonmonotonic objective with no critical constraints.

3.7 Find if the following problem is well constrained:

$$\max f = x_1 - x_2$$
$$\text{subject to } g_1 = 2x_1 + 3x_2 - 10 \le 0,$$
$$g_2 = -5x_1 - 2x_2 + 2 \le 0,$$
$$g_3 = -2x_1 + 7x_2 - 8 \le 0.$$

3.8 Solve:

$$\min f = (x_1 - 3)^2 + (x_2 - 3)^2$$
$$\text{subject to } x_1 + x_2 \le 0; x_1, x_2 \ge 0.$$

3.9 Using both monotonicity principles, solve the following for positive finite $x_i, i = 1, 2, 3$:

$$\max x_1$$
$$\text{subject to } \exp(x_1) \le x_2, \exp(x_2) \le x_3, x_3 \le 10.$$

3.10 *Explosive-Actuated Cylinder* (Siddall 1972) A quick-action cylinder is to be designed so that it is powered by an explosive cartridge rather than by hydraulic pressure. The general configuration is shown in the figure. The cartridge explodes in the chamber with fixed volume and the gas expands through the vent into the cylinder, pushing the piston. We are primarily concerned with the size of the cylinder, because it is part of a mechanism requiring a minimum total length for the cylinder.

The design must satisfy certain specifications arising from other system considerations and availability of materials. These specifications are:

Maximum allowable cylinder outside diameter, $D_{\max} = 1.0''$.

Maximum overall length, $L_{\max} = 2''$.

Fixed chamber volume, $V_c = 0.084 \text{ in}^3$.

Kinetic energy to be delivered, $W_{\min} = 600$ lb-in.

Maximum piston force, $F_{\max} = 700$ lb.

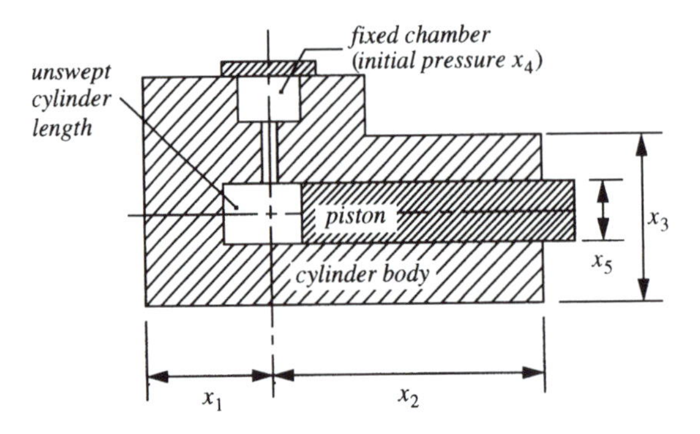

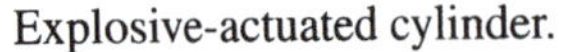
Explosive-actuated cylinder.

Yield strength in tension of material, $S_{yt} = 125$ kpsi.

Factor of safety for strength, $N = 3$.

All the above specifications give values for the problem's design parameters. The design variables are as follows:

x_1 = unswept cylinder length (inches),

x_2 = working stroke of piston (inches),

x_3 = outside diameter of cylinder (inches),

x_4 = initial pressure (psi),

x_5 = piston diameter (inches).

The objective function is the total length of the cylinder. Neglecting the thickness of the wall at the end of the stroke we have

$$\min f = x_1 + x_2.$$

The first constraint involves the kinetic energy requirement and is expressed by

$$\frac{(10^3)x_4 v_1}{1-\gamma}\left(v_2^{1-\gamma} - v_1^{1-\gamma}\right) \geq W_{\min},$$

where

$$v_1 = V_c + (\pi/4)x_1 x_5^2,$$
$$v_2 = v_1 + (\pi/4)x_2 x_5^2,$$

with v_1 and v_2 being the initial and final volume of combustion and $\gamma = 1.2$ being the ratio of specific heats.

The piston force constraint is expressed by

$$(1000\pi/4)x_4 x_5^2 \leq F_{\max}.$$

The wall stress constraint can be written as

$$\sigma_e \leq S_{yt}/N,$$

where the equivalent stress σ_e is given by the failure criterion. Using the maximum shear stress (Guest) criterion for simplicity, we have

$$\sigma_e = \sigma_1 - \sigma_2,$$

with the principal stresses σ_1 and σ_2 given by

$$\sigma_1 = \frac{x_4\left(x_3^2 + x_5^2\right)}{x_3^2 - x_5^2},$$
$$\sigma_2 = -x_4.$$

Finally, the geometric constraints are

$$x_3 \le D_{\max},$$

$$x_1 + x_2 \le L_{\max},$$

$$x_5 < x_3 \text{(strict inequality)}.$$

All variables are positive.

(a) Prove that the model is reduced to the form below with the variable monotonicities as shown:

$$\min f(x_1^+, x_2^+)$$

subject to

$$g_1(x_1^-, x_2^-, x_4^-, x_5^-) \le 1 \quad \text{(kinetic energy)},$$

$$g_2(x_4^+, x_5^+) \le 1 \quad \text{(piston force)},$$

$$g_3(x_3^-, x_4^+, x_5^+) \le 1 \quad \text{(wall stress)},$$

$$\left.\begin{array}{l} g_4(x_3^+) \le 1 \\ g_5(x_1^+, x_2^+) \le 1 \end{array}\right\} \quad \text{(geometry)}.$$

(b) Using this chapter's principles, derive the following rules: (1) The kinetic energy requirement is always critical. (2) The piston force and/or the wall stress requirement must be critical. (3) If the wall stress requirement is critical, then the outside diameter of the cylinder must be set at its maximum allowable value. (4) The maximum length constraint is uncritical.

3.11 Design a flat head air tank with double the internal capacity of the example.

3.12 (From Alice Agogino, University of California, Berkeley.) Use monotonicity analysis and consider several cases to solve

$$\min f = 100x_3^2 + x_4^2$$

$$\text{subject to } x_3 = x_2 - x_1^2,$$

$$x_4 = 1 - x_1,$$

where $x_i, i = 1, \dots, 4$ are real, although not necessarily positive.

3.13 Examine the problem

$$\min f = 2x_3 - x_4$$

subject to

$$(1)\ x_3 - 3x_4 \le 4, \qquad (5)\ x_2 + x_3 = 6,$$

$$(2)\ 3x_3 - 2x_4 \le 3, \qquad (6)\ x_3 + x_4 \le 7,$$

$$(3)\ -x_1 + x_2 - x_3 + x_4 = 2, \qquad (7)\ x_3 + 3x_4 \le 5.$$

$$(4)\ x_2 - x_3 \le 2,$$

Answer with brief justification the following. (a) Which variables are relevant, irrelevant? (b) Which constraints form conditionally critical sets? (c) Which constraints are uncritical? (d) Which constraints are dominant? (e) Rewrite the model with irrelevant variables deleted and dominated constraints relaxed, and indicate critical constraints. (f) Is this reduced problem constraint bound? (g) Does the reduced problem have multiply critical constraints? (h) Solve the original problem.

3.14 Apply regional monotonicity to the problem with $x_1, x_2 \geq 0$:

$$\min f = x_1^2 + x_2^2 - 2x_1 - 4x_2$$

$$\text{subject to } g_1 = x_1 + 4x_2 - 5 \leq 0,$$

$$g_2 = 2x_1 + 3x_2 - 6 \leq 0.$$

3.15 Apply regional monotonicity to the problem (Rao 1978):

$$\min (1/3)(x_1 + 1)^3 + x_2$$

$$\text{subject to } g_1 = -x_1 + 1 \leq 0,$$

$$g_2 = -x_2 \leq 0.$$

3.16 Study the problem

$$\min f = x$$

$$\text{subject to } g_1 = x^2 - 5x + 4 \leq 0.$$

(a) analytically with the monotonicity principles; (b) graphically as in Figure 3.7.

4

Interior Optima

The difficulties of the slopes we have overcome. Now we have to face the difficulties of the valleys.

Bertold Brecht (1898–1956)

Design problems rarely have no constraints. If the number of active constraints, equalities and inequalities, is less than the number of design variables, degrees of freedom still remain. Suppose that we are able to eliminate explicitly all active constraints, while dropping all inactive ones. The reduced problem would have only an objective function depending on the remaining variables, and no constraints. The number of design variables left undetermined would be equal to the number of degrees of freedom and the problem would be still unsolved. The following example shows how this situation may be addressed.

Example 4.1 Consider the design of a round, solid shaft subjected to a steady torque T and a completely reversed bending moment M. It has been decided that fatigue failure is the only constraint of interest. The cost function should represent some trade-off between the quality and the quantity of the material used. Oversimplifying, we may assume a generic steel material and take the ultimate strength s_u and the diameter d as design variables, with objective being the cost per unit length:

$$C(d, s_u) = C_1 d^2 + C_2 s_u. \tag{4.1}$$

The cost coefficients C_1 and C_2 are measured in dollars per unit area and dollars per unit strength, respectively. The fatigue strength constraint is $\sigma_a \leq s_n$ where σ_a is an alternating equivalent stress and s_n is the endurance limit. Using the von Mises criterion, we may set

$$\sigma_a = \left(\sigma_{xa}^2 + 3\tau_{xya}^2\right)^{1/2},$$

where $\sigma_{xa} = 32M/\pi d^3$ and $\tau_{xya} = 16T/\pi d^3$. Following usual practice (Juvinall 1983), we may set $s_n = K s_u$, where the constant K represents various correction factors. This is an empirical relation between ultimate and endurance strength for steels and uses the (not strictly true) assumption that correction factors do not depend on the diameter d.

The problem is stated after some rearrangement as

$$\begin{aligned}&\text{minimize } C = C_1 d^2 + C_2 s_u\\&\text{subject to } s_u d^3 \geq C_3,\end{aligned} \tag{4.2}$$

where $C_3 = (K\pi)^{-1}(1{,}024M^2 + 768T^2)^{1/2}$, a positive parameter. Since the objective requires lower bounds for both design variables, the constraint must be active. Eliminating s_u, we get

$$\min C = C_1 d^2 + C_2 C_3 d^{-3}. \tag{4.3}$$

This reduced problem has one degree of freedom and no constraints, except the obvious limitation $d > 0$. From elementary calculus, a solution for (4.3) can be found by setting the first derivative of C with respect to d equal to zero and solving for d. Then, to verify a minimum, the second derivative must be positive for that value of d. So here we have

$$\partial C/\partial d = 2C_1 d - 3C_2 C_3 d^{-4} = 0,$$

with the solution $d_\dagger = (3C_2C_3/2C_1)^{1/5}$. Also,

$$\partial^2 C/\partial d^2 = 2C_1 + 12C_2C_3 d^{-5} > 0 \quad \text{for } d > 0.$$

Thus, the point $d_\dagger$ is the minimum d_*. The value of s_{u*} is found from the active constraint. ■

The above one-dimensional unconstrained problem (4.3) was solved with the assumption that the function is continuous and differentiable. This type of assumption about the behavior of the function is generally necessary for deriving operationally useful results in multidimensional situations.

Formally, the *unconstrained problem* is stated as

$$\begin{aligned}&\text{minimize } f(\mathbf{x})\\&\text{subject to } \mathbf{x} \in \mathcal{X} \subseteq \Re^n,\end{aligned} \tag{4.4}$$

where $\mathcal{X}$ is a *set constraint.* This deemphasizes the usual explicit constraints by assuming that they are all accounted for in an appropriate selection of $\mathcal{X}$. If $\mathcal{X}$ is an *open* set and a solution exists, that solution will be an *interior optimum*. In the shaft design above, the set $\mathcal{X}$ was simply $\mathcal{X} = \{d \mid d > 0\}$.

Note that in the theory developed in this chapter no assumption is made about positive values of the variables, as was done in Chapter 3.

4.1 Existence

Before we start looking for methods to locate interior minima we need to have some idea about when a function will indeed have a minimum. These topics are handled formally in the mathematical analysis of functions. Here we review some basic existence concepts to motivate the need for caution when we apply the theory of this chapter.

The Weierstrass Theorem

In the previous chapter we saw that well boundedness is a necessary condition for the existence of properly defined optima. This was *necessary* for monotonic functions, where the optimum would occur at the boundary. A function can only have an interior optimum if it is nonmonotonic. In this case existence can be associated with another function property: continuity. But now *sufficient* rather than necessary conditions can be stated. This result, in the one-dimensional case, is the *Weierstrass Theorem*, well known in real analysis:

> *A continuous function defined on a closed finite interval attains its maximum and minimum in that interval.*

Its proof for the case of a maximum will be outlined below to encourage some appreciation for the delicacy of certain existence conditions.

Recall that continuity in a function means that as we approach a point in the domain of the function, we also approach the corresponding point in the range of the function. Formally, a function $f(x)$, defined in $\mathfrak{R}$, is continuous in the interval $a \leq x \leq b$, if and only if $|f(x_0) - f(x)| < \delta$ for every x such that $|x_0 - x| < \varepsilon$, where x_0 is any point in the interval and ε and δ are the usual small positive numbers. Recall also that finite intervals on $\mathfrak{R}$ possess *cluster* (or *accumulation*) points. By definition, a point p in a subset $\mathcal{S}$ of a metric space $\mathcal{E}$ is a cluster point of $\mathcal{S}$ if any open ball with center p contains an infinite number of points in $\mathcal{S}$. Here the metric space is $\mathfrak{R}$ and the subset is a finite interval in $\mathfrak{R}$. Every subinterval containing a cluster point will also contain infinitely many other points of the interval. Now, to prove the theorem about the existence of a maximum we must prove that the values of f form a bounded set, which, therefore, possesses a supremum. We will prove that by contradiction. Assume that there is a sequence $\{x_n\}$ of points in $a \leq x \leq b$ for which $f(x_n)$ increases without limit, meaning that $f(x_n)$ can become infinitely large. This sequence will have a cluster point x_c in the interval, with $f(x_c)$ being finite. Near x_c there will always be values x_n such that $|f(x_c) - f(x_n)|$ is infinitely large. But then f will have a discontinuity at x_c, a contradiction. Thus f has a supremum, say U. To complete the proof we must show that there exists an x_U such that $f(x_U) = U$. We can always create a sequence $\{x_n\}$, such that $f(x_n) \rightarrow U$ as $n \rightarrow \infty$, and crossing out some terms we can create a convergent subsequence $\{x_{n,k}\}$ with x_U as its limit. Thus $f(x_{n,k}) \rightarrow f(x_U)$, and $k \rightarrow \infty$, because of continuity, while $f(x_{n,k}) \rightarrow U$ because of the original construction of the sequence $\{x_n\}$. Therefore, $f(x_U) = U$.

Sufficiency

The Weierstrass Theorem can be generalized in $\mathfrak{R}^n$ if we replace the closed, finite interval with a *closed and bounded set*. Such a set is called *compact*. The generalized theorem is:

> *A function continuous on a compact set in $\mathfrak{R}^n$ attains its maximum and minimum in that set.*

Note carefully that this existence theorem is a *sufficient* condition. A function may have extrema even though it is neither continuous nor defined on a compact set. For example, consider the function

$$f = \begin{cases} 2x^2 + x^{-2} & x \neq 0, \\ 0 & x = 0, \end{cases}$$

with domain the open interval $(-1, \infty)$. The construction above allows the function to be finite at $x = 0$. A local minimum occurs at $x = 2^{-1/4}$, and the global one is at zero. Yet f is neither continuous, nor bounded, nor defined on a compact set.

In design problems, demanding continuity can be troublesome. Although functions such as weight or stress may be continuous, the nature of the design variables is often discrete, for example, in standard sizes. Solving optimization problems directly with discrete variables is generally difficult – with some exceptions. Usually we solve the problem with continuous variables, where the above theory would apply, and then try to locate the discrete solution in the vicinity of the continuous one. This may involve more than just rounding up or down, as discussed in Chapter 6.

Compactness can be easier to handle. The existence theorems do not imply that the optimum is an interior one. It could be on the boundary, which explains the need for a closed constraint set. The discussion in Chapter 3 showed how to detect open unbounded constraint sets. Those arguments give a simple and rigorous procedure for verifying the appropriateness of the model, that is, for creating the mathematical conditions that allow the existence and subsequent detection of the optimal design.

4.2 Local Approximation

Knowing that an optimum exists is only useful if we have an *operational* way of finding it. For a function of one variable, such as in Example 4.1, the direct solution of the optimality conditions of zero first derivative and positive second derivative is an operational way of finding a minimum. Extending the optimality conditions to functions of many variables is not difficult, and it involves a concept fundamental in the study of optimization problems: *local approximation* of functions. In this section we will present the familiar idea of the *Taylor series* approximation and develop some notation used throughout the rest of the book.

Taylor Series

If an infinite power series converges in some interval, then the sum (or limit) of the series has a value for each x in the interval. The series

$$\sum_{n=0}^{\infty} \alpha_n (x - x_0)^n, \quad -a < x - x_0 < a, \; a > 0$$

is convergent in the stated interval and can be used to define a function of x, since $f(x)$ is defined uniquely for each x:

$$f(x) = \sum_{n=0}^{\infty} \alpha_n (x - x_0)^n, \quad x_0 - a < x < x_0 + a. \tag{4.5}$$

Inverting the argument, we see that a given function can be represented by an infinite power series, if we can calculate the correct coefficients α_n. One way to do this is to assume that the function has derivatives of any order and create the *Taylor series expansion of f about the point x_0*, that is,

$$f(x) = \sum_{n=0}^{\infty} \frac{f^{(n)}(x_0)}{n!}(x - x_0)^n, \tag{4.6}$$

where $f^{(n)}(x_0)$ is the nth-order derivative at x_0. It should be emphasized that the expansion holds within an interval of convergence containing the point. Note also that not all functions can be represented by power series, even if all necessary derivatives can be computed. Loosely speaking, a function may change so rapidly that a polynomial representation cannot follow it. As an example, consider the function

$$f(x) = \begin{pmatrix} e^{-1/x^2}, & x \neq 0 \\ 0, & x = 0 \end{pmatrix} \quad x \in \Re. \tag{4.7}$$

All its derivatives of any order vanish at the origin, that is, $f^{(n)}(0) = 0$ for all positive integers n. The absolute value of the function in the immediate neighborhood of the origin is smaller than any arbitrary power term and a Taylor series cannot be constructed (see also Hancock, 1917).

The expansion (4.6) is exact but requires an infinite number of terms. Taylor's Theorem says that a finite number of terms, N, plus a *remainder* depending on N can be used instead. The remainder can be bounded from above using the Schwartz inequality $|xy| \leq |x| \cdot |y|$ and the following expression results:

$$f(x) = f(x_0) + \sum_{n=1}^{N} \frac{f^{(n)}(x_0)}{n!}(x - x_0)^n + o(|x - x_0|^N). \tag{4.8}$$

The *order symbol o* (lowercase omicron) means that all terms with $n > N$ will tend to zero *faster* than $|x - x_0|^N$, as x approaches x_0, that is, terms with $n > N$ are small compared to $|x - x_0|^N$.

From (4.8) we can obtain local approximations to a function: A *first-order* or *linear* approximation is

$$f(x) = f(x_0) + \frac{df(x_0)}{dx}(x - x_0) \tag{4.9}$$

and a *second-order* or *quadratic* approximation is

$$f(x) = f(x_0) + \frac{df(x_0)}{dx}(x - x_0) + \frac{1}{2}\frac{d^2 f(x_0)}{dx^2}(x - x_0)^2. \tag{4.10}$$

These approximations are good whenever the higher order terms can be legitimately neglected. Although the accuracy improves by adding more terms, the effort required for calculating higher order derivatives makes the linear and quadratic approximations the only practical ones.

The Taylor series approximation can be extended to functions of several variables. The absolute value measure of length, $|x|$, is replaced by the *Euclidean Norm* $\|\bullet\|$, that is, the length of a vector $\mathbf{x}$ is given by

$$\|\mathbf{x}\| \triangleq \left(\sum_{i=1}^{n} x_i^2\right)^{1/2}. \tag{4.11}$$

The Taylor series linear and quadratic approximations for $f(\mathbf{x})$ about $\mathbf{x}_0$ are now

$$f(\mathbf{x}) = f(\mathbf{x}_0) + \sum_{i=1}^{n} \frac{\partial f(\mathbf{x}_0)}{\partial x_i}(x_i - x_{i0}) + o(\|\mathbf{x} - \mathbf{x}_0\|) \tag{4.12}$$

and

$$\begin{aligned} f(\mathbf{x}) = f(\mathbf{x}_0) &+ \sum_{i=1}^{n} \frac{\partial f(\mathbf{x}_0)}{\partial x_i}(x_i - x_{i0}) \\ &+ \frac{1}{2}\sum_{i=1}^{n}\sum_{j=1}^{n} \frac{\partial^2 f(\mathbf{x}_0)}{\partial x_i \partial x_j}(x_i - x_{i0})(x_j - x_{j0}) + o(\|\mathbf{x} - \mathbf{x}_0\|^2), \end{aligned} \tag{4.13}$$

where $\mathbf{x} = (x_1, x_2, \ldots, x_n)^T$ and $\mathbf{x}_0 = (x_{10}, x_{20}, \ldots, x_{n0})^T$. If we employ vector notation, we can obtain more compact expressions that are easier to manipulate algebraically in subsequent derivations of formulas. We define the *gradient* vector ∇f to be the *row* vector of the first partial derivatives of f:

$$\nabla f \triangleq (\partial f/\partial x_1, \partial f/\partial x_2, \ldots, \partial f/\partial x_n). \tag{4.14}$$

Some alternative symbols for ∇f are $\nabla f_{\mathbf{x}}$, $\partial f/\partial \mathbf{x}$, and $f_{\mathbf{x}}$.

We define the *Hessian* matrix $\mathbf{H}$ of f to be the *square, symmetric* matrix of the second derivatives of f:

$$\mathbf{H} \triangleq \begin{pmatrix} \partial^2 f/\partial x_1^2 & \ldots & \partial^2 f/\partial x_1 \partial x_n \\ \vdots & & \vdots \\ \partial^2 f/\partial x_n \partial x_1 & \ldots & \partial^2 f/\partial x_n^2 \end{pmatrix}. \tag{4.15}$$

Alternative symbols for the Hessian are $\nabla^2 F$, $\partial^2 f/\partial \mathbf{x}^2$, and $f_{\mathbf{xx}}$. Next we define the *perturbation vector* $\partial \mathbf{x} \triangleq \mathbf{x} - \mathbf{x}_0$, with components $x_i - x_{i0}$, and the resulting *function perturbation*

$$\partial f \triangleq f(\mathbf{x}) - f(\mathbf{x}_0).$$

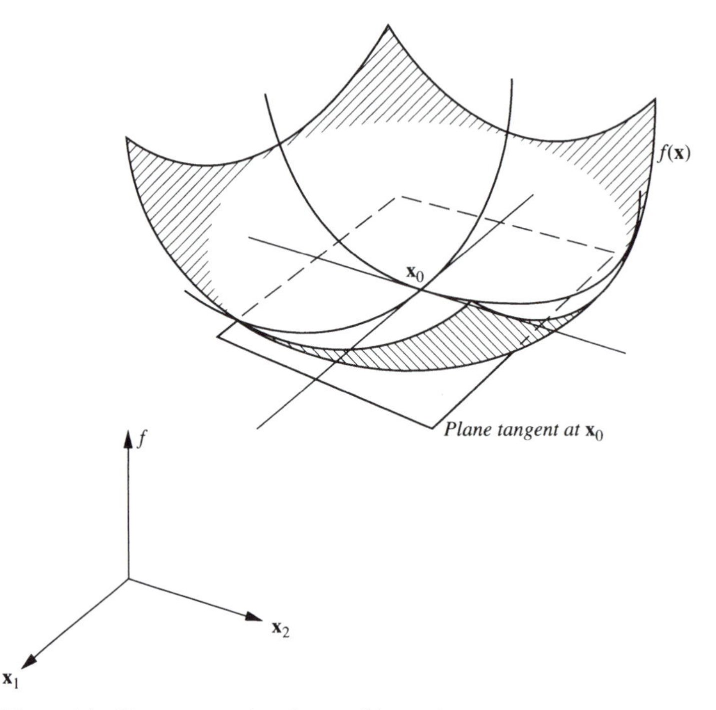

Figure 4.1. Planar approximation to $f(x_1, x_2)$ at $\mathbf{x}_0$.

Now we can write Equations (4.12) and (4.13) in the compact vector form

$$\partial f = \nabla f(\mathbf{x}_0)\partial\mathbf{x} + o(\|\partial\mathbf{x}\|), \tag{4.16}$$

$$\partial f = \nabla f(\mathbf{x}_0)\partial\mathbf{x} + \tfrac{1}{2}\partial\mathbf{x}^T\mathbf{H}(\mathbf{x}_0)\partial\mathbf{x} + o(\|\partial\mathbf{x}\|^2). \tag{4.17}$$

Geometrically, the linear approximation in the two-dimensional case is a planar approximation of the two-dimensional surface, that is, a plane tangent to the surface $f(x_1, x_2)$ at $(x_{10}, x_{20})^T$, as in Figure 4.1.

The above two equations give the linear and quadratic approximations, respectively, for function perturbations that result from perturbations in the variables, assuming the higher order terms are negligible. These approximations are local and provide a tool for analyzing the behavior of a function near a point. The *differential theory* of multivariable optimization uses these approximations to derive local properties of optimality.

Quadratic Functions

This special class of functions can provide useful insights for general functions locally approximated by quadratics.

Example 4.2 Consider the function

$$f = x_1^2 - 3x_1x_2 + 4x_2^2 + x_1 - x_2,$$

which has $\partial f/\partial x_1 = 2x_1 - 3x_2 + 1$, $\partial f/\partial x_2 = -3x_1 + 8x_2 - 1$, $\partial^2 f/\partial x_1^2 = 2$, $\partial^2 f/\partial x_2^2 = 8$, and $\partial^2 f/\partial x_1 \partial x_2 = -3$. Therefore, the gradient and Hessian at any point are

$$\nabla f = (2x_1 - 3x_2 + 1, -3x_1 + 8x_2 - 1),$$
$$\mathbf{H} = \begin{pmatrix} 2 & -3 \\ -3 & 8 \end{pmatrix}.$$

Note that since the function is a quadratic polynomial in x_1 and x_2, the Hessian is independent of the variables. In the Taylor expansion the second-order approximation will be exact, with the higher order terms always zero. Consider the points $\mathbf{x}_0 = (0, 0)^T$ and $\mathbf{x} = (1, 1)^T$, where $f(\mathbf{x}_0) = 0$ and $f(\mathbf{x}) = 2$ so that the exact perturbation is $\partial f = f(\mathbf{x}) - f(\mathbf{x}_0) = 2$. The second-order Taylor expansion is

$$\begin{aligned}\partial f &= (2x_{10} - 3x_{20} + 1, -3x_{10} + 8x_{20} - 1)(\partial x_1, \partial x_2)^T \\ &\quad + \frac{1}{2}\begin{pmatrix} \partial x_1 \\ \partial x_2 \end{pmatrix}^T \begin{pmatrix} 2 & -3 \\ -3 & 8 \end{pmatrix} \begin{pmatrix} \partial x_1 \\ \partial x_2 \end{pmatrix} \\ &= (2x_{10} - 3x_{20} + 1)\partial x_1 + (-3x_{10} + 8x_{20} - 1)\partial x_2 \\ &\quad + \frac{1}{2}(2\partial x_1^2 - 6\partial x_1 \partial x_2 + 8\partial_2^2).\end{aligned}$$

Taking $x_{10} = x_{20} = 0$ and $\partial x_1 = \partial x_2 = 1$, this expression gives $\partial f = 2$. ■

Example 4.3 Consider the function

$$f = (3 - x_1)^2 + (4 - x_2)^2,$$

which has $\partial f/\partial x_1 = -6 + 2x_1$, $\partial f/\partial x_2 = -8 + 2x_2$, $\partial^2 f/\partial x_1^2 = 2$, $\partial^2 f/\partial x_2^2 = 2$, and $\partial^2 f/\partial x_1 \partial x_2 = 0$. Therefore at any point

$$\nabla f = (-6 + 2x_1, -8 + 2x_2),$$
$$\mathbf{H} = \begin{pmatrix} 2 & 0 \\ 0 & 2 \end{pmatrix}.$$

Here the Hessian is again independent of the variables, but it is also *diagonal*. This occurs because the function is *separable*, that is, it is composed of separate functions that each depend on only one of the variables. We could write

$$f = f_1(x_1) + f_2(x_2),$$

where $f_1(x_1) = (3 - x_1)^2$ and $f_2(x_2) = (4 - x_2)^2$. Separable functions are easier to optimize because we can look at only one variable at a time. ■

Both the preceding examples dealt with *quadratic* functions. A general form for quadratic functions is

$$f(\mathbf{x}) = c + \mathbf{b}^T\mathbf{x} + \tfrac{1}{2}\mathbf{x}^T\mathbf{A}\mathbf{x}, \tag{4.18}$$

where c is a scalar, $\mathbf{b}$ is an n-vector, and $\mathbf{A}$ is a symmetric $n \times n$ matrix whose elements a_{ij} are the coefficients of the quadratic terms $x_i x_j$. The gradient and the Hessian are given by

$$\nabla f = \mathbf{b}^T + \mathbf{x}^T \mathbf{A}, \quad \mathbf{H} = \mathbf{A}, \tag{4.19}$$

and they are related in a simple but useful way, namely,

$$\nabla f(\mathbf{x}_2) - \nabla f(\mathbf{x}_1) = (\mathbf{x}_2 - \mathbf{x}_1)^T \mathbf{H}, \tag{4.20}$$

where $\mathbf{x}_1$ and $\mathbf{x}_2$ are two distinct points in the domain of f. The product $\mathbf{x}^T\mathbf{A}\mathbf{x}$ is called a *quadratic form*, a special case of the bilinear form $\mathbf{x}_1^T \mathbf{A} \mathbf{x}_2$.

The assumption of symmetry for $\mathbf{A}$ in (4.18) does not imply lack of generality. A nonsymmetric square matrix can easily be transformed into an equivalent symmetric one. In fact,

$$\mathbf{x}^T\mathbf{A}\mathbf{x} = \tfrac{1}{2}\mathbf{x}^T\mathbf{A}\mathbf{x} + \tfrac{1}{2}\mathbf{x}^T\mathbf{A}\mathbf{x} = \tfrac{1}{2}\mathbf{x}^T\mathbf{A}\mathbf{x} + \tfrac{1}{2}\mathbf{x}^T\mathbf{A}^T\mathbf{x} = \mathbf{x}^T\left[\tfrac{1}{2}(\mathbf{A} + \mathbf{A}^T)\right]\mathbf{x}.$$

The matrix $(\mathbf{A} + \mathbf{A}^T)$ is symmetric, as shown by using the definition

$$(\mathbf{A} + \mathbf{A}^T)^T = \mathbf{A} + \mathbf{A}^T.$$

Example 4.4 Consider a quadratic form with matrix

$$\mathbf{A} = \begin{pmatrix} 3 & 4 \\ 1 & 2 \end{pmatrix}.$$

The equivalent symmetric matrix is found from

$$\frac{1}{2}(\mathbf{A} + \mathbf{A}^T) = \frac{1}{2}\left[\begin{pmatrix} 3 & 4 \\ 1 & 2 \end{pmatrix} + \begin{pmatrix} 3 & 1 \\ 4 & 2 \end{pmatrix}\right] = \frac{1}{2}\begin{pmatrix} 6 & 5 \\ 5 & 4 \end{pmatrix}.$$

The quadratic function corresponding to $\mathbf{A}$ is

$$\begin{aligned} f &= \frac{1}{2}\mathbf{x}^T\mathbf{A}\mathbf{x} = \frac{1}{2}(x_1, x_2)\begin{pmatrix} 3 & 4 \\ 1 & 2 \end{pmatrix}\begin{pmatrix} x_1 \\ x_2 \end{pmatrix} = \frac{1}{2}\left(3x_1^2 + x_1x_2 + 4x_1x_2 + 2x_2^2\right) \\ &= \frac{1}{2}\left(3x_1^2 + 5x_1x_2 + 2x_2^2\right). \end{aligned}$$

The Hessian is easily found to be the same as $(\frac{1}{2})(\mathbf{A} + \mathbf{A}^T)$. ■

Vector Functions

This idea of local approximation can be extended to *vector* functions $\mathbf{f}(\mathbf{x}) = [f_1(\mathbf{x}), \ldots, f_m(\mathbf{x})]^T$. Here we will mention only that for a vector function, the gradient vector is simply replaced by the *Jacobian* matrix of all first partial derivatives

$$\mathbf{J} \triangleq \begin{bmatrix} \partial f_1/\partial x_1, \ldots, \partial f_1/\partial x_n \\ \vdots \qquad\qquad \vdots \\ \partial f_m/\partial x_1, \ldots, \partial f_m/\partial x_n \end{bmatrix}. \tag{4.21}$$

An alternative symbol used is $\partial \mathbf{f}/\partial \mathbf{x}$. The linear approximation of $\mathbf{f}$ is given by

$$\mathbf{f}(\mathbf{x}) = \mathbf{f}(\mathbf{x}_0) + \frac{\partial \mathbf{f}(\mathbf{x}_0)}{\partial \mathbf{x}}\partial \mathbf{x}. \tag{4.22}$$

Example 4.5 The Jacobian may be viewed as a column vector whose elements are the gradients of the components of $\mathbf{f}$:

$$(\partial \mathbf{f})/(\partial \mathbf{x}) = (\nabla f_1, \nabla f_2, \ldots, \nabla f_m)^T.$$

If $\mathbf{f}$ has components $f_1 = \mathbf{a}^T\mathbf{x} + b$, $f_2 = \frac{1}{2}\mathbf{x}^T\mathbf{B}\mathbf{x} + \mathbf{d}^T\mathbf{x}$ the Jacobian is given by

$$\frac{\partial \mathbf{f}}{\partial \mathbf{x}} = \begin{pmatrix} \mathbf{a}^T \\ \mathbf{x}^T\mathbf{B} + \mathbf{d}^T \end{pmatrix}.$$

Vector functions will be used later for representation of constraint sets. ■

4.3 Optimality

First-Order Necessity

Suppose that $f(\mathbf{x})$ has a minimum f_* at $\mathbf{x}_*$. Locally at $\mathbf{x}_*$, any perturbations in $\mathbf{x}$ must result in higher values of f by definition of $\mathbf{x}_*$, that is, to first order

$$\partial f_* = \nabla f(\mathbf{x}_*)\partial \mathbf{x}_* \geq 0. \tag{4.23}$$

This implies that for all $\partial \mathbf{x}_* \neq \mathbf{0}$, the gradient $\nabla f(\mathbf{x}_*) = \mathbf{0}^T$, that is, *all* partial derivatives $\partial f/\partial x_i$ at $\mathbf{x}_*$ must be zero. We can see why this is true by contradiction; in component form Equation (4.23) is written as

$$\partial f = \sum_{i=1}^{n} \frac{\partial f(\mathbf{x}_*)}{\partial x_i}\partial x_i \geq 0, \tag{4.24}$$

where the subscript $_*$ is dropped for convenience. Assume that there is a j such that $(\partial f/\partial x_j) \neq 0$. Then choose a component perturbation ∂x_j with sign opposite to that of the derivative so that the jth component's contribution to ∂f will be $(\partial f/\partial x_j)\partial x_j < 0$. Next, hold all other component perturbations $\partial x_i = 0$, $i \neq j$, so that the total change will be

$$\partial f = (\partial f/\partial x_j)\partial x_j < 0, \tag{4.25}$$

which contradicts (4.24) and the hypothesis of a nonzero component of the gradient at the minimum.

Note that in the derivation above there is an implicit assumption that the points $\mathbf{x}_* + \partial\mathbf{x}$ belong to the feasible set $\mathcal{X}$ for the problem as posed in (4.4). This is guaranteed only if $\mathbf{x}_*$ is in the interior of $\mathcal{X}$. We will see in Chapter 5 how the result changes for $\mathbf{x}_*$ on the boundary of the feasible set. Here we have derived a *first-order necessary condition for an unconstrained (or interior) local minimum:*

> *If $f(\mathbf{x})$, $x \in \mathcal{X} \subseteq \Re^n$, has a local minimum at an interior point $\mathbf{x}_*$ of the set $\mathcal{X}$ and if $f(\mathbf{x})$ is continuously differentiable at $\mathbf{x}_*$, then $\nabla f(\mathbf{x}_*) = \mathbf{0}^T$.*

This result is a *necessary* condition because it may also be true at points that are not local minima. The gradient will be zero also at a local maximum and at a saddlepoint (see Figure 4.2). The saddlepoint may be particularly tricky to rule out because it

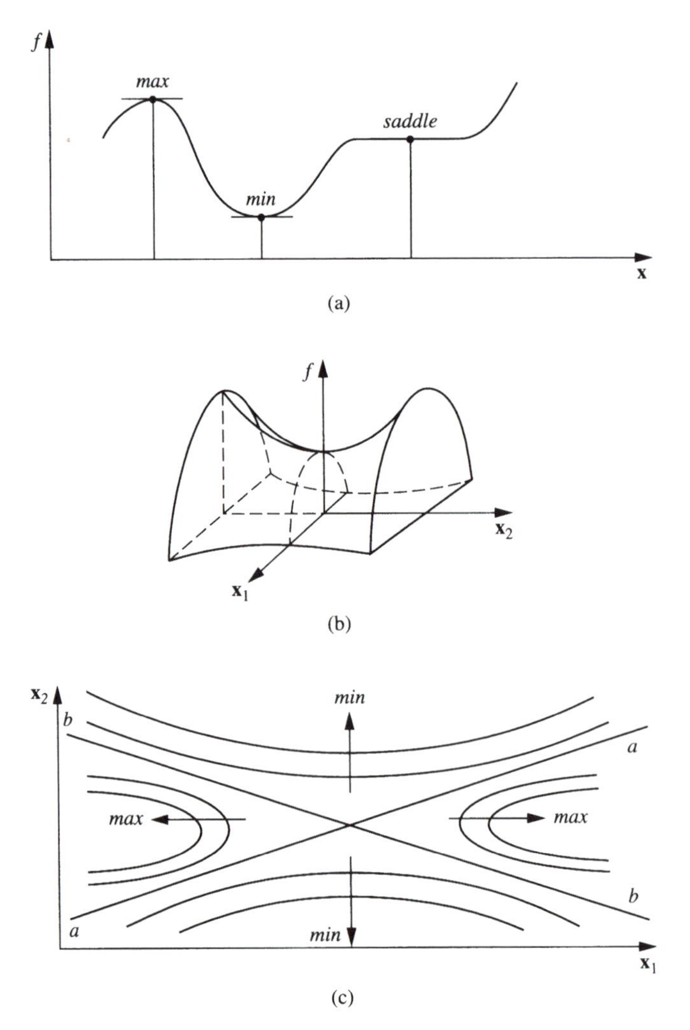

Figure 4.2. Zero gradients at saddlepoints.

appears to be a minimum if one approaches it from only certain directions. Yet both ascending and descending directions lead away from it. All points at which the gradient is zero are collectively called *stationary points*, and the above necessary condition is often called the *stationarity condition*.

The value for f_* with $\nabla f(\mathbf{x}_*) = \mathbf{0}^T$ may not be uniquely defined by a single $\mathbf{x}_*$, but by more, possibly infinitely many $\mathbf{x}_*$s giving the same minimum value for f. In other words, $f(\mathbf{x}_*) = f_*$ for a single value of f_* but several distinct $\mathbf{x}_*$. A *valley* is a line or plane whose points correspond to the same minimum of the function. A *ridge* would correspond to a maximum. We will see some examples later in this section.

Second-Order Sufficiency

The first-order local approximation (4.23) gives a useful but inconclusive result. Using second-order information, we should be able to make more firm conclusions. If $\mathbf{x}_\dagger$ is a stationary point of f, then Equation (4.17) gives

$$\partial f_\dagger = \tfrac{1}{2}\partial\mathbf{x}^T\mathbf{H}(\mathbf{x}_\dagger)\partial\mathbf{x}, \tag{4.26}$$

where the higher order terms have been neglected and $\nabla f(\mathbf{x}_\dagger) = \mathbf{0}^T$ has been accounted for. The sign of $\partial f_\dagger$ depends on the sign of the differential quadratic form $\partial\mathbf{x}^T\mathbf{H}_\dagger\partial\mathbf{x}$, which is a scalar. If this quadratic form is strictly positive for all $\partial\mathbf{x} \neq \mathbf{0}$, then $\mathbf{x}_\dagger$ is definitely a local minimum. This is true because then the higher order terms can be legitimately neglected, and a *sufficient* condition has been identified. A real, symmetric matrix whose quadratic form is strictly positive is called *positive-definite*. With this terminology, the *second-order sufficiency condition for an unconstrained local minimum* is as follows:

> *If the Hessian matrix of $f(\mathbf{x})$ is positive-definite at a stationary point $\mathbf{x}_\dagger$, then $\mathbf{x}_\dagger$ is a local minimum.*

Example 4.6 Consider the function

$$f(x_1, x_2) = 2x_1 + x_1^{-2} + 2x_2 + x_2^{-2},$$

which has

$$\nabla f = (2 - 2x_1^{-3}, 2 - 2x_2^{-3}),$$

$$\mathbf{H} = \begin{pmatrix} 6x_1^{-4} & 0 \\ 0 & 6x_2^{-4} \end{pmatrix},$$

and a stationary point $\mathbf{x}_\dagger = (1, 1)^T$. The differential quadratic form is at every point

$$\partial\mathbf{x}^T\mathbf{H}\,\partial\mathbf{x} = 6x_1^{-4}\partial x_1^2 + 6x_2^{-4}\partial x_2^2 > 0,$$

except at $(0, 0)^T$. The Hessian is positive-definite at $(1, 1)^T$, which then is a local minimum.

Consider also the function from Example 4.3,

$$f = (3 - x_1)^2 + (4 - x_2)^2,$$

which has a stationary point $\mathbf{x}_\dagger = (3, 4)^T$ and a Hessian positive-definite everywhere, since

$$\partial \mathbf{x}^T \mathbf{H}\, \partial \mathbf{x} = 2\partial x_1^2 + 2\partial x_2^2 > 0$$

for all nonzero perturbations. Thus $(3, 4)^T$ is a local minimum.

Both these functions, being separable, have diagonal Hessians and their differential quadratic forms involve only *sums of squares* of perturbation components and no cross-product terms $\partial x_1 \partial x_2$. Therefore, the possible sign of the quadratic form can be found by just looking at the signs of the diagonal elements of the Hessian. If they are all strictly positive, as in these functions, the Hessian will be positive-definite. ■

The example motivates the idea that any practical way of applying the second-order sufficiency condition will involve some form of diagonalization of the Hessian. This is the basis of all practical tests for positive-definiteness. Here are three familiar tests from linear algebra.

POSITIVE-DEFINITE MATRIX TESTS

A square, symmetric matrix is positive-definite if and only if any of the following is true:

1. All its eigenvalues are positive.
2. All determinants of its leading principal minors are positive; that is, if $\mathbf{A}$ has elements a_{ij}, then all the determinants

$$|a_{11}|, \quad \begin{vmatrix} a_{11} & a_{12} \\ a_{21} & a_{22} \end{vmatrix}, \quad \begin{vmatrix} a_{11} & a_{12} & a_{13} \\ a_{21} & a_{22} & a_{23} \\ a_{31} & a_{32} & a_{33} \end{vmatrix}, \ldots, \det(\mathbf{A})$$

must be positive.

3. All the pivots are positive when $\mathbf{A}$ is reduced to row-echelon form, working systematically along the main diagonal.

The third test is equivalent to "completing the square" and getting positive signs for each square term. This test essentially utilizes a Gauss elimination process (see, e.g., Wilde and Beightler, 1967, or Reklaitis, Ravindran, and Ragsdell, 1983). Here we illustrate it in some examples.

Example 4.7 Consider the quadratic function

$$f = -4x_1 + 2x_2 + 4x_1^2 - 4x_1x_2 + x_2^2,$$

which has

$$\nabla f = (-4 + 8x_1 - 4x_2, 2 - 4x_1 + 2x_2),$$

$$\mathbf{H} = \begin{pmatrix} 8 & -4 \\ -4 & 2 \end{pmatrix}.$$

The two components of the gradient are linearly dependent so that setting them equal to zero gives an infinity of stationary points on the line

$$2x_{1\dagger} - x_{2\dagger} = 1.$$

Moreover, the Hessian is *singular* at every point since the second row is a multiple of the first. Looking at the quadratic form, we have

$$\begin{aligned} \partial\mathbf{x}^T\mathbf{H}\,\partial\mathbf{x} &= (\partial x_1, \partial x_2)\begin{pmatrix} 8 & -4 \\ -4 & 2 \end{pmatrix}\begin{pmatrix} \partial x_1 \\ \partial x_2 \end{pmatrix} \\ &= (8\partial x_1 - 4\partial x_2)\partial x_1 + (-4\partial x_1 + 2\partial x_2)\partial x_2 \\ &= 8\partial x_1^2 - 8\partial x_1\partial x_2 + 2\partial x_2^2 = 2(2\partial x_1 - \partial x_2)^2. \end{aligned}$$

Therefore, at any stationary point we have

$$\partial f_\dagger = \tfrac{1}{2}\partial\mathbf{x}^T\mathbf{H}\,\partial\mathbf{x} = (2\partial x_1 - \partial x_2)^2 \geq 0.$$

The perturbation is zero only if $2\partial x_1 - \partial x_2 = 0$. A stationary point could be $\mathbf{x}_\dagger = (1, 1)^T$. Then $\partial x_1 = x_1 - 1$, $\partial x_2 = x_2 - 1$, and the zero second-order perturbations will occur along the line $2(x_1 - 1) - (x_2 - 1) = 0$ or $2x_1 - x_2 = 1$, which is exactly the line of stationary points.

Since the second-order approximation is exact for a quadratic function, we conclude that the minimum $f_* = -1$ occurs along the *straight valley* $2x_{1*} - x_{2*} = 1$, as in Figure 4.3. ■

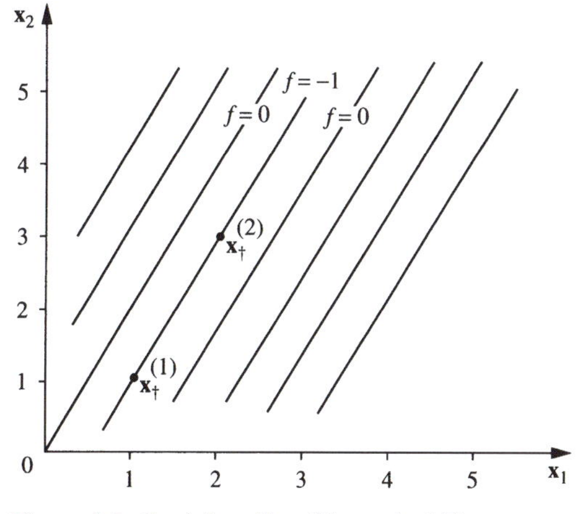

Figure 4.3. Straight valley (Example 4.7).

Nature of Stationary Points

The singularity of the Hessian gave a quadratic form that was not strictly positive but only nonnegative. This could make a big difference in assessing optimality. When the second-order terms are zero at a stationary point, the higher order terms will in general be needed for a conclusive study. The condition $\partial\mathbf{x}^T\mathbf{H}\,\partial\mathbf{x} \geq 0$ is no longer sufficient but only *necessary*. The associated matrix is called *positive-semidefinite*. Identifying semidefinite Hessians at stationary points of functions higher than quadratic should be a signal for extreme caution in reaching optimality conclusions. Some illustrative cases are examined in the exercises.

The terminology and *sufficiency* conditions for determining the nature of a stationary point are summarized below.

Quadratic Form	Hessian Matrix	Nature of x
positive	positive-definite	local minimum
negative	negative-definite	local maximum
nonnegative	positive-semidefinite	probable valley
nonpositive	negative-semidefinite	probable ridge
any sign	indefinite	saddlepoint

A symmetric matrix with negative eigenvalues will be negative-definite; if some eigenvalues are zero and the others have the same sign, it will be semidefinite; if the eigenvalue signs are mixed, it will be indefinite.

Example 4.8 Consider the function

$$f(x_1, x_2) = x_1^4 - 2x_1^2x_2 + x_2^2.$$

The gradient and Hessian are given by

$$\nabla f = \left(4x_1^3 - 4x_1x_2, -2x_1^2 + 2x_2\right) = 2\left(x_1^2 - x_2\right)(2x_1, -1),$$

$$\mathbf{H} = \begin{pmatrix} 12x_1^2 - 4x_2 & -4x_1 \\ -4x_1 & 2 \end{pmatrix}.$$

Since $(2x_1, -1) \neq \mathbf{0}^T$, all stationary points must lie on the parabola,

$$x_{1\dagger}^2 - x_{2\dagger} = 0.$$

At any such stationary point the Hessian is singular:

$$\mathbf{H}_\dagger = 2\begin{pmatrix} 4x_1^2 & -2x_1 \\ -2x_1 & 1 \end{pmatrix}.$$

The second-order terms, after completing the square, give

$$\partial f_\dagger = \tfrac{1}{2}\partial\mathbf{x}^T\mathbf{H}_\dagger\partial\mathbf{x} = (2x_{1\dagger}\partial x_1 - \partial x_2)^2 \geq 0.$$

The quadratic form is nonnegative at every stationary point but this does not yet prove a valley, since the higher order terms in the Taylor expansion may be important along directions where both gradient and quadratic forms vanish. Note, however, that such directions must yield a new point $\mathbf{x}_{\dagger}^{(2)}$ such that

$$\left[x_{1\dagger}^{(2)}\right]^2 - x_{2\dagger}^{(2)} = 0,$$
$$2x_{1\dagger}^{(1)}\left(x_{1\dagger}^{(2)} - x_{1\dagger}^{(1)}\right) - \left(x_{2\dagger}^{(2)} - x_{2\dagger}^{(1)}\right) = 0,$$

where $\mathbf{x}_{\dagger}^{(1)}$ is the current point. Solving these equations, we see that the only solution possible is $\mathbf{x}_{\dagger}^{(2)} = \mathbf{x}_{\dagger}^{(1)}$, that is, there are no directions where both gradient and Hessian vanish simultaneously. The valley $x_1^2 - x_2 = 0$ is a flat curved one, and no further investigation is necessary.

If we apply a transformation

$$\hat{x} \triangleq x_1^2 - x_2$$

the function becomes after rearrangement

$$f(x_1, x_2) \equiv f(\hat{x})\left[= \left(x_1^2 - x_2\right)^2\right] = \hat{x}^2,$$

with a unique minimum at $\hat{x} = 0$; that is, $x_1^2 = x_2$ represents a family of minima for the original space. ■

Such transformations could occasionally substantially reduce the effort required for identifying the true nature of stationary points with semidefinite Hessians.

4.4 Convexity

The optimality conditions in the previous section were derived algebraically. There is a nice geometric meaning that can give further insight. Let us examine first a function of one variable. A zero first derivative means that the tangent to the graph of the function at x_* must have zero slope, for example, line 1 in Figure 4.4. A positive second derivative means that the graph must curl up away from the point. The tangent lines at points near x_* must have increasing slopes, that is, the *curvature* must be positive.

Convex Sets and Functions

Positive curvature can be expressed geometrically in two other equivalent ways. A line tangent at any point $f(x_2)$ will never cross the graph of the function, for example, line 2 in Figure 4.4. Also, a line connecting any two points $f(x_1)$, $f(x_2)$ on the graph will be entirely above the graph of the function between the points x_1, x_2, for example, line 3 in Figure 4.4. A function that exhibits this behavior is called *convex*. The geometric property of a function in the neighborhood of a stationary point that will ensure that the point is a minimum is called *local convexity*. The concept of

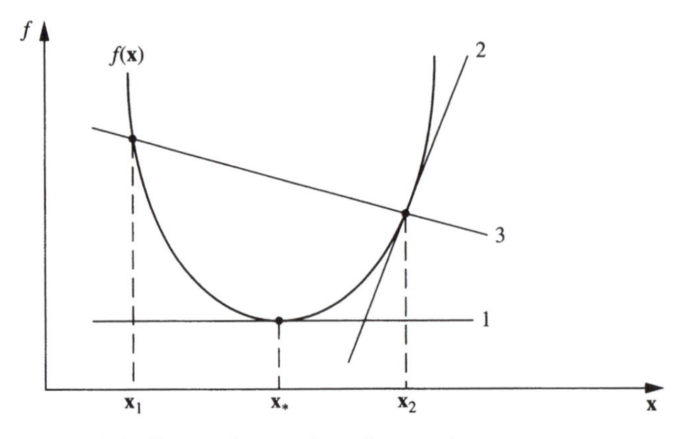

Figure 4.4. Geometric meaning of convexity.

convexity can be defined more rigorously and be generalized to sets and to functions of many variables. We will discuss these next. The most important result we will reach is that just as local convexity guarantees a local minimum, so *global* convexity will guarantee a *global* minimum.

A set $\mathcal{S} \subseteq \Re^n$ is convex if, for every point $\mathbf{x}_1$, $\mathbf{x}_2$ in $\mathcal{S}$, the point

$$\mathbf{x}(\lambda) = \lambda\mathbf{x}_2 + (1 - \lambda)\mathbf{x}_1, \quad 0 \leq \lambda \leq 1 \tag{4.27}$$

belongs also to the set. The geometric meaning of convexity of a set is that a line between two points of the set contains only points that belong to the set (Figure 4.5).

Convexity is a desirable property for the set constraint of an unconstrained optimization problem, or more generally for the feasible domain of a constrained problem. Roughly speaking, this is true because convex sets exclude difficult nonlinearities that could confound most elegant optimization theory and algorithms. This will become more evident when examining the local methods of Chapter 7.

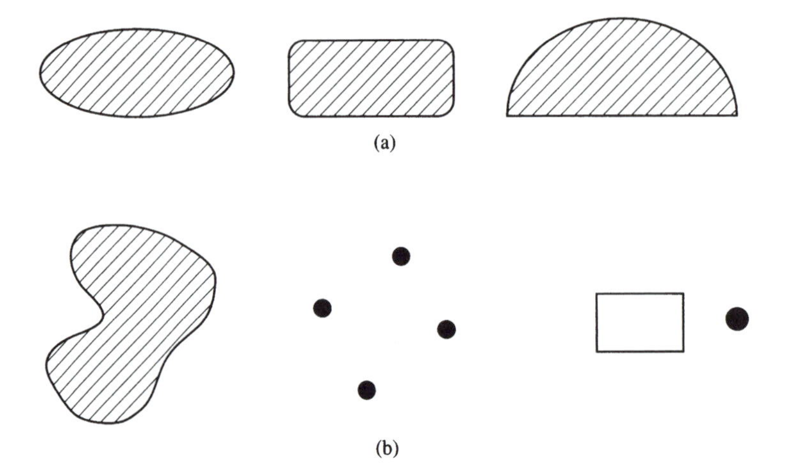

Figure 4.5. (a) Convex sets; (b) nonconvex sets.

An example of a useful convex set is the *hyperplane*, which can be defined as the set

$$\mathcal{H} = \{\mathbf{x} \mid \mathbf{x} \in \Re^n, \mathbf{a}^T\mathbf{x} = c\}, \tag{4.28}$$

where $\mathbf{a}$ is a nonzero vector of real numbers and c is a scalar. The hyperplane is the n-dimensional generalization of the usual plane in the three-dimensional space. The vectors $\mathbf{x}$ represent the set of solutions for a single linear equation. So in $\Re$ the hyperplane is just the point $x = c/a$; in $\Re^2$ it is the line $a_1x_1 + a_2x_2 = c$; in $\Re^3$ it is the plane $a_1x_1 + a_2x_2 + a_3x_3 = c$.

A function $f: \mathcal{X} \to \Re$, $\mathcal{X} \subseteq \Re^n$ defined on a nonempty convex set $\mathcal{X}$ is called *convex* on $\mathcal{X}$ if and only if, for every $\mathbf{x}_1, \mathbf{x}_2 \in \mathcal{X}$:

$$f(\lambda\mathbf{x}_2 + (1-\lambda)\mathbf{x}_1) \leq \lambda f(\mathbf{x}_2) + (1-\lambda)f(\mathbf{x}_1), \tag{4.29}$$

where $0 \leq \lambda \leq 1$. This gives an algebraic definition of the geometric meaning of convexity described earlier, that is, that the graph of a convex function between any two points $\mathbf{x}_1, \mathbf{x}_2$ will lie on or below the line connecting the two points $f(\mathbf{x}_1)$, $f(\mathbf{x}_2)$. If (4.29) holds as a strict inequality for all $\mathbf{x}_1 \neq \mathbf{x}_2$ $(0 < \lambda < 1)$, then $f(\mathbf{x})$ is *strictly convex*. A function $f(\mathbf{x})$ is (strictly) concave if $-f(\mathbf{x})$ is (strictly) convex. Concave sets are not defined.

CONVEXITY PROPERTIES

Here are some useful properties that follow immediately from the above definitions of convexity.

1. If $\mathcal{S}$ is convex and $\alpha \in \Re$, then the set $\alpha\mathcal{S}$ is convex.
2. If $\mathcal{S}_1, \mathcal{S}_2$ are convex sets, then the set $\mathcal{S}_1 + \mathcal{S}_2 = \{\mathbf{y} \mid \mathbf{y} = \mathbf{x}_1 + \mathbf{x}_2, \mathbf{x}_1 \in \mathcal{S}_1, \mathbf{x}_2 \in \mathcal{S}_2\}$ is convex.
3. The intersection of convex sets is convex.
4. The union of convex sets is usually nonconvex.
5. If f is a convex function and $\alpha \geq 0$, then αf is a convex function.
6. If f_1, f_2 are convex on a set $\mathcal{S}$, then $f_1 + f_2$ is also convex on $\mathcal{S}$.

These properties will be useful in the study of constrained problems.

Let us now consider a convex function $f(\mathbf{x})$ on a convex set $\mathcal{S}$ and define the set $\mathcal{S}_c$:

$$\mathcal{S}_c \triangleq \{\mathbf{x} \mid \mathbf{x} \in \mathcal{X}, f(\mathbf{x}) \leq c\},$$

where c is a scalar. Suppose that $\mathbf{x}_1, \mathbf{x}_2 \in \mathcal{S}_c$ so that $f(\mathbf{x}_1) \leq c$ and $f(\mathbf{x}_2) \leq c$. Taking a point $\lambda\mathbf{x}_2 + (1-\lambda)\mathbf{x}_1$, $0 \leq \lambda \leq 1$, we have

$$f(\lambda\mathbf{x}_2 + (1-\lambda)\mathbf{x}_1) \leq \lambda f(\mathbf{x}_2) + (1-\lambda)f(\mathbf{x}_1) \leq \lambda c + (1-\lambda)c = c.$$

Therefore, $\lambda \mathbf{x}_2 + (1 - \lambda)\mathbf{x}_1$ belongs to $\mathcal{S}_c$, which is then convex. This simple result shows an intimate relation between convex functions and convex sets defined by inequalities:

If $f(\mathbf{x})$ is convex, then the set $\mathcal{S}_c \subseteq \Re^n$ defined by $\{\mathbf{x} \mid f(\mathbf{x}_1) \leq c\}$ is convex for every real number c.

This property provides a way to determine convexity of the feasible domain when it is described by explicit inequalities, $g_j(\mathbf{x}) \leq 0$, $j = 1, \ldots, m$. In fact, if all the g_js are convex, the intersection of the sets $\{\mathbf{x} \mid g_j(\mathbf{x}) \leq 0\}$, $j = 1, \ldots, m$, is the feasible domain and it is convex. This result is discussed further in Chapter 5.

Differentiable Functions

Verifying convexity of a function from the definition is often very cumbersome. To get an operationally useful characterization we will assume that the function is also differentiable. For a function of one variable we associated convexity with a positive second derivative. The expected generalization for convex differentiable functions of several variables should be positive-definite Hessian. We will prove this result next.

First, recognize that the definition of a convex function is equivalent to the geometric statement that the tangent at any point of the graph does not cross the graph. Analytically, this is stated as

$$f(\mathbf{x}_1) \geq f(\mathbf{x}_0) + \nabla f(\mathbf{x}_0)(\mathbf{x}_1 - \mathbf{x}_0). \tag{4.30}$$

The proof is left as an exercise (Exercise 4.19). Next we apply Taylor's Theorem with the remainder term being a quadratic calculated at point $\mathbf{x}(\lambda) = \lambda \mathbf{x}_1 + (1 - \lambda)\mathbf{x}_0$ between $\mathbf{x}_0$ and $\mathbf{x}_1$ $(0 \leq \lambda \leq 1)$:

$$\begin{aligned} f(\mathbf{x}_1) = & f(\mathbf{x}_0) + \nabla f(\mathbf{x}_0)(\mathbf{x}_1 - \mathbf{x}_0) \\ & + \tfrac{1}{2}(\mathbf{x}_1 - \mathbf{x}_0)^T \mathbf{H}[\mathbf{x}(\lambda)](\mathbf{x}_1 - \mathbf{x}_0). \end{aligned} \tag{4.31}$$

If $\mathbf{H}[\mathbf{x}(\lambda)]$ is positive-semidefinite for all λs, Equation (4.31) implies (4.30), and so f is convex. Conversely, if f is convex, then (4.30) must hold. Now, if the Hessian is not positive-semidefinite everywhere, the quadratic term in (4.31) could be negative for some λ, which would imply $f(\mathbf{x}_1) < f(\mathbf{x}_0) + \nabla f(\mathbf{x}_0)(\mathbf{x}_1 - \mathbf{x}_0)$, contradicting the convexity assumption. Thus we have reached the following result:

A differentiable function is convex if and only if its Hessian is positive-semidefinite in its entire convex domain.

Example 4.9 The set $\mathcal{G} = \{\mathbf{x} \mid \mathbf{x} \in \Re^2, g = (3 - x_1)^2 + (4 - x_2)^2 \leq 1\}$ is convex, because the Hessian of g is positive-definite (which is stronger than positive-semidefinite). ■

Example 4.10 The set constraint described by the inequalities

$$g_1(\mathbf{x}) = \mathbf{d}^T\mathbf{x} \le c_1, \quad g_2(\mathbf{x}) = \mathbf{x}^T\mathbf{Q}\mathbf{x} + \mathbf{a}^T\mathbf{x} + b \le c_2$$

is convex, if the matrix $\mathbf{Q}$ is positive-semidefinite, since then it will be the intersection of two convex sets. ■

It should be noted that the convex domain $\mathcal{D}$ of f above is assumed open, so that all points $\mathbf{x}$ are interior. Otherwise, convexity may not imply positivity of the Hessian. For example, consider a function $f(x)$, $x \in \mathfrak{R}$, that is convex in $\mathcal{D}$ but has an inflection point on the boundary of $\mathcal{D}$. Also, a positive-definite Hessian implies *strict* convexity, but the converse is generally not true.

Example 4.11 The function $f(\mathbf{x}) = 4x_1^4 + 4x_2^4$ is strictly convex but the Hessian is singular at the origin. ■

The most practical result of convexity follows immediately from the definition of convexity by inequality (4.30). If the minimizer $\mathbf{x}_*$ of a convex function exists, then

$$f(\mathbf{x}) \ge f(\mathbf{x}_*) + \nabla f(\mathbf{x}_*)(\mathbf{x} - \mathbf{x}_*). \tag{4.32}$$

If, in addition, $\mathbf{x}_*$ is an interior point and the function is differentiable, then $\nabla f(\mathbf{x}_*) = \mathbf{0}^T$ is not only necessary but also sufficient for optimality. Global validity of (4.32) will imply that $\mathbf{x}_*$ is the global minimizer. In summary,

If a differentiable (strictly) convex function with a convex open domain has a stationary point, this point will be the (unique) global minimum.

Example 4.12 Consider the function

$$f = x_1^2 + 2x_2^2 + 3x_3^2 + 3x_1x_2 + 4x_1x_3 - 3x_2x_3.$$

The function is quadratic of the form $f = \mathbf{x}^T\mathbf{Q}\mathbf{x}$ where

$$\mathbf{Q} = \begin{pmatrix} 1 & 0 & 0 \\ 3 & 2 & -4 \\ 4 & 1 & 3 \end{pmatrix}, \quad \mathbf{A} = \frac{1}{2}(\mathbf{Q} + \mathbf{Q}^T) = \frac{1}{2}\begin{pmatrix} 2 & 3 & 4 \\ 3 & 4 & -3 \\ 4 & -3 & 6 \end{pmatrix}$$

so that $f = \mathbf{x}^T\mathbf{A}\mathbf{x}$. Since $\mathbf{A}$ is symmetric, it is also the Hessian of f. Since the diagonal elements are positive but the leading 2×2 principal minor is negative, the matrix $\mathbf{A}$ is indefinite and the function is nonconvex, with a saddle at the origin.

The function $f_1 = 2x_1^2 + 3x_2^2 + 3x_1x_2$ is again of the form $\mathbf{x}^T\mathbf{Q}_1\mathbf{x}$ where

$$\mathbf{Q}_1 = \begin{pmatrix} 2 & 0 \\ 3 & 3 \end{pmatrix}, \quad \mathbf{A}_1 = \frac{1}{2}(\mathbf{Q}_1 + \mathbf{Q}_1^T) = \frac{1}{2}\begin{pmatrix} 4 & 3 \\ 3 & 6 \end{pmatrix},$$

but $\mathbf{A}_1$ is now positive-definite. Consequently, f_1 is convex with a global minimum at the origin. ■

Example 4.13 Consider the general sum of two positive power functions

$$f(\mathbf{x}) = c_1 x_1^{\alpha_{11}} x_2^{\alpha_{12}} + c_2 x_1^{\alpha_{21}} x_2^{\alpha_{22}},$$

where the cs are positive constants, the αs are real numbers, and the xs are strictly positive. It is rather involved to see if such a function is convex by direct differentiation. Another way is to use the monotonic transformation of variables $x_i = \exp y_i$, $i = 1, 2$ and the shorthand t_1, t_2 for the terms of the function, that is,

$$f(\mathbf{y}) = t_1 + t_2 = c_1 \exp(\alpha_{11} y_1 + \alpha_{12} y_2) + c_2 \exp(\alpha_{21} y_1 + \alpha_{22} y_2).$$

Note that $\partial t_i / \partial y_j = \alpha_{ij} t_i$, so that we find easily

$$\begin{aligned}
\partial f / \partial y_1 &= \alpha_{11} t_1 + \alpha_{21} t_2, \\
\partial f / \partial y_2 &= \alpha_{12} t_1 + \alpha_{22} t_2, \\
\partial^2 f / \partial y_i \partial y_j &= \alpha_{1i} \alpha_{1j} t_1 + \alpha_{2i} \alpha_{2j} t_2.
\end{aligned}$$

The determinants of the leading principal minors of the Hessian are

$$\begin{aligned}
&\partial^2 f / \partial y_1^2 = \alpha_{11}^2 t_1 + \alpha_{21}^2 t_2 > 0, \\
&(\partial^2 f / \partial y_1^2)(\partial^2 f / \partial y_2^2) - (\partial^2 f / \partial y_1 \partial y_2)^2 \\
&\quad = (\alpha_{11}^2 t_1 + \alpha_{21}^2 t_2)(\alpha_{12}^2 t_1 + \alpha_{22}^2 t^2) - (\alpha_{11}\alpha_{12} t_1 + \alpha_{21}\alpha_{22} t_2)^2 \\
&\quad = t_1 t_2 (\alpha_{11}\alpha_{22} - \alpha_{21}\alpha_{12})^2 \geq 0,
\end{aligned}$$

where the positivity is guaranteed from $t_1, t_2 > 0$ by definition of f. Thus, the function is convex with respect to y_1 and y_2, and its stationary point should be a global minimum. If $\alpha_{11}\alpha_{22} - \alpha_{21}\alpha_{12} \neq 0$, this point should be unique, with the function $f(y_1, y_2)$ being strictly convex. The stationary point, however, *does not exist*, since there we must have $t_1 = t_2 = 0$ or, $x_1^{\alpha_{11}} x_2^{\alpha_{12}} = 0$, $x_1^{\alpha_{21}} x_2^{\alpha_{22}} = 0$ simultaneously. This is ruled out by the strict positivity of x_1, x_2 . The function is not *well bounded*, yet it is convex.

This becomes obvious by looking at a special case:

$$f = x_1^2 x_2 + x_1^{-1} x_2^{-1}.$$

Here we have

$$\begin{aligned}
\partial f / \partial x_1 &= 2x_1 x_2 - x_1^{-2} x_2^{-1} = (2x_1^3 x_2^2 - 1)/x_1^2 x_2, \\
\partial f / \partial x_2 &= x_1^2 - x_1^{-1} x_2^{-2} = (x_1^3 x_2^2 - 1)/x_1 x_2^2.
\end{aligned}$$

If $x_1^3 x_2^2 > 1$, then f is increasing wrt both x_1 and x_2, and so it is not well bounded from below (i.e., it is not bounded away from zero). No $\mathbf{x} \neq \mathbf{0}$ can satisfy the stationarity conditions. In fact, the problem can be shown to be monotonic by making the transformation $x_3 = x_1^3 x_2^2$. Then, $f_1(x_1, x_3) = x_1^{1/2}(x_3^{1/2} + x_3^{-1/2})$, which is monotonic in x_1. Hence, the problem has no positive minimum.

The general problem can be bounded if we add another term:

$$f = c_1 x_1^{\alpha_{11}} x_2^{\alpha_{12}} + c_2 x_1^{\alpha_{21}} x_2^{\alpha_{22}} + c_3 x_1^{\alpha_{31}} x_2^{\alpha_{32}}.$$

Applying the exponential transformation, we can show that f is convex in the y_1, y_2 space. In fact, being the (positive) sum of the convex exponential functions it must be convex. A stationary point can be found quickly employing special procedures. Models with positive sums of power functions of many positive variables are called *Geometric Programming* problems (Duffin et al., 1967). They have been studied extensively as a special class of design optimization models (Beightler and Phillips, 1976). Note that although sums of positive power functions are convex under an exponential transformation, they are not well bounded unless the number of terms is greater than the number of variables. ■

4.5 Local Exploration

Functions that are not of a simple polynomial form or that may be defined implicitly through a computer program subroutine will have optimality conditions that are either difficult to solve for $\mathbf{x}_*$ or impossible to write in an explicit form. Many practical design models will have such functions.

Example 4.14 Engineering functions often contain exponential terms or other transcendental functions. These problems will almost always lead to nonlinear stationarity conditions that cannot be solved explicitly. For example, if

$$f = x_1 + x_2 + x_1 \exp(-x_2) + x_2^2 \exp(-x_1)$$

the stationarity conditions will be

$$\partial f/\partial x_1 = 1 + \exp(-x_2) - x_2^2 \exp(-x_1) = 0,$$

$$\partial f/\partial x_2 = 1 - x_1 \exp(-x_2) + 2x_2 \exp(-x_1) = 0,$$

which cannot be solved explicitly. ■

Gradient Descent

When optimality conditions cannot be manipulated to yield an explicit solution, an *iterative* procedure can be sought. One may start from an initial point where the function value is calculated and then take a step in a downward direction, where the function value will be lower. To make such a step, one utilizes local information and explores the immediate vicinity of the current point; hence, all iterative methods perform some kind of *local exploration*. There is direct equivalence with a design procedure, where an existing design is used to generate information about developing an improved one. If the iteration scheme converges, the process will end at a stationary point where no further improvement is possible provided that a *descent* step was indeed performed at each iteration.

An obvious immediate concern is how to find a descent direction. The first-order perturbation of f at $\mathbf{x}_0$ shows that a descent move is found if

$$\partial f = \nabla f(\mathbf{x}_0)\partial\mathbf{x} < 0. \tag{4.33}$$

Therefore, descent perturbation $\partial\mathbf{x}$ will be one where

$$\partial\mathbf{x} = -\nabla f^T(\mathbf{x}_0). \tag{4.34}$$

Whatever the sign of the gradient vector, a move in the direction of the negative gradient will be a descent one, for then

$$\partial f = -|\nabla f(\mathbf{x}_0)|^2 < 0. \tag{4.35}$$

To develop an iterative procedure, let us set $\partial\mathbf{x}_k = \mathbf{x}_{k+1} - \mathbf{x}_k$, where k is the current iteration, and define for convenience $\mathbf{g}(\mathbf{x}) = \nabla f^T(\mathbf{x})$ and $f(\mathbf{x}_k) = f_k$. Then, from $\partial\mathbf{x}_k = -\nabla f^T(\mathbf{x}_k)$, we get

$$\mathbf{x}_{k+1} = \mathbf{x}_k - \mathbf{g}_k, \tag{4.36}$$

which should give a simple iterative method of local improvement.

Example 4.15 For the function of Example 4.8,

$$f = x_1^4 - 2x_1^2 x_2 + x_2^2,$$

the iteration (4.36) would give

$$\begin{pmatrix} x_{1,k+1} \\ x_{2,k+1} \end{pmatrix} = \begin{pmatrix} x_{1,k} \\ x_{2,k} \end{pmatrix} - 2\left(x_{1,k}^2 - x_{2,k}\right)\begin{pmatrix} 2x_{1,k} \\ -1 \end{pmatrix}.$$

Starting at a nonstationary point, take $\mathbf{x}_0 = (1.1, 1)^T$ with $f_0 = 44.1(10^{-3})$ and calculate

$$\mathbf{x}_1 = \begin{pmatrix} 1.1 \\ 1 \end{pmatrix} - 2(1.21 - 1)\begin{pmatrix} 2.2 \\ -1 \end{pmatrix} = \begin{pmatrix} 0.176 \\ 1.420 \end{pmatrix}, \quad f_1 = 1.929,$$

$$\mathbf{x}_2 = \begin{pmatrix} 0.176 \\ 1.420 \end{pmatrix} - 2(0.031 - 1.420)\begin{pmatrix} 0.352 \\ -1 \end{pmatrix} = \begin{pmatrix} 1.154 \\ -1.358 \end{pmatrix}, \quad f_2 = 7.235.$$

The expected descent property is not materialized, although the point is close to the valley point $(1, 1)^T$. Let us try another starting point $\mathbf{x}_0' = (5, 2)^T$ with $f_0' = 529$. Now calculate again

$$\mathbf{x}_1' = \begin{pmatrix} 5 \\ 2 \end{pmatrix} - 2(25 - 2)\begin{pmatrix} 10 \\ -1 \end{pmatrix} = \begin{pmatrix} -455 \\ 48 \end{pmatrix}, \quad f_1' = 4.2(10^{10}).$$

We see again that immediate divergence occurs. Clearly, the iteration (4.36) is not good. Since we know that the direction we move in is a descent one, the explanation for what happens must be that the step length is too large. We may control this length by providing an adjusting parameter α, that is, modify (4.36) as

$$\mathbf{x}_{k+1} = \mathbf{x}_k - \alpha\mathbf{g}_k,$$

where α is suitably chosen. Taking $\mathbf{x}_0'' = (5, 2)^T$, let us choose $\alpha = 0.01$ since the gradient norm there is very large. Then

$$\mathbf{x}_1'' = \begin{pmatrix} 5 \\ 2 \end{pmatrix} - (0.01)(2)(23)\begin{pmatrix} 10 \\ -1 \end{pmatrix} = \begin{pmatrix} 0.4 \\ 2.46 \end{pmatrix}, \quad f_1'' = 5.29.$$

Continuing with $\alpha = 0.1$, since the gradient norm is becoming smaller, we get

$$\mathbf{x}_2'' = \begin{pmatrix} 0.4 \\ 2.46 \end{pmatrix} - (0.1)(2)(-2.3)\begin{pmatrix} 0.8 \\ -1 \end{pmatrix} = \begin{pmatrix} 0.768 \\ 2 \end{pmatrix}, \quad f_2'' = 1.988,$$

$$\mathbf{x}_3'' = \begin{pmatrix} 0.768 \\ 2 \end{pmatrix} - (0.1)(2)(-1.41)\begin{pmatrix} 1.536 \\ -1 \end{pmatrix} = \begin{pmatrix} 1.201 \\ 1.718 \end{pmatrix}, \quad f_3'' = 0.076.$$

Some success has been achieved now, since the function value f_3'' approximates the optimal one within two significant digits. We may continue the iterations adjusting the step length again. ■

The example shows clearly that a gradient method must have an adjustable step length that is, in general, dependent on the iteration itself. Thus (4.36) must be revised as

$$\mathbf{x}_{k+1} = \mathbf{x}_k - \alpha_k \mathbf{g}_k. \tag{4.37}$$

The step length α_k can be determined heuristically, but a rigorous procedure is possible and in fact rather obvious. How this may be done will be examined in the next section.

Newton's Method

The simple gradient-based iteration (4.37) is nothing but a local linear approximation to the function applied successively to each new point. The approximation to the stationary point $\mathbf{x}_k$ is corrected at every step by subtracting the vector $\alpha_k \mathbf{g}_k$ from $\mathbf{x}_k$ to get the new approximation $\mathbf{x}_{k+1}$. Clearly, when we are close to a stationary point this correction can be very small and progress toward the solution will be very slow. The linear approximation is then not very efficient and higher order terms are significant. Let us approximate the function with a quadratic one using the Taylor expansion

$$f_{k+1} = f_k + \nabla f_k \partial \mathbf{x}_k + \left(\tfrac{1}{2}\right) \partial \mathbf{x}_k^T \mathbf{H}_k \partial \mathbf{x}_k. \tag{4.38}$$

The minimizer $\mathbf{x}_{k+1}$ can be found from the stationarity condition

$$\nabla f_k^T + \mathbf{H}_k \partial \mathbf{x}_k = \mathbf{0}. \tag{4.39}$$

Assuming that $\mathbf{H}_k$ is invertible, we can rewrite (4.39) as

$$\mathbf{x}_{k+1} = \mathbf{x}_k - \mathbf{H}_k^{-1} \mathbf{g}_k. \tag{4.40}$$

If the function is locally strictly convex, the Hessian will be positive-definite, the quadratic term in (4.38) will be strictly positive, and the iteration (4.40) will yield a lower function value. However, the method will move to higher function values if the Hessian is negative-definite and it may stall at singular points where the Hessian is semidefinite.

The iterative scheme (4.40) uses successive quadratic approximations to the function and is the purest form of *Newton's method.* The correction to the approximate minimizer is now $-\mathbf{H}_k^{-1}\mathbf{g}_k$, with the inverse Hessian multiplication serving as an acceleration factor to the simple gradient correction. Newton's method will move efficiently in the neighborhood of a local minimum where local convexity is present.

Example 4.16 Consider the function

$$f = 4x_1^2 + 3x_1x_2 + x_2^2$$

with gradient and Hessian given by

$$\mathbf{g}(\mathbf{x}) = \begin{pmatrix} 8x_1 + 3x_2 \\ 3x_1 + 2x_2 \end{pmatrix}, \quad \mathbf{H} = \begin{pmatrix} 8 & 3 \\ 3 & 2 \end{pmatrix}.$$

The Hessian is positive-definite everywhere and the function is strictly convex. The inverse of $\mathbf{H}$ is found simply from

$$\mathbf{H}^{-1} = \frac{1}{(16-9)}\begin{pmatrix} 2 & -3 \\ -3 & 8 \end{pmatrix} = \begin{pmatrix} \frac{2}{7} & -\frac{3}{7} \\ -\frac{3}{7} & \frac{8}{7} \end{pmatrix}.$$

Starting at $\mathbf{x}_0 = (1, 1)^T$, Newton's method gives

$$\mathbf{x}_1 = \begin{pmatrix} 1 \\ 1 \end{pmatrix} - \begin{pmatrix} \frac{2}{7} & -\frac{3}{7} \\ -\frac{3}{7} & \frac{8}{7} \end{pmatrix}\begin{pmatrix} 11 \\ 5 \end{pmatrix} = \begin{pmatrix} 0 \\ 0 \end{pmatrix}.$$

The minimizer is obviously $(0, 0)^T$ and Newton's method approximating a quadratic function exactly will reach the solution in *one step* from any starting point. This happens because the Hessian is fixed for quadratic functions. Since the function is strictly convex, the stationary point will be a global minimum. ■

Example 4.17 Consider again Example 4.8, where the function $f = x_1^4 - 2x_1^2x_2 + x_2^2$ has the valley $x_{1*}^2 = x_{2*}$. The correction vector, $\mathbf{s}$, according to Newton's method is

$$\mathbf{s} = -\mathbf{H}^{-1}\mathbf{g} = -\left[8(x_1^2 - x_2)\right]^{-1}\begin{pmatrix} 2 & 4x_1 \\ 4x_1 & 12x_1^2 - 4x_2 \end{pmatrix}\begin{pmatrix} 4x_1(x_1^2 - x_2) \\ -2(x_1^2 - x_2) \end{pmatrix}$$

$$= -\frac{2(x_1^2 - x_2)}{8(x_1^2 - x_2)}\begin{pmatrix} 0 \\ 8x_1^2 - 12x_1^2 + 4x_2 \end{pmatrix} = \begin{pmatrix} 0 \\ x_1^2 - x_2 \end{pmatrix}.$$

Therefore,

$$\mathbf{x}_{k+1} = \begin{pmatrix} x_1 \\ x_2 \end{pmatrix}_k + \begin{pmatrix} 0 \\ x_1^2 - x_2 \end{pmatrix}_k = \begin{pmatrix} x_1 \\ x_1^2 \end{pmatrix}_k.$$

Any starting point will lead into the valley in one iteration. This nice result is coincidental, since f is not quadratic. Note, however, that in an actual numerical implementation, the algebraic simplifications above would not be possible. If an iteration is attempted after the first one, an error will occur since the quantity $(x_1^2 - x_2)$ in the denominator of the elements in $\mathbf{H}^{-1}$ will be zero. ■

Example 4.18 Consider the function $f = x^4 - 32x$, which has $\partial f/\partial x = 4x^3 - 32$ and $\partial^2 f/\partial x^2 = 12x^2 > 0$. The minimum occurs at $x_* = 2$, $f_* = -48$. The Newton iteration gives

$$x_{k+1} = x_k - [(4x^3 - 32)/12x^2]_k = 0.6667x_k + 2.6667x_k^{-2}.$$

Based on this iteration and starting at $x_0 = 1$ we find the sequence of points $\{x_k\} = \{1, 3.3333, 2.4622, 2.0813, 2.0032, \ldots\}$ and function values $\{f_k\} = \{-31, 16.79, -42.0371, -47.8370, -47.9998, \ldots\}$.

Convergence occurred very quickly, but note that the function value went *up* at the first iteration, although the Hessian is positive-definite there. This is an unpleasant surprise because it means that positive-definiteness of the Hessian is not sufficient to guarantee descent when a step length of one is used. This may be the case for functions that are not quadratic so that the second-order terms are not enough to approximate the function correctly. What can happen is that between two successive points the curvature of the function is steeper than the quadratic approximation and the Newton step goes into an area where the function value is larger than before. ■

Example 4.19 Consider the function

$$f = \tfrac{1}{3}x_1^3 + x_1x_2 + \tfrac{1}{2}x_2^2 + 2x_2 - \left(\tfrac{2}{3}\right),$$

which has $\mathbf{g} = (x_1^2 + x_2, x_1 + x_2 + 2)^T$ with the stationary points $(2, -4)^T$ and $(-1, -1)^T$. The Hessian is

$$H = \begin{pmatrix} 2x_1 & 1 \\ 1 & 1 \end{pmatrix}$$

and the Newton correction direction is

$$\begin{aligned} \mathbf{s} &= -(2x_1 - 1)^{-1}\begin{pmatrix} 1 & -1 \\ -1 & 2x_1 \end{pmatrix}\begin{pmatrix} x_1^2 + x_2 \\ x_1 + x_2 + 2 \end{pmatrix} \\ &= -(2x_1 - 1)^{-1}\left(x_1^2 - x_1 - 2, x_1^2 + 2x_1x_2 + 4x_1 - x_2\right)^T. \end{aligned}$$

Applying Newton's method starting at $(1, 1)^T$ we have

$$\mathbf{x}_0 = \begin{pmatrix} 1 \\ 1 \end{pmatrix}, \quad \mathbf{s}_0 = \begin{pmatrix} 2 \\ -6 \end{pmatrix}, \quad f_0 = 3.1667,$$

$$\mathbf{x}_1 = \begin{pmatrix} 1 \\ 1 \end{pmatrix} + \begin{pmatrix} 2 \\ -6 \end{pmatrix} = \begin{pmatrix} 3 \\ -5 \end{pmatrix}, \quad \mathbf{s}_1 = \begin{pmatrix} -0.8 \\ 0.8 \end{pmatrix}, \quad f_1 = -4.1667,$$

$$\mathbf{x}_2 = \begin{pmatrix} 3 \\ -5 \end{pmatrix} + \begin{pmatrix} -0.8 \\ 0.8 \end{pmatrix} = \begin{pmatrix} 2.2 \\ -4.2 \end{pmatrix}, \quad \mathbf{s}_2 = \begin{pmatrix} -0.1882 \\ 0.1882 \end{pmatrix}, \quad f_2 = -5.9373,$$

$$\mathbf{x}_3 = \begin{pmatrix} 2.2 \\ -4.2 \end{pmatrix} + \begin{pmatrix} -0.1882 \\ 0.1882 \end{pmatrix} = \begin{pmatrix} 2.0118 \\ -4.0118 \end{pmatrix}, \quad f_3 = -5.9998.$$

The point $(2, -4)^T$ with $f_* = -6$ is in fact a local minimum, since $\mathbf{H}$ is positive-definite there.

Suppose that the initial (or some intermediate) point is $\mathbf{x}_0 = (-1, -1)^T$. Now $\mathbf{s}_0 = (0, 0)^T$ and no search direction is defined. Although we know that a minimizer exists, this happens because the point $(-1, -1)^T$ is a *saddlepoint*, with $\mathbf{H}$ being indefinite there. ■

The examples demonstrate that the basic local exploration methods (gradient method and Newton's method) are not as effective as one might desire. The gradient method may be hopelessly inefficient, while Newton's method may go astray too easily for a general nonlinear function. Modifications and extensions of the basic methods have been developed to overcome most difficulties. Some of these ideas will be examined in the next two sections of this chapter and in Chapter 7. We should keep in mind that many design models possess characteristics defying the basic iterative strategies, and so a good understanding of their limitations is necessary for avoiding early disappointments in the solution effort. Even advanced iterative methods can fail, as we will point out further in Chapter 7.

4.6 Searching along a Line

The local exploration idea of the previous section gave us two descent directions: the negative gradient $-\mathbf{g}_k$ and the Newton $-\mathbf{H}_k^{-1}\mathbf{g}_k$. We saw that the gradient is not useful unless we control how far we move along that direction. The same may be true for the Newton direction. The addition of an adjustable step length, as in (4.37), is an effective remedy. The question is how to determine a good value for the step size α_k. To do this, let us think of the local exploration as a general iterative procedure

$$\mathbf{x}_{k+1} = \mathbf{x}_k + \alpha_k \mathbf{s}_k, \tag{4.41}$$

where $\mathbf{s}_k$ is a *search direction vector*. Searching along the line determined by $\mathbf{s}_k$, we would like to stop at the point that gives the smallest value of the function on that line. That new point $\mathbf{x}_{k+1}$ will be at a distance $\alpha_k \mathbf{s}_k$ from $\mathbf{x}_k$. Thus, α_k is simply the solution to the problem

$$\min_{\alpha} f(\mathbf{x}_k + \alpha \mathbf{s}_k); \quad 0 \leq \alpha < \infty, \tag{4.42}$$

which we write formally as

$$\alpha_k = \arg \min_{0 \leq \alpha < \infty} f(\mathbf{x}_k + \alpha \mathbf{s}_k). \tag{4.43}$$

Positivity of α_k is the only requirement because $\mathbf{s}_k$ is already assumed to be a descent direction of $\mathbf{x}_k$. The general local exploration procedure will have two phases: a direction-finding phase to determine $\mathbf{s}_k$ and a *line search* along $\mathbf{s}_k$ to determine α_k. Most numerical (iterative) optimization techniques are essentially of this nature. The differences among practical algorithms are in the theoretical selection of $\mathbf{s}_k$ and α_k, and in the numerical calculation of their values.

Gradient Method

Here we can assume that an *exact* line search is possible, that is, the value $\alpha_* = \alpha_k$ of problem (4.43) can be found precisely and the corresponding value f_* exists (which would not be true for an unbounded problem). Then, if we select $\mathbf{s}_k = -\mathbf{g}_k$, we define the method known as the *gradient method*. We may terminate the iterations, when the gradient becomes sufficiently small, say, $\|\mathbf{g}_k\| < \varepsilon, \varepsilon > 0$. Thus, an algorithm may be as follows.

GRADIENT ALGORITHM

1. For $k = 0$: Select $\mathbf{x}_0$, ε.
2. For $k > 0$: Set $\mathbf{s}_k = -\mathbf{g}_k$.
   Compute $\alpha_k = \arg\min_\alpha f(\mathbf{x}_k - \alpha\mathbf{g}_k)$.
   Set $\mathbf{x}_{k+1} = \mathbf{x}_k - \alpha_k\mathbf{g}_k$.
3. For $k = k + 1$: Stop, if $\|\mathbf{g}_{k+1}\| < \varepsilon$. Otherwise repeat step 2.

Note that the *termination criterion* $\|\mathbf{g}_k\| < \varepsilon$ is only one example of several that may be used depending on the problem and the meaning of its solution. Other examples of termination criteria may be $\|\mathbf{x}_{k+1} - \mathbf{x}_k\| < \varepsilon$ or $(f_{k+1} - f_k) < -\varepsilon$, or a combination of them.

The line search subproblem is an important one to solve correctly and efficiently because the overall algorithmic performance may depend on it. Discussion on this is deferred until Chapter 7. As an illustration here, let us use a simple quadratic approximation for the line search and apply it to the gradient method. The Taylor expansion at $\mathbf{x}_k$ is given by

$$f(\mathbf{x}_k - \alpha\mathbf{g}_k) = f(\mathbf{x}_k) - \alpha\nabla f(\mathbf{x}_k)\mathbf{g}_k + \left(\tfrac{1}{2}\right)\alpha^2\mathbf{g}_k^T\mathbf{H}(\mathbf{x}_k)\mathbf{g}_k \tag{4.44}$$

or, in condensed form,

$$f_{k+1} = f_k - \alpha\mathbf{g}_k^T\mathbf{g}_k + \left(\tfrac{1}{2}\right)\alpha^2\mathbf{g}_k^T\mathbf{H}_k\mathbf{g}_k. \tag{4.45}$$

Differentiating with respect to α and setting the result equal to zero, we get

$$-\mathbf{g}_k^T\mathbf{g}_k + \alpha\mathbf{g}_k^T\mathbf{H}_k\mathbf{g}_k = \mathbf{0}, \tag{4.46}$$

which solved for α gives $\alpha_k = \mathbf{g}_k^T\mathbf{g}_k/\mathbf{g}_k^T\mathbf{H}_k\mathbf{g}_k$. This expression will work for positive-definite $\mathbf{H}_k$ and is equivalent to taking a single Newton step to predict the mimimum along the gradient direction. The iteration step 2 in the general gradient algorithm will then be

$$\mathbf{x}_{k+1} = \mathbf{x}_k - (\mathbf{g}_k^T \mathbf{g}_k / \mathbf{g}_k^T \mathbf{H}_k \mathbf{g}_k)\mathbf{g}_k. \tag{4.47}$$

The iterative Newton method, used fully for this one-dimensional search, would require solving (4.46) iteratively until the minimum wrt α is found.

Example 4.20 For quadratic functions the Hessian will be fixed and will not depend on the iteration k. This makes the iteration (4.47) much easier to apply. For the function $f = 4x_1^2 + 3x_1x_2 + x_2^2$ of Example 4.16, (4.47) applied to $\mathbf{x}_0 = (1, 1)^T$ gives the first iteration

$$\mathbf{x}_1 = \begin{pmatrix} 1 \\ 1 \end{pmatrix} - \left[\frac{(11, 5)\binom{11}{5}}{(11, 5)\begin{pmatrix} 8 & 3 \\ 3 & 2 \end{pmatrix}\binom{11}{5}} \right] \begin{pmatrix} 11 \\ 5 \end{pmatrix} = \begin{pmatrix} 1 \\ 1 \end{pmatrix} - 0.1083 \begin{pmatrix} 11 \\ 5 \end{pmatrix},$$

or $\mathbf{x}_1 = (-0.191, 0.458)^T$ with $f_1 = 0.093$. Although this is a great improvement from $f_0 = 8$, it is not as good as that obtained from the full Newton step of Example 4.16.

For nonquadratic functions the Hessian will vary with k. For the function $f = x_1^4 - 2x_1^2x_2 + x_2^2$ of Examples 4.15 and 4.17, (4.47) applied to $\mathbf{x}_0 = (1.1, 1)^T$ gives

$$\mathbf{x}_1 = \begin{pmatrix} 1.1 \\ 1 \end{pmatrix} - \left[\frac{(0.924, -0.42)\binom{0.924}{-0.42}}{(0.924, -0.42)\begin{pmatrix} 10.52 & -4.4 \\ -4.4 & 2 \end{pmatrix}\binom{0.924}{-0.42}} \right] \begin{pmatrix} 0.924 \\ -0.42 \end{pmatrix}$$

$$= (1.1, 1)^T - 0.0808\,(0.924, -0.42)^T = (1.025, 1.034)^T$$

with $f_1 = 2.7(10^{-4})$, an improvement from $f_0 = 44.1(10^{-3})$. ■

Modified Newton's Method

The effect of the step size on the general iteration (4.41) can be discovered quickly if we apply to (4.41) an approach similar to the one we used for deriving (4.47). Assuming perturbations $\partial\mathbf{x}_k = \alpha\mathbf{s}_k$, the Taylor expansion at $\mathbf{x}_k$ will be

$$f(\mathbf{x}_k + \alpha\mathbf{s}_k) = f_k + \alpha\mathbf{g}_k^T\mathbf{s}_k + \left(\tfrac{1}{2}\right)\alpha^2\mathbf{s}_k^T\mathbf{H}_k\mathbf{s}_k + \alpha^3 o(\|\mathbf{s}_k\|^2), \tag{4.48}$$

where we include the order terms to indicate an exact expansion. The value of α that minimizes this function is taken as the step size. The stationary point along $\mathbf{s}_k$ is found from taking the derivative of (4.48) with respect to α, setting it equal to zero, and solving for $\alpha_* = \alpha_k$. Namely, we solve for α_k the equation

$$\mathbf{g}_k^T\mathbf{s}_k + \alpha_k\mathbf{s}_k^T\mathbf{H}_k\mathbf{s}_k + \alpha_k^2 o(\|\mathbf{s}_k\|^2) = 0. \tag{4.49}$$

Note that at points $\mathbf{x}_k$ where the norm $\|\mathbf{s}_k\|$ is not zero, the higher order term may be significant in estimating α_k. For example, in Newton's method, we take $\mathbf{s}_k = -\mathbf{H}_k^{-1}\mathbf{g}_k$

and (4.49) will be rewritten as

$$(\alpha_k - 1)\mathbf{g}_k^T \mathbf{H}_k^{-1} \mathbf{g}_k + \alpha_k^2 o\left(\|\mathbf{H}_k^{-1}\mathbf{g}_k\|^2\right) = 0. \tag{4.50}$$

For $\mathbf{H}_k^{-1}\mathbf{g}_k \neq 0$, taking $\alpha_k = 1$ as is the case with the simple Newton's method will not satisfy the stationarity condition and may give a poor choice of α_k. The function value f_{k+1} may be larger than f_k. Therefore, a suggested modification would be

$$\mathbf{x}_{k+1} = \mathbf{x}_k - \alpha_k \mathbf{H}_k^{-1}\mathbf{g}_k, \tag{4.51}$$

$$\alpha_k = \arg \min_{0 \le \alpha < \infty} f\left(\mathbf{x}_k - \alpha \mathbf{H}_k^{-1}\mathbf{g}_k\right). \tag{4.52}$$

This modification will work well for positive-definite Hessians.

4.7 Stabilization

Given a good choice for α_k, the otherwise efficient Newton's method will be unstable when the Hessian is not positive-definite at points $\mathbf{x}_k$ during iterations; given a symmetric matrix $\mathbf{M}_k$, the gradient and Newton's methods can be classed together by the general iteration

$$\mathbf{x}_{k+1} = \mathbf{x}_k - \alpha_k \mathbf{M}_k \mathbf{g}_k, \tag{4.53}$$

where $\mathbf{M}_k = \mathbf{I}$ for gradient descent and $\mathbf{M}_k = \mathbf{H}_k^{-1}$ for Newton's method. A positive-definite $\mathbf{M}_k$ is sometimes called a *metric*. Equation (4.53) suggests a perturbation in the direction $-\mathbf{M}_k\mathbf{g}_k$. The first-order Taylor approximation gives

$$f_{k+1} - f_k = -\alpha_k \mathbf{g}_k^T \mathbf{M}_k \mathbf{g}_k, \tag{4.54}$$

so that descent is accomplished for $\mathbf{M}_k$ positive-definite. Thus, the gradient method with $\mathbf{M}_k = \mathbf{I}$ will satisfy this condition. But Newton's method will usually fail if the nearest stationary point is not a minimum. So we need a compromise that retains some of the efficiency of the second-order approximation while remaining stable and avoiding nonoptimal points. We can do this by constructing a matrix

$$\mathbf{M}_k = (\mathbf{H}_k + \mu_k \mathbf{I})^{-1} \tag{4.55}$$

and selecting a positive scalar μ_k so that $\mathbf{M}_k$ is always positive-definite.

Modified Cholesky Factorization

One way to construct $\mathbf{M}_k$ is to perform a spectral decomposition of $\mathbf{H}_k$ and add μ_k s to all eigenvalues until all of them are positive. This method has not been fully explored and is considered rather expensive computationally. Another way proposed by Gill and Murray (1974) is based on the *Cholesky factorization* of $\mathbf{H}_k$. We will summarize this method here, particularly for demonstrating how to get away from a saddlepoint in a practical way.

The Cholesky factorization of a symmetric matrix $\mathbf{H}$ is the product $\mathbf{H} = \mathbf{LDL}^T$ where $\mathbf{L}$ is a lower triangular matrix with unit diagonal elements and $\mathbf{D}$ is a diagonal matrix. This factorization is a special case of a diagonalization procedure based on Gauss elimination, so that the solution of a symmetric linear system would take about half the computational effort. If the elements of **H**, **L**, and **D** are h_{ij}, l_{ij}, and d_{jj}, respectively, the Cholesky factors can be calculated column by column from the formulas

$$d_{jj} = h_{jj} - \sum_{s=1}^{j-1} d_{ss} l_{js}^2, \tag{4.56}$$

$$l_{ij} = (1/d_{jj}) \left(h_{ij} - \sum_{s=1}^{j-1} d_{ss} l_{js} l_{is} \right), \quad i = j+1, \ldots, j+n$$

for all $j = 1, \ldots, n$. The factorization gives an efficient solution procedure of a linear system. For example, the Newton direction $\mathbf{s}_k = -\mathbf{H}_k^{-1}\mathbf{g}_k$ is the solution of the system

$$\mathbf{H}_k \mathbf{s}_k = -\mathbf{g}_k, \tag{4.57}$$

which is equivalent to solving the system

$$\begin{aligned} \mathbf{L}_k \mathbf{p}_k &= -\mathbf{g}_k, \\ \mathbf{L}_k^T \mathbf{s}_k &= \mathbf{D}_k^{-1} \mathbf{p}_k, \end{aligned} \tag{4.58}$$

where $\mathbf{H}_k = \mathbf{L}_k \mathbf{D}_k \mathbf{L}_k^T$. The Cholesky factorization is well defined for positive-definite matrices but not for indefinite ones. The *modified Cholesky factorization* by Gill and Murray requires construction of a modified matrix $\overline{\mathbf{H}}$ decomposable to $\overline{\mathbf{D}}$ and $\overline{\mathbf{L}}$ so that all elements of the factors are uniformly bounded and all elements of $\overline{\mathbf{D}}$ are positive:

$$\overline{d}_{jj} > \delta; \quad \left| l_{ij}^2 \overline{d}_{jj} \right| \le \beta^2 \quad \text{for } i > j, \tag{4.59}$$

where δ is a small positive number and β is estimated from $\beta^2 = \max\{|d_{jj}|, h_{jj}/(n^2-1)^{0.5}, \varepsilon_M\}$, with ε_M being the machine precision. The rationale for this choice is elaborated further in the cited source.

The construction of $\overline{\mathbf{H}}$ proceeds column by column. Assuming the $j-1$ columns have been computed, for the jth column we compute

$$\overline{d}_{jj} = \max\{|d_{jj}|, \delta\}. \tag{4.60}$$

Using this in place of d_{jj}, we check the second of the (4.59) conditions. If it is satisfied, the l_{ij}s of the jth column are kept as calculated. If not, then d_{jj} is corrected as

$$\overline{d}_{jj} = h_{jj} + e_{jj} - \sum_{s=1}^{j-1} d_{ss} l_{js}^2, \tag{4.61}$$

where the positive scalar e_{jj} is chosen so that $\max |l_{ij}^2 \overline{d}_{jj}| = \beta^2, i > j$. When the process is completed, we get a matrix

$$\overline{\mathbf{H}}_k = \overline{\mathbf{L}}_k \overline{\mathbf{D}}_k \overline{\mathbf{L}}_k^T = \mathbf{H}_k + \mathbf{E}_k \tag{4.62}$$

where $\overline{\mathbf{H}}_k$ is positive-definite and $\mathbf{E}_k$ is a nonnegative diagonal matrix with elements e_{jj}. These will have either the values required for (4.61) or the values $2|d_{jj}|$ or $|d_{jj}|+\delta$ depending on the choice in (4.60).

Given the modified factorization (4.62), it can be shown that a *direction of negative curvature*, that is, a direction such that $\partial\mathbf{x}^T\mathbf{H}\partial\mathbf{x} < 0$, at a saddlepoint can be found by solving the system

$$\overline{\mathbf{L}}_k^T \partial\mathbf{x}_k = \mathbf{e}_s, \tag{4.63}$$

where $\mathbf{e}_s$ is the unit vector of the sth coordinate. This coordinate corresponds to the smallest of the quantities $(\overline{d}_{jj} - e_{jj})$ above. For an indefinite matrix, the value of $(-\overline{d}_{ss} - e_{ss})$ will be negative and the solution of (4.63) will give a descent vector $\partial\mathbf{x}_k$.

Example 4.21 Consider the function $f = 2x_1^2 - 4x_1x_2 + 1.5x_2^2 + x_2$, which is a quadratic of the form $f = \frac{1}{2}\mathbf{x}^T\mathbf{A}\mathbf{x} + x_2$ with

$$\mathbf{A} = \begin{pmatrix} 4 & -4 \\ -4 & 3 \end{pmatrix},$$

so that $\mathbf{A}$ is also the Hessian of f and is indefinite (see Exercise 4.10). Applying the factorization above, we have

$$\begin{aligned} j=1\colon\ & d_{11} = h_{11} = 4, l_{11} = 1, l_{21} = d_{11}^{-1}h_{21} = -1, \\ & \beta^2 = \max\{4, 4/\sqrt{3}, \ldots\} = 4, \\ & l_{21}^2 d_{11} \le \beta^2 \text{ implies } 4 \le 4 \text{ (i.e., satisfied)}, \\ j=2\colon\ & d_{22} = h_{22} - d_{11}l_{21}^2 = -1, \overline{d}_{22} = 1, \\ & l_{12} = 0, l_{22} = 1\ (i > 2 \text{ does not exist}). \end{aligned}$$

Thus,

$$\begin{aligned} \overline{\mathbf{A}} = \overline{\mathbf{L}}\overline{\mathbf{D}}\overline{\mathbf{L}}^T &= \begin{pmatrix} 1 & 0 \\ -1 & 1 \end{pmatrix}\begin{pmatrix} 4 & 0 \\ 0 & 1 \end{pmatrix}\begin{pmatrix} 1 & -1 \\ 0 & 1 \end{pmatrix} = \begin{pmatrix} 4 & -4 \\ -4 & 5 \end{pmatrix} \\ &= \begin{pmatrix} 4 & -4 \\ -4 & 3 \end{pmatrix} + \begin{pmatrix} 0 & 0 \\ 0 & 2 \end{pmatrix} = \mathbf{A} + \mathbf{E}, \end{aligned}$$

where $\mathbf{E}$ has $e_{22} = 2|d_{22}|$. From (4.63) we get

$$\begin{pmatrix} 1 & -1 \\ 0 & 1 \end{pmatrix}\begin{pmatrix} \partial x_1 \\ \partial x_2 \end{pmatrix} = \begin{pmatrix} 0 \\ 1 \end{pmatrix}$$

since $(\overline{d}_{ss} - e_{ss}) = (\overline{d}_{22} - e_{22}) = (1 - 2) = -1$. Solving for $\partial\mathbf{x}$, we get $\partial\mathbf{x} = (1, 1)^T$. The reader may compare this with results from Exercise 4.10. ∎

The modification (4.62) may sometimes result in substantial modification to the Hessian, and this could have a bad influence on the actual performance of the method (Fletcher 1980).

4.8 Trust Regions

The algorithms discussed in the previous sections perform two distinct tasks during every iteration: They find a search direction $\mathbf{s}_k$, and they determine how far to move in the direction $\mathbf{s}_k$ by performing a line search. The discussion focused mostly on the quality of the search direction. What happens during the line search may also have significant influence on the performance of the algorithm. We will discuss line search types in Chapter 7; here we want to show that thinking about how far we can step in a direction leads to a different class of algorithms.

Moving with Trust

Of the two popular search directions, gradient/steepest descent and Newton, the Newton direction is generally considered to be more powerful and faster. However, this statement is made under the assumption that few line searches are needed. If the search length $\|\alpha(\mathbf{s}_k)\|$ is really short or if the Hessian at $\mathbf{x}_k$ is very different from the Hessian at $\mathbf{x}_*$, then the steepest descent direction may be a better direction to search.

This can be seen by examining Figures 4.6 and 4.7. The contour lines of the function

$$f(x_1, x_2) = (x_1 - 2)^2 + (x_1 - 2)^2 x_2^2 + (x_2 + 1)^2 \tag{4.64}$$

are drawn in Figure 4.6. Rays representing the steepest descent direction and the Newton direction are drawn from point A [1.7, −0.7], near the solution, and B [1.5, 1.5] further away. In Figure 4.7 the objective function is plotted along the two different directions. Near point A (Figure 4.7a), the steepest descent direction is slightly better

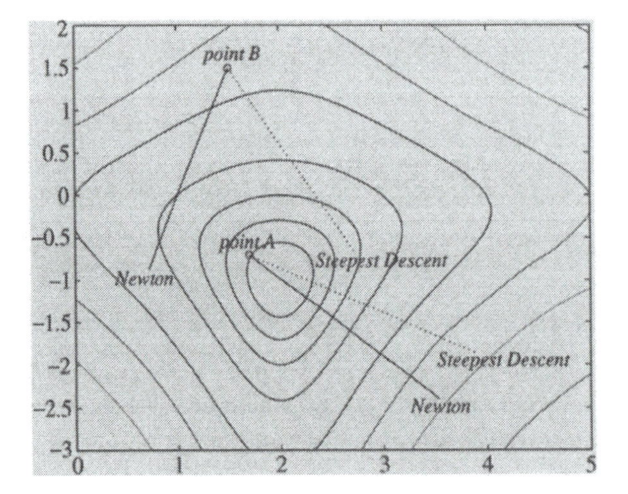

Figure 4.6. Contour lines of Equation (4.64).

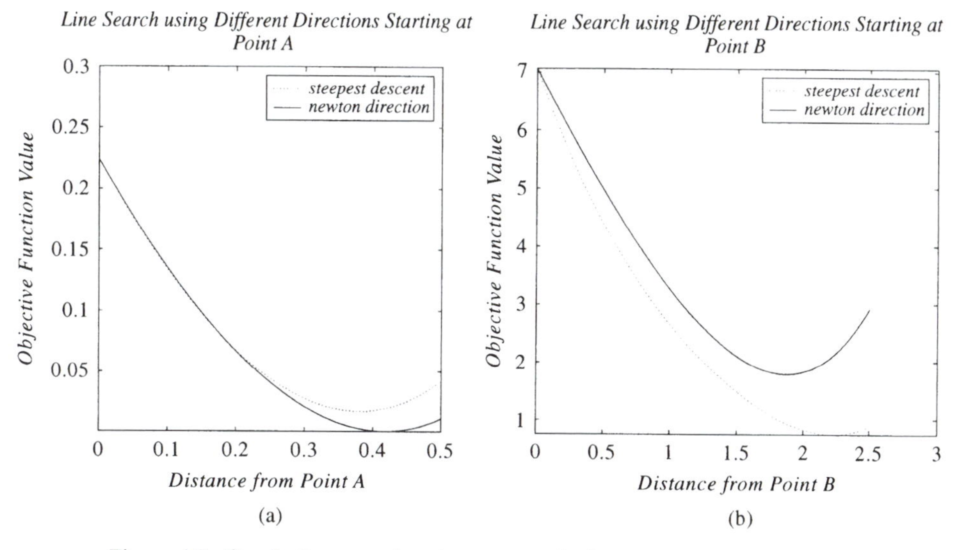

Figure 4.7. Quadratic approximation may not be better.

than the Newton direction, that is, $f(\mathbf{x}_A + \alpha\mathbf{s}_{\text{gradient}}) < f(\mathbf{x}_A + \alpha\mathbf{s}_{\text{newton}})$ for small values of α. In the region around point A the function is nearly quadratic. As we move away from A the Newton direction provides a better function value and a new point that is closer to $\mathbf{x}_*$. Near point B (Figure 4.7b) the objective function deviates from a quadratic. As we move away the steepest descent direction is better than the Newton. A subsequent line search would yield a much better point.

Experience with line search algorithms shows that line searches are typically much shorter during the early iterations, when the iterates are far from the solution. As the algorithm progresses, if the function is approximately quadratic in the neighborhood of the solution, the line search eventually becomes unnecessary, namely, the step size will be always $\alpha = 1$.

These insights should motivate the idea that, when we select a certain local approximation form for a function, say quadratic, we do not need to perform a line search as long as we stay within an area where we trust our approximation, the *trust region*.

Trust Region Algorithm

A trust region radius, denoted Δ, is defined as a limit on the size of a step we may take in a given direction of search. We can use a quadratic model, but we restrict the search problem so that the step we take will always stay within the trust region. Formally, we can achieve this by solving the constrained minimization problem

$$\begin{aligned} &\min f \approx \mathbf{g}_k^T\mathbf{s} + \tfrac{1}{2}\mathbf{s}^T\mathbf{H}_k\mathbf{s} \\ &\text{subject to } \|\mathbf{s}\| \leq \Delta \end{aligned} \tag{4.65}$$

to calculate the step $\mathbf{s}_k$. The inequality $\|\mathbf{s}\| \leq \Delta$ implies two possibilities. If $\|\mathbf{s}\| < \Delta$ then $\mathbf{s}_k$ is given by the equation

$$\mathbf{s}_k = -\mathbf{H}_k\mathbf{g}_k \quad \text{if } \|\mathbf{H}_k\mathbf{g}_k\| < \Delta. \tag{4.66}$$

If $\|\mathbf{s}_k\| = \Delta$ then $\mathbf{s}_k$ is given by the system

$$\mathbf{s}_k = -(\mathbf{H}_k + \mu\mathbf{I})\mathbf{g}_k, \quad \mu > 0, \text{ and } \|\mathbf{s}_k\| = \Delta. \tag{4.67}$$

The additional quantity μ is associated with the trust region constraint. (It is the *Lagrange multiplier*, defined in Chapter 5.)

The first trust region algorithms (Levenberg 1944, Marquardt 1963) ignore Δ and manipulate μ directly to ensure that $\mathbf{H}_k + \mu\mathbf{I}$ is positive definite. Conceptually, if an estimate of μ is given a priori, it is easier to solve the matrix equation $\mathbf{s}_k = -(\mathbf{H}_k + \mu\mathbf{I})\mathbf{g}_k$ than it is to solve (4.65). However, here we will focus on the direct use of a trust region that allows an intuitive extension to trust region algorithms for constrained problems, which will be discussed in Section 7.8.

In a typical trust region algorithm a "full step" $\mathbf{s}_k$ is performed and it is accepted only if $\mathbf{x}_k + \mathbf{s}_k$ is a better point than $\mathbf{x}_k$, that is, if $f(\mathbf{x}_k + \mathbf{s}_k) < f(\mathbf{x}_k)$. If $\mathbf{x}_k + \mathbf{s}_k$ is not a better point, then the trust region is made smaller, so that $\mathbf{s}$ becomes increasingly more like a steepest descent direction.

As iterations progress, it is necessary to update Δ in a more sophisticated manner than simply shrinking it when steps are unacceptable. To do this, we need a measure of how well (4.65) approximates the original function. We define the ratio

$$r_k = \frac{\text{actual reduction}}{\text{predicted reduction}}, \tag{4.68}$$

where

$$\text{actual reduction} = f(\mathbf{x}_k) - f(\mathbf{x}_k + \mathbf{s}_k), \tag{4.69}$$

$$\text{predicted reduction} = -(\mathbf{g}_k^T \mathbf{s}_k + \mathbf{s}_k^T \mathbf{H}_k \mathbf{s}_k). \tag{4.70}$$

A "good step" occurs when the predicted reduction from the quadratic model is close to the actual reduction experienced (e.g., $r_k > 0.75$). The trust region radius is then increased. A "bad step" would be for $r_k < 0.25$ and the trust region radius would be decreased. A prototype algorithm follows.

TRUST REGION ALGORITHM

1. Start with some point $\mathbf{x}_1$ and a trust region radius $\Delta_1 > 0$. Set the iteration counter $k = 1$.
2. Calculate the gradient $\mathbf{g}_k$ and Hessian $\mathbf{H}_k$ at $\mathbf{x}_k$.
3. Calculate the step $\mathbf{s}_k$ by solving (4.65).
4. Calculate the value $f(\mathbf{x}_k + \mathbf{s}_k)$ along with the predicted reduction, actual reduction, and ratio r_k.
5. (Unacceptable Step) If $f(\mathbf{x}_k + \mathbf{s}_k) \geq f(\mathbf{x}_k)$, then do not accept the new point. Set $\Delta_{k+1} = \Delta_k/2$, $\mathbf{x}_{k+1} = \mathbf{x}_k$, $k = k + 1$ and go to Step 3.
6. (Altering the Trust Region) Set $\mathbf{x}_{k+1} = \mathbf{x}_k + \mathbf{s}_k$. If $r_k < 0.25$, then set $\Delta_{k+1} = \Delta_k/2$; if $r_k > 0.75$, then set $\Delta_{k+1} = 2\Delta_k$; otherwise set $\Delta_{k+1} = \Delta_k$. Set $k = k + 1$ and go to Step 2.

The theoretical framework of trust region algorithms keeps candidate steps bound providing advantages over line search methods. One advantage is the ability to model and use negative curvature. As we saw in the previous section on stabilization, a good search method will force the Hessian approximation to be positive-definite in order to guarantee a descent direction. If the actual problem has negative curvature the algorithms will not work well. In the trust region approach no such restriction is required. Another advantage is in the use of second-order information. If second-order estimates are inaccurate, line search algorithms will use a bad search direction resulting in very small movement, too many iterations, or even convergence failure in the line search. A trust region algorithm will reduce the domain over which the approximate model is believed accurate, effectively decreasing the detrimental effects of inaccurate second-order estimates. This property of the trust region algorithms has spurred much recent research into models that are not based on quadratic approximations.

There are also some disadvantages. When the trust region is too large, a large approximation is usually re-solved after shrinking the trust region, which is more expensive computationally than simply shortening a line search. A large approximate problem (4.65) is also more difficult to solve when the Hessian is possibly indefinite.

4.9 Summary

The theory for interior optima developed in this chapter is the basis for a variety of local iterative techniques. The gradient and Newton's methods are the most obvious and pure methods. The gradient method is generally trustworthy but slow, while Newton's method is fast but temperamental. We examined one modification to Newton's method based on modifying the Cholesky factorization of the Hessian. Other Newton-like methods will be studied in Chapter 7.

Optimality results for unconstrained problems are extended to constrained ones in Chapter 5. The same approach is used for differentiable functions that can be approximated locally by a Taylor series. Hence the mathematical foundations laid in this chapter are useful for the remainder of the book.

Even for problems that do not satisfy the assumptions under which the theory was developed, the concepts presented are still useful and necessary before pursuing more specialized techniques. For example, *nondifferentiable* or *nonsmooth optimization* methods use the concept of a *subgradient* defined for convex functions through a relation analogous to (4.34).

Notes

The material in this chapter is classical optimization theory. A careful and readable introduction to differentiable functions optimization is given by Russell (1970). Close to the presentation here is the one in Wilde and Beightler (1967). The classical text by Hancock (1917) is still the best source for studying some often overseen complexities such as the Genocchi and Peano counterexample of Exercise 4.14.

Several texts on nonlinear programming could serve for additional information. The text by Luenberger (1973, 1984) is particularly well written, while the book by Gill, Murray, and Wright (1981) provides a wealth of information on several subjects. Both these two references, as well as the very readable books by Fletcher (1980, 1981), could be of help for extensions of many topics in the present book, particularly of Chapters 5 and 7.

All the above references were consulted for the presentation of this chapter. Some readers may need to refresh some basic calculus and linear algebra background. The texts by Protter and Morey (1964) and Noble (1969) could serve this purpose.

Exercises

4.1 The Taylor expansion for a multivariable function is a *multiple series* expansion. Its general representation in coordinate form employs a special symbol for *multiple sums* as follows:

$$\sum_{1 \le \sum_{i=1}^{n} r_i \le N} (\,).$$

This operator means that the sum of the terms in parentheses is taken over all possible combinations of r_is that add up to a number between 1 and N, with the r_is all nonnegative. If an equality is used for the summation index, that is, $\sum r_i = N$, it means that only the combinations of r_is adding up to N exactly should be used. The generalization of (4.13) is thus given by

$$f(\mathbf{x}) = f(\mathbf{x}_0) + \sum_{1 \le \sum_{i=1}^{n} r_i \le N} \frac{\partial^{(\sum r_i)} f(\mathbf{x}_0)}{\partial x_1^{r_1} \cdots \partial x_n^{r_n}} \prod_{i=1}^{n} \frac{(x_i - x_{i0})^{r_i}}{r_i!}$$

$$+ o(\|\mathbf{x} - \mathbf{x}_0\|^N).$$

Using this expression, verify (4.13) for $n = 2$ and $N = 2$.

4.2 Rewrite the generalized Taylor series expansion of Exercise 4.1 in terms of perturbations. Verify by direct comparison that an alternative way of representing this generalization is

$$\partial f = \sum_{p=1}^{N} d^p f(\mathbf{x}_0)/p! + o(\|\partial \mathbf{x}\|^N),$$

where the operator d^p is defined by

$$d^{(p)} f(\mathbf{x}) = \underbrace{\sum_{i=1}^{n} \sum_{j=1}^{n} \cdots \sum_{k=1}^{n}}_{p \text{ summations}} \frac{\partial^{(p)} f(\mathbf{x})}{\partial x_i \partial x_j \cdots \partial x_k} \partial x_i \partial x_j \cdots \partial x_k.$$

4.3 Using the results from Exercise 4.2, derive a complete expression for the *third-order* approximation to ∂f, for $n = 2$.

4.4 Show that the Hessian matrix of a separable function is always diagonal.

4.5 Prove the expressions (4.19) for the gradient and Hessian of a general quadratic function. Use them to verify the results in Examples 4.2 and 4.3.

4.6 Using the methods of this chapter, find the minimum of the function

$$f = (1 - x_1)^2 + 100\left(x_2 - x_1^2\right)^2.$$

This is the well-known Rosenbrock's "banana" function, a test function for numerical optimization algorithms.

4.7 Find the global minimum of the function

$$f = 2x_1^2 + x_1x_2 + x_2^2 + x_2x_3 + x_3^2 - 6x_1 - 7x_2 - 8x_3 + 19.$$

4.8 Prove by completing the square that if a function $f(x_1, x_2)$ has a stationary point, then this point is

(a) a local minimum, if

$$\left(\partial^2 f/\partial x_1^2\right)\left(\partial^2 f/\partial x_2^2\right) - \left(\partial^2 f/\partial x_1 \partial x_2\right)^2 > 0 \quad \text{and} \quad \partial^2 f/\partial x_1^2 > 0;$$

(b) a local maximum, if

$$\left(\partial^2 f/\partial x_1^2\right)\left(\partial^2 f/\partial x_2^2\right) - \left(\partial^2 f/\partial x_1 \partial x_2\right)^2 > 0 \quad \text{and} \quad \partial^2 f/\partial x_1^2 < 0;$$

(c) a saddlepoint, if

$$\left(\partial^2 f/\partial x_1^2\right)\left(\partial^2 f/\partial x_2^2\right) - \left(\partial^2 f/\partial x_1 \partial x_2\right)^2 < 0.$$

4.9 Find the nature of the stationary point(s) of

$$f = -4x_1 + 2x_2 + 4x_1^2 - 4x_1x_2 + 101x_2^2 - 200x_1^2x_2 + 100x_1^4.$$

Hint: Try the transformation $\hat{x}_1 = 2x_1 - x_2$ and $\hat{x}_2 = x_1^2 - x_2$.

4.10 Show that the stationary point of the function

$$f = 2x_1^2 - 4x_1x_2 + 1.5x_2^2 + x_2$$

is a saddle. Find the directions of downslopes away from the saddle using the differential quadratic form.

4.11 Find the nature of the stationary points of the function

$$f = x_1^2 + 4x_2^2 + 4x_3^2 + 4x_1x_2 + 4x_1x_3 + 16x_2x_3.$$

4.12 Find the point in the plane $x_1 + 2x_2 + 3x_3 = 1$ in $\Re^3$ that is nearest to the point $(-1, 0, 1)^T$.

4.13 Prove the *Maclaurin Theorem* for functions of one variable stated as follows: If the function $f(x), x \in \mathcal{X} \subseteq \Re$, has an interior stationary point and its lowest order nonvanishing derivative is positive and of *even* order, then the stationary point is a minimum.

4.14 Lagrange (1736–1813) incorrectly stated an extension to Maclaurin's Theorem of Exercise 4.13 using the differential operator $d^P f(\mathbf{x})$ of Exercise 4.2 instead of the derivative, for functions of several variables. Genocchi and Peano gave a counterexample a century later using the function

$$f = \left(x_1 - a_1^2 x_2^2\right)\left(x_1 - a_2^2 x_2^2\right)$$

with a_1 and a_2 constants.

(a) Show that Lagrange's extension identifies a minimum for f at the origin.

(b) Show that the origin is in fact a maximum for all points on the curve $x_1 = \left(\frac{1}{2}\right)(a_1^2 + a_2^2)x_2^2$, if $a_1 \neq a_2$.

4.15 Consider the function

$$f = -x_2 + 2x_1x_2 + x_1^2 + x_2^2 - 3x_1^2x_2 - 2x_1^3 + 2x_1^4.$$

(a) Show that the point $(1, 1)^T$ is stationary and that the Hessian is positive-semidefinite there.

(b) Find a straight line along which the second-order perturbation ∂f is zero.

(c) Examine the sign of third- and fourth-order perturbations along the line found above.

(d) Identify the nature of $(1, 1)^T$ according to Lagrange (Exercise 4.14) and disprove its minimality by calculating the function values at, say, $(0, 0)^T$, $(0, 0.25)^T$.

(e) Make the transformation $\hat{x}_1 = x_1^2 - x_2$ and $\hat{x}_2 = 2x_1^2 - 2x_1 - x_2 + 1$, which gives $f = \hat{x}_1\hat{x}_2$, where $\hat{x}_1, \hat{x}_2$ are unrestricted in sign. Show that f is unbounded below by selecting values of $\hat{x}_1, \hat{x}_2$ that give real values for x_1, x_2.

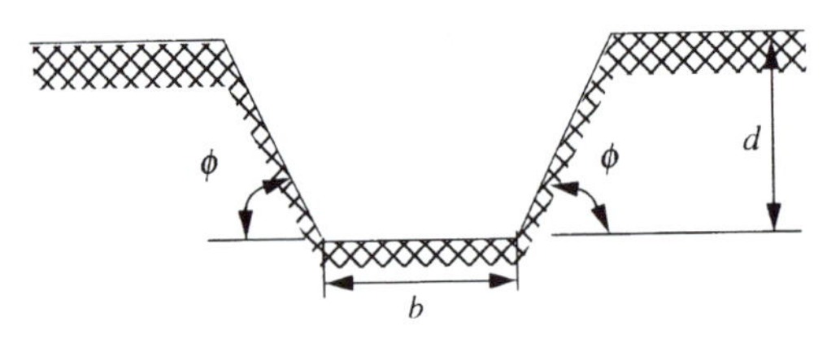

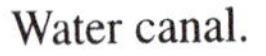
Water canal.

4.16 A water canal with a fixed cross-sectional area must be designed so that its discharge capacity (i.e., flow rate) is maximized (see figure). The design variables are the height d, the width of base b, and the angle of the sides ϕ. It can be shown that the flow rate is proportional to the inverse of the so-called wetted perimeter p, given by $p = b + (2d/\sin\phi)$. The area is easily calculated as $A = db + d^2 \cot\phi$. For a given $A = 100$ ft^2, find and prove the globally optimal design. Can you extend this result for all values of the parameter A? (From Stark and Nicholls 1972.)

4.17 Prove that a hyperplane is a convex set.

4.18 Show that if $\mathbf{x}(\lambda) = \lambda\mathbf{x}_1 + (1 - \lambda)\mathbf{x}_2$ and $F(\lambda) = f[\mathbf{x}(\lambda)]$, then

$$dF/d\lambda = \nabla f[\mathbf{x}(\lambda)](\mathbf{x}_1 - \mathbf{x}_2),$$

$$d^2F/d\lambda^2 = (\mathbf{x}_1 - \mathbf{x}_2)^T(\partial^2 f/\partial\mathbf{x}^2)(\mathbf{x}_1 - \mathbf{x}_2), \quad 0 \leq \lambda \leq 1.$$

4.19 Prove that $f(\mathbf{x})$ is convex, if and only if

$$f(\mathbf{x}) \geq f(\mathbf{x}_0) + \nabla f(\mathbf{x}_0)(\mathbf{x} - \mathbf{x}_0)$$

for all $\mathbf{x}$ and $\mathbf{x}_0$ (in a convex set).
Hint: For the "only if" part, use the definition and examine the case $\lambda \to 0$. For the "if" part, apply the inequality for $\mathbf{x} = \mathbf{x}_1$, $\mathbf{x} = \mathbf{x}_2$ and construct the definition.

4.20 Minimize the function

$$f = x_1^2 + 2x_1x_2 + 4x_1x_3 + 3x_2^2 + 2x_2x_3 + 5x_3^2$$

using the gradient method and Newton's method. Perform a few iterations with each method starting from the point $\mathbf{x}_0 = (1, 1, 1)^T$. Confirm results analytically.

4.21 Minimize the function

$$f = 5x_1^2 + 12x_1x_2 - 16x_1x_3 + 10x_2^2 - 26x_2x_3 + 17x_3^2 - 2x_1 - 4x_2 - 6x_3$$

starting a local exploration from the base case. Apply the gradient method and Newton's method. Compare results for at least a few iterations.

4.22 Show that the least-squares fit for a line $y = a + bt$, with a and b constants, is given by the solution of the system $\mathbf{A}^T\mathbf{A}\mathbf{x} = \mathbf{A}^T\mathbf{y}$ where $\mathbf{y} = (y_1, y_2, \ldots, y_m)^T$, $\mathbf{x} = (a, b)^T$, m is the number of observations, and

$$\mathbf{A} = \begin{bmatrix} 1 & t_1 \\ 1 & t_2 \\ \vdots & \vdots \\ 1 & t_m \end{bmatrix}.$$

4.23 For the function $f = 2x_1^2 - 3x_1x_2 + 8x_2^2 + x_1 - x_2$ find and prove the minimum analytically. Then find explicitly the value α_k that is the solution to an exact line search $\mathbf{x}_{k+1} = \mathbf{x}_k + \alpha_k\mathbf{s}_k$. Perform iterations using $\mathbf{s}_k$ from the gradient and Newton's methods, starting from $\mathbf{x}_0 = (1, 1)^T$. Compare results.

4.24 Apply the gradient method to minimize the function $f = (x_1 - x_2)^2 + (x_2 - 1)^2$ starting at $(0, 0)^T$. Repeat with Newton's method. Compare.

4.25 Minimize the function $f = x_1^2 + x_1^4x_2^2 + x_3^2$. Use analysis. Then make some iterations starting near a solution point and use both the gradient and Newton's method.

4.26 Find the stationary points of the function

$$f = x_1 + x_1^{-1} + x_2 + x_2^{-1}.$$

5

Boundary Optima

The contrary (is) a benefit and from the differences (we find) the most beautiful harmony.
Heraclitus (6th century B.C.)

The minimizer of a function that has an open domain of definition will be an interior point, if it exists. For a differentiable function, the minimizer will be also a stationary point with $\nabla f(\mathbf{x}_*) = \mathbf{0}^T$. The obvious question is what happens if there are constraints and the function has a minimum at the boundary. We saw in Chapter 3 that design models will often have boundary optima because of frequent monotonic behavior of the objective function. Monotonicity analysis can be used to identify active constraints, but this cannot always be done without iterations. Moreover, when equality constraints are present, direct elimination of the constraints and reduction to a form without equalities will be possible only if an explicit solution with respect to enough variables can be found. Thus, we would like to have optimality conditions for constrained problems that can be operationally useful without explicit elimination of constraints. These conditions lead to computational methods for identifying constrained optima. This is the main subject here. As in Chapter 4, the theory in this chapter does not require the assumption of positivity for the variables.

5.1 Feasible Directions

Recalling the iterative process of reaching a minimum in unconstrained problems, $\mathbf{x}_{k+1} = \mathbf{x}_k + \alpha_k \mathbf{s}_k$, we recognize that there was an implicit assumption that $\mathbf{x}_{k+1}$ was still an interior point. However, if the set $\mathcal{X}$ is closed, an iterant $\mathbf{x}_k$ may be on the boundary or close to it. Then there will be directions $\mathbf{s}_k$ and step lengths α_k that could yield an infeasible $\mathbf{x}_{k+1}$. In other words, it is possible to have infeasible perturbations $\partial\mathbf{x}_k = \alpha_k \mathbf{s}_k$. This leads us to the following definitions: Given an $\mathbf{x} \in \mathcal{X}$, a perturbation $\partial\mathbf{x}$ is a *feasible perturbation*, if and only if $\mathbf{x} + \partial\mathbf{x} \in \mathcal{X}$. Similarly, a *feasible direction* $\mathbf{s}$ at $\mathbf{x}$ is defined if and only if there exists a scalar $\alpha_u > 0$, such that $\mathbf{x} + \alpha\mathbf{s} \in \mathcal{X}$ for all $0 \leq \alpha \leq \alpha_u$. From the definition of a local minimum, it is now evident that the necessary condition for a local minimum that may be on the boundary of $\mathcal{X}$ is simply

$$\nabla f(\mathbf{x}_*)\mathbf{s} \geq 0 \quad \text{for all feasible } \mathbf{s}. \tag{5.1}$$

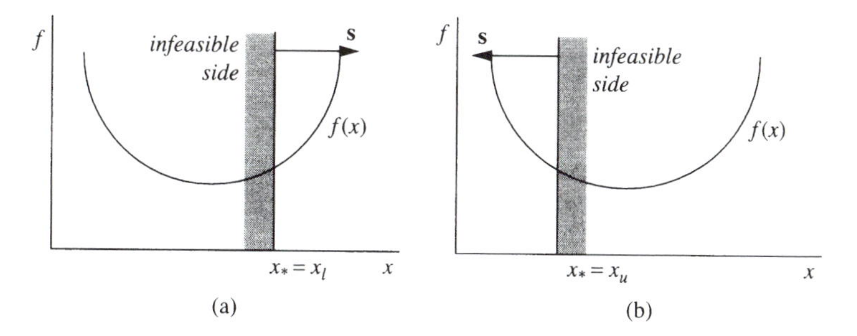

Figure 5.1. Local minima on the boundary (one-dimensional problem).

This condition says that any allowable movements away from $\mathbf{x}_*$ must be in directions of increasing f. In the one-dimensional case, Figure 5.1, we see that at a local boundary minimum the slope of the function there and the allowable change in x must have the same sign. For a local boundary maximum they should have opposite signs. Looking at Figure 5.1(a), we see that $df/dx > 0$ and $x_L - x \leq 0$ must be true at x_L, which is also the minimizer. Locally, the problem is $\{\min f(x^+)$, subject to $g(x^-) = x_L - x \leq 0\}$. The first monotonicity principle applied locally will give $x_* = x_L$, and that is a nice interpretation of the necessary condition.

In two dimensions the geometric meaning of (5.1) is essentially that at a minimum no feasible directions should exist that are at an angle of more than 90° from the gradient direction (Figure 5.2). In the figure, cases (a) and (b) are local minima, but (c) and (d) are not. Note that the gradient vector in the two-dimensional case of $f(x_1, x_2)$ is given by

$$\nabla f = (\partial f/\partial x_1)\mathbf{e}_1 + (\partial f/\partial x_2)\mathbf{e}_2, \tag{5.2}$$

where $\mathbf{e}_1$ and $\mathbf{e}_2$ are unit vectors in the x_1 and x_2 coordinate directions, respectively. In the two-dimensional contour plot representation of $f(x_1, x_2)$, the gradient at a point can be drawn by taking the line tangent to the contour at that point and bringing a perpendicular to the tangent. The direction of ∇f will be the direction of increasing f.

Condition (5.1) implies a fundamental way for constructing iterative procedures to solve constrained problems. If $\mathbf{x}_k$ is a nonoptimal point, a move in the feasible direction $\mathbf{s}_k$ should be made so that $\nabla f(\mathbf{x}_k)\mathbf{s}_k < 0$. The step length α_k is found from solving the problem

$$\min_{0 \leq \alpha < \infty} f(\mathbf{x}_k + \alpha \mathbf{s}_k), \quad \text{subject to } \mathbf{x}_k + \alpha \mathbf{s}_k \in \mathcal{X}. \tag{5.3}$$

Obviously, the major question is how to enforce the feasibility requirement in an operationally useful manner.

Example 5.1 Consider minimizing the function

$$f = (x_1 - 2)^2 + (x_2 - 2)^2 \quad \text{where } (x_1, x_2)^T \text{ belongs to the set } \mathcal{X} \subseteq \Re^n$$

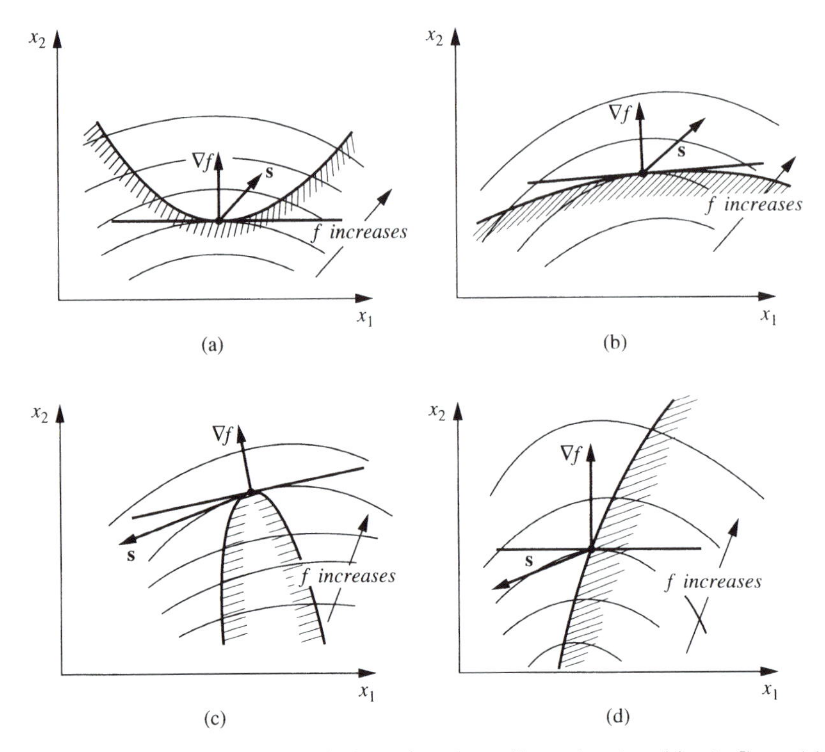

Figure 5.2. Local minima on the boundary (two-dimensional problem). Cases (a) and (b) are minima; cases (c) and (d) are not.

defined by

$$\mathcal{X} = \{\mathbf{x} \mid -x_1 + x_2 \leq 1, 2x_1 + 3x_2 \leq 8, \; x_1 \geq 0, \; x_2 \geq 0\}.$$

The set is shown in Figure 5.3. The unconstrained minimum is at $(2, 2)^T$, an infeasible point since it violates $2x_1 + 3x_2 \leq 8$. From the figure it is seen that for feasibility, $x_2 \leq 2$. This makes f *decreasing* wrt x_2 *in the feasible domain*, and at least one constraint will be active to bound x_2 from above. Evidently the second constraint is active, so the constrained minimum will be found on the line $2x_1 + 3x_2 = 8$. In fact, for $x_1 = (8 - 3x_2)/2$ we get $f = (2 - 1.5x_2)^2 + (x_2 - 2)^2$, which gives $x_{2*} = 1.54$ and, therefore, $x_{1*} = 1.69$. At this point the local optimality condition

$$[\nabla f(1.69, 1.54)][(x_1, x_2)^T - (1.69, 1.54)^T] \geq 0$$

says that the vector $\nabla f(1.69, 1.54) = (-0.62, -0.92)$ must make an acute angle with all vectors $[(x_1 - 1.69), (x_2 - 1.54)]^T$, with x_1, x_2 feasible, which the figure verifies. However, at the point $(4, 0)^T$ we have

$$\begin{aligned}[\nabla f(4, 0)][(x_1, x_2)^T - (4, 0)^T] &= (4, -4)(x_1 - 4, x_2)^T \\ &= 4(x_1 - x_2 - 4) \leq 0 \quad \text{for all feasible } x_1, x_2.\end{aligned}$$

The necessary condition (5.1) for minimality is violated at this obvious maximizer. ∎

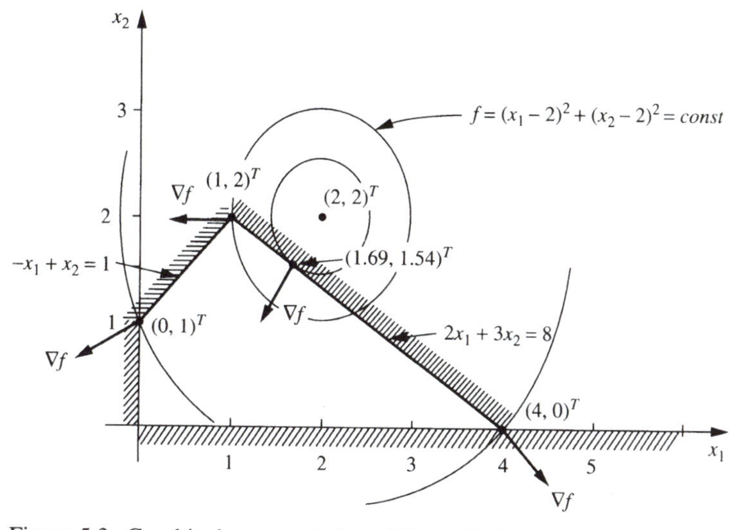

Figure 5.3. Graphical representation of Example 5.1.

5.2 Describing the Constraint Surface

In Chapter 4 we focused on problems of the form {min $f(\mathbf{x})$ subject to $\mathbf{x} \in \mathcal{X} \subseteq \Re^n$}. The constraints were implicitly accounted for in the *set constraint* $\mathcal{X}$. Defining f over an *open* $\mathcal{X}$, we assured that our results held for interior points. Boundary points require the use of a feasible direction vector, but how is such feasibility assured? To derive practical results we need more information about the constraints. So now we will turn our attention to problems of the explicit form

$$\begin{aligned} &\min f(\mathbf{x}) \\ &\text{subject to } \mathbf{h}(\mathbf{x}) = \mathbf{0}, \quad \mathbf{g}(\mathbf{x}) \leq \mathbf{0}, \quad \mathbf{x} \in \mathcal{X} \subseteq \Re^n, \end{aligned} \tag{5.4}$$

where the set constraint will be assumed to have a secondary role, that is, the feasible space described by the equality and inequality constraints will be assumed to be imbedded in the set $\mathcal{X}$. The minimum of $f(\mathbf{x})$ that also satisfies the equality and inequality constraints will be referred to as a *boundary* or *constrained minimum*. For simplicity of presentation, all equality constraints in this chapter will be assumed active. Yet the cautions about inactive equalities in Chapter 3 should be kept in mind.

If the problem has only inequality constraints, it is possible that the minimum is *interior* to the feasible domain, that is, all the constraints are inactive and the usual conditions for an unconstrained minimum would apply. This would be the case in Example 5.1 if the center of the circles representing the objective function was moved inside the convex polyhedron representing the constraints. However, if (active) equality constraints exist, they must be satisfied strictly. Consider m equality constraints that are *functionally independent*, that is, none of them can be derived from the others through algebraic operations. Together they represent a *hypersurface* of dimension $n - m$, which is a subset of $\Re^n$ and on which the minimum of the objective must lie. For example, in a two-dimensional problem, a constraint $h_1(x_1, x_2) = 0$ will represent a curve in the (x_1, x_2) plane on which $\mathbf{x}_*$ must lie. If there is another $h_2(x_1, x_2) = 0$, then $\mathbf{x}_*$ will be simply the only feasible point, namely, the intersection of the two curves (if it exists). In a three-dimensional problem with two constraints, the intersection will be a curve in space on which $\mathbf{x}_*$ must lie.

Regularity

To create a local first-order theory for such problems it would be enough to work with the linear approximation of the hypersurface near a given point, rather than the surface itself. For a scalar function $f: \mathcal{R}^n \to \mathcal{R}$ we did this using the gradient of the function, as indicated by the first-order term of the Taylor expansion. The linear approximation was obtained by using a plane orthogonal to the gradient (i.e., a tangent plane at the point) and checking for a minimum with respect to small displacements along the tangent plane. We would like to be able to do the same for vector functions $\mathbf{h}: \mathcal{R}^n \to \mathcal{R}^m$, used to describe hypersurfaces of the form $\mathbf{h}(\mathbf{x}) = \mathbf{0}$. In other words, if the constraint set is described by

$$\mathcal{S} = \{\mathbf{x} \in \mathcal{X} \subseteq \mathcal{R}^n : h_j(\mathbf{x}) = 0,\ j = 1, \ldots, m\} \tag{5.5}$$

what should its nature be so that we can define tangent planes properly?

To see the difficulty, let us look at the following situations.

(a) The set $\mathcal{S}_1 = \{(x_1, x_2, x_3)^T \in \mathcal{R}^3 : x_1^2 + x_2^2 + x_3^2 - 1 = 0\}$ has $n = 3, m = 1$; so it represents a surface of dimension $3 - 1 = 2$ in $\mathcal{R}^3$. Indeed, it represents a sphere in 3-D space.

(b) The set $\mathcal{S}_2 = \{(x_1, x_2, x_3)^T \in \mathcal{R}^3 : x_1^2 + x_2^2 + x_3^2 - 1 = 0,\ x_1 + x_2 + x_3 - 1 = 0\}$ has $n = 3$, $m = 2$; so it represents a surface of dimension $3 - 2 = 1$, that is, a curve in R^3. Indeed, it represents a circle that is the intersection of a sphere and a plane in 3-D space.

(c) The set $\mathcal{S}_3 = \{\mathbf{x} \in \mathcal{R}^4 : x_1 + x_2 + x_3 + x_4 = 0,\ x_1 - x_2 + x_3 - x_4 = 0,\ x_1 + x_3 = 0\}$ has $n = 4$, $m = 3$; so it should represent a surface of dimension one. However, it actually represents a *plane surface of dimension two*. Notice that the three hyperplanes $\mathbf{a}_j^T \mathbf{x} = 0$, $j = 1, 2, 3$ have normal vectors $\mathbf{a}_1^T = (1, 1, 1, 1)$, $\mathbf{a}_2^T = (1, -1, 1, -1)$, $\mathbf{a}_3^T = (1, 0, 1, 0)$, which are *linearly dependent*. Now suppose that these three hyperplanes actually corresponded to the tangent hyperplanes at a point $\mathbf{x}$ of three hypersurfaces having an intersection containing $\mathbf{x}$. Then a point $\mathbf{y}$ belonging to the *tangent plane of the intersection* at $\mathbf{x}$ would satisfy the relations $\mathbf{a}_j^T \mathbf{y} = 0$, $j = 1, 2, 3$. (This is understood intuitively from the one-dimensional case where the normal to a curve at a point is normal to the tangent at that point.) Now in $\mathcal{R}^4$ the intersection of three hypersurfaces will usually be a one-dimensional curve and so the tangent plane would be a line. However, since the above relations $\mathbf{a}_j^T \mathbf{y} = 0$ are not independent, they give a plane instead, and the tangent "line" is not well defined.

The conclusion of this third example is that the tangent plane of the intersection of hypersurfaces $h_j(\mathbf{x}_0) = 0$ will have the expected $n - m$ dimension if and only if the normals at any point are linearly independent, that is, the gradients $\nabla h_i(\mathbf{x}_0)$ are linearly independent.

Now we will formalize these ideas: A set of equality constraints on $\mathcal{R}^n$, $h_1(\mathbf{x}) = 0$, $h_2(\mathbf{x}) = 0, \ldots, h_m(\mathbf{x}) = 0$, defines a *hypersurface* of dimension $n - m$ if the constraints are functionally independent. This surface is represented by the set $\mathcal{S} = \{\mathbf{x} \in \mathcal{R}^n : \mathbf{h}(\mathbf{x}) = \mathbf{0}\}$. We assume that the functions $h_j(\mathbf{x})$, $j = 1, \ldots, m$ are continuous and differentiable so that the surface is smooth. Each point on this smooth surface has a tangent plane that is defined as containing the derivative of *any* differentiable curve on the surface $\mathcal{S}$, passing through the point (Figure 5.4(a)).

A point $\mathbf{x} \in \mathcal{S}$ is called a *regular* point if and only if the gradient vectors $\nabla h_1(\mathbf{x}), \nabla h_2(\mathbf{x}), \ldots, \nabla h_m(\mathbf{x})$ are linearly independent. The simplifying assumption that any point we considered must be regular is usually referred to as a *constraint qualification.*

Tangent and Normal Hyperplanes

The *normal plane* (or subspace) for $\mathcal{S}$ at a regular point $\mathbf{x}$ is the subspace $\mathcal{N}(\mathbf{x})$ of $\mathcal{R}^n$ spanned by the gradient vectors $\nabla h_j(\mathbf{x})$:

$$\mathcal{N}(\mathbf{x}) = \big\{\mathbf{z} \in \mathcal{R}^n : \mathbf{z} = \alpha_1 \nabla h_1^T(\mathbf{x}) + \cdots + \alpha_m \nabla h_m^T(\mathbf{x}); \\ \alpha_1, \ldots, \alpha_m \in \mathcal{R}\big\}. \tag{5.6}$$

The *tangent plane* (or subspace) for $\mathcal{S}$ at a regular point $\mathbf{x}$ is the subspace $\mathcal{T}(\mathbf{x})$ of $\mathcal{R}^n$, orthogonal to the normal space, that is,

$$\mathcal{T}(\mathbf{x}) = \{\mathbf{y} \in \mathcal{R}^n : \nabla \mathbf{h}(\mathbf{x})\mathbf{y} = \mathbf{0}\}. \tag{5.7}$$

Note that the expression (5.7) is not a definition but rather a representation of the tangent plane at regular points. The condition of regularity is not imposed on the constraint surface itself but on its representation in terms of an $\mathbf{h}(\mathbf{x})$ (Luenberger 1973). Once again we must recognize that the properties of the optimization problem are intimately dependent upon the model we construct to represent the design.

The above definitions of normal and tangent subspaces require that they pass through the origin. It is conceptually better to think that they pass through the point $\mathbf{x}$. This can be affected by a simple translation. Moreover, it can be shown rigorously (Russell 1970) that in the neighborhood of a point $\mathbf{x}_0$ we can represent points $\mathbf{x}_1$ by moving along the tangent and normal spaces, as in Figure 5.4(b).

We have arrived at our goal, which was to develop a modeling machinery for applying local theory to explicitly constrained problems. We described only equality constraints, since any inequality constraints of interest will have to be active at a minimum and presumably tight (i.e., satisfied as equalities).

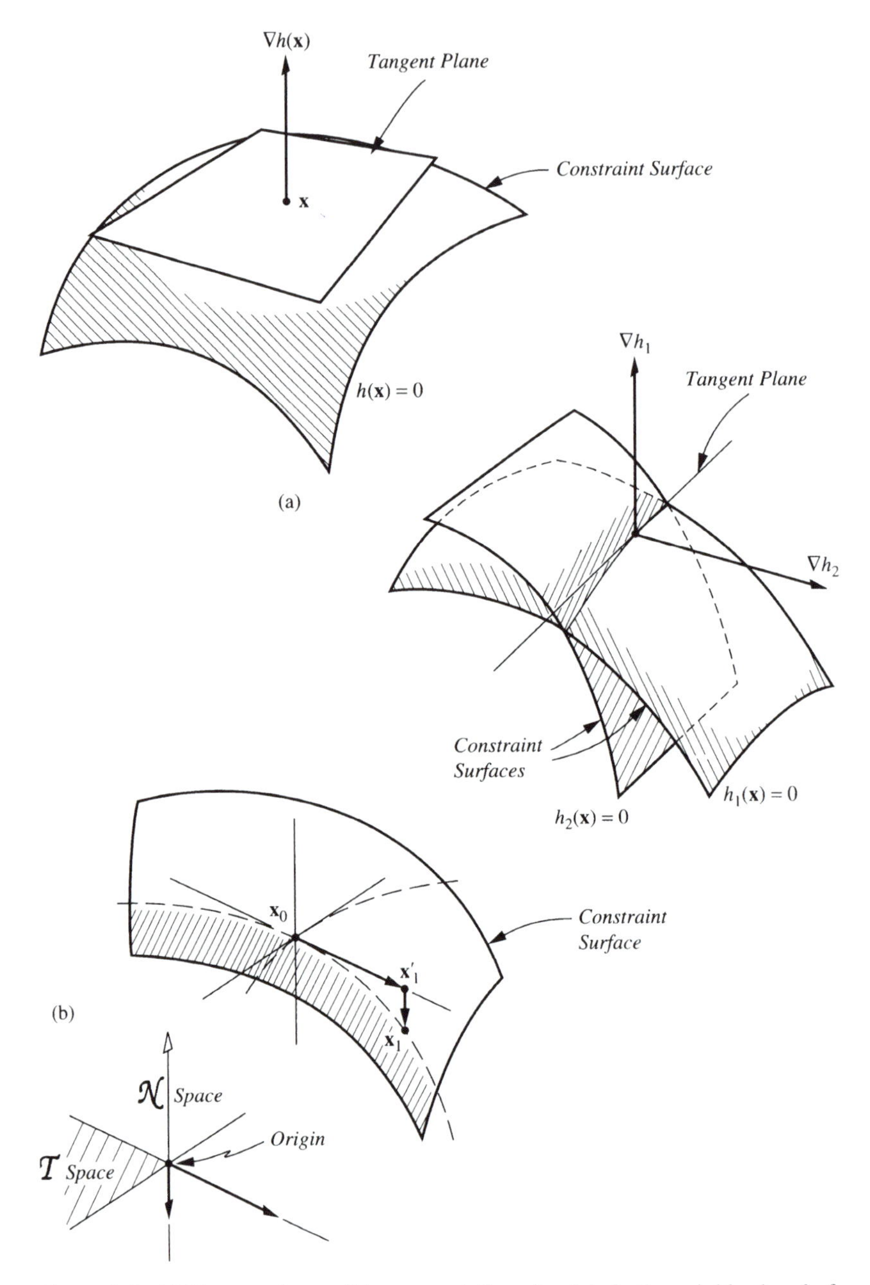

Figure 5.4. (a) Tangent planes; (b) representation of points in the neighborhood of $\mathbf{x}_0$.

5.3 Equality Constraints

The simplest way to deal with a single equality constraint would be to use *direct elimination*: Solve the equality for one of the variables and substitute for it throughout the entire model. In this way, one variable and one constraint would be eliminated from the problem. The degrees of freedom would remain the same but the size would be reduced, with the equality accounted for. When many equalities are present, this method would require an explicit solution of the system of simultaneous equations, giving closed-form expressions for a number of variables (equal to the number of constraints) in terms of the remaining variables. Substitution would then result in a reduced problem. Although this method can be effective, it is too often impossible to apply. Thus, in the case where constraints are too difficult to solve explicitly, some other method must be used.

Reduced (Constrained) Gradient

Local optimality conditions would probably require derivative information to express feasible perturbations about a point. With this in mind, we examine the problem

$$\begin{aligned}&\min f(\mathbf{x}) \\ &\text{subject to } h_j(\mathbf{x}) = 0, \quad j = 1, 2, \ldots, m.\end{aligned} \tag{5.8}$$

Any feasible point $\mathbf{x}$ must satisfy the constraints. A small perturbation $\partial\mathbf{x}$ about this point will result in perturbations ∂h_j of the constraints. Clearly, the point $\mathbf{x} + \partial\mathbf{x}$ will be infeasible if $\partial h_j \neq 0$ for some j. Therefore, feasible perturbations are assured if $\partial h_j = 0$, $j = 1, 2, \ldots, m$.

The first-order approximations of the perturbations for objective and constraint functions are

$$\begin{aligned}\partial f &= \nabla f \partial\mathbf{x} = \sum_{i=1}^{n} (\partial f/\partial x_i)\partial x_i, \\ \partial h_j &= \nabla h_j \partial\mathbf{x} = \sum_{i=1}^{n} (\partial h_j/\partial x_i)\partial x_i = 0, \quad j = 1, 2, \ldots, m.\end{aligned} \tag{5.9}$$

If the values of the derivatives are known at the point $\mathbf{x}$, the above relations give $m + 1$ linear equations in $n + 1$ unknowns. The case of $m > n$ is ruled out by assuming linear independence of the constraint equations. The case $m = n$ is of no interest since the solution is the trivial one, $\partial\mathbf{x} = \mathbf{0}$, meaning $\mathbf{x}$ is the only feasible point that exists. Thus, $m < n$ is assumed, so that the degrees of freedom of the system are $p = n - m$.

We can rearrange (5.9) as follows:

$$\begin{aligned}-\partial f + \sum_{i=1}^{m} (\partial f/\partial x_i)\partial x_i &= -\sum_{i=m+1}^{n} (\partial f/\partial x_i)\partial x_i, \\ \sum_{i=1}^{m} (\partial h_j/\partial x_i)\partial x_i &= -\sum_{i=m+1}^{n} (\partial h_j/\partial x_i)\partial x_i, \quad j = 1, \ldots, m.\end{aligned} \tag{5.10}$$

This simply means that we select m out of n independent variables, set them on the left side, and call them *solution* or *state variables*:

$$s_i \triangleq x_i; \quad i = 1, \ldots, m. \tag{5.11}$$

The remaining variables $(n - m)$ we call *decision variables*:

$$d_i \triangleq x_i \quad i = m+1, \ldots, n \quad \text{or} \quad i = 1, \ldots, p. \tag{5.12}$$

The number of decision variables is equal to the number of degrees of freedom. Decision variables can have arbitrary perturbations ∂d_i while the state variable perturbations ∂s_i must conform to feasibility according to Equations (5.10) rewritten now as

$$-\partial f + \sum_{i=1}^{m} (\partial f/\partial s_i)\partial s_i = -\sum_{i=1}^{p} (\partial f/\partial d_i)\partial d_i, \tag{5.13}$$
$$\sum_{i=1}^{m} (\partial h_j/\partial s_i)\partial s_i = -\sum_{i=1}^{p} (\partial h_j/\partial d_i)\partial d_i, \quad j = 1, \ldots, m.$$

In vector notation, each of (5.13) is written as

$$-\partial f + (\partial f/\partial \mathbf{s})\partial \mathbf{s} = -(\partial f/\partial \mathbf{d})\partial \mathbf{d}, \tag{5.14a}$$
$$(\partial \mathbf{h}/\partial \mathbf{s})\partial \mathbf{s} = -(\partial \mathbf{h}/\partial \mathbf{d})\partial \mathbf{d}, \tag{5.14b}$$

where $\partial \mathbf{h}/\partial \mathbf{s}$ and $\partial \mathbf{h}/\partial \mathbf{d}$ are the $m \times m$ and $m \times p$ Jacobian matrices of the vector function $\mathbf{h}(\mathbf{x})$. Now assuming that $\partial \mathbf{h}/\partial \mathbf{s}$ was constructed so that the gradients $\partial h_j/\partial \mathbf{s}$, $j = 1, \ldots, m$ are linearly independent, we can use its inverse to solve (5.14b) and substitute in (5.14a):

$$\partial \mathbf{s} = -(\partial \mathbf{h}/\partial \mathbf{s})^{-1}(\partial \mathbf{h}/\partial \mathbf{d})\partial \mathbf{d}, \tag{5.15a}$$
$$\begin{aligned} \partial f &= (\partial f/\partial \mathbf{d})\partial \mathbf{d} + (\partial f/\partial \mathbf{s})\partial \mathbf{s} \\ &= [(\partial f/\partial \mathbf{d}) - (\partial f/\partial \mathbf{s})(\partial \mathbf{h}/\partial \mathbf{s})^{-1}(\partial \mathbf{h}/\partial \mathbf{d})]\partial \mathbf{d}. \end{aligned} \tag{5.15b}$$

The quantity in brackets can be thought of as the gradient of a new *unconstrained* function $z(\mathbf{d})$, which would be equivalent to the original objective function f if the solution variables had been eliminated. Thus, we can define a quantity

$$\partial z/\partial \mathbf{d} \triangleq (\partial f/\partial \mathbf{d}) - (\partial f/\partial \mathbf{s})(\partial \mathbf{h}/\partial \mathbf{s})^{-1}(\partial \mathbf{h}/\partial \mathbf{d}), \tag{5.16}$$

which we call the *constrained* or *reduced gradient* of the function f. Its components are called *constrained derivatives*. Then we write

$$\partial z \equiv \partial f = (\partial z/\partial \mathbf{d})\partial \mathbf{d} \tag{5.17}$$

to denote perturbations of the objective function that are also feasible. The feasible domain of z is in the p-dimensional space; the function z is considered unconstrained since we assume $\mathbf{d}$ to be an interior point of the p-dimensional feasible domain. Thus, the obvious condition for a (constrained) stationary point $\mathbf{x}_\dagger = (\mathbf{d}_\dagger, \mathbf{s}_\dagger)^T$ is that

$$(\partial z/\partial \mathbf{d})_\dagger = \mathbf{0}^T, \tag{5.18}$$

that is, the constrained derivatives must all vanish. This result suggests that the theory of unconstrained local optima can be applied to equality constrained optima, if we work in the reduced space. If $\mathbf{x}_\dagger$ is a regular point, then the construction of constrained derivatives is guaranteed.

Lagrange Multipliers

The necessary condition (5.18) for the reduced gradient to vanish at a stationary point can be rewritten as

$$\left(\frac{\partial f}{\partial \mathbf{d}}\right)_\dagger - \left(\frac{\partial f}{\partial \mathbf{s}}\right)_\dagger \left(\frac{\partial \mathbf{h}}{\partial \mathbf{s}}\right)_\dagger^{-1} \left(\frac{\partial \mathbf{h}}{\partial \mathbf{d}}\right)_\dagger = \mathbf{0}^T, \tag{5.19}$$

with all quantities being evaluated at $\mathbf{x}_\dagger$. Now we may define

$$\boldsymbol{\lambda}^T \triangleq -\left(\frac{\partial f}{\partial \mathbf{s}}\right)_\dagger \left(\frac{\partial \mathbf{h}}{\partial \mathbf{s}}\right)_\dagger^{-1}. \tag{5.20}$$

With some rearrangement, (5.19) and (5.20) are written as

$$\left(\frac{\partial f}{\partial \mathbf{d}}\right)_\dagger + \boldsymbol{\lambda}^T \left(\frac{\partial \mathbf{h}}{\partial \mathbf{d}}\right)_\dagger = \mathbf{0}^T, \tag{5.21}$$

$$\left(\frac{\partial f}{\partial \mathbf{s}}\right)_\dagger + \boldsymbol{\lambda}^T \left(\frac{\partial \mathbf{h}}{\partial \mathbf{s}}\right)_\dagger = \mathbf{0}^T.$$

From (5.21) it is clear that the gradients $(\partial h_j/\partial \mathbf{s})$ that represent the rows of $(\partial \mathbf{h}/\partial \mathbf{s})$ must be linearly independent. Thus, we expect that the values of $\boldsymbol{\lambda}^T$ at $\mathbf{x}_\dagger$ are uniquely defined. Then (5.21) will have a uniquely defined $\boldsymbol{\lambda}^T$ solution, if the gradients $(\partial h_j/\partial \mathbf{d})$ representing the rows of $(\partial \mathbf{h}/\partial \mathbf{d})_\dagger$ are also linearly independent. In other words, the point $\mathbf{x}_\dagger$ is assumed to be a regular point. The stationarity conditions (5.21) can now be rewritten in terms of the m-vector $\boldsymbol{\lambda}$ and the original vector $\mathbf{x} = (\mathbf{d}, \mathbf{s})^T$ as

$$\nabla f(\mathbf{x}_\dagger) + \boldsymbol{\lambda}^T \nabla \mathbf{h}(\mathbf{x}_\dagger) = \mathbf{0}^T. \tag{5.22}$$

This relation shows that the *necessary condition* for a minimum is that *the gradient of the objective must be a linear combination of the gradients of the constraints, at this minimizing point.* Notice that there is no restriction on the sign of the components of the vector $\boldsymbol{\lambda}$. The significance of $\boldsymbol{\lambda}$ will be discussed in Section 5.9. It is emphasized, however, that Equation (5.20) defines $\boldsymbol{\lambda}$ only at constrained stationary points.

The stationarity condition (5.22) is often expressed in terms of a special function, the *Lagrangian* function defined by

$$L(\mathbf{x}, \boldsymbol{\lambda}) \triangleq f(\mathbf{x}) + \boldsymbol{\lambda}^T \mathbf{h}(\mathbf{x}), \tag{5.23}$$

where $\boldsymbol{\lambda}$ is a vector of new variables called *Lagrange multipliers*. This name should be reserved for the values of $\boldsymbol{\lambda}$ that satisfy (5.22), in which case the definition will coincide with the one in (5.20). It is easy to verify that (5.22) and $\mathbf{h}(\mathbf{x}) = \mathbf{0}$ comprise the *stationarity conditions of the Lagrangian*. Values of $\boldsymbol{\lambda}$ other than those satisfying stationarity should be considered *estimates* of the Lagrange multipliers. This distinction is not always observed in the literature. The use of multiplier estimates in numerical procedures will be examined in Chapter 7. Note that sometimes the multipliers are defined with a sign opposite to the one in Equation (5.20), in which case the Lagrangian is also defined by

$$L(\mathbf{x}, \boldsymbol{\lambda}) = f(\mathbf{x}) - \boldsymbol{\lambda}^T \mathbf{h}(\mathbf{x}). \tag{5.23$'$}$$

We will see in Section 5.6 that the choice of sign convention is related to whether we use the positive or negative (null) form for the model (see also Section 1.2).

Example 5.2 Consider the problem

$$\begin{aligned} &\min f(x_1, x_2) = (x_1 - 2)^2 + (x_2 - 2)^2 \\ &\text{subject to } h_1(x_1, x_2) = x_1^2 + x_2^2 - 1 = 0, \end{aligned}$$

where $x_i > 0$. The function f represents a cone with a minimum (apex) at $(2, 2)^T$ and the constraint is a circle centered at $(0, 0)^T$ with radius 1. Let us define $x_1 = s_1$ and $x_2 = d_1$ and calculate

$$\begin{aligned} \partial f/\partial \mathbf{s} &= \partial f/\partial s_1 = 2(s_1 - 2), \\ \partial f/\partial \mathbf{d} &= \partial f/\partial d_1 = 2(d_1 - 2), \\ \partial \mathbf{h}/\partial \mathbf{s} &= \partial h_1/\partial s_1 = 2s_1, \\ \partial \mathbf{h}/\partial \mathbf{d} &= \partial h_1/\partial d_1 = 2d_1, \\ \partial z/\partial \mathbf{d} &= (\partial f/\partial d_1) - (\partial f/\partial s_1)(\partial h_1/\partial s_1)^{-1}(\partial h_1/\partial d_1) \\ &= 2(d_1 - 2) - 2(s_1 - 2)(2s_1)^{-1}(2d_1) = -4 + 4d_1 s_1^{-1}. \end{aligned}$$

The constrained stationary point is found from setting $\partial z/\partial \mathbf{d} = \mathbf{0}^T$ and combining with the equality constraint to solve the system

$$-4 + 4d_1 s_1^{-1} = 0, \qquad d_1^2 + s_1^2 = 1.$$

This gives $s_1 = d_1 = 1/\sqrt{2} = 0.707$, the intersection of the line $s_1 = d_1 (x_1 = x_2)$ with the circle $d_1^2 + s_1^2 = 1 (x_1^2 + x_2^2 = 1)$. If we use the stationarity condition (5.22) in terms of the original variables, we get the system

$$\begin{aligned} &(2x_1 - 4, 2x_2 - 4) + \lambda_1(2x_1, 2x_2) = (0, 0), \\ &x_1^2 + x_2^2 - 1 = 0, \end{aligned}$$

which is solved for $x_1 = x_2 = 1/\sqrt{2}$, $\lambda_1 = 2\sqrt{2} - 1$. The two methods give the same result. The point is indeed a *global* minimum, although the local stationarity condition alone does not guarantee that. ■

Example 5.3 Consider the problem

$$\begin{aligned} &\min f = x_1^2 + x_2^2 - x_3^2 \\ &\text{subject to } h_1 = 5x_1^2 + 4x_2^2 + x_3^2 - 20 = 0, \\ &\qquad\qquad h_2 = x_1 + x_2 - x_3 = 0, \end{aligned}$$

for which we have

$$\partial f/\partial \mathbf{x} = (2x_1, 2x_2, -2x_3),$$
$$\partial \mathbf{h}/\partial \mathbf{x} = \begin{pmatrix} 10x_1 & 8x_2 & 2x_3 \\ 1 & 1 & -1 \end{pmatrix}.$$

To apply the reduced gradient method we define $x_1 = s_1$, $x_2 = s_2$, and $x_3 = d_1$, and therefore we have from Equation (5.16)

$$\begin{aligned}
\partial z/\partial \mathbf{d} &= \partial f/\partial d_1 - \begin{pmatrix} \partial f/\partial s_1 \\ \partial f/\partial s_2 \end{pmatrix}^T \begin{pmatrix} \partial h_1/\partial s_1 & \partial h_1/\partial s_2 \\ \partial h_2/\partial s_1 & \partial h_2/\partial s_2 \end{pmatrix}^{-1} \begin{pmatrix} \partial h_1/\partial d_1 \\ \partial h_2/\partial d_1 \end{pmatrix} \\
&= -2d_1 - (2s_1, 2s_2) \begin{pmatrix} 10s_1 & 8s_2 \\ 1 & 1 \end{pmatrix}^{-1} \begin{pmatrix} 2d_1 \\ -1 \end{pmatrix} \\
&= -2d_1 - (10s_1 - 8s_2)^{-1}(2s_1, 2s_2) \begin{pmatrix} 1 & -8s_2 \\ -1 & 10s_1 \end{pmatrix} \begin{pmatrix} 2d_1 \\ -1 \end{pmatrix} \\
&= -2d_1 - \frac{4d_1(s_1 - s_2) - 4s_1 s_2}{10s_1 - 8s_2}.
\end{aligned}$$

Setting $\partial z/\partial \mathbf{d} = 0$ and combining with the constraints, we arrive at the system

$$\begin{aligned}
&s_1 s_2/(6s_1 - 5s_2) = d_1, \\
&s_1 + s_2 = d_1, \\
&5s_1^2 + 4s_2^2 = 20 - d_1^2,
\end{aligned}$$

which in general would be solved iteratively. Here, combining the first two equations, we find the explicit relation $s_1/s_2 = \pm(\frac{5}{6})^{1/2} = \pm 0.9129$. Eliminating d_1 from the third equation, we get $(12 \pm 2.1908)s_1^2 = 20$ with the following solutions: $s_1 = \pm 1.1872$, $s_1 = \pm 1.4279$. Thus there are four constrained stationary points:

$$\begin{aligned}
&(1.1872, 1.30, 2.4872)^T, \quad (1.4279, -1.5641, -0.1362)^T, \\
&(-1.1872, -1.30, -2.4872)^T, \quad (-1.4279, 1.5641, 0.1362)^T.
\end{aligned}$$

Although we can calculate the function values at these points, the smallest one may not necessarily be the minimizer. To determine that, we would need second-order information in the reduced space, as we will describe in the next section.

Using the $\boldsymbol{\lambda}$ approach of Equation (5.22), we get

$$\begin{pmatrix} 2x_1 \\ 2x_2 \\ -2x_3 \end{pmatrix}^T + (\lambda_1, \lambda_2) \begin{pmatrix} 10x_1 & 8x_2 & 2x_3 \\ 1 & 1 & -1 \end{pmatrix} = \begin{pmatrix} 0 \\ 0 \\ 0 \end{pmatrix}^T,$$

or

$$\begin{aligned}
&2x_1 + 10\lambda_1 x_1 + \lambda_2 = 0, \\
&2x_2 + 8\lambda_1 x_2 + \lambda_2 = 0, \\
&-2x_3 + 2\lambda_1 x_3 - \lambda_2 = 0,
\end{aligned}$$

which together with $h_1 = 0$, $h_2 = 0$ give a system of five equations in five unknowns. The solutions obtained should be the same as those above. ■

5.4 Curvature at the Boundary

For an unconstrained function the nature of the stationary points is determined by the curvature at that point as expressed by the Hessian matrix. The same should be possible for a constrained stationary point, where now not only the curvature of the function but also that of the active constraints would be important. As was the case with the gradient, some form of a *constrained Hessian* should be determined in the reduced space of the decision variables $\mathbf{d}$.

Constrained Hessian

In the following discussion, the matrix symbol $(\partial^2\mathbf{y}/\partial\mathbf{x}^2)$ means a vector of Hessians

$$(\partial^2\mathbf{y}/\partial\mathbf{x}^2) \triangleq \left(\frac{\partial^2 y_1}{\partial\mathbf{x}^2}, \ldots, \frac{\partial^2 y_m}{\partial\mathbf{x}^2}\right)^T \quad \text{for } \mathbf{y} = (y_1, \ldots, y_m)^T. \tag{5.24}$$

The use of this notation is shown below:

$$(\partial f/\partial\mathbf{s})(\partial^2\mathbf{s}/\partial\mathbf{x}^2) = \sum_{i=1}^{m}(\partial f/\partial s_i)(\partial^2 s_i/\partial\mathbf{x}^2).$$

Recall that the reduced gradient is given by

$$(\partial z/\partial\mathbf{d}) = (\partial f/\partial\mathbf{d}) + (\partial f/\partial\mathbf{s})(\partial\mathbf{s}/\partial\mathbf{d}), \tag{5.25}$$

where $\partial\mathbf{s}/\partial\mathbf{d} = -(\partial\mathbf{h}/\partial\mathbf{s})^{-1}(\partial\mathbf{h}/\partial\mathbf{d})$. Differentiating (5.25) wrt $\mathbf{d}$, we have

$$\begin{aligned}
\frac{\partial^2 z}{\partial\mathbf{d}^2} &= \frac{\partial}{\partial\mathbf{d}}\left(\frac{\partial z}{\partial\mathbf{d}}\right)^T = \frac{\partial}{\partial\mathbf{d}}\left(\frac{\partial f}{\partial\mathbf{d}}\right)^T + \frac{\partial}{\partial\mathbf{d}}\left[\left(\frac{\partial f}{\partial\mathbf{s}}\right)\left(\frac{\partial\mathbf{s}}{\partial\mathbf{d}}\right)\right]^T \\
&= \frac{\partial^2 f}{\partial\mathbf{d}^2} + \left(\frac{\partial^2 f}{\partial\mathbf{d}\,\partial\mathbf{s}}\right)\left(\frac{\partial\mathbf{s}}{\partial\mathbf{d}}\right) + \left(\frac{\partial\mathbf{s}}{\partial\mathbf{d}}\right)^T\left[\frac{\partial}{\partial\mathbf{d}}\left(\frac{\partial f}{\partial\mathbf{s}}\right)^T\right] \\
&\quad + \left(\frac{\partial f}{\partial\mathbf{s}}\right)\left[\frac{\partial}{\partial\mathbf{d}}\left(\frac{\partial\mathbf{s}}{\partial\mathbf{d}}\right)^T\right]^T \\
&= \frac{\partial^2 f}{\partial\mathbf{d}^2} + \left(\frac{\partial^2 f}{\partial\mathbf{d}\,\partial\mathbf{s}}\right)\left(\frac{\partial\mathbf{s}}{\partial\mathbf{d}}\right) + \left(\frac{\partial\mathbf{s}}{\partial\mathbf{d}}\right)^T \\
&\quad \times\left[\frac{\partial^2 f}{\partial\mathbf{s}\,\partial\mathbf{d}} + \left(\frac{\partial^2 f}{\partial\mathbf{s}^2}\right)\left(\frac{\partial\mathbf{s}}{\partial\mathbf{d}}\right)\right] + \left(\frac{\partial f}{\partial\mathbf{s}}\right)\left(\frac{\partial^2\mathbf{s}}{\partial\mathbf{d}^2}\right).
\end{aligned}$$

So the matrix of second constrained derivatives is given by

$$\frac{\partial^2 z}{\partial \mathbf{d}^2} = \frac{\partial^2 f}{\partial \mathbf{d}^2} + \left(\frac{\partial^2 f}{\partial \mathbf{d}\,\partial \mathbf{s}}\right)\left(\frac{\partial \mathbf{s}}{\partial \mathbf{d}}\right) + \left(\frac{\partial \mathbf{s}}{\partial \mathbf{d}}\right)^T \left(\frac{\partial^2 f}{\partial \mathbf{s}\,\partial \mathbf{d}}\right)$$
$$+ \left(\frac{\partial \mathbf{s}}{\partial \mathbf{d}}\right)^T \left(\frac{\partial^2 f}{\partial \mathbf{s}^2}\right)\left(\frac{\partial \mathbf{s}}{\partial \mathbf{d}}\right) + \left(\frac{\partial f}{\partial \mathbf{s}}\right)\left(\frac{\partial^2 \mathbf{s}}{\partial \mathbf{d}^2}\right). \tag{5.26}$$

In this expression all quantities are computable except $\partial^2\mathbf{s}/\partial\mathbf{d}^2$, which is still unknown. An expression for this matrix can be found by setting the second-order perturbation of the constraints equal to zero, as required by feasibility:

$$\frac{\partial}{\partial \mathbf{d}}\left(\frac{\partial \mathbf{h}}{\partial \mathbf{d}}\right)^T = \frac{\partial}{\partial \mathbf{d}}\left[\left(\frac{\partial \mathbf{h}}{\partial \mathbf{d}}\right) + \left(\frac{\partial \mathbf{h}}{\partial \mathbf{s}}\right)\left(\frac{\partial \mathbf{s}}{\partial \mathbf{d}}\right)\right]^T = \mathbf{0}. \tag{5.27}$$

The left-hand side is exactly the same as the one we used to derive (5.26), with f replaced by $\mathbf{h}$. Therefore, we easily arrive at the equation

$$\frac{\partial^2 \mathbf{h}}{\partial \mathbf{d}^2} + \left(\frac{\partial^2 \mathbf{h}}{\partial \mathbf{d}\,\partial \mathbf{s}}\right)\left(\frac{\partial \mathbf{s}}{\partial \mathbf{d}}\right) + \left(\frac{\partial \mathbf{s}}{\partial \mathbf{d}}\right)^T \left(\frac{\partial^2 \mathbf{h}}{\partial \mathbf{d}\,\partial \mathbf{s}}\right)^T$$
$$+ \left(\frac{\partial \mathbf{s}}{\partial \mathbf{d}}\right)^T \left(\frac{\partial^2 \mathbf{h}}{\partial \mathbf{s}^2}\right)\left(\frac{\partial \mathbf{s}}{\partial \mathbf{d}}\right) + \left(\frac{\partial \mathbf{h}}{\partial \mathbf{s}}\right)\left(\frac{\partial^2 \mathbf{s}}{\partial \mathbf{d}^2}\right) = \mathbf{0}. \tag{5.28}$$

Equation (5.28) can be solved for $(\partial^2\mathbf{s}/\partial\mathbf{d}^2)$ in terms of computable quantities.

We can obtain somewhat more compact expressions by observing that (5.26) can be written as

$$\frac{\partial^2 z}{\partial \mathbf{d}^2} = (\mathbf{I}, (\partial\mathbf{s}/\partial\mathbf{d})^T)\begin{pmatrix} \dfrac{\partial^2 f}{\partial \mathbf{d}^2} & \dfrac{\partial^2 f}{\partial \mathbf{d}\,\partial \mathbf{s}} \\ \dfrac{\partial^2 f}{\partial \mathbf{s}\,\partial \mathbf{d}} & \dfrac{\partial^2 f}{\partial \mathbf{s}^2} \end{pmatrix}\begin{pmatrix} \mathrm{I} \\ \dfrac{\partial \mathbf{s}}{\partial \mathbf{d}} \end{pmatrix} + \left(\frac{\partial f}{\partial \mathbf{s}}\right)\left(\frac{\partial^2 \mathbf{s}}{\partial \mathbf{d}^2}\right),$$

or

$$\frac{\partial^2 z}{\partial \mathbf{d}^2} = (\mathbf{I}, (\partial\mathbf{s}/\partial\mathbf{d})^T) f_{\mathbf{xx}}(\mathbf{I}, (\partial\mathbf{s}/\partial\mathbf{d}))^T + (\partial f/\partial\mathbf{s})(\partial^2\mathbf{s}/\partial\mathbf{d}^2), \tag{5.29}$$

where $f_{\mathbf{xx}}$ is the Hessian of $f(\mathbf{x})$ partitioned in terms of decision and state variables, namely,

$$f_{\mathbf{xx}} = \begin{pmatrix} \partial^2 f/\partial\mathbf{d}^2 & \partial^2 f/\partial\mathbf{d}\,\partial\mathbf{s} \\ \partial^2 f/\partial\mathbf{s}\,\partial\mathbf{d} & \partial^2 f/\partial\mathbf{s}^2 \end{pmatrix}, \tag{5.30}$$

and $\mathbf{I}$ is the identity matrix. Note that the mixed-variable Hessians are not square in general and that $\partial^2 f/\partial\mathbf{s}\,\partial\mathbf{d} = (\partial^2 f/\partial\mathbf{d}\,\partial\mathbf{s})^T$. This can be verified by viewing them as Jacobians of the gradient vector functions in column form (recall Example 4.5). With this notation the calculation of the constrained Hessian in the $\mathbf{d}$-space is summarized as follows:

1. Calculate $\partial \mathbf{s}/\partial \mathbf{d}$ by solving the system

$$(\partial \mathbf{h}/\partial \mathbf{s})(\partial \mathbf{s}/\partial \mathbf{d}) = -(\partial \mathbf{h}/\partial \mathbf{d}). \tag{5.31}$$

2. Calculate $\partial^2 \mathbf{s}/\partial \mathbf{d}^2$ by solving the system

$$(\partial \mathbf{h}/\partial \mathbf{s})(\partial^2 \mathbf{s}/\partial \mathbf{d}^2) = -(\mathbf{I}, (\partial \mathbf{s}/\partial \mathbf{d})^T)\mathbf{h}_{\mathbf{xx}}(\mathbf{I}, (\partial \mathbf{s}/\partial \mathbf{d}))^T. \tag{5.32}$$

3. Calculate $\partial^2 z/\partial \mathbf{d}^2$ from (5.29).

The above calculations can always be performed at any given point provided that the Jacobian $\partial \mathbf{h}/\partial \mathbf{s}$ has full rank. This requirement would be satisfied at regular points. The matrix $\mathbf{h}_{\mathbf{xx}}$ is assumed partitioned in the same way as $f_{\mathbf{xx}}$ in (5.30).

Example 5.4 We may test the point $(s_1, d_1)_\dagger^T = (0.707, 0.707)^T$ for Example 5.2. The calculations are very easy to perform because no matrix inversions are required; we will find first symbolic expressions and then substitute numerical values.

1. Calculate the Jacobian of the solution function $\mathbf{s}(\mathbf{d})$ from (5.31):

$$\partial \mathbf{s}/\partial \mathbf{d} = -(\partial \mathbf{h}/\partial \mathbf{s})^{-1}(\partial \mathbf{h}/\partial \mathbf{d}) = -s_1^{-1} d.$$

2. Calculate the Hessian of $\mathbf{s}(\mathbf{d})$ from (5.32):

$$\partial^2 \mathbf{s}/\partial \mathbf{d}^2 = -(\partial \mathbf{h}/\partial \mathbf{s})^{-1} \left[(1, -d_1/s_1) \begin{pmatrix} \frac{\partial^2 h_1}{\partial d_1^2} & \frac{\partial^2 h_1}{\partial d_1 \partial s_1} \\ \frac{\partial^2 h_1}{\partial s_1 \partial d_1} & \frac{\partial^2 h_1}{\partial s_1^2} \end{pmatrix} \begin{pmatrix} 1 \\ -d_1/s_1 \end{pmatrix} \right]$$

$$= -(1/2s_1)(1, -d_1/s_1) \begin{pmatrix} 2 & 0 \\ 0 & 2 \end{pmatrix} \begin{pmatrix} 1 \\ -d_1/s_1 \end{pmatrix} = -(s_1^2 + d_1^2)/s_1^3.$$

3. Calculate the constrained derivative from (5.29):

$$\frac{\partial^2 z}{\partial d_1^2} = (1, -d_1/s_1) \begin{pmatrix} \frac{\partial^2 f}{\partial d_1^2} & \frac{\partial^2 f}{\partial d_1 \partial s_1} \\ \frac{\partial^2 f}{\partial s_1 \partial d_1} & \frac{\partial^2 f}{\partial s_1^2} \end{pmatrix} \begin{pmatrix} 1 \\ -\frac{d_1}{s_1} \end{pmatrix} - \frac{\partial f}{\partial s_1} \left(\frac{s_1^2 + d_1^2}{s_1^3} \right)$$

$$= \left[2 + 2d_1^2 s_1^{-2}\right] - 2(s_1 - 2)\left(s_1^2 + d_1^2\right)s_1^{-3}$$

$$= 4\left(s_1^2 + d_1^2\right)s_1^{-3}.$$

At the point $d_{1\dagger} = s_{1\dagger} = 0.707$, $\partial^2 z/\partial d_1^2 = 11.315 > 0$ and so we have a constrained minimum. If $\partial^2 z/\partial d_1^2$ is positive everywhere, the point is a global minimum (try to prove it in Exercise 5.21). ∎

Second-Order Sufficiency

The example demonstrated an application of the more general *sufficiency condition* for a constrained minimum:

A feasible point $\mathbf{x}_* = (\mathbf{d}_*, \mathbf{s}_*)^T$ *that satisfies the conditions* $(\partial z/\partial \mathbf{d})_* = \mathbf{0}^T$ *and* $\partial \mathbf{d}^T(\partial^2 z/\partial \mathbf{d}^2)_* \partial \mathbf{d} > 0$ *is a local constrained minimum.*

Another statement of sufficiency useful both theoretically and computationally can be reached through the Lagrangian. To derive the alternate expression we will use the shorthand symbols

$$\mathbf{S_d} = \partial \mathbf{s}/\partial \mathbf{d}, \quad L_{\mathbf{dd}} = \partial^2 L/\partial \mathbf{d}^2, \quad L_{\mathbf{ds}} = \partial^2 L/\partial \mathbf{d}\, \partial \mathbf{s}$$

in the same spirit as in previous derivations.

We start by solving (5.32) for $\partial^2 \mathbf{s}/\partial \mathbf{d}^2$ and substituting in (5.29):

$$\begin{aligned} \frac{\partial^2 z}{\partial \mathbf{d}^2} &= (\mathbf{I}, \mathbf{S_d}^T)\, f_{\mathbf{xx}}\, (\mathbf{I}, \mathbf{S_d})^T - \left(\frac{\partial f}{\partial \mathbf{s}}\right)\left(\frac{\partial \mathbf{h}}{\partial \mathbf{s}}\right)^{-1} \left[(\mathbf{I}, \mathbf{S_d}^T)\mathbf{h_{xx}}(\mathbf{I}, \mathbf{S_d})^T\right] \\ &= (\mathbf{I}, \mathbf{S_d}^T)\, f_{\mathbf{xx}}(\mathbf{I}, \mathbf{S_d})^T + \boldsymbol{\lambda}^T\left[(\mathbf{I}, \mathbf{S_d}^T)\mathbf{h_{xx}}(\mathbf{I}, \mathbf{S_d})^T\right] \\ &= (\mathbf{I}, \mathbf{S_d}^T)(f_{\mathbf{xx}} + \boldsymbol{\lambda}^T \mathbf{h_{xx}})(\mathbf{I}, \mathbf{S_d})^T, \end{aligned} \tag{5.33}$$

where $\boldsymbol{\lambda}^T$ was substituted from (5.20). Noting that $L_{\mathbf{xx}} = f_{\mathbf{xx}} + \boldsymbol{\lambda}^T \mathbf{h_{xx}}$, we create a partition similar to (5.30), which now gives

$$\begin{aligned} \frac{\partial^2 z}{\partial \mathbf{d}^2} &= (\mathbf{I}, \mathbf{S_d}^T)\begin{pmatrix} L_{\mathbf{dd}} & L_{\mathbf{ds}} \\ L_{\mathbf{sd}} & L_{\mathbf{ss}} \end{pmatrix}\begin{pmatrix} \mathbf{I} \\ \mathbf{S_d} \end{pmatrix} \\ &= L_{\mathbf{dd}} + \mathbf{S_d}^T L_{\mathbf{sd}} + L_{\mathbf{ds}}\mathbf{S_d} + \mathbf{S_d}^T L_{\mathbf{ss}}\mathbf{S_d}. \end{aligned} \tag{5.34}$$

Now we develop the quadratic form

$$\begin{aligned} \partial \mathbf{d}^T(\partial^2 z/\partial \mathbf{d}^2)\partial \mathbf{d} &= \partial \mathbf{d}^T\left(L_{\mathbf{dd}} + \mathbf{S_d}^T L_{\mathbf{sd}} + L_{\mathbf{ds}}\mathbf{S_d} + \mathbf{S_d}^T L_{\mathbf{ss}}\mathbf{S_d}\right)\partial \mathbf{d} \\ &= \partial \mathbf{d}^T L_{\mathbf{dd}}\partial \mathbf{d} + \partial \mathbf{s}^T L_{\mathbf{sd}}\partial \mathbf{d} + \partial \mathbf{d}^T L_{\mathbf{ds}}\partial \mathbf{s} + \partial \mathbf{s}^T L_{\mathbf{ss}}\partial \mathbf{s} \\ &= (\partial \mathbf{d}^T, \partial \mathbf{s}^T)\begin{pmatrix} L_{\mathbf{dd}} & L_{\mathbf{ds}} \\ L_{\mathbf{sd}} & L_{\mathbf{ss}} \end{pmatrix}\begin{pmatrix} \partial \mathbf{d} \\ \partial \mathbf{s} \end{pmatrix} = \partial \mathbf{x}^T L_{\mathbf{xx}}\partial \mathbf{x}, \end{aligned} \tag{5.35}$$

where $\partial \mathbf{s} = \mathbf{S_d}\partial \mathbf{d}$ was used.

A very elegant result has been discovered, namely, that the differential quadratic form of the reduced function is equal to the differential quadratic form of the Lagrangian. This form can be evaluated without variable partitioning. Moreover, the perturbations $\partial \mathbf{d}$ and $\partial \mathbf{s}$ applied to the evaluation of $\partial^2 z/\partial \mathbf{d}^2$ conform to the requirement of maintaining feasibility, that is, $(\partial \mathbf{h}/\partial \mathbf{d})\partial \mathbf{d} + (\partial \mathbf{h}/\partial \mathbf{s})\partial \mathbf{s} = \mathbf{0}$, which implies $\nabla \mathbf{h}\partial \mathbf{x} = \mathbf{0}$ in the $\mathbf{x}$-space. In other words, the perturbations $\partial \mathbf{x}$ in (5.35) are taken *only on the tangent plane*, as given in (5.7). This very important result allows a restatement of the sufficiency conditions:

If a feasible point $\mathbf{x}_*$ *exists together with a vector* $\boldsymbol{\lambda}$ *such that* $\nabla f(\mathbf{x}_*) + \boldsymbol{\lambda}^T \nabla \mathbf{h}(\mathbf{x}_*) = \mathbf{0}$ *and the Hessian of the Lagrangian with respect to* $\mathbf{x}$ *is positive definite on the subspace tangent to* $\mathbf{h}(\mathbf{x})$ *at* $\mathbf{x}_*$, *then* $\mathbf{x}_*$ *is a local constrained minimum.*

Note that the sufficiency condition requires only the calculation, at $\mathbf{x}_*$, of the form

$$\partial\mathbf{x}^T(\partial^2 L/\partial\mathbf{x}^2)\partial\mathbf{x} = \partial\mathbf{x}^T(\partial^2 f/\partial\mathbf{x}^2)\partial\mathbf{x} + \boldsymbol{\lambda}^T\partial\mathbf{x}^T(\partial^2\mathbf{h}/\partial\mathbf{x}^2)\partial\mathbf{x} \tag{5.36}$$

for $(\partial\mathbf{h}/\partial\mathbf{x})_*\partial\mathbf{x} = \mathbf{0}$. This is a weaker condition than requiring positive-definiteness of $\partial^2 L/\partial\mathbf{x}^2$ for all $\partial\mathbf{x}$. In the above condition we assume as usual that the point $\mathbf{x}_*$ is regular and that all equality constraints are active. Zero values for some multipliers could pose a problem (see Section 5.6).

The Lagrangian formulation will often offer an advantage in algorithmic theory, but not necessarily in actual computations.

Example 5.5 Recall the problem in Example 5.2:

$$\begin{aligned}&\min f = (x_1 - 2)^2 + (x_2 - 2)^2\\&\text{subject to } h_1 = x_1^2 + x_2^2 - 1 = 0,\end{aligned}$$

for which the Lagrangian function is

$$L = (x_1 - 2)^2 + (x_2 - 2)^2 + \lambda_1\left(x_1^2 + x_2^2 - 1\right)$$

and the stationarity conditions for the Lagrangian are

$$\begin{aligned}\partial L/\partial x_1 &= (2x_1 - 4) + \lambda_1(2x_1) = 0,\\ \partial L/\partial x_2 &= (2x_2 - 4) + \lambda_1(2x_2) = 0,\\ \partial L/\partial\lambda_1 &= x_1^2 + x_2^2 - 1 = 0.\end{aligned}$$

Their solution was found to be $(x_1, x_2, \lambda_1)^T = (0.707, 0.707, 1.828)^T$. The Hessian of the Lagrangian wrt $\mathbf{x}$ is

$$L_{\mathbf{xx}} = \begin{pmatrix}2 & 0\\ 0 & 2\end{pmatrix} + \lambda_1\begin{pmatrix}2 & 0\\ 0 & 2\end{pmatrix} = \begin{pmatrix}2 + 2\lambda_1 & 0\\ 0 & 2 + 2\lambda_1\end{pmatrix}.$$

Any value of $\lambda_1 > -1$ and any $\partial\mathbf{x}$ (including those on the tangent subspace of h_1) will give a positive-definite matrix. ■

Example 5.6 Consider the problem with $x_i > 0$:

$$\begin{aligned}&\max f = x_1^2 x_2 + x_2^2 x_3 + x_1 x_3^2\\&\text{subject to } h = x_1^2 + x_2^2 + x_3^2 - 3 = 0.\end{aligned}$$

The stationarity conditions for the Lagrangian are

$$\begin{aligned}\partial L/\partial x_1 &= x_3^2 + 2x_1x_2 + 2x_1\lambda = 0,\\ \partial L/\partial x_2 &= x_1^2 + 2x_2x_3 + 2x_2\lambda = 0,\\ \partial L/\partial x_3 &= x_2^2 + 2x_1x_3 + 2x_3\lambda = 0,\\ \partial L/\partial\lambda &= x_1^2 + x_2^2 + x_3^2 - 3 = 0,\end{aligned}$$

The symmetry of the problem implies a possible solution with $x_1 = x_2 = x_3$. Making use of that, we find easily $x_1 = -2\lambda/3$, $\lambda = \pm\frac{3}{2}$. Since the problem asks for $x_i > 0$, we select $\lambda = -\frac{3}{2}$ and $\mathbf{x}_\dagger = (1, 1, 1)^T$. The Hessian of the Lagrangian wrt $\mathbf{x}$ is given by

$$L_{\mathbf{xx}} = 2\begin{pmatrix} x_2+\lambda & x_1 & x_3 \\ x_1 & x_3+\lambda & x_2 \\ x_3 & x_2 & x_1+\lambda \end{pmatrix}.$$

At the above selected stationary point the differential quadratic form is

$$\left(\tfrac{1}{2}\right)\partial\mathbf{x}^T L_{\mathbf{xx}}\partial\mathbf{x} = \partial\mathbf{x}^T \begin{pmatrix} -\frac{1}{2} & 1 & 1 \\ 1 & -\frac{1}{2} & 1 \\ 1 & 1 & -\frac{1}{2} \end{pmatrix}\partial\mathbf{x}$$
$$= -\tfrac{1}{2}\left(\partial x_1^2 + \partial x_2^2 + \partial x_3^2\right) + 2(\partial x_1\partial x_2 + \partial x_1\partial x_3 + \partial x_2\partial x_3).$$

Perturbations on the plane tangent to h are found from

$$\nabla h\partial\mathbf{x} = (2x_1, 2x_2, 2x_3)(\partial x_1, \partial x_2, \partial x_3)^T = 0,$$

or $x_1\partial x_1 + x_2\partial x_2 + x_3\partial x_3 = 0$, which at $\mathbf{x}_\dagger$ gives $\partial x_1 = -\partial x_2 - \partial x_3$. Substituting this in the above expression of the quadratic form and reducing the results to a sum of squares, we get

$$\left(\tfrac{1}{2}\right)\partial\mathbf{x}^T L_{\mathbf{xx}}\partial\mathbf{x} = -3\left[(\partial x_2 + \partial x_3/2)^2 + 3\partial x_3^2/4\right],$$

which means that $L_{\mathbf{xx}}$, on the plane $\nabla h\partial\mathbf{x} = 0$, is negative-definite and the point $(1, 1, 1)^T$ is a local maximum. It is left as an exercise to see if the negative stationary point $(-1, -1, -1)^T$ would be a minimum (Exercise 5.21). ■

Bordered Hessians

There is a simple test to determine if the matrix $L_{\mathbf{xx}}$ is positive-definite on the tangent subspace (Luenberger 1984). We form a *"bordered" Hessian* matrix

$$\mathbf{B} \triangleq \begin{bmatrix} \mathbf{0} & \nabla\mathbf{h} \\ \nabla\mathbf{h}^T & L_{\mathbf{xx}} \end{bmatrix}. \tag{5.37}$$

Then we compute the sign of the last $n - m$ principal minors of $\mathbf{B}$, where n and m are the numbers of variables and active constraints, respectively. The matrix $L_{\mathbf{xx}}$ is positive-definite on $\nabla\mathbf{h}\partial\mathbf{x} = \mathbf{0}$, if and only if all these minors have sign $(-1)^m$. The proof of this result can be found in the reference cited above.

Example 5.7 Consider the problem (Luenberger 1984)

$$\min f = x_1 + x_2^2 + x_2x_3 + 2x_3^2$$
$$\text{subject to } h = \tfrac{1}{2}\left(x_1^2 + x_2^2 + x_3^2\right) - \tfrac{1}{2} = 0.$$

Here $\mathbf{x}_\dagger = (1, 0, 0)^T$ and $\lambda = -1$, as can be found from the stationarity conditions. At that point

$$\nabla h = (1, 0, 0), \quad L_{\mathbf{xx}} = \begin{pmatrix} -1 & 0 & 0 \\ 0 & 1 & 1 \\ 0 & 1 & 3 \end{pmatrix},$$

so that the bordered Hessian is

$$\mathbf{B} = \left(\begin{array}{c|ccc} 0 & 1 & 0 & 0 \\ \hline 1 & -1 & 0 & 0 \\ 0 & 0 & 1 & 1 \\ 0 & 0 & 1 & 3 \end{array}\right).$$

Checking the last $3 - 1 = 2$ principal minors, we find that they have

$$\det \begin{pmatrix} 0 & 1 & 0 \\ 1 & -1 & 0 \\ 0 & 0 & 1 \end{pmatrix} = -1, \quad \det(\mathbf{B}) = -2.$$

Thus both have $(-1)^1 = -1$ sign, and $L_{\mathbf{xx}}$ will be positive-definite on the tangent subspace. ■

5.5 Feasible Iterations

In Section 5.1 we introduced the idea of feasible perturbations about a point $\mathbf{x}$ on a constraint surface and in Section 5.3 we developed a theory for maintaining first-order feasibility (i.e., satisfaction of linearized constraints). In the present section we will examine two minimization methods that generate feasible perturbations while decreasing the objective function at the same time. The first method uses the reduced space approach described in the previous sections; the second employs projections on the subspace tangent to the constraint surface. Both methods are now classical and have been successfully implemented in algorithmic procedures (see also Chapter 7).

Generalized Reduced Gradient Method

Applying the optimality conditions in the reduced space of the decision variables will encounter in practice the same difficulties as for unconstrained problems. Direct solution of $\partial z/\partial \mathbf{d} = \mathbf{0}^T$ may be impossible and local explorations in the reduced space will be necessary. Moreover, whereas in the unconstrained case we generously assumed that iterant $\mathbf{x}_k$s will remain feasible, in the presence of equalities any movement in the $\mathbf{d}$-space must be accompanied by adjustments in the $\mathbf{s}$-space. This can be accomplished by solving the constraints $\mathbf{h}(\mathbf{d}, \mathbf{s}) = \mathbf{0}$ for $\mathbf{s}$ given a $\mathbf{d}$.

Let us be more concrete and consider a gradient iteration with respect to the decision variables

$$\mathbf{d}_{k+1} = \mathbf{d}_k - \alpha_k(\partial z/\partial \mathbf{d})_k^T. \tag{5.38}$$

The corresponding state variables can be found from (5.15a):

$$\begin{aligned} \mathbf{s}'_{k+1} &= \mathbf{s}_k - (\partial \mathbf{h}/\partial \mathbf{s})_k^{-1}(\partial \mathbf{h}/\partial \mathbf{d})_k \partial \mathbf{d}_k \\ &= \mathbf{s}_k + \alpha_k (\partial \mathbf{h}/\partial \mathbf{s})_k^{-1}(\partial \mathbf{h}/\partial \mathbf{d})_k (\partial z/\partial \mathbf{d})_k^T. \end{aligned} \tag{5.39}$$

This calculation is based on the linearization of the constraints and it will not satisfy the constraints exactly unless they are all linear. However, a solution to the nonlinear system

$$\mathbf{h}(\mathbf{d}_{k+1}, \mathbf{s}_{k+1}) = \mathbf{0}, \tag{5.40}$$

given $\mathbf{d}_{k+1}$, can be found iteratively using $\mathbf{s}'_{k+1}$ as an initial guess. Linearizing (5.40), for $\mathbf{d}_{k+1}$ fixed, we get the following "inner" iteration

$$[\mathbf{s}_{k+1}]_{j+1} = \left[\mathbf{s}_{k+1} - (\partial \mathbf{h}/\partial \mathbf{s})_{k+1}^{-1} \mathbf{h}(\mathbf{d}_{k+1}, \mathbf{s}_{k+1})\right]_j, \tag{5.41}$$

where $\mathbf{s}_{k+1} = \mathbf{s}'_{k+1}$ for $j = 0$. In most circumstances this procedure will converge and the new feasible point $\mathbf{x}_{k+1} = (\mathbf{d}_{k+1}, \mathbf{s}_{k+1})^T$ will be determined.

This general strategy is used in a class of local iterative methods under the name *generalized reduced gradient* (GRG) methods, with the term "generalized" attached historically because the first methods of this type were implemented for linear constraints. A typical GRG iteration is shown in Figure 5.5. A move $(\mathbf{d}_{k+1} - \mathbf{d}_k)$ is

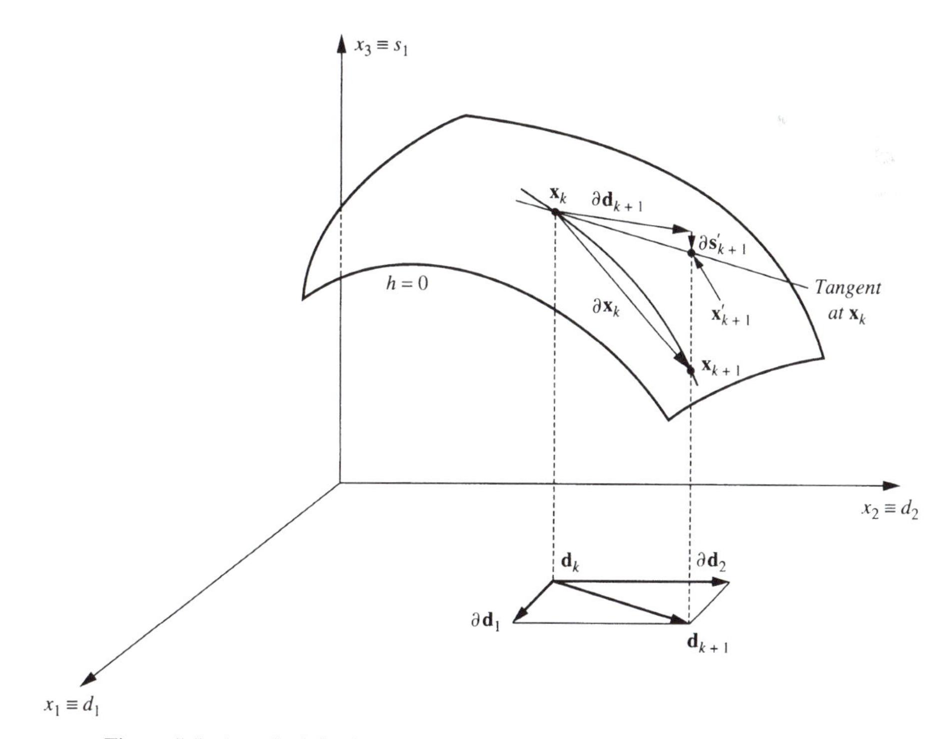

Figure 5.5. A typical GRG move; in three dimensions and for one constraint surface, there will be two decision variables $\mathbf{d} = (d_1, d_2)^T$ and one state variable s_1; the surface $h = 0$ may be also an active inequality $g \leqq 0$.

performed in the reduced decision space to decrease the objective function, through an iteration such as (5.38). Once the **d**-space move is complete, an automatic adjustment in the **s**-space is also enforced because the reduced gradient was constructed so that first-order feasibility is maintained, that is, $\partial\mathbf{h} = \nabla\mathbf{h}\partial\mathbf{x} = 0$. This **s**-space adjustment, given by Equation (5.39), brings the iteration to a point $\mathbf{x}'_{k+1}$ on the plane tangent to the constraint surface at $\mathbf{x}_k$. Unless the constraints are linear, the point $\mathbf{x}'_{k+1}$ will be infeasible and the state variable vector $\mathbf{s}'_{k+1}$ must be adjusted further to return to the constraint surface. This requires solving the nonlinear system of equations $\mathbf{h}(\mathbf{x}) = \mathbf{0}$ with $\mathbf{d}_{k+1}$ fixed and $\mathbf{s}'_{k+1}$ as an initial guess, as in Equation (5.41) above.

The iteration on the decision variables, such as Equation (5.38), may be performed based on Newton's method rather than on the gradient method, using the results from Section 5.4, that is,

$$\mathbf{d}_{k+1} = \mathbf{d}_k - \alpha_k(\partial^2 z/\partial\mathbf{d}^2)_k^{-1}(\partial z/\partial\mathbf{d})_k^T. \tag{5.42}$$

The state variables are calculated next using the quadratic approximation

$$\partial\mathbf{s}'_k = (\partial\mathbf{s}/\partial\mathbf{d})_k\partial\mathbf{d}_k + \left(\tfrac{1}{2}\right)\partial\mathbf{d}_k^T(\partial^2\mathbf{s}/\partial\mathbf{d}^2)_k\partial\mathbf{d}_k, \tag{5.43}$$

where $\partial\mathbf{s}'_k = \mathbf{s}'_{k+1} - \mathbf{s}_k$ and where $(\partial\mathbf{s}/\partial\mathbf{d})_k$ and $(\partial^2\mathbf{s}/\partial\mathbf{d}^2)_k$ are calculated from (5.31) and (5.32), respectively. Then, iteration (5.41) can be used to return to feasibility. In general, this return to the constraint surface will not be exact, but within a prespecified acceptable error in the satisfaction of $\mathbf{h}(\mathbf{x}) = \mathbf{0}$.

The constraint surface may consist of active equalities and/or inequalities. In complete algorithmic implementations care must be taken in deciding which constraints are active as we move toward the optimum and in maintaining feasibility with respect to other locally inactive constraints. These issues are examined in Chapter 7.

Example 5.8 Consider the problem

$$\begin{aligned}
&\min f = (x_1 - 1)^2 + (x_2 - 1)^2 + (x_3 - 1)^2 + (x_4 - 1)^2 \\
&\text{subject to } h_1 = x_1 - 4x_2^2 + 2x_2x_3 - x_3^2 = 0, \\
&\qquad\qquad h_2 = x_1 - x_4 = 0,
\end{aligned}$$

which has two degrees of freedom (Beightler, Phillips, and Wilde 1979). We define $x_1 = s_1$, $x_2 = d_1$, $x_3 = d_2$, and $x_4 = s_2$ (this choice will be discussed later) and calculate the following:

$$\partial f/\partial\mathbf{d} = (2d_1 - 2, 2d_2 - 2), \quad \partial f/\partial\mathbf{s} = (2s_1 - 2, 2s_2 - 2),$$

$$\partial\mathbf{h}/\partial\mathbf{d} = \begin{pmatrix} -8d_1 + 2d_2 & 2d_1 - 2d_2 \\ 0 & 0 \end{pmatrix}, \quad \partial\mathbf{h}/\partial\mathbf{s} = \begin{pmatrix} 1 & 0 \\ 1 & -1 \end{pmatrix},$$

$$\partial^2 f/\partial\mathbf{d}^2 = \begin{pmatrix} 2 & 0 \\ 0 & 2 \end{pmatrix}, \quad \partial^2 f/\partial\mathbf{s}^2 = \begin{pmatrix} 2 & 0 \\ 0 & 2 \end{pmatrix},$$

$$\partial^2 f/\partial \mathbf{d}\,\partial \mathbf{s} = \mathbf{0},\ \partial^2 \mathbf{h}/\partial \mathbf{d}\,\partial \mathbf{s} = \mathbf{0},$$

$$\partial^2 \mathbf{h}/\partial \mathbf{d}^2 = \left[\begin{pmatrix} -8 & 2 \\ 2 & -2 \end{pmatrix}, \begin{pmatrix} 0 & 0 \\ 0 & 0 \end{pmatrix}\right]^T, \qquad \partial^2 \mathbf{h}/\partial \mathbf{s}^2 = \mathbf{0}.$$

Now we can calculate the state quantities

$$\begin{aligned} \partial \mathbf{s}/\partial \mathbf{d} &= \begin{pmatrix} -1 & 0 \\ -1 & 1 \end{pmatrix} \begin{pmatrix} -8d_1 + 2d_2 & 2d_1 - 2d_2 \\ 0 & 0 \end{pmatrix} \\ &= \begin{pmatrix} 8d_1 - 2d_2 & -2d_1 + 2d_2 \\ 8d_1 - 2d_2 & -2d_1 + 2d_2 \end{pmatrix}, \end{aligned}$$

$$\partial^2 \mathbf{h}/\partial \mathbf{d}^2 + (\partial \mathbf{h}/\partial \mathbf{s})(\partial^2 \mathbf{s}/\partial \mathbf{d}^2) = \mathbf{0};$$

the second equation is (5.28). This matrix equation expanded becomes

$$\left[\begin{pmatrix} -8 & 2 \\ 2 & -2 \end{pmatrix}, \begin{pmatrix} 0 & 0 \\ 0 & 0 \end{pmatrix}\right]^T + \begin{pmatrix} 1 & 0 \\ 1 & -1 \end{pmatrix} \left[\frac{\partial^2 s_1}{\partial \mathbf{d}^2}, \frac{\partial^2 s_2}{\partial \mathbf{d}^2}\right]^T = \mathbf{0},$$

giving

$$\left[\begin{pmatrix} -8 & 2 \\ 2 & -2 \end{pmatrix}, \begin{pmatrix} 0 & 0 \\ 0 & 0 \end{pmatrix}\right]^T + \left[\left(\frac{\partial^2 s_1}{\partial \mathbf{d}^2}\right), \left(\frac{\partial^2 s_1}{\partial \mathbf{d}^2} - \frac{\partial^2 s_2}{\partial \mathbf{d}^2}\right)\right]^T = \mathbf{0},$$

from which it follows that

$$(\partial^2 s_1/\partial \mathbf{d}^2) = (\partial^2 s_2/\partial \mathbf{d}^2) = \begin{pmatrix} 8 & -2 \\ -2 & 2 \end{pmatrix}.$$

Now we evaluate the reduced gradient and Hessian:

$$\partial z/\partial \mathbf{d} = (2d_1 - 2, 2d_2 - 2) + (2s_1 - 2, 2s_2 - 2)(\partial \mathbf{s}/\partial \mathbf{d}),$$

$$\begin{aligned} \partial^2 z/\partial \mathbf{d}^2 &= \begin{pmatrix} 2 & 0 \\ 0 & 2 \end{pmatrix} + \left(\frac{\partial \mathbf{s}}{\partial \mathbf{d}}\right)^T \begin{pmatrix} 2 & 0 \\ 0 & 2 \end{pmatrix} \left(\frac{\partial \mathbf{s}}{\partial \mathbf{d}}\right) \\ &\quad + \begin{pmatrix} 2s_1 - 2 \\ 2s_2 - 2 \end{pmatrix}^T \left[\left(\frac{\partial^2 s_1}{\partial \mathbf{d}^2}\right), \left(\frac{\partial^2 s_2}{\partial \mathbf{d}^2}\right)\right]^T, \end{aligned}$$

where $\partial \mathbf{s}/\partial \mathbf{d}$ and $\partial^2 \mathbf{s}/\partial \mathbf{d}^2$ were not substituted for economy of space.

All we need now is a starting point for the iterations. Taking $\mathbf{d}_0 = (1, 1)^T$ corresponding to $\mathbf{s}_0 = (3, 3)^T$, we perform a Newton step according to (5.42) with $\alpha_k = 1$. First find

$$\partial z/\partial \mathbf{d} = (0, 0) + (4, 4) \begin{pmatrix} 6 & 0 \\ 6 & 0 \end{pmatrix} = (48, 0),$$

$$\begin{aligned} \partial^2 z/\partial \mathbf{d}^2 &= \begin{pmatrix} 2 & 0 \\ 0 & 2 \end{pmatrix} + \begin{pmatrix} 6 & 6 \\ 0 & 0 \end{pmatrix} \begin{pmatrix} 2 & 0 \\ 0 & 2 \end{pmatrix} \begin{pmatrix} 6 & 0 \\ 6 & 0 \end{pmatrix} + 8 \begin{pmatrix} 8 & -2 \\ -2 & 2 \end{pmatrix} \\ &= \begin{pmatrix} 210 & -16 \\ -16 & 18 \end{pmatrix}. \end{aligned}$$

Next calculate the step

$$\mathbf{d}_1 = \begin{pmatrix}1\\1\end{pmatrix} - \frac{1}{3524}\begin{pmatrix}18 & 16\\16 & 210\end{pmatrix}\begin{pmatrix}48\\0\end{pmatrix} = \begin{pmatrix}0.7548\\0.7821\end{pmatrix}$$

and the corresponding $(\partial\mathbf{s})_0'$ from (5.43),

$$(\partial s_1)_0' = \begin{pmatrix}6\\0\end{pmatrix}^T \begin{pmatrix}-0.2452\\-0.2179\end{pmatrix} + \frac{1}{2}\begin{pmatrix}-0.2452\\-0.2179\end{pmatrix}^T \begin{pmatrix}8 & -2\\-2 & 2\end{pmatrix}\begin{pmatrix}-0.2452\\-0.2179\end{pmatrix}$$
$$= -1.2900 = (\partial s_2)_0'.$$

Therefore, $\mathbf{s}_1' = (1.7100, 1.7100)^T$, which corresponds to $(h_1)_1 = 0.0001$ and $(h_2)_1 = 0.0000$. Taking this as $j = 0$, we use the linearization (5.41) to get a better estimate of the state variables:

$$[\mathbf{s}_1]_1 = \begin{pmatrix}1.7100\\1.7100\end{pmatrix} + \begin{pmatrix}-1 & 0\\-1 & 1\end{pmatrix}\begin{pmatrix}0.0001\\0.0000\end{pmatrix} = \begin{pmatrix}1.7099\\1.7099\end{pmatrix}.$$

This estimate is very accurate. In fact, even $\mathbf{s}_1'$ was already feasible up to the fourth decimal place of the constraint value.

We have completed the first feasible iteration giving the new point $\mathbf{x}_1 = (1.7099, 0.7548, 0.7821, 1.7099)^T$ with $f_1 = 1.1155$, an improvement from $f_0 = 8$. More iterations can be performed in the same way. ■

Some comments should be made about the selection of state and decision variables. A lot of computational effort is often spent in nonlinear equation solving to return to feasibility. How much computation is required often depends on the choice of state and decision variables. Poor choices may lead to singular or near-singular matrices that could stall the search. This need for partitioning of variables is, in fact, a drawback of the reduced space approach. A simple procedure that may be followed before the iteration begins is to renumber all the equations systematically (if possible) so that each state variable is solvable by one equation, progressing along the diagonal as shown below:

$$\begin{aligned} h_1 &= s_1 + \phi_1(\mathbf{d}),\\ h_2 &= s_2 + \phi_2(\mathbf{d}; s_1),\\ &\vdots\\ h_m &= s_m + \phi_m(\mathbf{d}; s_1, s_2, \ldots, s_{m-1}). \end{aligned} \tag{5.44}$$

This procedure is further discussed in Beightler, Phillips, and Wilde (1979).

Gradient Projection Method

A very similar idea, which historically was developed separately and earlier than GRG-type algorithms, is to operate without variable partitioning using the linear approximation of the constraint surface at a point $\mathbf{x}_k$, that is, the tangent subspace of the constraints at $\mathbf{x}_k$, defined again by $\nabla\mathbf{h}\partial\mathbf{x} = \mathbf{0}$. The idea is illustrated in Figure 5.6.

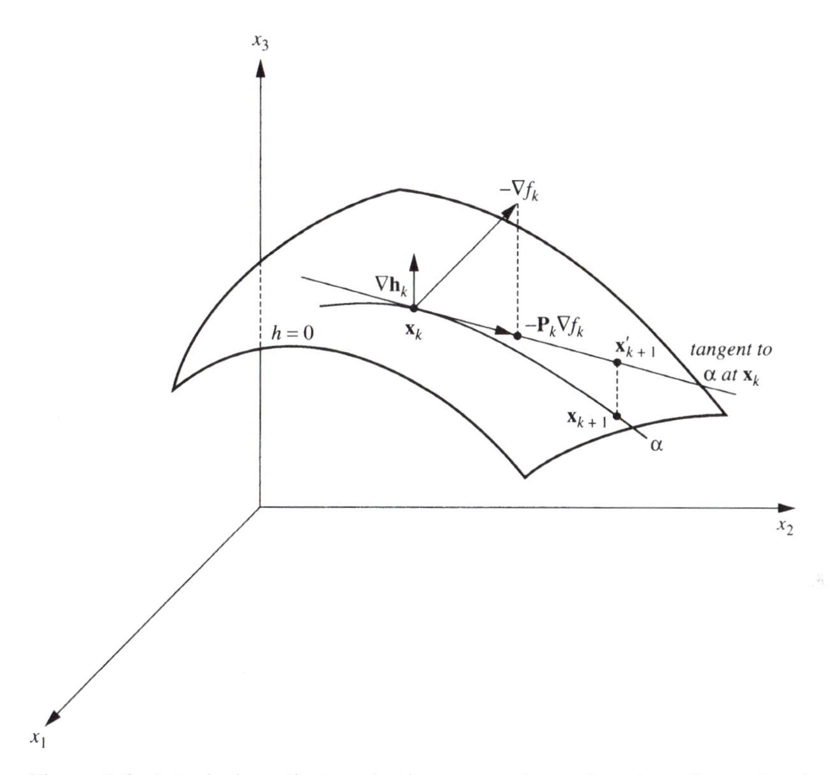

Figure 5.6. A typical gradient projection move; the surface $h = 0$ may be also an active inequality $g \leqq 0$.

The negative gradient at $\mathbf{x}_k$ is *projected* on the tangent subspace, so that a vector $-\mathbf{P}_k \nabla f_k$ is defined, with $\mathbf{P}_k$ a matrix corresponding to the projection transformation. This is then taken as a search direction $\mathbf{s}_k$ in the usual iteration

$$\mathbf{x}_{k+1} = \mathbf{x}_k + \alpha_k \mathbf{s}_k.$$

A proper selection of α_k gives a new point $\mathbf{x}'_{k+1}$ on the tangent subspace, which is infeasible except if the constraints are linear. Then a correction step is taken to return to the constraint surface. Usually, a direction orthogonal to $-\mathbf{P}_k \nabla f_k$ is selected and the intersection point $\mathbf{x}_{k+1}$ with the constraint surface is the final result of the iteration. This type of algorithm has been known as a *gradient projection method.*

As mentioned earlier, projection appears to have an advantage over the reduced space approach, because no partitioning of the variables is required. The projection method can be seen as a "symmetric" one, since it does not discriminate among variables as the "asymmetric" reduced space does.

To illustrate how the projection is computed, let us examine the case with linear equality constraints only:

$$\min f(\mathbf{x})$$
$$\text{subject to } \mathbf{h}(\mathbf{x}) = \mathbf{A}\mathbf{x} - \mathbf{b} = \mathbf{0}.$$

Since here $\mathbf{A} = \nabla\mathbf{h}$, the tangent subspace at a point $\mathbf{x}$ is described by

$$\mathcal{T} = \{\mathbf{s} \in \mathcal{R}^n: \mathbf{As} = \mathbf{0}\},$$

which is the first-order feasibility requirement. The normal subspace at a point on the constraint surface is given by

$$\mathcal{N} = \{\mathbf{z} \in \mathcal{R}^n: \mathbf{A}^T\mathbf{p} = \mathbf{z}, \mathbf{p} \in \mathcal{R}^m\},$$

where $\mathbf{p}$ is a vector of (as yet unspecified) parameters, and m is the dimension of $\mathbf{h}$, that is, the number of equality constraints. The two spaces $\mathcal{T}$ and $\mathcal{N}$ are orthogonal and complementary with respect to the original space; therefore, any vector $\mathbf{x}$ can be expressed as a sum of vectors from each of the two spaces $\mathcal{T}$ and $\mathcal{N}$. Our goal then is to find such an expression for the negative gradient of f at a given iterant $\mathbf{x}_k$. To this end we set $\mathbf{g}_k = \nabla f_k^T$ and let

$$-\mathbf{g}_k = \mathbf{s}_k + \mathbf{z}_k = \mathbf{s}_k + \mathbf{A}^T\mathbf{p}_k \tag{5.45}$$

with $\mathbf{s}_k \in \mathcal{T}(\mathbf{x}_k)$ and $\mathbf{z}_k \in \mathcal{N}(\mathbf{x}_k)$. Premultiplying (5.45) by $\mathbf{A}$ we get

$$-\mathbf{Ag}_k = \mathbf{As}_k + \mathbf{AA}^T\mathbf{p}_k. \tag{5.46}$$

Since $\mathbf{As}_k = \mathbf{0}$ by definition and also $\mathbf{x}_k$ is assumed regular, Equation (5.46) can be solved for $\mathbf{p}_k$, giving

$$\mathbf{p}_k = -(\mathbf{AA}^T)^{-1}\mathbf{A}\nabla f_k^T \tag{5.47}$$

Substituting in Equation (5.45) and solving for $\mathbf{s}_k$, we get

$$\mathbf{s}_k = -[\mathbf{I} - \mathbf{A}^T(\mathbf{AA}^T)^{-1}\mathbf{A}]\nabla f_k^T = -\mathbf{P}\nabla f_k^T, \tag{5.48}$$

where $\mathbf{I}$ is the identity matrix and $\mathbf{P}$ is the *projection matrix*

$$\mathbf{P} \triangleq \mathbf{I} - \mathbf{A}^T(\mathbf{AA}^T)^{-1}\mathbf{A}, \tag{5.49}$$

which projects $-\nabla f_k$ on the subspace $\mathbf{As} = \mathbf{0}$.

In the case of linear constraints the subspace of the constraints is the same as the tangent subspace (translated by $\mathbf{b}$, as mentioned in Section 5.2), and so the projection matrix is constant throughout the iterations, at least as long as the same active constraints are used. For nonlinear constraints the projection matrix will change with k, with $\mathbf{A}$ replaced by the Jacobian $\nabla\mathbf{h}(\mathbf{x}_k) = \nabla\mathbf{h}_k$, or $\nabla\mathbf{g}_k$ for active inequalities. Thus, Equation (5.49) is modified to

$$\mathbf{P}_k = \mathbf{I} - \nabla\mathbf{h}_k^T\left(\nabla\mathbf{h}_k\nabla\mathbf{h}_k^T\right)^{-1}\nabla\mathbf{h}_k, \tag{5.50}$$

which is computed at each iteration. If a gradient method is used, the iteration

$$\mathbf{x}'_{k+1} = \mathbf{x}_k - \alpha_k\mathbf{P}_k\nabla f_k^T \tag{5.51}$$

gives the new point on the tangent subspace. To return to the constraint surface using a direction $\mathbf{s}'_k$ orthogonal to the tangent subspace, that is, on the normal subspace at

$\mathbf{x}'_{k+1}$, we set

$$\mathbf{s}'_k = (\nabla \mathbf{h}'_{k+1})^T \mathbf{c}_k, \tag{5.52}$$

with the unspecified vector $\mathbf{c}$ determined, for example, as follows. Approximate $\mathbf{h}(\mathbf{x}_{k+1})$ by setting $\mathbf{x}_{k+1} = \mathbf{x}'_{k+1} + \mathbf{s}'_k$ and using the Taylor expansion

$$\mathbf{h}(\mathbf{x}'_{k+1} + \mathbf{s}'_k) = \mathbf{h}(\mathbf{x}'_{k+1}) + \nabla \mathbf{h}(\mathbf{x}'_{k+1})\mathbf{s}'_k.$$

Assuming Newton–Raphson iterations and using Equation (5.52), we get

$$\mathbf{h}(\mathbf{x}'_{k+1}) + \nabla \mathbf{h}'_{k+1}(\nabla \mathbf{h}'_{k+1})^T \mathbf{c}_k = \mathbf{0},$$

or, for $\mathbf{x}'_{k+1}$ being regular,

$$\mathbf{c}_k = -[\nabla \mathbf{h}'_{k+1}(\nabla \mathbf{h}'_{k+1})^T]^{-1}\mathbf{h}'_{k+1}. \tag{5.53}$$

Thus, the iteration formula for getting from $\mathbf{x}'_{k+1}$ to $\mathbf{x}_{k+1}$ will be

$$[\mathbf{x}_{k+1}]_{j+1} = \left[\mathbf{x}_{k+1} - (\nabla \mathbf{h}_{k+1})^T \left(\nabla \mathbf{h}_{k+1} \nabla \mathbf{h}^T_{k+1}\right)^{-1} \mathbf{h}_{k+1}\right]_j, \tag{5.54}$$

with $\mathbf{x}_{k+1} = \mathbf{x}'_{k+1}$ at $j = 0$ and all the Jacobians computed once at $\mathbf{x}'_{k+1}$. This is the same idea as Equation (5.41) for a GRG-type gradient descent.

Example 5.9 Consider the problem (Figure 5.7)

$$\min f = x_1^2 + (x_2 - 3)^2$$
$$\text{subject to } g = x_2^2 - 2x_1 \le 0, \quad x_1 \ge 0, \quad x_2 \ge 0.$$

We have $\nabla f = (2x_1, 2x_2 - 6)$, $\nabla g = (-2, 2x_2)$. Because the unconstrained minimum violates the constraint, g will be active. Assume an initial point $\mathbf{x}_0 = (0.5, 1)^T$, where $\nabla f_0 = (1, -4)$, $\nabla g_0 = (-2, 2)$. The projection matrix at $\mathbf{x}_0$ is [Equation (5.50)]

$$\mathbf{P}_0 = \begin{pmatrix} 1 & 0 \\ 0 & 1 \end{pmatrix} - (-2, 2)^T \left[\begin{pmatrix} -2 \\ 2 \end{pmatrix}^T \begin{pmatrix} -2 \\ 2 \end{pmatrix} \right]^{-1} (-2, 2) = \frac{1}{2} \begin{pmatrix} 1 & 1 \\ 1 & 1 \end{pmatrix}.$$

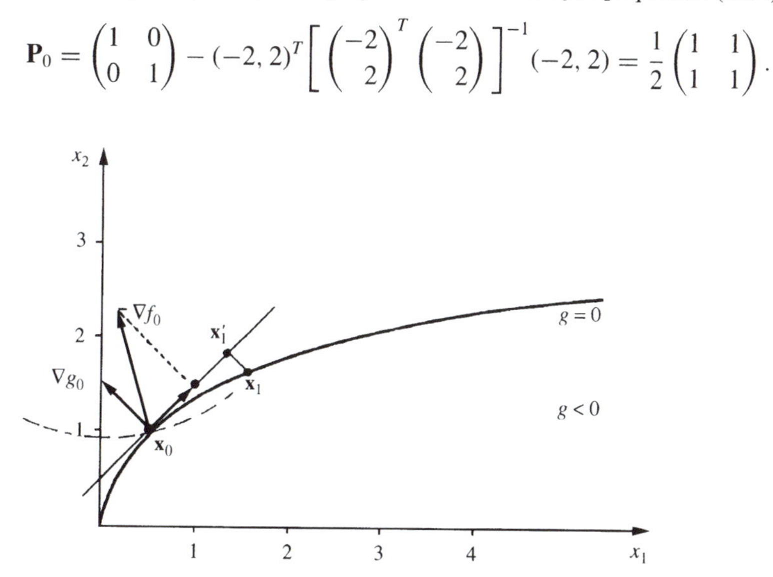

Figure 5.7. A gradient projection iteration for Example 5.9.

The iterant $\mathbf{x}_1'$ on the subspace $\partial g = 0$ [Equation (5.51)] is

$$\mathbf{x}_1' = \begin{pmatrix} 0.5 \\ 1 \end{pmatrix} - (0.4) \begin{pmatrix} 0.5 & 0.5 \\ 0.5 & 0.5 \end{pmatrix} \begin{pmatrix} 1 \\ -4 \end{pmatrix} = \begin{pmatrix} 1.1 \\ 1.6 \end{pmatrix}.$$

To return to the surface, Equation (5.54) may be used:

$$[\mathbf{x}_1]_1 = \begin{pmatrix} 1.1 \\ 1.6 \end{pmatrix} - \left[\begin{pmatrix} -2 \\ 3.2 \end{pmatrix} \left[\begin{pmatrix} -2 \\ 3.2 \end{pmatrix}^T \begin{pmatrix} -2 \\ 3.2 \end{pmatrix} \right]^{-1} \right] (0.36) = \begin{pmatrix} 1.15 \\ 1.52 \end{pmatrix},$$

which gives $g_1 = 0.0104$. (For a rigorous justification of $\alpha_0 = 0.4$ above, see Exercise 7.18.) If this constraint satisfaction accuracy is acceptable, the point is taken as the final $\mathbf{x}_1$. More iterations can be performed in a similar way, although the point is almost optimal. We may check for termination using the condition $\|\mathbf{P}_k \nabla f_k^T\| < \varepsilon$. ■

A discussion on algorithmic implementation of these two feasible direction methods is deferred until Chapter 7.

5.6 Inequality Constraints

We now examine the problem with only inequality constraints in the standard negative null form

$$\begin{aligned} &\min f(\mathbf{x}) \\ &\text{subject to } \mathbf{g}(\mathbf{x}) \leq \mathbf{0}. \end{aligned} \tag{5.55}$$

If there were any equalities, let us assume that they have been deleted, explicitly or implicitly. At a constrained minimum, only the *active* constraints will be significant. In general, only some of the g_js will be active at $\mathbf{x}_*$; so let us represent by $\bar{\mathbf{g}}$ the vector of active inequalities. At $\mathbf{x}_*$ no difference exists between $\bar{\mathbf{g}}$ and equality constraints $\mathbf{h}$. Hence the first-order optimality condition (5.22) will apply to the active inequalities:

$$\nabla f(\mathbf{x}_*) + \boldsymbol{\mu}^T \nabla \bar{\mathbf{g}}(\mathbf{x}_*) = \mathbf{0}^T. \tag{5.56}$$

Here $\boldsymbol{\mu}$ is the vector of Lagrange multipliers associated with the active inequalities, defined exactly as in (5.20), that is, $\boldsymbol{\mu}^T = -(\partial f/\partial \mathbf{s})_*(\partial \bar{\mathbf{g}}/\partial \mathbf{s})_*^{-1}$, if a partition $\mathbf{x} = (\mathbf{d}, \mathbf{s})^T$ was effected. This partition is not necessary and $\boldsymbol{\mu}$ could be defined directly from (5.56). If $\mathbf{x}_*$ exists, the multipliers will be well defined provided $\mathbf{x}_*$ is a regular point.

There is a difference, however, between $\boldsymbol{\lambda}$ and $\boldsymbol{\mu}$. Recall that the components of $\boldsymbol{\lambda}$ were unrestricted in sign. For inequality constrained problems in the form of (5.55) the first-order perturbations must satisfy at $\mathbf{x}_*$, for *all* $\partial \mathbf{x}_*$,

$$\partial f_* = \nabla f_* \partial \mathbf{x}_* \geq 0 \quad \text{(optimality)}, \tag{5.57}$$

$$\partial \bar{\mathbf{g}}_* = \nabla \bar{\mathbf{g}}_* \partial \mathbf{x}_* \leq \mathbf{0} \quad \text{(feasibility)}. \tag{5.58}$$

Combining (5.56)–(5.58) and assuming perturbations $\partial \mathbf{x}_* \neq \mathbf{0}$, we get $\boldsymbol{\mu}^T \partial \bar{\mathbf{g}} \leq \mathbf{0}$, or since $\partial \bar{\mathbf{g}} \leq \mathbf{0}$ for feasibility, $\boldsymbol{\mu} \geq \mathbf{0}$. The multipliers $\boldsymbol{\mu}$ are *restricted in sign* to be nonnegative. In fact, at regular points multipliers associated with active inequalities must be strictly positive. Active constraints with zero multipliers are possible, for example, if the matrix $\nabla\bar{\mathbf{g}}(\mathbf{x}_*)$ in (5.56) is singular, meaning that $\mathbf{x}_*$ is not a regular point. This situation is usually referred to as *degeneracy* and could defy iterative procedures designed on the assumption of regularity for all iterant $\mathbf{x}_k$, including the optimizer. Measures can be taken in degenerate cases, but they will not be discussed here. For now, we will assume nondegeneracy, so that zero multipliers can be associated only with inactive constraints. Thus, the distinction between active and inactive constraints can be dropped by generalizing the optimality condition (5.56) as follows:

$$\nabla f_* + \boldsymbol{\mu}^T \nabla \mathbf{g}_* = \mathbf{0}^T, \quad \boldsymbol{\mu}^T \mathbf{g} = 0, \quad \boldsymbol{\mu} \geq \mathbf{0}. \tag{5.59}$$

The condition $\boldsymbol{\mu}^T\mathbf{g} = 0$ is called the *complementary slackness* or *transversality* condition, implying that if $g_i < 0$, then $\mu_i = 0$ and vice versa.

Karush–Kuhn–Tucker Conditions

It is now a simple matter to state the necessary optimality conditions for a problem with both equalities and inequalities. They will be a combination of the conditions (5.22) and (5.59). The problem is stated as

$$\begin{aligned} &\min f(\mathbf{x}) \\ &\text{subject to } \mathbf{h}(\mathbf{x}) = \mathbf{0}, \quad \mathbf{g}(\mathbf{x}) \leq \mathbf{0}, \end{aligned} \tag{5.60}$$

and the *necessary* conditions, known as the *Karush–Kuhn–Tucker* (KKT) conditions are

1. $\mathbf{h}(\mathbf{x}_*) = \mathbf{0}$, $\mathbf{g}(\mathbf{x}_*) \leq \mathbf{0}$;
2. $\nabla f_* + \boldsymbol{\lambda}^T \nabla \mathbf{h}_* + \boldsymbol{\mu}^T \nabla \mathbf{g}_* = \mathbf{0}^T$, where $\boldsymbol{\lambda} \neq \mathbf{0}$, $\boldsymbol{\mu} \geq \mathbf{0}$, $\boldsymbol{\mu}^T \mathbf{g} = 0$. (5.61)

Here $\mathbf{x}_*$, the minimizer, is assumed to be a regular point. A point that satisfies the KKT conditions (5.61) is called a *KKT point* and may not be a minimum since the conditions are not sufficient. Second-order information is necessary to verify the nature of a KKT point. Again the procedure would be the same as for equalities but now the active inequalities must be included. The second-order sufficiency conditions are then as follows:

> *If a KKT point $\mathbf{x}_*$ exists, such that the Hessian of the Lagrangian on the subspace tangent to the active constraints (equalities and inequalities) is positive-definite at $\mathbf{x}_*$, then $\mathbf{x}_*$ is a local constrained minimum.*

There are several variations of the KKT conditions depending on the explicit statement of the problem (5.60). For example, in computational procedures it may be desirable to handle simple upper and lower bound constraints separately (see Exercises 5.5 and 5.6).

The difference between the multipliers $\boldsymbol{\lambda}$ and $\boldsymbol{\mu}$ is the sign restriction for $\boldsymbol{\mu}$. One should see readily that directed equalities such as those discovered by monotonicity analysis will also have a sign restriction on the corresponding λs. This could be useful when trying to find a solution for the KKT conditions, which in general is a complicated task.

Example 5.10 Consider the problem with $x_1, x_2 > 0$:

$$\begin{aligned}&\min f = 8x_1^2 - 8x_1x_2 + 3x_2^2\\&\text{subject to } g_1 = x_1 - 4x_2 + 3 \le 0,\\&\qquad\qquad g_2 = -x_1 + 2x_2 \le 0.\end{aligned}$$

We will find all KKT points and test them for minimality. But first let us quickly find the correct solution by monotonicity analysis. The objective is written $f = 8x_1(x_1 - x_2) + 3x_2^2$ and since g_2 states that $x_1 \ge 2x_2$, f is increasing wrt x_1 in the feasible domain. Then g_2 must be active since it is the only constraint decreasing wrt x_1. Thus, $x_{1*} = 2x_{2*}$. Elimination of x_1 yields the problem $\{\min f = 19x_2^2$, subject to $x_2 \ge \frac{3}{2}\}$. Obviously, $x_{2*} = \frac{3}{2}$. Both constraints are active.

Let us now write the KKT conditions, in the special case of only inequalities, Equation (5.59):

$$\begin{aligned}&16x_1 - 8x_2 + \mu_1 - \mu_2 = 0,\\&-8x_1 + 6x_2 - 4\mu_1 + 2\mu_2 = 0,\\&\mu_1(x_1 - 4x_2 + 3) = 0, \quad \mu_1 \ge 0, \quad x_1 - 4x_2 + 3 \le 0,\\&\mu_2(-x_1 + 2x_2) = 0, \quad \mu_2 \ge 0, \quad -x_1 + 2x_2 \le 0.\end{aligned}$$

This is a system of equalities and inequalities, requiring a case-by-case examination of possible solutions. The possibilities are as follows:

1. $\mu_2 = 0$, $\mu_1 \ne 0$ so that $g_1 = 0$. Then we have

$$\begin{aligned}&16x_1 - 8x_2 + \mu_1 = 0, \quad x_1 - 4x_2 + 3 = 0,\\&-8x_1 + 6x_2 - 4\mu_1 = 0, \quad -x_1 + 2x_2 \le 0,\end{aligned}$$

   which imply $x_1 = 0.464x_2$; this violates g_2, so the case is abandoned.

2. $\mu_2 = 0$, $\mu_1 = 0$ so that an interior solution is expected. The system resulting is

$$\begin{aligned}&16x_1 - 8x_2 = 0, \quad x_1 - 4x_2 + 3 \le 0,\\&-8x_1 + 6x_2 = 0, \quad -x_1 + 2x_2 \le 0,\end{aligned}$$

   which clearly has no solution.

3. $\mu_2 \ne 0$, $\mu_1 \ne 0$ so that $g_1 = 0$, $g_2 = 0$. Solving the two active constraints, we find $x_1 = 2x_2 = 3$. The stationarity conditions are then solved to find the positive values $\mu_1 = 28.5$, $\mu_2 = 64.5$. This is the solution identified by monotonicity analysis.

4. $\mu_2 \ne 0$, $\mu_1 = 0$ so that $g_2 = 0$. Then we have

$$\begin{aligned}&16x_1 - 8x_2 - \mu_2 = 0, \quad x_1 - 4x_2 + 3 \le 0,\\&-8x_1 + 6x_2 + 2\mu_2 = 0, \quad -x_1 + 2x_2 = 0,\end{aligned}$$

   which can be easily seen to have no solution.

Thus, the KKT point identified is $\mathbf{x} = (3, \frac{3}{2})^T$ and $\boldsymbol{\mu} = (28.5, 64.5)^T$. To test for minimality we must evaluate the differential quadratic form of the Lagrangian. Because the constraints are linear, the Hessian of the Lagrangian is the same as the Hessian of the objective, which is easily found to be positive-definite. So the point is indeed a minimizer.

Since monotonicity analysis identified the solution as a global one, we might be able to prove the same based on the sufficient conditions. In fact, the Hessian is *globally* positive-definite, which means that f is convex, while the solution is the (convex) intersection of the hyperplanes $g_1 = 0$, $g_2 = 0$. Since these hyperplanes are linearly independent, their intersection is a unique point. Thus, the unique minimum is guaranteed to be global. ■

Lagrangian Standard Forms

On several occasions so far we have made the observation that the "standard" form of optimality conditions and related properties has a corresponding "standard" form of the original model. This can be a source of confusion if one reads the literature carelessly. As we pointed out in Chapter 1, the most often used standard forms are the negative and positive null forms. The expressions we developed in the present section for the KKT conditions were based on the negative null form. It is sometimes desirable to work with the positive null form but *maintain the property that the multipliers associated with active inequalities be nonnegative*. To do this, one must define the Lagrangian function by subtracting the constraint terms, as was pointed out already in Equation (5.23′).

We summarize these results here for easy reference.

NEGATIVE NULL FORM

Model: min f, subject to $\mathbf{h} = \mathbf{0}, \mathbf{g} \leq \mathbf{0}$.

Lagrangian: $L = f + \boldsymbol{\lambda}^T\mathbf{h} + \boldsymbol{\mu}^T\mathbf{g}$.

KKT conditions:
$$\nabla f + \boldsymbol{\lambda}^T \nabla\mathbf{h} + \boldsymbol{\mu}^T \nabla\mathbf{g} = \mathbf{0}^T,$$
$$\mathbf{h} = \mathbf{0}, \mathbf{g} \leq \mathbf{0},$$
$$\boldsymbol{\lambda} \neq \mathbf{0}, \boldsymbol{\mu} \geq \mathbf{0}, \boldsymbol{\mu}^T\mathbf{g} = 0.$$

POSITIVE NULL FORM

Model: min f, subject to $\mathbf{h} = \mathbf{0}, \mathbf{g} \geq \mathbf{0}$.

Lagrangian: $L = f - \boldsymbol{\lambda}^T\mathbf{h} - \boldsymbol{\mu}^T\mathbf{g}$.

KKT conditions:
$$\nabla f - \boldsymbol{\lambda}^T \nabla\mathbf{h} - \boldsymbol{\mu}^T \nabla\mathbf{g} = \mathbf{0}^T, \tag{5.61′}$$
$$\mathbf{h} = \mathbf{0}, \mathbf{g} \geq \mathbf{0},$$
$$\boldsymbol{\lambda} \neq \mathbf{0}, \boldsymbol{\mu} \geq \mathbf{0}, \boldsymbol{\mu}^T\mathbf{g} = 0.$$

The proof of the conditions for the positive null form is entirely analogous to the one for the negative form given in this section. Other standard forms will lead to other

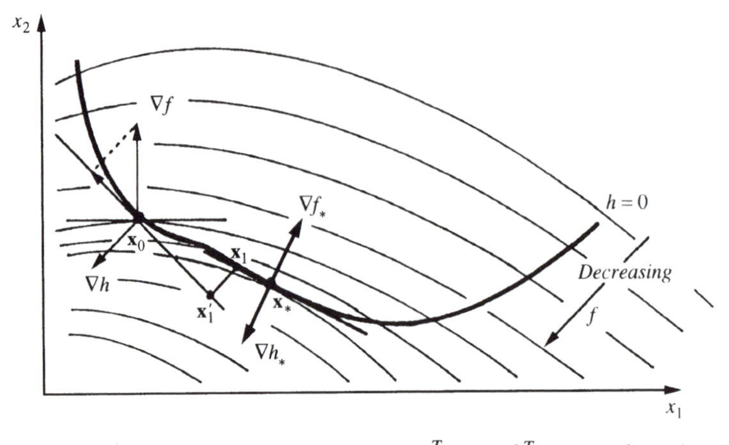

Figure 5.8. Geometric meaning of $\nabla f + \boldsymbol{\lambda}^T \nabla h = \mathbf{0}^T$ at $\mathbf{x}_*$ and $\mathbf{x}_0 \neq \mathbf{x}_*$.

variations of the above conditions (see Exercises 5.5 and 5.6). See also Section 5.9 on the interpretation of Lagrange multipliers.

5.7 Geometry of Boundary Optima

The derivation of optimality conditions on the boundary, presented in the previous section, was basically algebraic. There is a nice geometric interpretation of these conditions that provides further insight. This is the topic of the present section.

Interpretation of KKT Conditions

Recall the observation in Section 5.1 that at a local constrained minimum, all feasible directions must form an angle with the gradient direction that is less than or equal to 90° (Figure 5.2). Looking now at Figure 5.8, where one equality constraint is depicted together with the objective function in two dimensions, we see that any feasible direction at $\mathbf{x}_*$ is *exactly* orthogonal to ∇f. Moreover, any of the (two) possible feasible directions, being on the tangent to $h(\mathbf{x})$ at $\mathbf{x}_*$, will be also orthogonal to ∇h. Thus, ∇f_* and ∇h_* are colinear, that is, linearly dependent according to $\nabla f_* + \lambda \nabla h_* = \mathbf{0}^T$. This can be generalized to higher dimensions if we imagine that the feasible directions will lie on the hyperplane tangent to the intersection of all the constraint hypersurfaces given by (5.7), that is, $\nabla \mathbf{h}(\mathbf{x}_*)\mathbf{s} = \mathbf{0}$. Then the gradient of f being orthogonal to $\mathbf{s}$ will lie on the normal space of the constraint set at $\mathbf{x}_*$ and will be expressed as a linear combination of the ∇h_is according to (5.6).

At points other than optimal, such as $\mathbf{x}_0$ in Figure 5.8, the feasible direction, being on the tangent, will make an angle other than 90° with the gradient of the objective. A feasible *descent* direction will have an angle larger than 90° and a move in this direction will yield a point $\mathbf{x}_1'$ that will be infeasible, unless h is linear. Thus, a

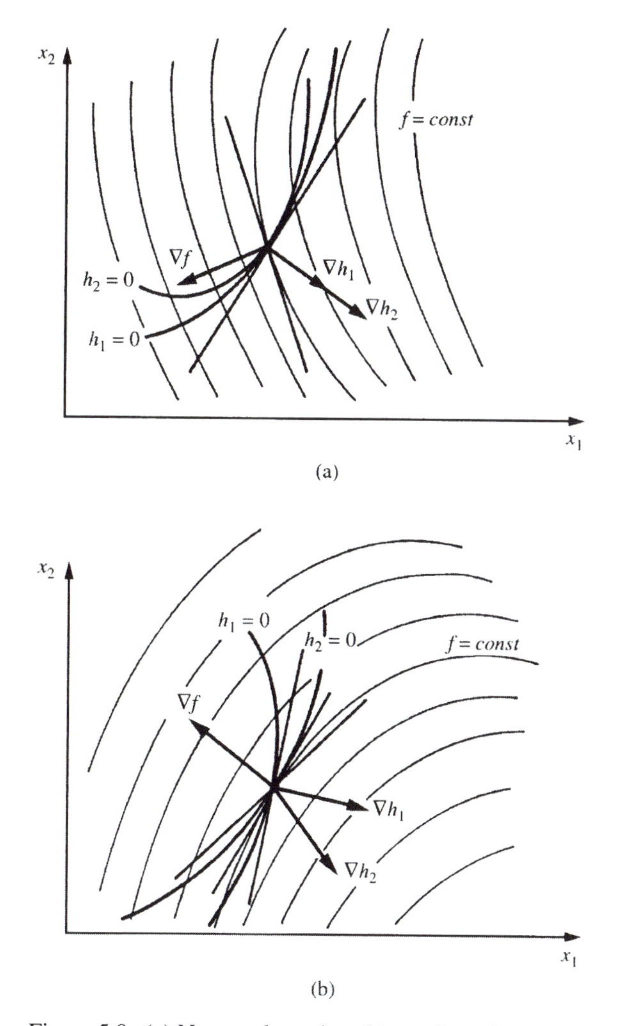

Figure 5.9. (a) Nonregular point; (b) regular point.

restoration step is needed to get back on the constraint at $\mathbf{x}_1$. This is in fact the basic idea for the gradient projection method of Section 5.5.

The theoretical need for the regularity of $\mathbf{x}_*$ is now easily understood from Figure 5.9. In (a), the point is not regular since ∇h_1 and ∇h_2 are linearly dependent. The point $\mathbf{x}_*$ is obviously optimal, being the only feasible one, but it cannot satisfy the optimality condition $\nabla f_* = -\lambda_1 \nabla h_{1*} - \lambda_2 \nabla h_{2*}$. In (b), where $\mathbf{x}_*$ is regular, the condition holds.

Turning to inequalities, the same observations as for equalities are made for the active inequalities (Figure 5.10). The difference is that now the feasible directions exist within a cone, say, at point $\mathbf{x}_0$. The existence of a feasible side corresponds to the sign restriction on the multipliers at $\mathbf{x}_*$. Thus, $\mathbf{x}_*$ in Figure 5.10 is a KKT point with ∇f_* expressed as a linear combination of the gradients ∇g_1 and ∇g_2 of the active constraints with μ_1 and μ_2 being strictly positive and feasibility satisfied.

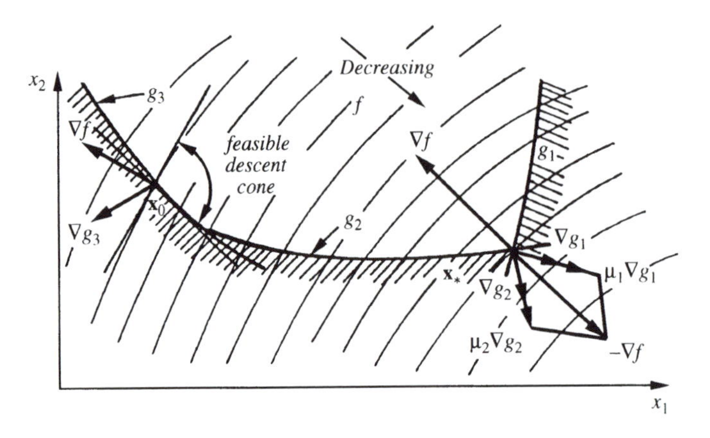

Figure 5.10. Geometric meaning of the KKT conditions.

Interpretation of Sufficiency Conditions

Since a KKT point may not be a minimum, we should examine the geometric meaning of the sufficiency condition $\partial \mathbf{x}^T(\partial^2 L/\partial \mathbf{x}^2)\partial \mathbf{x}$ at a KKT point. From the condition for only (active) inequalities,

$$\partial \mathbf{x}^T(\partial^2 L/\partial \mathbf{x}^2)\partial \mathbf{x} = \partial \mathbf{x}^T(\partial^2 f/\partial \mathbf{x}^2)\partial \mathbf{x} + \sum_i \mu_i[\partial \mathbf{x}^T(\partial^2 g_i/\partial \mathbf{x}^2)\partial \mathbf{x}], \quad (5.62)$$

we see that the Hessian of the Lagrangian is positive-definite if the functions f and g_i (all i) are convex, because then their differential quadratic forms will be positive-definite and $\mu_i > 0$ (all i). Recall that if g_i is convex, then the inequality $g_i \leq 0$ represents a convex set of points $\mathbf{x} \in \mathcal{X}$ that are feasible wrt g_i. Simultaneous satisfaction of all g_is will occur at the intersection set of all sets $g_i \leq 0$, which will also be convex. But that is exactly the feasible space defined by $\mathbf{g}(\mathbf{x}) \leq \mathbf{0}$. Thus, geometrically the sufficient condition says that *locally*, at a KKT point, the objective function must be convex and the feasible space must be also convex. In Figure 5.11, the three KKT points $\mathbf{x}_1$, $\mathbf{x}_2$, and $\mathbf{x}_3$ are shown. At $\mathbf{x}_1$ and $\mathbf{x}_3$, f and g are locally

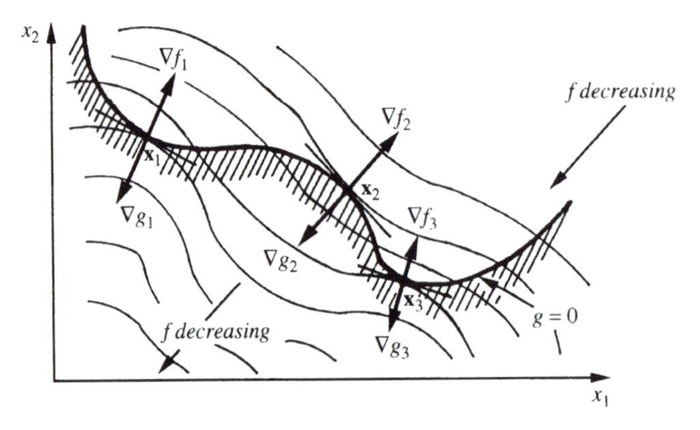

Figure 5.11. Nonsufficiency of KKT conditions.

convex, but at $\mathbf{x}_2$ they are locally concave. Only $\mathbf{x}_1$ and $\mathbf{x}_3$ are minima, with the global one being $\mathbf{x}_1$.

If convexity is a global property, then just as in the unconstrained case, the minimizer will be unique and global. This happy situation gives the following optimality condition:

If a KKT point exists for a convex function subject to a convex constraint set, then this point is a unique global minimizer.

This statement for a global minimum is actually stronger than necessary. Milder forms of convexity such as pseudoconvexity or quasiconvexity of f would do as well. Proving these more generalized results is not of particular interest here and the reader may consult several of the references (e.g., Avriel 1976 or Bazaraa and Shetty, 1979) about this. The point, however, is that global solutions can sometimes occur even if the function f is not perfectly convex.

The above sufficiency interpretation can be extended to equality constraints in the original $\mathcal{X}$-space, if we take into account multiplier signs as can be best seen from the following example.

Example 5.11 Consider the problem

$$\min f = (x_1 - 2)^2 + (x_2 - a)^2$$
$$\text{subject to } h = x_2^2 - 2x_1 = 0,$$

where a is a parameter and x_1 and x_2 are taken positive. The stationarity conditions are

$$2(x_1 - 2) - 2\lambda = 0, \quad 2(x_2 - a) + 2\lambda x_2 = 0.$$

Solving in terms of λ, we find $x_1 = 2 + \lambda$, $x_2 = a/(1 + \lambda)$. Substituting in the constraint after elimination of λ and x_1, we get $x_2^3 - 2x_2 - 2a = 0$. The solution of this cubic equation can be found explicitly in terms of a (see, for example, Spiegel, 1968). Let us examine two cases:

a. For $a = 3$ we find $x_1 = 2.38$, $x_2 = 2.18$, $\lambda = 0.38$. The second-order condition is satisfied, since

$$\partial^2 f/\partial \mathbf{x}^2 + \lambda(\partial^2 h/\partial \mathbf{x}^2) = \begin{pmatrix} 2 & 0 \\ 0 & 2 \end{pmatrix} + 0.38 \begin{pmatrix} 0 & 0 \\ 0 & 2 \end{pmatrix} = \begin{pmatrix} 2 & 0 \\ 0 & 2.76 \end{pmatrix}$$

is a positive-definite matrix.

b. For $a = 1$ we find $x_1 = 1.56$, $x_2 = 1.77$, $\lambda = -0.44$. The second-order condition is again satisfied since

$$\partial^2 f/\partial \mathbf{x}^2 + \lambda(\partial^2 h/\partial \mathbf{x}^2) = \begin{pmatrix} 2 & 0 \\ 0 & 2 \end{pmatrix} - 0.44 \begin{pmatrix} 0 & 0 \\ 0 & 2 \end{pmatrix} = \begin{pmatrix} 2 & 0 \\ 0 & 1.12 \end{pmatrix}.$$

Since $\partial \mathbf{d}^T(\partial^2 z/\partial \mathbf{d}^2)\partial \mathbf{d} = \partial \mathbf{x}^T(\partial^2 L/\partial \mathbf{x}^2)\partial \mathbf{x}$, the *reduced* objective is locally convex (on the surface $h(\mathbf{x})$) in both cases. But in the original space the situation is as shown in

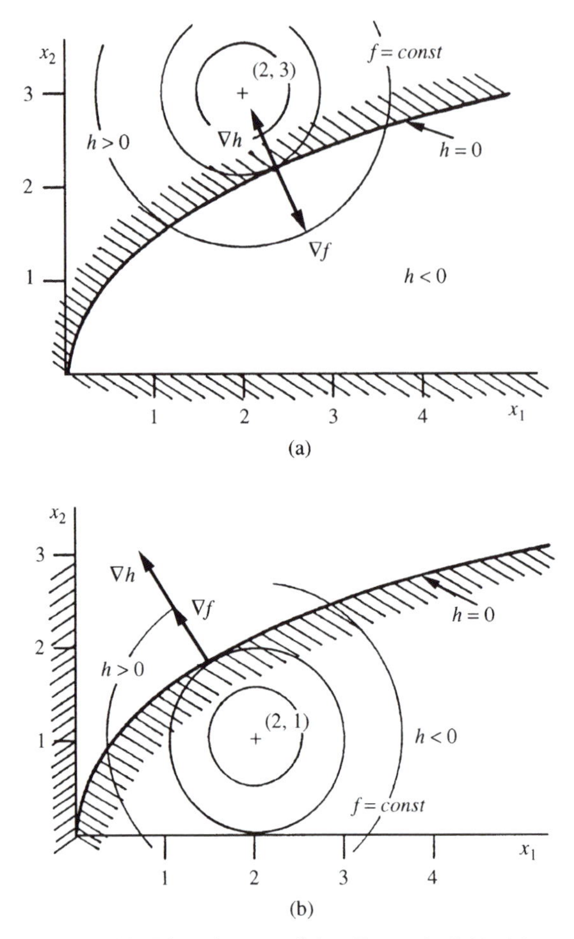

Figure 5.12. Directing equalities. Example 5.11: (a) $a = 3$, (b) $a = 1$.

Figure 5.12. If we think of h as a directed equality, it would act in the same way as an active inequality that has the unconstrained minimum on its infeasible side. Thus, in (a) we have $h \equiv <0$ and the feasible space, bounded by $x_2 = 0$ and $h = 0$, is convex. In (b) we have $h \equiv >0$ and the feasible space, bounded by $x_1 = 0$ and $h = 0$, is not convex. In the convex case with $a = 3$, the multiplier is positive. In the nonconvex case with $a = 1$ the multiplier is negative, but now $h \equiv >0$ is not in the negative null form, so if we reverse the inequality as $-h \equiv <0$, then λ will have the same magnitude but a positive sign. Fortunately, although the feasible set is nonconvex, the KKT point is the global minimum.

We see then that a possible simple way to handle sufficiency geometrically in the original feasible space in the presence of (active) equalities is to direct these equalities properly and then try to prove convexity. Of course, lack of convexity will not disprove the optimality of the KKT point. Proper directions for the equalities can be found by applying the monotonicity principles. If the point tested for optimality has $x_1 \geq 2$, then f is increasing wrt x_1 and h must be decreasing wrt x_1; therefore, $h = x_2^2 - 2x_1 \equiv <0$ is the correct direction for proper bounding of x_1. Similarly, if $x_1 \leq 2$, we have $h \equiv >0$. This is exactly what we found above. ■

5.8 Linear Programming

A problem having objective and constraint functions that are all linear is called a *linear programming* (LP) problem. Although design models are rarely linear, solution strategies for this special case are sufficiently important to discuss them here. The traditional method for solving linear programs is the *Simplex Method* and its revisions, originally developed by Dantzig (1963). The solution strategy is presented here as a specialization of the methods we developed for nonlinear models and is equivalent to the Simplex Method although the morphology of the manipulations appears different. Our exposition follows closely the one by Best and Ritter (1985). The important idea of a *local active set strategy* is introduced here (see also Section 7.4).

The general statement of a linear programming model is

$$\begin{aligned} \min f &= \mathbf{c}^T\mathbf{x} && (n \text{ variables}) \\ \text{subject to } \mathbf{h} &= \mathbf{A}_1\mathbf{x} - \mathbf{b}_1 = \mathbf{0} && (m_1 \text{ equalities}), \\ \mathbf{g} &= \mathbf{A}_2\mathbf{x} - \mathbf{b}_2 \leq \mathbf{0} && (m_2 \text{ inequalities}), \end{aligned} \tag{5.63}$$

where the vectors $\mathbf{b}_1$, $\mathbf{b}_2$, and $\mathbf{c}$ and the matrices $\mathbf{A}_1$ and $\mathbf{A}_2$ are all parameters. Their dimensions match the dimensions of $\mathbf{x}$, $\mathbf{h}$, and $\mathbf{g}$. Before we proceed with discussing the general problem (5.63), we will look at an example and study the geometry of such problems in two dimensions.

Example 5.12 Consider the problem

$$\begin{aligned} &\max f = 2x_1 + x_2 \\ &\text{subject to } g_1\colon x_1 + 2x_2 \leq 8, \\ &\qquad g_2\colon x_1 - x_2 \leq \tfrac{3}{2}, \\ &\qquad g_3\colon 2x_1 \geq 1, \\ &\qquad g_4\colon 2x_2 \geq 1. \end{aligned} \tag{a}$$

The problem is reformulated in the standard negative null form

$$\begin{aligned} &\min f = -2x_1 - x_2 \\ &\text{subject to } g_1 = x_1 + 2x_2 - 8 \leq 0, \\ &\qquad g_2 = x_1 - x_2 - 1.5 \leq 0, \\ &\qquad g_3 = -2x_1 + 1 \leq 0, \\ &\qquad g_4 = -2x_2 + 1 \leq 0, \end{aligned} \tag{b}$$

so that $\mathbf{c} = (-2, -1)^T$, $\mathbf{b}_2 = (8, 1.5, -1, -1)^T$, and

$$\mathbf{A}_2^T = \begin{pmatrix} 1 & 1 & -2 & 0 \\ 2 & -1 & 0 & -2 \end{pmatrix}.$$

No equalities exist, and so $\mathbf{b}_1 = \mathbf{0}$ and $\mathbf{A}_1 = \mathbf{0}$. The geometric representation of the feasible space and the contours of f are shown in Figure 5.13. The constraint surface

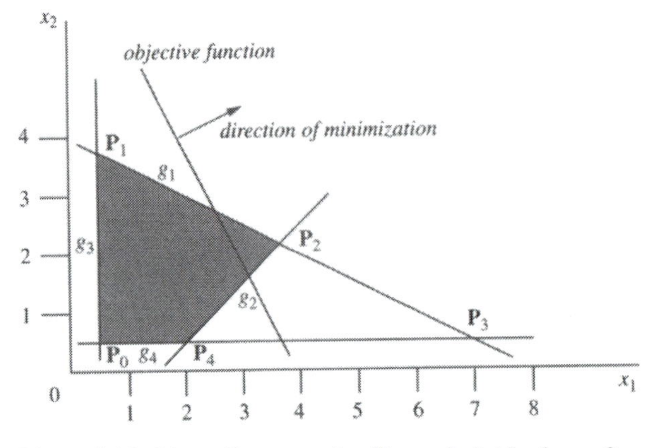

Figure 5.13. Linear Programming Example 5.12; $f = -2x_1 - x_2$.

is comprised of intersecting straight lines. The objective function contours are straight parallel lines. It is geometrically evident that the solution will be found by moving the line $f = \text{const}$ up and to the right, to the furthest point of the feasible domain, that is, point $\mathbf{P}_2$. This point is the intersection of the active constraints $g_1 = 0$ and $g_2 = 0$, and so the optimum is constraint bound.

Let us now use monotonicity analysis to solve this problem. In symbolic form we have

$$
\begin{aligned}
&\min f(x_1^-, x_2^-) \\
&\text{subject to } g_1(x_1^+, x_2^+) \le 0, \\
&\qquad g_2(x_1^+, x_2^-) \le 0, \\
&\qquad g_3(x_1^-) \le 0, \\
&\qquad g_4(x_2^-) \le 0.
\end{aligned} \tag{c}
$$

To bound x_2 from above, g_1 must be active (it is critical). Therefore, set $x_1 = 8 - 2x_2$ and substitute in (b). After rearrangement, we get

$$
\begin{aligned}
&\min f(x_2^+) = -16 + 3x_2 \\
&\text{subject to } g_2(x_2^-) = 2.17 - x_2 \le 0, \\
&\qquad g_3(x_2^+) = x_2 - 3.75 \le 0, \\
&\qquad g_4(x_2^-) = 0.5 - x_2 \le 0.
\end{aligned} \tag{d}
$$

A lower bound for x_2 is provided by both g_2 and g_4, but g_2 is dominant and therefore active. Thus $x_{2*} = 2.17$ and $x_{1*} = 3.67$. This is again point $\mathbf{P}_2$, the constraint-bound solution we found before. ■

Some immediate general remarks can be made based on this example. In a linear model, the objective and constraint functions are always monotonic. If equalities exist, we can assume, without loss of generality, that they have been eliminated,

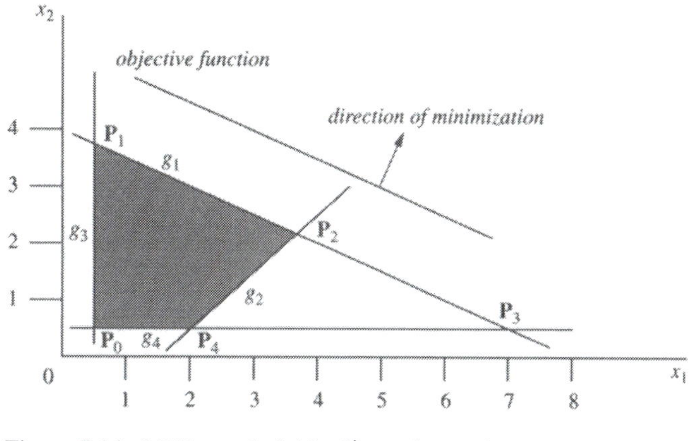

Figure 5.14. LP Example 5.13; $f' = -2x_1 - 4x_2$.

explicitly or implicitly, so that the resulting reduced problem will be monotonic. From the first monotonicity principle, there will be always at least one active constraint, identified possibly with the aid of dominance. Subsequent elimination of active constraints will always yield a new monotonic problem. The process will continue as long as activity can be proven, until no variables remain in the objective. The solution reached usually will be at a vertex of the feasible space, which is the intersection of as many active constraint surfaces as there are variables. The only other possibility is to have variables left in the constraints that do not appear in the objective function. In this case, the second monotonicity principle would indicate the existence of an infinite number of optima along the edge or face whose normal matches that of the objective function gradient. The limiting optimal values will be at the corners of the "*optimal face*" that correspond to upper and lower bounds on the variable not appearing in the objective function.

Example 5.13 Examine the same problem as in Example 5.12, but with different objective function (see Figure 5.14):

$$\begin{aligned}
&\min f' = -2x_1 - 4x_2\\
&\text{subject to } g_1 = x_1 + 2x_2 - 8 \le 0,\\
&\qquad g_2 = x_1 - x_2 - 1.5 \le 0,\\
&\qquad g_3 = -2x_1 + 1 \le 0,\\
&\qquad g_4 = -2x_2 + 1 \le 0.
\end{aligned}$$

As before, g_1 is active and elimination of x_1 results in

$$\begin{aligned}
&\min f = -16\\
&\text{subject to } g_2\colon x_2 \ge 2.17,\\
&\qquad g_3\colon x_2 \le 3.75,\\
&\qquad g_4\colon x_2 \ge 0.5.
\end{aligned}$$

The solution is now given by $2.17 \le x_{2*} \le 3.75$ and $x_{1*} + 2x_{2*} = 8$. Geometrically, the optimal face, an edge in this case, is the line segment $g_1 = 0$ limited by the points $\mathbf{P}_1$ and $\mathbf{P}_2$ corresponding to $x_{2*} = 2.17$ (g_2 active) and $x_{2*} = 3.75$ (g_3 active). ■

Optimality Conditions

In the n-dimensional space $\mathcal{R}^n$, a hyperplane $\mathbf{a}^T\mathbf{x} = b$ can be thought of as separating two *half-spaces* defined by $\mathbf{a}^T\mathbf{x} \le b$ and $\mathbf{a}^T\mathbf{x} \ge b$. A half-space $\mathbf{a}^T\mathbf{x} \le b$ is a convex set. The constraint set $\{\mathbf{a}_1^T\mathbf{x} \le b_1, \mathbf{a}_2^T\mathbf{x} \le b_2, \ldots, \mathbf{a}_m^T\mathbf{x} \le b_m\}$, or $\mathbf{Ax} \le \mathbf{b}$, is the intersection of the convex half-spaces and is therefore also convex. Thus the LP problem involves minimization of a convex (linear) function over a convex constraint set. Therefore, the KKT optimality conditions, applied to LP models, will be both necessary and sufficient. Moreover, if the KKT point is unique it will be the global minimizer. Recalling that the gradient of $f = \mathbf{a}^T\mathbf{x}$ is simply $\mathbf{a}^T$, we can state the KKT conditions for the LP problem (5.63) as follows:

A regular point $\mathbf{x}_*$ is an optimal solution for the LP problem if and only if there exist vectors $\boldsymbol{\lambda}$ and $\boldsymbol{\mu}$ such that

$$\begin{aligned}
&1.\ \mathbf{A}_1\mathbf{x}_* = \mathbf{b}_1, \quad \mathbf{A}_2\mathbf{x}_* \le \mathbf{b}_2;\\
&2.\ \mathbf{c}^T + \boldsymbol{\lambda}^T\mathbf{A}_1 + \boldsymbol{\mu}^T\mathbf{A}_2 = \mathbf{0}^T, \quad \boldsymbol{\mu} \ge \mathbf{0}, \quad \boldsymbol{\lambda} \ne \mathbf{0}, \qquad (5.64)\\
&3.\ \boldsymbol{\mu}^T(\mathbf{A}_2\mathbf{x}_* - \mathbf{b}_2) = 0.
\end{aligned}$$

This is derived directly from the KKT conditions (5.61) applied to a convex problem. Note that regularity of $\mathbf{x}$ everywhere means that the constraints are linearly independent.

In the linear programming terminology, a corner point where at least n (the dimension of $\mathbf{x}$) constraints are tight with linearly independent gradients is called an *extreme* point. If these constraints are exactly equal to n, then we have a *nondegenerate extreme point*, which is equivalent to our previous definition of a regular constraint-bound point. In the nonlinear programming (NLP) terminology, nondegeneracy refers to association of active constraints with *strictly* positive multipliers. This in LP is expressed by saying that $\mathbf{x}_*$ must satisfy the *strict complementary slackness condition*: $\mathbf{a}_j^T\mathbf{x} \le b_j$ implies $\mu_j > 0$ for all $j = 1, \ldots, m_2$. We prefer to adhere to the usual NLP terminology in our present discussion and the reader should recall these nuances when consulting various references.

Example 5.14 Consider the problem of Example 5.12 with an additional constraint g_5: $x_1 \le 3.67$ (see Figure 5.15). The optimality conditions apart from feasibility are

$$\begin{pmatrix}-2\\-1\end{pmatrix} + \mu_1\begin{pmatrix}1\\2\end{pmatrix} + \mu_2\begin{pmatrix}1\\-1\end{pmatrix} + \mu_3\begin{pmatrix}-2\\0\end{pmatrix} + \mu_4\begin{pmatrix}0\\-2\end{pmatrix} + \mu_5\begin{pmatrix}1\\0\end{pmatrix} = \begin{pmatrix}0\\0\end{pmatrix},$$

$$\mu_1(x_1 + 2x_2 - 8) = 0, \quad \mu_1 \ge 0,$$

$$\mu_2(x_1 - x_2 - 1.5) = 0, \quad \mu_2 \ge 0,$$

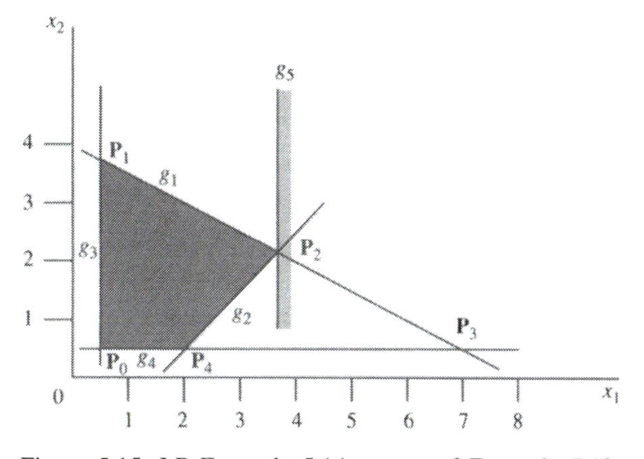

Figure 5.15. LP Example 5.14; repeat of Example 5.12 with additional constraint g_5: $x_1 \leq 3.67$.

$$\mu_3(-2x_1 + 1) = 0, \qquad \mu_3 \geq 0,$$

$$\mu_4(-2x_2 + 1) = 0, \qquad \mu_4 \geq 0,$$

$$\mu_5(x_1 - 3.67) = 0, \qquad \mu_5 \geq 0.$$

At the point $\mathbf{x} = (3.67, 2.17)^T$, constraints g_1, g_2, and g_5 are satisfied as strict equalities while g_3 and g_4 are inactive; so let $\mu_3 = \mu_4 = 0$ and write out the gradient equations:

$$\mu_1 + \mu_2 + \mu_5 = 2, \quad 2\mu_1 - \mu_2 = 1.$$

From the second one, $\mu_2 = 2\mu_1 - 1 \geq 0$ or $\mu_1 \geq \frac{1}{2}$. The first one then gives $3\mu_1 + \mu_5 = 3$ and $\mu_1 = (3 - \mu_5)/3 \geq \frac{1}{2}$ or $\mu_5 \leq \frac{3}{2}$. Thus, any $\mu_5 \leq \frac{3}{2}$ including $\mu_5 = 0$ will satisfy the optimality requirements. The situation is shown in Figure 5.15. Point $\mathbf{P}_2$ is nonregular, which explains why there are values $\mu_j > 0$ associated with $g_j = 0$, $j = 1, 2, 5$.

Next reconsider the problem stated in Example 5.13 (without the added constraint g_5) (Figure 5.14). Optimality conditions, other than feasibility, are

$$\begin{pmatrix} -2 \\ -4 \end{pmatrix} + \mu_1 \begin{pmatrix} 1 \\ 2 \end{pmatrix} + \mu_2 \begin{pmatrix} 1 \\ -1 \end{pmatrix} + \mu_3 \begin{pmatrix} -2 \\ 0 \end{pmatrix} + \mu_4 \begin{pmatrix} 0 \\ -2 \end{pmatrix} = \begin{pmatrix} 0 \\ 0 \end{pmatrix},$$

$$\mu_1(x_1 + 2x_2 - 8) = 0, \qquad \mu_1 \geq 0,$$

$$\mu_2(x_1 - x_2 - 1.5) = 0, \qquad \mu_2 \geq 0,$$

$$\mu_3(-2x_1 + 1) = 0, \qquad \mu_3 \geq 0,$$

$$\mu_4(-2x_2 + 1) = 0, \qquad \mu_4 \geq 0.$$

At the point $\mathbf{P}_2$ (Figure 5.14), constraints g_1 and g_2 are active and g_3 and g_4 are inactive with $\mu_3 = \mu_4 = 0$. The gradient equations there are $\mu_1 + \mu_2 = 2$ and $2\mu_1 - \mu_2 = 4$. The unique solution for these is $\mu_1 = 2$ and $\mu_2 = 0$. Thus, $\mathbf{P}_2$ is degenerate. Note that this degeneracy is associated with nonunique optimizers, which give the same value for the objective. The reader should verify that point $\mathbf{P}_1$ is also degenerate, but other optimizers on the edge $\mathbf{P}_1\mathbf{P}_2$ are not degenerate because only g_1 would be active and the multipliers for the inactive constraints would be zero. ■

The certain knowledge that the optimal solution of an LP problem is an extreme point (or face) motivates a special iterative procedure for reaching the optimum. Starting from a given (feasible) extreme point, we change the variables to move to an adjacent extreme point where the objective function has a lower value. This means movement along a selected edge emanating from the current extreme point. We continue in this way until an extreme point is reached where no further improvement in the objective function is possible. This is the basic LP algorithm that we will describe in this section and it is equivalent to the Simplex Method. Handling starting points that are not extreme and/or feasible, as well as handling nonregular and degenerate points, requires rather simple modifications of the basic algorithm.

Basic LP Algorithm

We examine first the case with only inequality constraints:

$$\min f = \mathbf{c}^T\mathbf{x} \qquad (5.65)$$
$$\text{subject to } g_j = \mathbf{a}_j^T\mathbf{x} \leq b_j, \quad j = 1, \ldots, m.$$

Assume that a regular extreme point $\mathbf{x}_0$ is known where the first $n(<m)$ constraints are active. To partition these active constraints, define

$$\mathbf{A}_0^T = (\mathbf{a}_1, \ldots, \mathbf{a}_n), \quad \mathbf{b}_0 = (b_1, \ldots, b_n)^T. \qquad (5.66)$$

The columns of the $n \times n$ matrix $\mathbf{A}_0^T$ are the gradients of the constraints active at $\mathbf{x}_0$ and, by the regularity assumption, $\mathbf{A}_0$ is nonsingular with an inverse defined as

$$\mathbf{D}_0 \triangleq \mathbf{A}_0^{-1} = (\mathbf{d}_1, \ldots, \mathbf{d}_n). \qquad (5.67)$$

In general, there will be n edges emanating from $\mathbf{x}_0$ that can be represented in the form $\mathbf{x}_0 - \alpha_0\mathbf{d}_i$ where $\alpha_0 > 0$ and $i = 1, \ldots, n$. In fact,

$$\mathbf{a}_j^T(\mathbf{x}_0 - \alpha_0\mathbf{d}_i) = \mathbf{a}_j^T\mathbf{x}_0 - \alpha_0\left(\mathbf{a}_j^T\mathbf{d}_i\right) = \begin{cases} b_j, & j \neq i, \\ b_i - \alpha_0 < b_i, & j = i, \end{cases} \qquad (5.68)$$

which simply means that moving along an edge requires continuing activity of $n-1$ constraints, with the remaining one of the n becoming inactive. Note that $\mathbf{a}_j^T\mathbf{d}_i = 0$ for $j \neq i$ and $\mathbf{a}_j^T\mathbf{d}_j = 1$, by definition of the inverse matrix.

A first iteration can be performed according to $\mathbf{x}_1 = \mathbf{x}_0 - \alpha_0\mathbf{s}_0$ where the search direction $\mathbf{s}_0$ and step size α_0 must be determined. A search direction for points along an edge will be a component of $\mathbf{D}_0$. Any such component will do provided a decrease in

f is guaranteed. Taking $f(\mathbf{x}_1) = \mathbf{c}^T(\mathbf{x}_0 - \alpha_0\mathbf{s}_0) = \mathbf{c}^T\mathbf{x}_0 - \alpha_0\mathbf{c}^T\mathbf{s}_0$, we need $\mathbf{c}^T\mathbf{s}_0 > 0$ for descent. A traditional way to do this is to set $\mathbf{s}_0 = \mathbf{d}_s$ such that

$$\mathbf{c}^T\mathbf{d}_s = \max_{1\le j\le n} \{\mathbf{c}^T\mathbf{d}_j : \mathbf{c}^T\mathbf{d}_j > 0\}. \tag{5.69}$$

Although this will give a direction of descent, the actual decrease in the objective function will depend on the step size. We need to find the maximum step size that maintains feasibility along an edge. Any feasible step size $\alpha \ge 0$ must satisfy the previously inactive constraints:

$$\mathbf{a}_j^T(\mathbf{x}_0 - \alpha\mathbf{s}_0) \le b_j, \quad j = n+1, \dots, m. \tag{5.70}$$

We know from (5.68) that the other constraints $j = 1, \dots, n$ will be satisfied by the choice $\mathbf{s}_0 = \mathbf{d}_s$. Since $\mathbf{a}_j^T\mathbf{x}_0 < b_j$ $(j = n+1, \dots, m)$, the feasibility requirement (5.70) need only be checked for those js where $\mathbf{a}_j^T\mathbf{s}_0 < 0$. Therefore, we select α_0 to be the least upper bound of the set of αs such that (5.70) is satisfied with $\mathbf{a}_j^T\mathbf{s}_0 < 0$. Thus, for $\mathbf{s}_0 = \mathbf{d}_s$ we have

$$\alpha_0 = \min_{n+1\le j\le m} \left\{ \frac{\mathbf{a}_j^T\mathbf{x}_0 - b_j}{\mathbf{a}_j^T\mathbf{d}_s} \ \text{with}\ \mathbf{a}_j^T\mathbf{d}_s < 0 \right\}. \tag{5.71}$$

This condition may give more than one j corresponding to α_0. This is what we call a *tie*. Resolving ties is important for handling degeneracy. Here we assume ad hoc that α_0 corresponds to a constraint with index $j = l$, with l the smallest index found from (5.71).

The selections suggested by (5.69) and (5.71) define an algorithm that utilizes an *active set strategy*, that is, a set of rules by which inequality constraints may be added or deleted from the set of the currently active ones in a way that promotes progress toward the solution. Active set strategies are also used in NLP algorithms and can be particularly helpful in solving design models with relatively many inequalities and few design variables. This subject will be discussed again in Chapters 6 and 7.

Before we look at the equalities, we will make an observation that has computational importance. An LP iteration replaces one active constraint $\mathbf{g}_s$ with another one $\mathbf{g}_l$ according to (5.69) and (5.71). To do this, we need the inverse $\mathbf{D}_0 = \mathbf{A}_0^{-1}$ at $\mathbf{x}_0$. Once point $\mathbf{x}_1$ is identified, we will need the inverse $\mathbf{D}_1 = \mathbf{A}_1^{-1}$ to repeat the process. The matrices $\mathbf{A}_0$ and $\mathbf{A}_1$ differ only in one column:

$$\begin{aligned} \mathbf{A}_0^T &= (\mathbf{a}_1, \dots, \mathbf{a}_{s-1}, \mathbf{a}_s, \mathbf{a}_{s+1}, \dots, \mathbf{a}_n), \\ \mathbf{A}_1^T &= (\mathbf{a}_1, \dots, \mathbf{a}_{s-1}, \mathbf{a}_l, \mathbf{a}_{s+1}, \dots, \mathbf{a}_n). \end{aligned} \tag{5.72}$$

We should expect that this special property would be exploited so that complete inversion of the matrices is not required at every iteration. This is in fact possible with the following procedure:

1. Assume that $\mathbf{A}_0^T$ and $\mathbf{A}_1^T$ are nonsingular and that $\mathbf{a}_l$ and $\mathbf{d}_s$ are not orthogonal, that is, $\mathbf{a}_l^T\mathbf{d}_s \neq 0$.
2. Define for simplicity of presentation

$$\begin{aligned} \mathbf{D}_0 &= \mathbf{A}_0^{-1} = (\mathbf{d}_1, \ldots, \mathbf{d}_{s-1}, \mathbf{d}_s, \mathbf{d}_{s+1}, \ldots, \mathbf{d}_n)_0, \\ \mathbf{D}_1 &= \mathbf{A}_1^{-1} = (\mathbf{d}_1, \ldots, \mathbf{d}_{s-1}, \mathbf{d}_s, \mathbf{d}_{s+1}, \ldots, \mathbf{d}_n)_1. \end{aligned} \tag{5.73}$$

3. Compute the elements of $\mathbf{D}_1$ according to

$$\begin{aligned} (\mathbf{d}_i)_1 &= (\mathbf{d}_i)_0 - \left[\frac{\mathbf{a}_l^T(\mathbf{d}_i)_0}{\mathbf{a}_l^T(\mathbf{d}_s)_0}\right](\mathbf{d}_s)_0, \quad i \neq s, \\ (\mathbf{d}_s)_1 &= (\mathbf{d}_s)_0/\mathbf{a}_l^T(\mathbf{d}_s)_0. \end{aligned} \tag{5.74}$$

These expressions are generalized for any iteration k by replacing 0 with k and 1 with $k+1$. Using them reduces the arithmetic operations required for inversion by a factor of n.

Let us now examine the case with equality and inequality constraints:

$$\begin{aligned} &\min f = \mathbf{c}^T\mathbf{x} \\ &\text{subject to } h_j = \mathbf{a}_j^T\mathbf{x} - b_j = 0, \quad j = 1, \ldots, m_1, \\ &\qquad\qquad g_j = \mathbf{a}_j^T\mathbf{x} - b_j \leq 0, \quad j = m_1 + 1, \ldots, m_1 + m_2. \end{aligned} \tag{5.75}$$

The only difference now is that the equality constraints (barring degeneracy) will remain active throughout the iterations, and so all matrices used above will be simply augmented by treating the equalities as active inequalities. In the active set strategy implemented by (5.69) and (5.70), the equalities will be *excluded*. Thus, the selection of the search index s will be made only among indices $j = m_1 + 1, \ldots, m_1 + m_2$.

A more formal exposition of the above can be given as we did for the inequalities. This is not particularly important here and we will omit it. The main task is proper bookkeeping of the currently active set.

These ideas now will be illustrated with some examples.

Example 5.15 Reconsider Example 5.12 shown in Figure 5.13 and stated again as:

$$\begin{aligned} &\min f = -2x_1 - x_2 \\ &\text{subject to } g_1 = x_1 + 2x_2 - 8 \leq 0, \quad g_3 = -2x_1 + 1 \leq 0, \\ &\qquad\qquad g_2 = x_1 - x_2 - 1.5 \leq 0, \quad g_4 = -2x_2 + 1 \leq 0. \end{aligned}$$

Take the point $\mathbf{P}_0$, $\mathbf{x}_0 = (0.5, 0.5)^T$, as the starting point. We will use the symbol $\mathcal{J}_k$ to indicate the indices of the active constraints in iteration k, to help us in bookkeeping.

INITIALIZATION $(k = 0)$

We have

$$\mathbf{x}_0 = (0.5, 0.5)^T, \quad \mathcal{J}_0 = (3, 4)$$

and

$$\mathbf{A}_0^T = \begin{pmatrix} -2 & 0 \\ 0 & -2 \end{pmatrix}, \quad \mathbf{D}_0 = \begin{pmatrix} -\frac{1}{2} & 0 \\ 0 & -\frac{1}{2} \end{pmatrix}.$$

To select the search direction $\mathbf{s}_0$ and leaving constraint g_s we evaluate (5.69):

$$\mathbf{c}^T\mathbf{d}_s = \max_{j=3,4}\left[(-2,-1)\begin{pmatrix} -\frac{1}{2} \\ 0 \end{pmatrix}, (-2,-1)\begin{pmatrix} 0 \\ -\frac{1}{2} \end{pmatrix}\right] = 1.$$

Therefore, $s = 3$ and $\mathbf{s}_0 = \mathbf{d}_3 = (-\frac{1}{2}, 0)^T$. The new point will lie on the edge $\mathbf{x}_0 - \alpha_0\mathbf{d}_3$, which is the boundary of the active constraint $g_4 = 0$. We must now select a value for α_0. We do this by using (5.71) for $j = 1, 2$. Calculate the values

$$\mathbf{a}_1^T\mathbf{d}_3 = (1, 2)\left(-\tfrac{1}{2}, 0\right)^T = -\tfrac{1}{2}, \quad \mathbf{a}_2^T\mathbf{d}_3 = (1, -1)\left(-\tfrac{1}{2}, 0\right)^T = -\tfrac{1}{2}.$$

Since both are negative, we next calculate

$$\alpha_0 = \min_{j=1,2}\left[\frac{(1,2)(0.5,0.5)^T - 8}{(-0.5)}, \frac{(1,-1)(0.5,0.5)^T - 1.5}{(-0.5)}\right] = 3,$$

which corresponds to $l = 2$ and the new entering active constraint g_2. Thus, the new point must be $\mathbf{P}_4$, with

$$\mathbf{x}_1 = \mathbf{x}_0 - \alpha_0\mathbf{s}_0 = (0.5, 0.5)^T - 3\left(-\tfrac{1}{2}, 0\right)^T = (2, 0.5)^T.$$

FIRST ITERATION $(k = 1)$

We have now

$$\mathbf{x}_1 = (2, 0.5)^T, \quad \mathcal{J}_1 = (2, 4)$$

and

$$\mathbf{A}_1^T = \begin{pmatrix} 1 & 0 \\ -1 & -2 \end{pmatrix}, \quad \mathbf{D}_1 = \begin{pmatrix} 1 & -0.5 \\ 0 & -0.5 \end{pmatrix}.$$

Note that $\mathbf{D}_1$ is calculated from (5.73) and (5.74) by letting

$$(\mathbf{d}_1)_0 = (-0.5, 0)^T = (\mathbf{d}_s)_0, \quad (\mathbf{d}_2)_0 = (0, -0.5)^T, \quad \text{and} \quad \mathbf{a}_l^T = (1, -1)$$

so that

$$(\mathbf{d}_2)_1 = \begin{pmatrix} 0 \\ -0.5 \end{pmatrix} - \left[\frac{(1,-1)\begin{pmatrix} 0 \\ -0.5 \end{pmatrix}}{(1,-1)\begin{pmatrix} -0.5 \\ 0 \end{pmatrix}}\right]\begin{pmatrix} -0.5 \\ 0 \end{pmatrix} = \begin{pmatrix} -0.5 \\ -0.5 \end{pmatrix},$$

$$(\mathbf{d}_s)_1 = (\mathbf{d}_1)_1 = \left[\frac{1}{(1,-1)\begin{pmatrix} -0.5 \\ 0 \end{pmatrix}}\right]\begin{pmatrix} -0.5 \\ 0 \end{pmatrix} = \begin{pmatrix} 1 \\ 0 \end{pmatrix}.$$

The new leaving constraint is selected from evaluating $\mathbf{c}^T\mathbf{d}_1 = (-2, -1)(1, 0)^T = -2$, $\mathbf{c}^T\mathbf{d}_2 = (-2, -1)(-0.5, -0.5)^T = 1.5$, which gives $\mathbf{s}_1 = \mathbf{d}_2$ with g_4 being deactivated

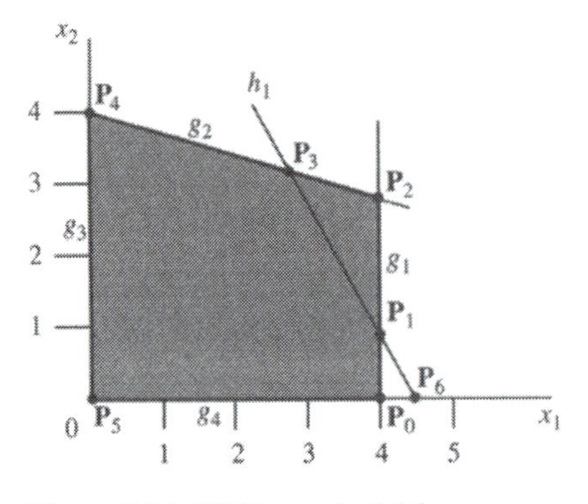

Figure 5.16. LP Example 5.16.

($s = 4$). So we will move along $g_2 = 0$. The value of α_1 is found again from (5.71) for $j = 1, 3$. Calculate the values:

$$\mathbf{a}_1^T(\mathbf{d}_2)_1 = (1, 2)(-0.5, -0.5)^T = -1.5,$$
$$\mathbf{a}_3^T(\mathbf{d}_2)_1 = (-2, 0)(-0.5, -0.5)^T = 1$$

Since only the first is negative, we calculate $\alpha_1 = [(1, 2)(2, 0.5)^T - 8]/(-1.5) = 3.33$, which corresponds to $l = 1$ and the new entering constraint g_1. The new point is $\mathbf{P}_2$ with $\mathbf{x}_2 = \mathbf{x}_1 - \alpha_1\mathbf{s}_1 = (2, 0)^T - 3.33(-0.5, -0.5)^T = (3.67, 2.17)^T$.

SECOND ITERATION ($k = 2$)

Repeating calculations as before, we find $\mathbf{D}_2$ and show that no positive $\mathbf{c}^T\mathbf{d}_j$ exists that gives a nonzero feasible step size. Therefore the point is optimal. ■

Example 5.16 Consider the problem shown in Figure 5.16 and stated as

$$\min f = -x_2$$
$$\text{subject to } h_1 = 2x_1 + x_2 - 9 = 0,$$
$$g_1 = x_1 - 4 \leq 0,$$
$$g_2 = x_1 + 3x_2 - 12 \leq 0,$$
$$g_3 = -x_1 \leq 0,$$
$$g_4 = -x_2 \leq 0.$$

In terms of the general statement (5.75), the problem has $m_1 = 1, m_2 = 4$. Any active set must include h_1. Recalling our discussion in Example 5.11, we could find an appropriate direction for h_1. If h_1 is absent, monotonicity analysis shows immediately that the optimum will be at the intersection of the active constraints g_2 and g_3, point $\mathbf{P}_4$. A directed h_1 should have this point in its infeasible half-space, that is, $h_1 \equiv >0$. The directed feasible domain would be the triangle $\mathbf{P}_1\mathbf{P}_2\mathbf{P}_3$. Note that with h_1 directed as above, monotonicity analysis immediately indicates g_2 active and $\mathbf{P}_3$ the optimum point.

For an iterative procedure, assume that we start at point $\mathbf{P}_0$, $\mathbf{x}_0 = (4, 0)^T$, and set $\mathcal{I}_0 = (g_1, g_4)$, where the symbol $\mathcal{I}_0$ is used for the set of active constraints, including equalities, instead of the set $\mathcal{J}_0$ that was defined to contain just the indices of the active inequality constraints (this is done to avoid more complicated notation). Thus,

$$\mathbf{A}_0^T = \begin{pmatrix} 1 & 0 \\ 0 & -1 \end{pmatrix}, \quad \mathbf{D}_0 = \mathbf{A}_0^{-1} = \begin{pmatrix} 1 & 0 \\ 0 & -1 \end{pmatrix}.$$

The search direction $\mathbf{s}_0$ and leaving constraint g_s is found from calculating

$$(\mathbf{c}^T\mathbf{d}_s)_0 = \max_{\mathcal{I}_0} \left\{ (0, -1)\begin{pmatrix} 1 \\ 0 \end{pmatrix}, (0, -1)\begin{pmatrix} 0 \\ -1 \end{pmatrix} \right\} = 1,$$

which gives $g_s = g_4$. The entering constraint must be h_1 and the corresponding step size is simply $\alpha_0 = (\mathbf{a}_j^T\mathbf{x}_0 - b_j)/\mathbf{a}_j^T\mathbf{d}_s$ for $j = m_1 = 1$, that is,

$$\alpha_0 = \frac{(2, 1)(4, 0)^T - 9}{(2, 1)(0, -1)^T} = 1.$$

The new point $\mathbf{x}_1 = \mathbf{x}_0 - \alpha_0(\mathbf{d}_s)_0 = (4, 0)^T - (0, -1)^T = (4, 1)^T$ is the extreme point $\mathbf{P}_1$ as expected.

To proceed with the next iteration we set $\mathcal{I}_1 = (g_1, h_1)$ and get

$$\mathbf{A}_1^T = \begin{pmatrix} 1 & 2 \\ 0 & 1 \end{pmatrix}, \quad \mathbf{D}_1 = \mathbf{A}_1^{-1} = \begin{pmatrix} 1 & 0 \\ -2 & 1 \end{pmatrix}.$$

The matrix $\mathbf{D}_1$ should be calculated by (5.74) in larger problems. Now we must retain h_1 in the active set, and so g_1 is the leaving constraint. If more candidates existed, as is the case in larger problems, the $(\mathbf{c}^T\mathbf{d}_j)$ values would be used.

Here we set $g_s = g_1$ and we calculate the values $\mathbf{a}_j^T\mathbf{d}_s$ for the inactive constraints, g_2: $(1, 3)(1, -2)^T = -5$, g_3: $(-1, 0)(1, -2)^T = -1$, and g_4: $(0, -1)(1, -2)^T = 2$. Only g_2 and g_3 are candidates for entering the active set. We then calculate

$$\alpha_1 = \min\left\{ \frac{(1, 3)(4, 1)^T - 12}{(-5)}, \frac{(-1, 0)(4, 1)^T - 0}{(-1)} \right\} = 1,$$

which indicates $g_l = g_2$. The new point $\mathbf{P}_3$ is the intersection of $h_1 = 0$ and $g_2 = 0$, calculated from $\mathbf{x}_2 = (\mathbf{x}_1 - \alpha_1(\mathbf{d}_s)_1) = (4, 1)^T - (1, -2)^T = (3, 3)^T$.

At $\mathbf{x}_2$, the active set is $\mathcal{I}_2 = (g_2, h_1)$ and we have

$$\mathbf{A}_2^T = \begin{pmatrix} 1 & 2 \\ 3 & 1 \end{pmatrix}, \quad \mathbf{D}_2 = \mathbf{A}_2^{-1} = \begin{pmatrix} -\frac{1}{5} & \frac{3}{5} \\ \frac{2}{5} & -\frac{1}{5} \end{pmatrix}.$$

Now we calculate $(0, -1)(-\frac{1}{5}, -\frac{2}{5})^T = -\frac{2}{5}$ and $(0, -1)(\frac{3}{5}, -\frac{1}{5})^T = \frac{1}{5}$. This indicates that improvement is possible only if we deactivate h_1, which is not acceptable. Therefore, the point $\mathbf{x}_2$ is optimal. ■

We conclude this section with one last observation, which is often applicable also to active set strategies for NLP problems. For the model {min $f = \mathbf{c}^T\mathbf{x}$, subject to $\mathbf{Ax} \leq \mathbf{b}$}, the optimality conditions (5.64) give multiplier values that satisfy $\mathbf{c}^T + \boldsymbol{\mu}^T\mathbf{A} = \mathbf{0}^T$, or $-\boldsymbol{\mu}^T = \mathbf{c}^T\mathbf{A}^{-1} = \mathbf{c}^T\mathbf{D}$, where $\mathbf{D}$ is just the inverse as defined in the basic LP algorithm. A single multiplier is thus given by

$$-\mu_j = \mathbf{c}^T\mathbf{d}_j, \quad j = 1, \ldots, m. \tag{5.76}$$

Then the rule for selecting the leaving constraint in the LP active set strategy is to deactivate the constraint with the *most negative estimate* of its associated Lagrange multiplier, as expressed by (5.69). The values $\mathbf{c}^T\mathbf{d}_j$ are indeed estimates of the true multipliers, which are properly defined only at the optimum, that is, only when the correct active set has been identified. (See also Section 7.4.)

5.9 Sensitivity

In our discussion of design optimization models in Chapters 1 and 2, we made a distinction among design variables, parameters, and constants. The parameters were treated as unchanging constants for the particular problem examined. This categorization is somewhat arbitrary in the sense that it depends on the modeler. For example, a parameter may be included among the variables if further knowledge from the model analysis so indicates. Moreover, the values given to parameters and/or constants have a degree of uncertainty because they are usually related to actual performance of the design and cannot be predicted accurately before operation begins. In some cases, accuracy cannot be obtained even then.

These observations point out that when an optimal solution has been identified, it would depend on the values of parameters used. Therefore, a *post-optimality* analysis would be necessary for a complete study of the problem and determination of the role of the parameters in the estimation of the optimal values. This analysis is often referred to as *parametric study*.

Sensitivity Coefficients

When *small* changes in the values of the parameters (or constants) are considered, we have a problem of *sensitivity analysis*. In this case, small changes in the values of the parameters produce a perturbation $\partial\mathbf{h}$ in the values of the constraint functions $\mathbf{h}$. In the standard form of nonlinear programming, such perturbations correspond to small deviations from the zero value of the constraints. Thus, sensitivity analysis studies the influence of perturbations $\partial\mathbf{h}$ about the point $\mathbf{h} = \mathbf{0}$, on the values $f(\mathbf{x}_*)$ and $\mathbf{x}_*$. Active inequalities may be included in the same way as equalities.

Let $\mathbf{x}_*$ be a feasible optimum and $\partial\mathbf{h}$ a vector of perturbations about $\mathbf{h}(\mathbf{x}_*) = \mathbf{0}$. Corresponding to $\partial\mathbf{h}$, there will be perturbations $\partial\mathbf{x}$ about $\mathbf{x}_*$. Using the partition of state and decision variables $\partial\mathbf{x} = (\partial\mathbf{d}, \partial\mathbf{s})^T$, we have

$$\partial\mathbf{h} = (\partial\mathbf{h}/\partial\mathbf{d})\partial\mathbf{d} + (\partial\mathbf{h}/\partial\mathbf{s})\partial\mathbf{s}, \tag{5.77}$$

where all the derivatives are calculated at $\mathbf{x}_*$. Multiplying through by $(\partial\mathbf{h}/\partial\mathbf{s})^{-1}$, since $\mathbf{x}_*$ is a regular point, and rearranging, we get

$$\partial\mathbf{s} = \left(\frac{\partial\mathbf{h}}{\partial\mathbf{s}}\right)^{-1}\partial\mathbf{h} - \left(\frac{\partial\mathbf{h}}{\partial\mathbf{s}}\right)^{-1}\left(\frac{\partial\mathbf{h}}{\partial\mathbf{d}}\right)\partial\mathbf{d}. \tag{5.78}$$

Using this expression, (5.14a) gives

$$\partial f = \partial z = \left(\frac{\partial f}{\partial\mathbf{s}}\right)\left(\frac{\partial\mathbf{h}}{\partial\mathbf{s}}\right)^{-1}\partial\mathbf{h} + \left(\frac{\partial z}{\partial\mathbf{d}}\right)\partial\mathbf{d}, \tag{5.79}$$

where $(\partial z/\partial\mathbf{d})$ is the reduced gradient as defined earlier; it must be zero since the perturbations are about $\mathbf{x}_*$. Thus, we can rewrite (5.79) as

$$\partial f(\mathbf{x}_*) = \partial z(\mathbf{x}_*) = \left(\frac{\partial z}{\partial\mathbf{h}}\right)_*\partial\mathbf{h}, \tag{5.80}$$

where

$$\left(\frac{\partial z}{\partial\mathbf{h}}\right)_* \triangleq \left(\frac{\partial f}{\partial\mathbf{s}}\right)_*\left(\frac{\partial\mathbf{h}}{\partial\mathbf{s}}\right)_*^{-1} \tag{5.81}$$

is an m-row vector whose elements are the partial derivatives of the objective function at the optimum with respect to perturbations $\partial\mathbf{h}$ in the constraint function values. The elements of this vector are usually called *sensitivity coefficients* and they give the rate of change of the optimal values of the objective relative to small changes in the values of the constraints. We stress the qualification *small* changes, because our analysis is based on first-order approximation valid locally near $\mathbf{x}_*$. If we compare (5.81) with (5.20), we see that at a constrained stationary point that is a minimum, that is, $\mathbf{x}_\dagger = \mathbf{x}_*$, we have

$$\left(\frac{\partial z}{\partial\mathbf{h}}\right)_* = -\boldsymbol{\lambda}^T. \tag{5.82}$$

Thus, the *sensitivity coefficients are given by the opposites of the elements of the multiplier vector* $\boldsymbol{\lambda}$.

Note that the above interpretation of the multipliers is based on using the negative null form of model. If the positive null form is used, that is, with $\mathbf{g}(\mathbf{x}) \geq \mathbf{0}$, then the definition of multipliers will be with a positive sign in Equation (5.20), namely,

$$\boldsymbol{\lambda}^T \triangleq (\partial f/\partial\mathbf{s})_{\mathbf{x}_\dagger}(\partial\mathbf{h}/\partial\mathbf{s})_{\mathbf{x}_\dagger}^{-1} \quad \text{(positive null form).} \tag{5.83}$$

The Lagrangian will be $L = f - \boldsymbol{\lambda}^T\mathbf{h} - \boldsymbol{\mu}^T\mathbf{g}$, as discussed in Section 5.6, and condition (5.82), which redefines the multipliers, will be

$$\left(\frac{\partial z}{\partial\mathbf{h}}\right)_* = \boldsymbol{\lambda}^T \quad \text{(positive null form).} \tag{5.84}$$

Thus, in the positive null form of model, the sensitivity coefficients are the multipliers themselves rather than their opposites. This should be kept in mind not only while reading this book but also when using standard computer codes. One should carefully check the exact model form used as input to the program and possibly also the definition of the Lagrangian.

Some confusion also exists in the newer literature concerning sensitivity terminology. In some cases, the term sensitivity analysis denotes the task of finding the derivatives $\partial f/\partial x_i$ at *any* design point, not the values $\partial z/\partial h_j$ at the optimum. Since we defined the λs as Lagrange multipliers at the optimum, our present terminology is consistent.

This interpretation of Lagrange multipliers is useful in understanding the rationale of active set strategies, as will be seen in Chapter 7.

5.10 Summary

We explored the basic ideas for optimality of constrained problems. The recurring theme in this book – that model affects solution – was once again encountered in our discussion on the need of regularity and nondegeneracy in the constraint set, in order to pose the optimality conditions in a straightforward way. This echoes the discussions in Chapter 3, where the clear distinction was made between constraints that are active and constraints that happen to be satisfied as equalities at the optimum.

Constrained optimality can be viewed in two ways. One requires the choice of state and decision variables and subsequent transformation of the problem into the reduced (decision) space. In the reduced space, the problem is treated as unconstrained and so optimality is verified by a zero reduced gradient and a positive-definite reduced Hessian. One potential advantage of this approach is that the reduced space is of smaller dimension so that operations required to achieve optimality are limited to a smaller number of variables (equal to the number of degrees of freedom). Another way is to work with the Lagrangian function, find its stationary points, and pick one for which the Hessian of the Lagrangian (possibly projected on the tangent subspace of the constraints) is positive-definite. Now, however, the variables are the original $\mathbf{x}$ plus the multipliers $\boldsymbol{\lambda}$, and so although the dimension has increased, the state and decision variable partition is avoided. In an iterative method, this means that we need not solve a nonlinear system to recover correct state values for each decision change – which is computationally expensive. This symmetric approach requiring no variable partitioning is realized in the gradient projection method.

The above comments are important because in practice the optimality conditions are only useful as a target of an iterative scheme. The methods of Chapter 7 are all motivated by the need to satisfy the optimality conditions at the end of the iterations.

Linear programming is a subject in its own right, particularly when problem size becomes very large. The active set strategy that was presented as a solution procedure is a special case of NLP active set strategies that are further explored in Chapter 7. One important point, however, is that the LP active set strategy can be rigorously proved to work under very mild conditions, while in the NLP case the strategy is by and large heuristic. The reasons will become apparent in the discussion of Section 7.4.

Notes

Much of the material in this chapter is classical optimization theory and as such can be found in many optimization texts in more or less detail. The mathematical requirements in modeling the constraint set are nicely described in Luenberger (1973, 1984) and Russell (1970). The KKT conditions are a special case of the Fritz John conditions where another multiplier for the objective function is included, one that is not necessarily positive but possibly zero. The assumption of regularity removes this last possibility and results in the KKT conditions. Occasionally, there are meaningful problems where a zero multiplier for the objective is possible. A more detailed discussion of optimality conditions and various constraint qualifications can be found in Bazaraa and Shetty (1979) or Mangasarian (1969).

The differential theory leading to the first- and second-order conditions in the reduced (constrained) space essentially follows the one in Beightler, Phillips, and Wilde (1979); the equivalence proof of the reduced space and projected Lagrangian differential quadratic forms, Equation (5.35), was new in the first edition. It should be noted that the optimality proofs given are based only on calculus and are straightforward, albeit somewhat tedious. Elegance can be achieved if convex theory is used, but this would have been an additional burden on the reader. Such proofs can be found in several of the NLP texts cited throughout this book.

The Karush–Kuhn–Tucker conditions are named after Karush (1939) and Kuhn and Tucker (1951) who derived these conditions first.

For the original ideas about using feasible directions in an iterative scheme see, for example, Zoutendijk (1960) and Rosen (1960, 1961), whose name is usually associated with the gradient projection method. The reduced (decision) space idea is usually attributed to Wolfe (1963) for linear constraints and its generalization is due to Wilde and Beightler (1967) and Abadie and Carpentier (1969). For a more comprehensive exposition see also Ben-Israel, Ben-Tal, and Zlobek (1981).

Linear programming became a field of study when Dantzig developed the Simplex Method in the late 1940s and early 1950s (see Dantzig, 1963). The method assumes a form of standard model (the "canonical" one) different from what we use in NLP, essentially one with only equalities. It also uses specialized terminology that is of no particular interest here. The active set approach we presented in Section 5.8 based on the text by Best and Ritter (1985) allows the reader to see LP as a special case of NLP strategies and to increase comprehension of both without additional formulations and terminology. Best and Ritter prove the equivalence of that and the Simplex Method. Among the very large literature on linear programming, we mention the classical texts by Luenberger (1984) and Murty (1983). The book by Murty (1986) ties many LP and NLP ideas together.

Exercises

5.1 Examine the corner points of the feasible domain in Example 5.1 and determine directions of feasible descent. Show the gradient directions at each point.

5.2 Sketch graphically the problem

$$\min f(\mathbf{x}) = (x_1 + 1)^2 + (x_2 - 2)^2$$

$$\text{subject to } g_1 = x_1 - 2 \leq 0, \quad g_3 = -x_1 \leq 0,$$
$$g_2 = x_2 - 1 \leq 0, \quad g_4 = -x_2 \leq 0.$$

Find the optimum graphically. Determine directions of feasible descent at the corner points of the feasible domain. Show the gradient directions of f and g_is at these points. Verify graphical results analytically using KKT conditions and monotonicity analysis.

5.3 Use the optimality conditions to solve the problem

$$\min f = (x_1 - 1)^2 + (x_2 - 1)^2$$
$$\text{subject to } g_1 = (x_1 + x_2 - 1)^3 \leq 0 \quad \text{and} \quad x_1 \geq 0, x_2 \geq 0.$$

Solve again, replacing g_1 with $g_1 = x_1 + x_2 - 1 \leq 0$. Compare. (From Bazaraa and Shetty, 1979.)

5.4 Graph the problem

$$\min f = -x_1, \text{ subject to}$$
$$g_1 = x_2 - (1 - x_1)^3 \leq 0 \quad \text{and} \quad x_2 \geq 0.$$

Find the solution graphically. Apply the optimality conditions and monotonicity rules. Discuss. (From Kuhn and Tucker, 1951.)

5.5 Show that the KKT conditions for a problem stated in the form

$$\min f(\mathbf{x})$$

$$\text{subject to } h_j(\mathbf{x}) = 0, \quad j = 1, \ldots, m,$$
$$g_j(\mathbf{x}) \leq 0, \quad j = 1, \ldots, m_1,$$
$$g_j(\mathbf{x}) \geq 0, \quad j = m_1 + 1, \ldots, m_1 + m_2$$

are as follows:

$$\nabla f + \boldsymbol{\lambda}^T \nabla \mathbf{h} + \boldsymbol{\mu}^T \nabla \mathbf{g} = \mathbf{0}^T,$$
$$\boldsymbol{\mu}^T \mathbf{g} = 0, \quad \mu_j \geq 0 \quad \text{for } j = 1, \ldots, m_1,$$
$$\mu_j \leq 0 \quad \text{for } j = m_1 + 1, \ldots, m_1 + m_2.$$

5.6 Show that the KKT conditions for a problem stated in the form

$$\min f(\mathbf{x})$$

$$\text{subject to } \mathbf{h}(\mathbf{x}) = \mathbf{0},$$
$$\mathbf{g}(\mathbf{x}) \leq \mathbf{0}$$
$$\mathbf{x} \geq \mathbf{0}$$

are as follows:

$$\nabla f + \boldsymbol{\lambda}^T \nabla \mathbf{h} + \boldsymbol{\mu}^T \nabla \mathbf{g} \geq \mathbf{0}^T,$$
$$[\nabla f + \boldsymbol{\lambda}^T \nabla \mathbf{h} + \boldsymbol{\mu}^T \nabla \mathbf{g}]\mathbf{x} = 0,$$
$$\boldsymbol{\mu}^T \mathbf{g} = 0, \quad \boldsymbol{\mu} \geq \mathbf{0}.$$

5.7 Solve the following problem using the appropriate form of the KKT conditions:

$$\min f = 6x_1^2 - 3x_2 - 4x_1x_2 + 4x_2^2$$
$$\text{subject to } g_1 = e^{x_1} + \left(\tfrac{1}{2}\right)x_1x_2 + \left(\tfrac{1}{2}\right)x_2^2 - 5 \leq 0,$$
$$x_1 \geq 0, \quad x_2 \geq 0.$$

5.8 Solve the following problem using KKT conditions:

$$\text{maximize } f = 3x_1 - x_2 + x_3^2$$
$$\text{subject to } g_1 = x_1 + x_2 + x_3 \leq 0$$
$$\text{and } h_1 = -x_1 + 2x_2 + x_3^2 = 0.$$

5.9 Derive optimality conditions for the problem

$$\min f = \mathbf{c}^T \mathbf{y}$$
$$\text{subject to } \mathbf{a}^T \mathbf{x} = b, \quad \mathbf{x} \geq \mathbf{0}$$
$$\text{and } \mathbf{y} = \left(x_1^{-1}, x_2^{-1}, \ldots, x_n^{-1}\right)^T, \quad \mathbf{x} = (x_1, x_2, \ldots, x_n)^T.$$

Find an expression for the point that satisfies these conditions. Assume all parameters are positive.

5.10 Find a local solution to the problem

$$\max f = x_1x_2 + x_2x_3 + x_1x_3$$
$$\text{subject to } h = x_1 + x_2 + x_3 - 3 = 0.$$

Use three methods: direct elimination, constrained derivatives, and Lagrange multipliers. Compare. Is the solution global?

5.11 Solve the problem

$$\min f = x_1^2 + x_2^2 + x_3^2$$
$$\text{subject to } h_1 = x_1^2/4 + x_2^2/5 + x_3^2/25 - 1 = 0$$
$$\text{and } h_2 = x_1 + x_2 - x_3 = 0.$$

Use the reduced gradient method and verify with the KKT conditions.

5.12 Use necessary and sufficient conditions to check if the point $\mathbf{x} = (1, 5)^T$ is a solution to the problem

$$\min f = x_1^2 - x_2$$
$$\text{subject to } h_1 = x_1 + x_2 - 6 = 0,$$
$$g_1 = x_1 - 1 \geq 0,$$
$$g_2 = 26 - x_1^2 - x_2^2 \geq 0.$$

5.13 For the problem

$$\min f = (1/2)\left[(x_1 - 1)^2 + x_2^2\right]$$
$$\text{subject to } h = -x_1 + bx_2^2 = 0,$$

where b is a fixed scalar, examine the values of b that make $\mathbf{x}_* = \mathbf{0}$. Explain graphically for $b = 1/4$ and $b = 1$. (From Fletcher, 1981.)

5.14 Consider the problem

$$\min f = x_1^2 + x_2^2 + x_3^2 + 40x_1 + 20x_2 - 3000$$
$$\text{subject to } g_1 = x_1 - 50 \geq 0,$$
$$g_2 = x_1 + x_2 - 100 \geq 0,$$
$$g_3 = x_1 + x_2 + x_3 - 150 \geq 0.$$

Solve using KKT conditions. Solve again using monotonicity analysis. (From Rao, 1978.)

5.15 Use KKT conditions to solve the problem

$$\min f = 0.044x_1^3x_2^{-1} + x_1^{-1} + 0.0592x_1x_2^{-3}$$
$$\text{subject to } g_1 = 1 - 8.62x_1^{-1}x_2^3 \geq 0$$
$$\text{and } 0 \leq x_1 \leq 5, \quad 0 \leq x_2 \leq 5.$$

Solve again using monotonicity analysis.

5.16 Repeat the above exercise for the problem

$$\max f = 0.020x_1^4x_2x_3^2/10^7$$
$$\text{subject to } 675 \geq x_1^2x_2, \quad 0.419(10^7) \geq x_1^2x_3^2,$$
$$0 \leq x_1 \leq 36, \quad 0 \leq x_2, \quad 0 \leq x_3 \leq 125.$$

5.17 A point $\mathbf{P}(x_1, x_2)$ lies on the curve $x_1^2 - x_1x_2 + x_2^2 = 1$. What is the minimum distance from $\mathbf{P}$ to the origin ($x_1 = 0, x_2 = 0$), as $\mathbf{P}$ moves along the curve? Explain analytically and graphically.

5.18 Use monotonicity arguments and constrained derivatives to find the value(s) of the parameter b for which the point $x_1 = 1$, $x_2 = 2$ is the solution to the problem

$$\max f = 2x_1 + bx_2$$
$$\text{subject to } g_1 = x_1^2 + x_2^2 - 5 \leq 0$$
$$\text{and } g_2 = x_1 - x_2 - 2 \leq 0.$$

5.19 Examine all possible parametric solutions to the problem

$$\min f = (x_1 - a)^2 + (x_2 - b)^2$$
$$\text{subject to } x_1 + x_2 \leq c$$
$$\text{and } x_1 \geq 0, \quad x_2 \geq 0,$$

where a, b, and c are real parameters. Explain using graphical representation. Derive analytically the parametric solutions.

5.20 Solve the problem

$$\min f = (1/2)(x_1^2 + x_2^2 + x_3^2) - (x_1 + x_2 + x_3)$$
$$\text{subject to } x_1 + x_2 + x_3 \le 1$$
$$\text{and } x_1 \ge 0, \quad x_2 \ge 0, \quad x_3 \ge 0.$$

Try first using the KKT conditions. Try again using monotonicity and taking advantage of symmetry.

5.21 Determine if the solution to the problem in Example 5.4 is a global one. Do the same for Example 5.6 with the bounds $x_i \ge 0$ removed.

5.22 Show that a maximization problem with inequalities will have sign restriction on the multipliers that is opposite to the minimization one.

5.23 Find the optimality conditions for the LP problem stated as

$$\min \mathbf{c}^T\mathbf{x}$$
$$\text{subject to } \mathbf{A}_1\mathbf{x} \le \mathbf{b}_1, \quad \mathbf{A}_2\mathbf{x} \le \mathbf{b}_2, \quad \mathbf{x} \le \mathbf{b}_3, \quad \mathbf{x} \ge \mathbf{b}_4.$$

5.24 Consider the LP problem

$$\min f = -3x_1 - 2x_2$$
$$\text{subject to } g_1 = 2x_1 + x_2 - 10 \le 0,$$
$$g_2 = x_1 + x_2 - 8 \le 0,$$
$$g_3 = x_1 - 4 \le 0,$$
$$x_1 \ge 0, \quad x_2 \ge 0.$$

Solve the problem analytically and graphically. Check which of the points $\mathbf{x}_1 = (4, 2)^T$, $\mathbf{x}_2 = (2, 6)^T$, $\mathbf{x}_3 = (4, 0)^T$ is a KKT point. Show the relevant gradient vectors at these points graphically and confirm what you found above.

5.25 Solve the LP problem

$$\max f = 10x_1 + 5x_2 + 8x_3 + 7x_4$$
$$\text{subject to } 6x_1 + 5x_2 + 4x_3 + 6x_4 \le 40,$$
$$3x_1 + x_2 \le 15,$$
$$x_1 + x_2 \le 10,$$
$$x_3 + 2x_4 \le 10,$$
$$2x_3 + x_4 \le 10.$$

5.26 Study the following LP problem using monotonicity analysis. Perform two iterations of the basic LP algorithm.

$$\max f = 8x_1 + 5x_2 + 6x_3 + 9x_4 + 7x_5 + 9x_6 + 6x_7 + 5x_8$$
$$\text{subject to } g_1\colon 5x_1 + 3x_2 + 2x_4 + 3x_6 + 4x_7 + 6x_8 \le 30,$$
$$g_2\colon 2x_1 + 4x_3 + 3x_4 + 7x_5 + x_7 \le 20,$$
$$g_3\colon 2x_1 + 4x_2 + 3x_3 \le 10,$$
$$g_4\colon 7x_1 + 3x_2 + 6x_3 \le 15,$$

$$g_5: 5x_1 + 3x_3 \le 12,$$
$$g_6: 3x_4 + x_5 + 2x_6 \le 7,$$
$$g_7: 2x_4 + 4x_5 + 3x_6 \le 9,$$
$$g_8: 8x_7 + 5x_8 \le 25,$$
$$g_9: 7x_7 + 9x_8 \le 30,$$
$$g_{10}: 6x_7 + 4x_8 \le 20.$$

Compare the structure of this problem with the one in Exercise 5.25. Explore any special way that may take advantage of this structure, for example, some form of decomposition. (From Hillier and Lieberman, 1967.)

5.27 Examine the problem

$$\max f = (8 + a)x_1 + (24 - 2a)x_2$$
$$\text{subject to } x_1 + 2x_2 \le 10,$$
$$2x_1 + x_2 \le 10, \quad x_1, x_2 \ge 0,$$

where a is a parameter in the range [0,10].

(a) Solve the problem for all a and determine the active constraints. Are the same constraints always active?

(b) Calculate the sensitivity coefficients as functions of a. Can you find an optimal value for the parameter a (if you could have a choice)?

6

Parametric and Discrete Optima

Many are called, but few are chosen.
Matthew XXII, 14 (c. A.D. 75)

In the previous two chapters we set out the theory necessary for locating local optima for unconstrained and constrained problems. From the optimality conditions we derived some basic search procedures that can be used to reach optima through numerical iterations. In Chapter 7 we will look at local search algorithms more closely. But before we do that we need to pick up the thread of inquiry for two earlier issues: the role of parameters and the presence of discrete variables.

One of the common themes in this book is to advocate "model reduction" whenever possible, that is, rigorously cutting down the number of constraint combinations and variables that could lead to the optimum – before too much numerical computation is done. This reduction has two motivations. The first is to seek a *particular* solution, that is, a numerical answer for a given set of parameter values. The second is to construct a specific *parametric optimization procedure*, which for any set of parameter values would directly generate the globally optimal solution with a minimum of iteration or searching.

If the variables are continuous, the most practical way to find a particular optimum for a single set of parameter values is simply to employ a numerical optimization code. When discrete variables are present, the resulting combinatorial problem can only be solved if we examine all possible solutions, directly or indirectly. A "local" solution has little meaning and gradient-based local optimization algorithms would not be applicable since there are no gradients to compute. One basic approach to solving such problems is to reduce the often astronomical number of possible solutions to a much smaller one that can be computed and compared directly. This reduction of the feasible domain has already been pointed out as profitable for continuous problems. It is essential for finding discrete optima.

The motivation to reduce models to construct optimized parametric design procedures comes from three applications. The first is to generate, without unnecessary iterative computation, the optimal design directly from a set of input parameter values. The second is to reoptimize specific equipment configurations in the face of changing parameter values. Designers call this "resizing." The third is to optimize large complicated systems by breaking them down into interacting components, each described by a component-level optimal parametric design procedure. Such decomposition makes possible the optimization of systems too large for the structure to be ignored.

Conversely, components with optimized parametric design procedures can be assembled into larger systems by making the outputs of some be the inputs of others, their various objectives being combined into a unifying system goal. Optimization of large-scale systems, their decomposition into smaller manageable subsystems, and coordinated solution of subsystems are important research and practical problems.

The chapter begins by resuming the discussion of monotonicity analysis, showing that the optimum air tank design examined in Chapter 3 can be generated for any new set of parameter values without repeating the analysis. Then it concentrates on how to use *monotonicity tables* to construct systematically a parametric procedure for locating the optimal designs of hypothetical hydraulic cylinders like that in Chapter 3.

An important distinction is made between two ways critical constraints can be used to eliminate variables and reduce the model. The easier method, employed throughout Chapter 3, is to solve the algebraic constraint expression by what we called "explicit algebraic elimination." But when the constraint is not solvable in closed form, variables can only be eliminated by an iterative numerical approach that we may call an "implicit numerical solution." Showing how to handle the latter situation greatly extends the applicability of monotonicity analysis to practical design problems involving, for example, kinematic simulation or stress analysis by finite element analysis.

Attention is then turned to examining how to handle variables that are discrete-valued instead of continuous, again using the air tank problem. The *branch-and-bound* technique employed was in fact introduced earlier in this chapter as a way of ruling out unpromising cases for the optimum. At that point it starts becoming obvious that the definitions of activity and optimality must be reexamined when discrete variables are present. Several definitions from the earlier chapters are extended and refined. Design of an electrical transformer is used to illustrate how the definitions can be used, along with systematic model reduction techniques.

As in Chapter 3, the present chapter considers the design variables to be strictly positive and finite as in most engineering models.

6.1 Parametric Solution

The first goal of an optimization study may well be finding the numerical solution to a particular problem, as in the air tank design of Chapter 3. Once this has been done, one often can generalize the analysis so that solution with new values of the parameters will be simple and quick. Thus, the solution method already obtained for steel should apply to aluminum, with appropriate changes to allow for the differing material strength.

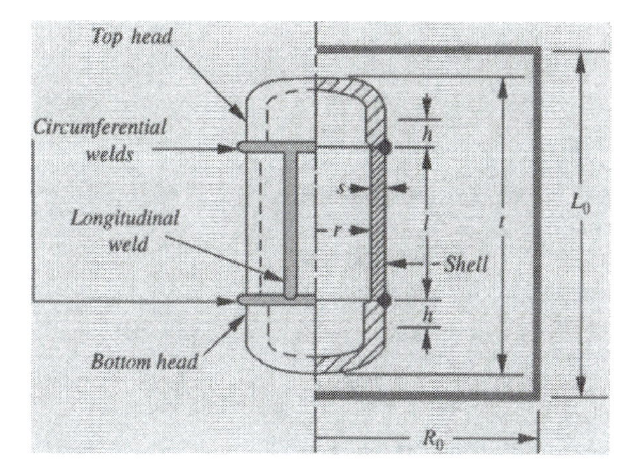

Figure 6.1. Air tank with dished heads.

The experience gained in solving one air tank problem would of course make it easy to solve another simply by repeating all the steps using the different numbers. Instead of this, however, we will show how to develop a procedure that, if given new data, will quickly generate the appropriate new optimum. In making this generalization, we also include the possibility of using the more common *dished* head design shown in Figure 6.1 as well as the more mathematically convenient flat head design already studied. Hemispherical heads are also included.

Particular Optimum and Parametric Procedures

Recall from Chapter 1 that one must always distinguish among constants, variables, and parameters. *Constants* (e.g., π or 2) never vary. *Variables* can be changed; there are in all six variables in the design of the exemplary air tank, symbolized by lowercase letters. *Parameters*, such as the tank's pressure rating, behave as constants for a particular design, but they can vary from one application to the next. Table 6.1 lists the ten parameters for the air tank example. There may also be *parametric functions*, depending entirely on the parameters, introduced into the

Table 6.1. Air Tank Design Parameters

C_h:	Head metal volume, dimensionless
D_0:	Maximum outside diameter, cm
E:	Joint efficiency, dimensionless
L_0:	Maximum total length, cm
L_l:	Minimum shell length, cm
L_u:	Maximum shell length, cm
P:	Pressure, atm
R_0:	Maximum outside radius, cm
S:	Allowable stress, Newtons/m^2
V:	Minimum capacity, liters

Table 6.2. Parametric Functions and Abbreviations

(a) Parametric Functions

K_h: Head thickness coefficient, dimensionless

$$K_h = \begin{cases} 2(CP/S)^{1/2} \text{ for flat heads} \\ P/(2S - 0.2P) \text{ for hemispherical heads} \end{cases}$$

K_l: Head depth coefficient, dimensionless

$$K_l = \begin{cases} 0 \text{ for flat heads} \\ \frac{4}{3} \text{ for hemispherical heads} \end{cases}$$

K_s: Shell thickness coefficient, dimensionless

$K_s = P/(2SE - 0.6P)$

K_v: Head volume coefficient, dimensionless

$$K_v = \begin{cases} 0 \text{ for flat heads} \\ 1 \text{ for hemispherical heads} \end{cases}$$

(b) Abbreviations

$K_0 = 2K_s + K_s^2$
$K_1 = 2C_hK_h - K_0K_v$
$K_2 = 1 + K_s$
$K_3 = K_v - 2K_l - 2K_h$

mathematical model as abbreviations or coefficients. Table 6.2 gives the four parametric functions (e.g., shell thickness coefficient). Parameters, parametric functions, and constants are represented by numbers or capital letters.

Monotonicity analysis of an optimization problem does not require any knowledge of parameter values other than their signs. Sometimes, as in the air tank design, so few possibilities remain for optimal cases that all of them can be evaluated and compared to determine the best. But a problem does not have to be large before the number of possibly optimal cases is too great for total enumeration.

Usually, as in the original statement of the air tank problem, one wishes to know the optimum for particular values of the parameters, given in advance. Such an optimum will be called a *particular optimum*. This is in contrast to an expression of the optimum as a function of the parameters, which will be called a *parametric optimum*. A parametric optimum is the result of a *parametric (optimization) procedure* identifying, for any given parameter values, the optimal case and the corresponding values of the decision variables.

Finding a parametric procedure may be justified when the same device is to be redesigned many times for various parameter values. But before undertaking such a task, it is still a good idea to find the particular optimum for the specific parameter values of immediate interest, as in the air tank design. Then any parametric procedure developed later can start by testing the optimal case for the particular solution. For parameter values near those for the particular optimum, such a procedure would usually avoid exploring all possible cases.

The simplest example of a parametric procedure is a design sizing formula, the kind found in engineering handbooks, which gives the optimizing arguments directly as easily calculated functions of problem parameter values. This can happen only in simple situations when the optimal case can be evaluated in advance. More generally, a parametric procedure must first use the parameter values given to pick the optimal case out of all the possibilities. Once this case is known, the appropriate sizing formulas are used to compute the optimum.

We now show how to generate systematically a parametric procedure for the air tank problem that gives, for any set of parameter values, the optimizing values of all variables. This will show how a parametric procedure, being for the most part noniterative, is more direct than the usual numerical optimization methods. In the example, the parametric procedure obtained can be easily programmed to find the globally optimal design for any combination of parameter values.

For easy reference, the air tank problem, for which a particular optimum has already been found, is restated in parametric form here:

$$\begin{aligned}
&\text{minimize } f = \pi(2rsl + s^2 l + 2C_h r^2 h) \text{ (metal volume)}\\
&\text{subject to}\\
&\quad h \geq K_h r && \text{(head thickness)},\\
&\quad s \geq K_s r && \text{(shell thickness)},\\
&\quad v = \pi r^2 (l + K_v r) && \text{(geometry)},\\
&\quad v \geq V && \text{(minimum capacity)},\\
&\quad l \geq L_l && \text{(minimum shell length)},\\
&\quad l \leq L_u && \text{(maximum shell length)},\\
&\quad t = l + 2K_l r + 2h && \text{(total length)},\\
&\quad r + s \leq R_0 && \text{(maximum outside radius)},\\
&\quad t \leq L_0 && \text{(maximum total length)}.
\end{aligned} \tag{6.1}$$

The monotonicities of all functions are exactly the same in this parametric representation as when specific parameter values are used. Hence, the first stage of the monotonicity analysis is the same as before, leading to the use of the constraints on head thickness, shell thickness, geometry, and minimum capacity to eliminate four variables: h, s, v, and l.

Notice that this time l is eliminated instead of r as before. This is because the multiply critical relation

$$v \equiv< \pi r^2 (l + K_v r),$$

which is slightly more complicated than the original one ($v \equiv< \pi r^2 l$), is solvable in closed form for l but not r. The solution for l is

$$l = V/\pi r^2 - K_v r.$$

Elimination of l (along with total length t) gives the following minimization problem in the single variable r:

$$\min f = VK_0 + \pi K_1 r^3$$

subject to

$$\begin{aligned}
&g_1: V/\pi r^2 - K_v r \le L_u && \text{(max. shell length)},\\
&g_2: V/\pi r^2 - K_v r \ge L_l && \text{(min. shell length)},\\
&g_3: V/\pi r^2 - K_3 r \le L_0 && \text{(max. total length)},\\
&g_4: \qquad\qquad K_2 r \le R_0 && \text{(max. outside radius)},
\end{aligned} \tag{6.2}$$

where the new positive parametric functions K_0, K_1, K_2, and K_3 are defined in Table 6.2. This reduction of a six-variable, nine-constraint problem down to one with a single variable and half as many constraints is typical in early engineering design models.

Branching

The original model has now been reduced to a problem with four inequalities in a single monotonic variable. One of the four must be active, unless by coincidence others happen also to be satisfied with strict equality. Subsequent analysis will show that one can tell in advance, after straightforward evaluation of certain formulas, exactly which constraint will bound r. In effect this analysis decomposes the original problem into four smaller ones, each with a simple solution.

Observe that the sense of the monotonicity of the reduced objective f in Equation (6.2) depends on the sign of K_1, the coefficient of r^3 in the only variable term. Thus, there are two major cases: one in which $K_1 > 0$, making f increase in r; the other in which $K_1 < 0$, making f decrease in r. Since each case will be completely solved in closed form, the proper case and its corresponding optimal solution can be determined merely by computing the parametric function K_1. Such subdivision of a problem into alternatives is called *branching*.

Let us first study the case where K_1 is positive so that f increases with r. By MP1, either g_1 or g_3 bounds r from below. Thus g_1 and g_3 together form a set of constraints conditionally critical for r. Since vessels bounded by either of these maximum length constraints would tend to have long shells, designs where $K_1 > 0$ will be referred to as *cylindrical*. The original (nonparametric) example gave a cylindrical design.

Graphical Interpretation

Figure 6.2 illustrates these concepts graphically. Since the objective increases in r, designs bounding r away from zero will lie on the left boundary of the feasible region. For every feasible l, minimum capacity is the only constraint on this left boundary, making it critical for r.

Along the feasible arc of the capacity constraint, the reduced objective has been shown to be monotonic in r after the elimination of l. If, as in the case under study, the reduced objective increases with r, the optimum must be at the extreme left

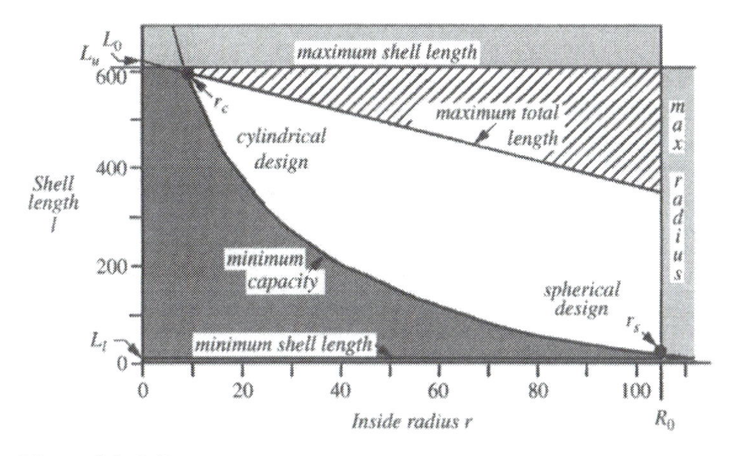

Figure 6.2. Effects of decreased total length constraints.

of this arc. This could be at the maximum total length constraint as shown. But if the maximum shell length constraint had been smaller, its intersection could have occurred far enough to the right to cut off the total length constraint. Thus, these two constraints together form a conditionally critical set.

If the reduced objective had been decreasing instead of increasing in r, the optimum would have been at the extreme right end of the capacity arc. Then the maximum radius and minimum length constraints would have been the conditionally critical set. Henceforth, such designs, for which $K_1 < 0$, will be called *spherical.*

Recall that a constraint bounding a monotonic variable x in the sense *opposite* to that required for criticality is said to be *uncritical* for x. Here, g_2 and g_4 are examples of uncritical constraints because each bounds the monotonic variable r from above, just opposite of what is needed to keep the problem well bounded, that is, $V/\pi r^2 - K_v r \not\geq L_l$ and $K_2 r \not\leq R_0$.

In Figure 6.2 these uncritical constraints are those at the lower right. Their feasible regions are left of their boundaries, that is, in the direction of improved values of the reduced objective.

Recall that this particular pattern of uncriticality and conditional criticality is appropriate only for an objective f increasing in r. When, however, the parametric coefficient K_1 is negative, so that f decreases with r, g_2 and g_4 would be conditionally critical instead of uncritical. Similarly, g_1 and g_3 would become uncritical.

One can solve the original problem by temporarily ignoring the uncritical constraints, provided the solution obtained also satisfies the uncritical constraints. The Uncriticality Theorem has shown that the critically relaxed minimum equals the true minimum if and only if the original constraints are consistent.

Parametric Tests

Once the constraints have been classified, parametric tests are constructed for each conditionally critical constraint set. In describing this construction it is convenient to let $R(l; K)$ represent the solution of the following generalization of the

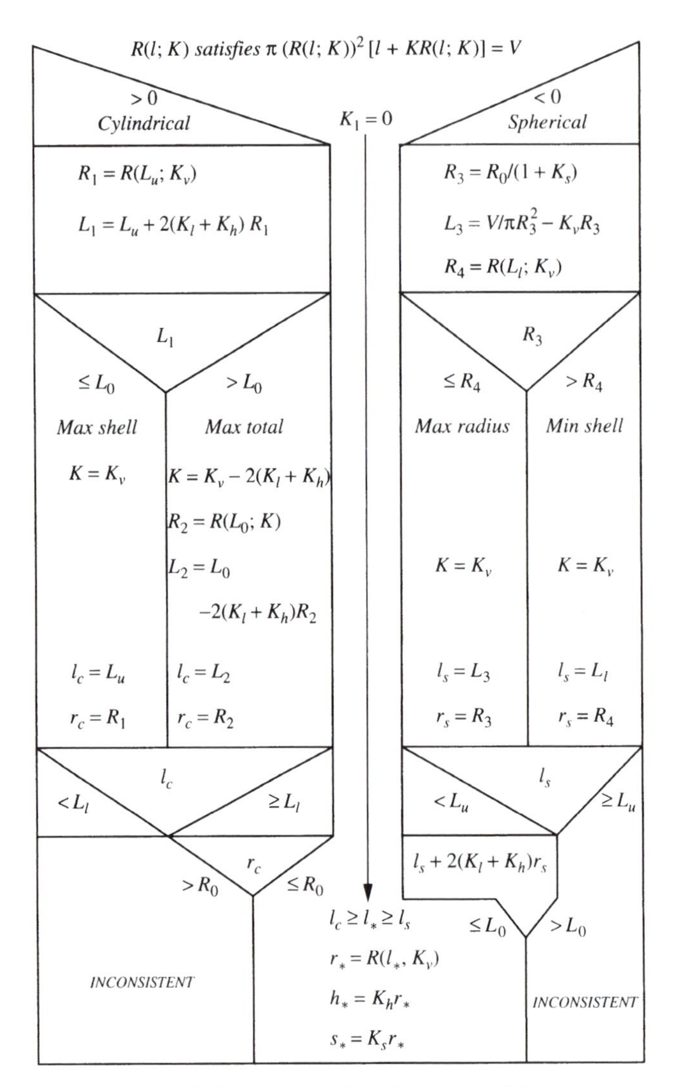

Figure 6.3. Logical decomposition chart (air tank).

minimum volume equation:

$$g_5\colon \pi\{R(l;K)\}^2(l + KR(l;K)) = v. \tag{6.3}$$

Here l is the shell length, the parameter K is the volume parameter K_v, and $R(l; K)$ is the radius giving minimum volume V for a given shell length l. This radius is an implicit function of l because it cannot be obtained in closed form.

Consider now the cylindrical case in which $K_1 > 0$ (Figure 6.3). There are two possibilities: Either the maximum shell length constraint or the total length constraint is active. In the first case, g_1 is a strict equality giving $l = L_u$; the corresponding radius is $R(L_u; K_v)$, abbreviated R_1. Notice that since L_u is a parameter, so is R_1. Since this design (R_1, L_u) satisfies the critical constraints on volume and shell length as strict equalities, it will be optimal if and only if it also satisfies the overall length constraint

g_3. Let L_1 denote the left member of constraint g_3, which is the overall length for the design (R_1, L_u). We have

$$L_1 = L_u + 2(K_l + K_h)R_1. \tag{6.4}$$

Then comparison of the parametric function L_1 with the maximum overall length parameter L_0 immediately tests for feasibility. For, if $L_1 \le L_0$, the design is optimal, confirming that maximum shell length rather than overall length is critical. Notice that this test depends entirely on parameters.

If, however, $L_1 > L_0$, the maximum shell length design is infeasible, indicating that it is overall length, not shell length, that determines the design. When this happens one must obtain the shell length as a function of radius by solving overall length constraint g_3:

$$l = L_0 - 2(K_l + K_h)r. \tag{6.5}$$

Using this to eliminate l from the minimum volume equation gives, after rearrangement, an equation of the same form as Equation (6.3), but with L_0 as length and with $K_v - 2(K_l + K_h)$ as parameter. Let R_2 represent this radius, implicitly determined, giving minimum volume for maximum overall length,

$$R_2 = R[L_0; K_v - 2(K_l + K_h)]. \tag{6.6}$$

The corresponding shell length L_2 is obtained by substituting this radius into Equation (6.5):

$$L_2 = L_0 - 2(K_l + K_h)R_2. \tag{6.7}$$

The left side of Figure 6.3 diagrams the logical decomposition for this cylindrical case. The subscript of r_c and l_c, the optimal design for $K_1 > 0$, refers to cylindrical vessels. In brief summary, if $L_1 \le L_0$, then $r_c = R_1$ and $l_c = L_u$; otherwise $r_c = R_2$ and $l_c = L_2$.

The solution must also be checked against the uncritical constraints on minimum shell length and maximum radius. Violation of either would prove inconsistency of the constraints, in which case no feasible solution would exist.

Of course the foregoing applies only in the cylindrical case. In the spherical case in which $K_1 < 0$, an appropriate set of parametric functions follows:

$$R_3 = R_0/(1 + K_s);\ L_3 = V/\pi R_3^2 - K_v R_3;\ R_4 = R(L_l; K_v). \tag{6.8}$$

Subscript s will be used in this spherical case by analogy to the way c was employed in the cylindrical case. The parametric test is that if $R_3 \le R_4$, then $r_s = R_3$ and $l_s = L_3$; otherwise $r_s = R_4$ and $l_s = L_l$. This is summarized on the right side of Figure 6.3.

In the borderline case where $K_1 = 0$, all feasible designs give the same objective value because the reduced objective does not depend on the radius. The procedures of Figure 6.3 then yield upper and lower bounds on the length:

$$l_s \le l_* \le l_c. \tag{6.9}$$

Table 6.3. Hydraulic Cylinder: Initial Monotonicity Table 0

Function Number	Functions	Variables d	i	t	f	p	s
(0)	d	+	.	.	.	.	.
(1)	$d - i - 2t = 0$	+*	−*	−*	.	.	.
(2)	$f - \pi i^2 p/4 = 0$	.	−*	.	+*	−*	.
(3)	$-f + F \leq 0$	.	.	.	−	.	.
(4)	$p - P \leq 0$	.	.	.	.	+	.
(5)	$s - ip/2t = 0$	.	−*	+*	.	−*	+*
(6)	$s - S \leq 0$	.	.	.	.	.	+
(7)	$-t + T \leq 0$	.	.	−	.	.	.

The radius r_* corresponding to a shell length l_* is obtained by solving the volume constraint equation, which becomes, for $l = l_*$,

$$r_* = R(l_*; K_v). \tag{6.10}$$

From r_*, the remaining s_* and h_* can be computed in the usual way. The uncritical constraints on maximum shell and overall length must also be checked to verify consistency, as shown in the center column of Figure 6.3.

Developing a parametric procedure becomes less possible as the number of variables and conditional constraints increases. Even the air tank problem took enough analysis and care that the effort was only barely justified. Thus, a parametric procedure should be considered as rare and valuable as any classic engineering formula. When a particular optimum has been found numerically, one might well ask if a parametric procedure can be developed. This possibility is especially strong when, as for the air tank, the particular optimum is constraint bound. The local optimization theory of Chapters 4 and 5 will be needed to handle the nonmonotonic situations that remain after monotonicity analysis has been completed.

6.2 The Monotonicity Table

Many principles, theorems, and lemmas to aid the monotonicity analysis and subsequent reduction of an optimization model were given in Chapter 3. Although each of the principles is simple, the process of applying them many times becomes complicated and confusing when the number of variables exceeds five or so. Since the result of one inference often becomes the premise of another, the analyst must be systematic in keeping track of the steps in the reasoning to avoid mistakes. Subsequent checking requires an adequate record of the inference process.

Such a record is provided by the *Monotonicity Table* (MT) to be described now (see Table 6.3). Roughly speaking, the MT has a column for each variable and a row for the objective function and each constraint. The elements of the resulting matrix indicate the incidence and monotonicity of each variable in each constraint.

Each column records a variable and the constraints that can bound it, whether from above, below, or both. As criticality and directions of constraints are established, the inequality and equality signs in the constraints are updated with the notation defined in Chapter 3.

Setting Up

A more general, less restricted, version of the hydraulic cylinder design problem of Section 3.7 will be used as an example of setting up and using a Monotonicity table to generate the possibly optimal cases needed to construct a procedure for parametric optimization. The original problem, having six variables and seven constraints, is eventually reduced to one constraint-bound case with a closed-form parametric solution and a set of three related cases solvable together by minimizing a single-variable function on a closed interval. In the parametric procedure, this minimization is not performed; instead, parametric tests for finding the optimum among these three cases are derived.

Table 6.3 shows the arrangement of rows and columns for this problem, in which it is desired to minimize the outside diameter d of the cylinder. The constraints, with numbers for reference, are listed to the left of each row. A column to the left of the constraints is reserved for recording the old constraints used in constructing those currently used.

The top row contains the monotonicities in the objective function of the variables heading the columns: signs "+" or "−" if increasing or decreasing respectively, a dot "·" if independent, and letters "n" if nonmonotonic and "u" if unknown. Monotonicities for the constraints are given in the matrix to the right of the constraints in a similar way. For the inequalities, these monotonicities refer to the constraint functions written in negative null form. The undirected equalities have the monotonicities recorded for whatever null form they have been arbitrarily given at the start. If and when they become directed, these initial monotonicities will be either confirmed or reversed. Until they have been directed, the monotonicities are marked with an asterisk.

First New Table: Reduction

Examination of the first variable column (d) shows that equality (1) is critical for d by MP1 if the initially assumed direction of (1) is reversed. Now well-constrained from below, d is eliminated by solving (1) for it and substituting for it wherever it appears, in this case only in the objective. This gives a reduced table (Table 6.4), which has the d column and constraint (1) deleted. The objective is modified to reflect the elimination of d. The constraint (1) eliminated, expressed as a properly directed critical inequality bounding d from below, is listed below the main monotonicity table as the first of the eliminated variables and constraints. This information will be useful for generating the optimal value of the eliminated variable d once the variables i and t, upon which it depends, are determined by later monotonicity analysis.

Table 6.4. Hydraulic Cylinder: Monotonicity Table 1 (Reduction)

Function Number	Functions	Variables i	t	f	p	s
(0, 1)	$i + 2t$	+	+	.	.	.
(2)	$f - \pi i^2 p/4 = 0$	$-^*$	.	$+^*$	$-^*$	.
(3)	$-f + F \leq 0$	.	.	−	.	.
(4)	$p - P \leq 0$	.	.	.	+	.
(5)	$s - ip/2t = 0$	$-^*$	$+^*$	.	$-^*$	$+^*$
(6)	$s - S \leq 0$	.	.	.	.	+
(7)	$-t + T \leq 0$	.	−	.	.	.

Eliminated Variables and Constraints

Constraint Number	Variable Eliminated
(1)	$d \geqq (i + 2t)$

Second New Table: Two Directions and Reductions

Examination of the first two variable columns i and t in first reduction Table 6.4 does not lead to any criticality conclusions because constraint (5) has not yet been directed. But MP2 confirms the initially assumed direction of equality (2) to bound f from above; inequality (3) is what bounds f from below. Both are semicritical rather than critical, and so neither should ordinarily be used to eliminate a variable. But since equality (2) is already known to be a strict equality, it makes sense to eliminate its simplest variable f with it. The force f and constraint (2) then joins the set of eliminated variables and constraints, and the deletion of the f column will accompany new monotonicities for i and p in inequality (3). In the next monotonicity table it will not be necessary to carry the dot "·" along with the new inequality (3, 2) indicating semicriticality for f, because f no longer appears explicitly. Thus (3, 2) constrains its variables i and p just like any other inequality, and f is automatically well constrained.

Before recording this, however, examine the s column. To constrain s properly MP2 requires equality (5) to be reversed from the direction initially assumed. Then both (5) and inequality (6) become semicritical for s, bounding it, respectively, below and above. As it was for f in the preceding paragraph, the strict equality (5) can be used to eliminate its simplest variable s. Notice that the monotonicity signs for (5) are opposite to those in Tables 6.3 and 6.4, information that will be reflected in the next reduction (Table 6.5). Now that all equalities have been directed, the monotonicity analysis can continue using Table 6.5, now free of asterisks and with variable s and constraint (5) moved to the head of the list of those eliminated. Newly eliminated variables and constraints go to the head, not the tail, of the list because this order will be followed in solving for the eliminated variables once the optimal values of the remaining variables, currently i, t, and p, have been found.

Table 6.5. Hydraulic Cylinder: Monotonicity Table 2 (2 Directions and Reductions)

Function Number	Functions	Variables		
		i	t	p
(0, 1)	$i + 2t$	+	+	.
(3, 2)	$-(\pi/4)i^2 p + F \leq 0$	−	.	−
(4)	$p - P \leq 0$	.	.	+
(6, 5)	$(1/2)ip/t - S \leq 0$	+	−	+
(7)	$-t + T \leq 0$	.	−	.

Eliminated Variables and Constraints

Constraint Number	Variable Eliminated
(5)	$s \geqq ((1/2)ip/t)$
(2)	$f \leq (\pi/4)i^2 p$
(1)	$d \geqq (i + 2t)$

Third New Table: Final Reduction

Table 6.5 shows that new inequality (3, 2) must be critical to constrain i from below. Since it also is a semicritical lower bound for p, it is in fact multiply critical; thus it should certainly be solved to eliminate a variable, the easier being p.

This gives a new table (Table 6.6), which no longer has a row for (3, 2) or a column for p. The solution (3, 2) for p at the head of the list of eliminated constraints has been substituted wherever p appeared in the solutions following, which greatly

Table 6.6. Hydraulic Cylinder: Monotonicity Table 3 (Final Reduction)

Function Number	Functions	Variables	
		i	t
(0, 1)	$i + 2t$	+	+
(4), (3, 2)	$/\pi)F/i^2 - P \leq 0$	−	.
(6, 5), (3, 2)	$(2F/\pi)/it - S \leq 0$	−	−
(7)	$-t + T \leq 0$	.	−

Eliminated Variables and Constraints

Constraint Number	Variable Eliminated
(3, 2)	$p \geqq ((4/\pi)F/i^2)$
(5), (3, 2)	$s \geqq ((2/\pi)F/it)$
2, (3, 2)[= 3]	$f \geq F$
(1)	$d \geqq (i + 2t)$

simplifies them. The third line is particularly interesting because in the preceding table it was constraint (2), just used to substitute p for f in constraint (3), thereby generating the new constraint (3, 2). Thus substituting (3, 2) back into (2), which is what must be done on the third line, amounts to what will be called a *desubstitution* of p for f. This of course regenerates the original constraint (3). The algebra is now shown to clarify this useful desubstitution concept, which will be employed several more times during the analysis. Its value is that merely by keeping track of the substitutions one can often avoid the unnecessary algebra of rederiving a previous constraint.

Here are the details for this desubstitution example. Constraint (3, 2), when added to constraint (2), gives the original constraint (3):

constraint (2):	$0 = f - (\pi/4)i^2 p$
constraint (3, 2):	$(\pi/4)i^2 p \geq F$
constraint (3):	$f \geq F$.

In general then, resubstitution generates a previous constraint without repeating algebra.

This is as far as the problem can be reduced without branching into cases. An interesting global fact emerges from the list of eliminated variables and constraints. This is that in all cases the optimizing force f must be at its specified lower limit F.

Branching by Conditional Criticality

Table 6.7 shows that all further criticality must be conditional, for a different pair of constraints conditionally bounds the two remaining variables i and t. That is, i is constrained critically from below by either (4, (3, 2)) or ((6, 5), (3, 2)); t by either ((6, 5), (3, 2)) or (7).

This lends itself to a straightforward branching procedure for beginning the case decomposition. Two sets of cases are generated depending on whether the simple inequality ((6, 5), (3, 2)) is tight or not at the optimum. If ((6, 5),(3, 2)) – labeled "stress limiting" because its right member is the maximum allowable stress – is relaxed, each variable is bounded by a single constraint as shown in Table 6.7. The list of eliminated variables and constraints is that for Table 6.6 with a "to check" constraint added as a reminder of the constraint relaxation. It is obtained by substituting preceding constraint (5, (3, 2)) for s into relaxed constraint ((6, 5), (3, 2)). Cancellation of the constraints (5, (3, 2)) leaves only the original stress limitation (6) to be checked. The resulting instruction is the first branch of the case decomposition.

Table 6.8 gives the resulting list of eliminated constraints and variables. There is no monotonicity table left, since now all variables have been eliminated. The list completely defines the optimizing solution for this relaxation. The long label (5, (3, 2)), (4, (3, 2), 7) for the stress computation merely reflects that the preceding variables i and t have been eliminated from the original stress relation (5, (3, 2)).

Table 6.7. Hydraulic Cylinder: Monotonicity Table 4 (Relaxation of (6, 5), (3, 2))

Function Number	Functions	Variables i	Variables t
(0, 1)	$i + 2t$	+	+
(4), (3, 2)	$(4/\pi)F/i^2 - P \leq 0$	–	.
(6, 5) (3, 2)	$(2/\pi)F/it - S < 0$	Relaxed	
(7)	$-t + T \leq 0$	.	–

Eliminated Variables and Constraints

Constraint Number	Variable Eliminated
(3, 2)	$p \geqq (4/\pi)F/i^2$
(5), (3, 2)	$s \geqq ((2/\pi)F/it)$
(6)	If $s - S < 0$, case is optimal. Else go to stress limited cases.
(1)	$d \geqq (i + 2t)$

This relaxed case's conditional (in contrast to global) facts are that (1) the optimizing thickness t must be at its specified lower limit T and (2) the optimizing pressure p must take its maximum allowed value P. Notice the systematic way in which this mildly complicated procedure branch has been generated.

The Stress-Bound Cases

If the relaxed stress constraint ((6, 5), (3, 2)) is violated, the Relaxation Theorem of Chapter 3 says the constraint must be active or semiactive. Here the solution is unique, making the theorem's corollary applicable. Therefore the constraint is in fact active and usable for eliminating another variable, say the wall thickness t this

Table 6.8. Hydraulic Cylinder: Relaxation

Eliminated Variables and Constraints

Constraint Number	Variable Eliminated
(4), (3, 2)	$i \geq 2\pi^{-1/2}(F/P)^{1/2}$
(7)	$t \geq T$
(4)	$p \geq P$
(5, (3, 2)), (4, (3, 2), 7)	$s \geqq (\pi^{-1/2}(FP)^{1/2}/T)$
(6)	If $s - S < 0$, case is optimal. Else go to stress limited cases.
(3)	$f \geq F$
(1)	$d \geqq (i + 2t)$

Table 6.9. Hydraulic Cylinder: Monotonicity Table 5 (Stress Restricted)

Function Number	Functions	Variable i
(0, 1), ((6, 5), (3, 2))	$i + (4F/\pi S)i^{-1}$	n
(4), (3, 2)	$(4/\pi)Fi^{-2} - P \leq 0$	$-$
(7), ((6, 5), (3, 2))	$i - (2F/\pi ST) \leq 0$	$+$

Eliminated Variables and Constraints

Number Restriction	Constraint Variable Eliminated
(6)	$s \leqq S$
(5, (3, 2)), (6)	$t \geqq (2F\pi S)/i$
(3, 2)	$p \geqq (4F/\pi)/i^2$
(3)	$f \geq F$
(1)	$d \geqq i + 2t$

time. The algebraic solution is put on the list of eliminated variables and constraints in Table 6.9, from which t has been eliminated from both the objective (0, 1) and constraint (7).

Monotonicity analysis is no longer applicable because the restricted objective function $f(i) = i + (4F/\pi S)i^{-1}$ is not monotonic, as indicated by the "n" on the objective line. Nevertheless, a simple closed-form branching process will now be constructed to test and generate any potentially optimal case that can arise.

The remaining independent variable i must lie in the interval

$$2(F/\pi P)^{1/2} \leq i \leq 2F/\pi ST$$

defined by the two remaining constraints. This interval is nonempty, that is, the minimization problem is consistent, if and only if

$$ST \leq (FP/\pi)^{1/2}.$$

Violation of this inequality signals the inconsistent case (henceforth labeled "case 0").

Since the second derivative of $f(i)$ is positive everywhere, the function is convex, with its unconstrained global minimum being at

$$i_\dagger = 2(F/\pi S)^{1/2},$$

the unique value where its first derivative vanishes. Thus there are three consistent cases: (1) $i_\dagger$ violates the lower (pressure) bound, which by the Relaxation Theorem must then be the constrained minimum, (2) $i_\dagger$ violates the upper (thickness) bound, which by the Relaxation Theorem must then be the constrained minimum, or (3) $i_\dagger$ satisfies both constraints and is therefore the unconstrained minimum.

All four cases are said to be *stress-bound*. In addition, cases 1 and 2 are also called *pressure-bound* and *thickness-bound*, respectively. Case 0 is the *inconsistent* one for which no feasible design exists.

Parametric Optimization Procedure

For the stress-bound cases the logic follows for either generating the minimizing internal diameter i_* or else finding the constraints to be inconsistent. Since Table 6.9 expresses all other variables as functions only of this value of i_*, the complete optimal design can be computed automatically. Two forms will be used. The first gives the branching tests in terms of inequalities on the unconstrained minimizer $i_\dagger$. Since both $i_\dagger$ and its bounds involve only parameters, not other problem variables, the second form involves only parameters. The full parametric solutions for each case, computed by substituting i_* for i in Table 6.9, are then listed.

ORIGINAL FORM

1. Consistency test (case 0):
   If $ST > (FP/\pi)^{1/2}$ then problem is inconsistent.
2. Unconstrained minimizer:
   Else compute $i_\dagger = 2(F/\pi S)^{1/2}$.
3. Lower bound test (case 1):
   If $i_\dagger < 2(F/(\pi P))^{1/2}$ (i.e., $P < S$) then $i_* = 2(F/(\pi P))^{1/2}$ (pressure-bound).
4. Upper bound test (case 2):
   Else if $i_\dagger > 2F/\pi ST$ then $i_* = 2F/\pi ST$ (thickness-bound).
5. Interior minimum (case 3):
   Else $i_* = i_\dagger = 2(F/\pi S)^{1/2}$.
6. Compute other optimized variables by setting $i = i_*$ in Table 6.9.

PARAMETRIC PROCEDURE

1. Consistency test (case 0):
   If $ST > (FP/\pi)^{1/2}$ then problem is inconsistent.
2. Stress-bound cases:
   Else $f_* = F$, $s_* = S$.
3. Lower bound test (case 1):
   If $P < S$ then pressure-bound (case 1):

$$
\begin{aligned}
i_* &= 2(F/(\pi P))^{1/2}, \\
t_* &= (FP/\pi)^{1/2}/S, \\
p_* &= P, \\
d_* &= i_* + 2t_*.
\end{aligned}
$$

4. Upper bound test (case 2):
   Else if $T > (F/\pi S)^{1/2}$ then thickness-bound (case 2):

$$i_* = 2F/\pi ST,$$
$$t_* = T,$$
$$p_* = \pi S^2 T^2/F.$$

5. Interior minimum (case 3):
   Else

$$i_* = 2(F/\pi S)^{1/2},$$
$$t_* = (F/\pi S)^{1/2},$$
$$p_* = S.$$

6. Minimum outside diameter:

$$d_* = i_* + 2t_*.$$

The second form, when combined with the fully parametric procedure of Table 6.6 for the relaxed understressed case, gives a completely parametric method for sizing the hydraulic cylinder optimally. Programmed for a computer, this scheme would appear to behave like an *optimal* expert system.

A warning needs to be issued: The hoop stress formula of constraint (5) is accurate only for *thin* walls, for which $i > 20t$, an example of a model validity constraint initially relaxed to ease the analysis. The reader can verify that this places the following bounds on the parameters. The relation of S to the lower bound represents the thin-wall pressure-bound optimum, while its relation to the upper bound represents the thickness-bound optimum:

$$10P < S < F/10\pi T.$$

These inequalities screen out parameter combinations whose optimal solutions violate the thin-wall assumption. Notice that the interior stress bound optimum *never* satisfies thin-wall theory because in this case $i_* = 2t_* < 20t_*$. Thus, the "optimal" solution for this approximate model may be merely a starting point for a more accurate stress analysis to obtain precise dimensions.

6.3 Functional Monotonicity Analysis

The hydraulic cylinder example of the preceding sections illustrates well how monotonicity analysis directs equations, identifies critical inequalities, and eliminates variables and constraints with them. But the example was overly academic in that every equation was algebraically simple enough to solve in closed form. More realistic engineering problems abound with functions unsolvable because they either are too complicated algebraically or are expressed as the result of a numerical procedure such as finite element analysis. Yet even then monotonicity can be identified and used to simplify a model just as if the equations could be solved explicitly.

Table 6.10. FEA Hydraulic Cylinder: Monotonicity Table 2 (after 3 Reductions and 2 Directions)

Function Number	Functions	Variables		
		i	t	p
(0, 1)	$i + 2t$	+	+	.
(3, 2)	$-(\pi/4)i^2 p + F \leq 0$	−	.	−
(4)	$p - P \leq 0$	.	.	+
(6, 5)	$s(i^+, t^-, p^+) - S \leq 0$	+	−	+

Eliminated Variables and Constraints

Constraint Number	Variable Eliminated
(5)	$s \geqq s(i^+, t^-, p^+)$
(2)	$f \geqq (\pi/4)i^2 p$
(1)	$d \geqq i + 2t$

The method for doing this is called *functional monotonicity analysis* and it relies on the implicit function theorem of classical calculus. This simple approach will be applied here to solve a more realistic variant of the hydraulic cylinder example. The key idea is that variables can still be eliminated, implicitly now instead of explicitly as before. Moreover, the reduced functions obtained, although now expressed functionally rather than algebraically, can often have their monotonicities identified and used for further monotonicity analysis. The outcome will be a precisely defined combination of algebraic operations and numerical procedures for generating the optimum, rather than a set of solved algebraic expressions linked by parametric inequalities.

Imagine then a new hydraulic cylinder model in which the minimum thickness constraint is deleted, and the hoop stress equation is replaced by a finite element analysis (FEA) code, which, given numerical values of the internal diameter i, wall thickness t, and hydraulic pressure p, computes combined stresses at various points in the cylinder wall. That is, the algebraic hoop stress equation $s = ip/2t$ is replaced by the functional equation

$$s = s(i^+, t^-, p^+),$$

the right-hand member representing the output of a finite element code. The superscripts "+" and "−" represent the monotonicities, here taken to be the same as those in the hoop stress equation.

Since all other functions are as in the earlier example, the first stages of the monotonicity analysis are just as shown in Tables 6.3 and 6.4, so they will not be repeated. The equivalent of MT2, showing the results of directing the equations and eliminating the three variables d, f, and s, is given in Table 6.10. It shows that not only is (3, 2) critical for i as before, but now (6, 5) involving the stress functional equation is also critical for t. The latter was only conditionally critical in the previous example because another constraint (7) also could bound t from below.

Explicit Algebraic Elimination

Both critical constraints will eventually be used to eliminate variables. To follow the order of the earlier example, let $p = (4F/\pi)i^{-2}$ be eliminated first. Notice that the solution was obtained in closed form by explicit algebraic manipulation. Its elimination from the stress function by substitution is symbolized by

$$s(i^+, t^-, p^+) = s(i^+, t^-, (4F/\pi)i^{-2})$$
$$= s(i^n, t^-).$$

The superscript "n" indicates that, with p eliminated, the function becomes nonmonotonic in i. This is because s increases with i when p is fixed, but it decreases with i when p is allowed to vary. To prove this statement, consider the total differential for s:

$$ds = (\partial s/\partial i)di + (\partial s/\partial t)dt + (\partial s/\partial p)dp$$
$$= (\partial s/\partial i)di + (\partial s/\partial t)dt + (\partial s/\partial p)(dp/di)di$$
$$= [(\partial s/\partial i) + (\partial s/\partial p)((-2)(4F/\pi)i^{-3})]di + (\partial s/\partial t)dt.$$

In the square brackets enclosing the coefficient of di, the first term is positive, and the second is the product of positive and negative factors, which is negative. Its sign is the monotonicity of s with respect to i, which is therefore negative for small i and positive for large i. For this reason s is nonmonotonic. This fact is not crucial to what follows; it is merely an example of how to analyze monotonicity of expressions that are only partly algebraic. The results are summarized in Table 6.11.

Table 6.11. FEA Hydraulic Cylinder: Monotonicity Table 3 (after All Algebraic Reductions)

Function Number	Functions	Variables i	t
(0, 1)	$i + 2t$	+	+
(4), (3, 2)	$(4F/\pi)i^{-2} - P \leq 0$	−	.
(6, 5) (3, 2)	$s(i^n, t^-) - S \leq 0$	n	−

Eliminated Variables and Constraints

Constraint Number	Variable Eliminated
(3, 2)	$p \geqq (4F/\pi)i^{-2}$
(5)	$s \geqq s(i^+, t^-, p^+)$
(2)	$f \geqq (\pi/4)i^2 p$
(1)	$d \geqq i + 2t$

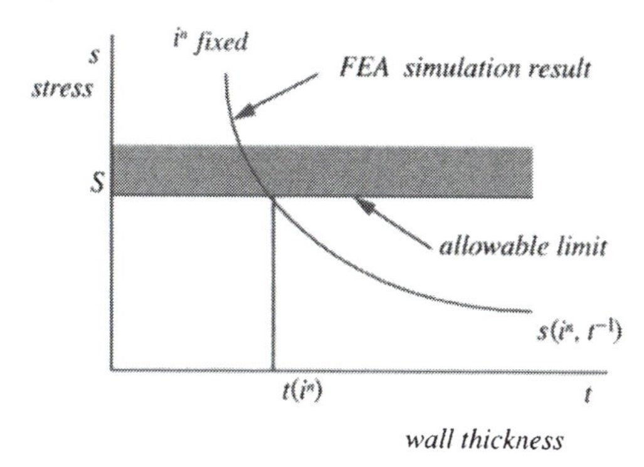

Figure 6.4. Finite element implicit numerical solution.

Implicit Numerical Solution

Table 6.11 shows that the reduced stress constraint [(6, 5), (3, 2)], although only known in functional form, must be critical for t by MP1. Since $s(i^n, t^-)$ is monotonic in t but not in i, the relation should be solved for the unique root t to avoid dealing with the multiple roots of i. The solution technique can only be numerical because $s(i^n, t^-)$ is not an algebraic expression, but instead the result of a finite element numerical computation.

Here is the situation. Any value of internal diameter i given is associated with a corresponding pressure p. With these two variables fixed, the FEA program generates, for any value of the wall thickness t, a set of stresses from which the largest is $s(i^+, t^-, p^+)$. There is only one value of t for which $s = S$, the allowable stress, and it must be found numerically by a root-finding procedure as illustrated in Figure 6.4. Since p cannot be chosen independently of i, from which it is calculated, the wall thickness producing the maximum allowable stress will be designated $t(i)$.

If i were varied over a wide range one could regard the corresponding values of $t(i)$ as an implicit function of i satisfying, over the range of i, the equation $s(i, t(i), (4F/\pi)i^{-2}) = S$. The implicit function theorem of classical mathematics guarantees that such an implicit function always exists for the differentiable functions common in design problems. In the problem at hand the entire function $t(i)$ is not needed. All that is required is the value of $t(i)$ corresponding to the as yet unknown optimizing value of the interior diameter. The iterative numerical process described for computing a few good values of $t(i)$ while seeking the optimal i will be called *implicit numerical solution*.

In Table 6.12, the final MT4, the only variable remaining is i, appearing nonmonotonically in the objective $i + 2t(i^n)$, constrained by a single inequality. Several remarks are in order. Notice first that in the list of eliminated variables t follows rather than precedes p, even though t would ordinarily go to the head of the list because it was eliminated after p. This is because the entire implicit function $t(i)$ is not being calculated – only a single numerical value for which p must be known before the search computation. Hence generating a solution requires determining the variables after i in the order p, t, s, etc.

Table 6.12. FEA Hydraulic Cylinder: Final Monotonicity Table 4

Function Number	Functions	Variable i
(0, 1)	$i + 2t(i^n)$	n
(4), (3, 2)	$(4F/\pi)i^{-2} - P \le 0$	–

Eliminated Variables and Constraints

Constraint Number	Variable Eliminated
(3, 2)	$p \geqq (4F/\pi)i^{-2}$
(6, 5), (3, 2)	$t(i)$: $s(i, t(i)) = S$
(5),((6, 5), (3, 2)), (3, 2)[= 6]	$s \geqq S$
(2)	$f \geqq F$
	$d \geqq i + 2t(i)$

Observe also that the expression for $t(i)$ is given as the numerical solution to an implicit equation rather than as an algebraic expression. Finally note how drastically the stress constraint identification [5, ((6, 5), (3, 2)), (3, 2)] simplifies to the original constraint (6) because of all the resubstitution. This of course still happens to the force constraint (2), just as in the example. Global facts of interest are: (1) The force f is the minimum specified, (2) the maximum stress is exactly that allowed, and (3) the optimal pressure is completely determined by the optimal internal diameter (and vice versa). The remaining task is to find this optimum i upon which everything else depends.

Optimization Using Finite Element Analysis

With the monotonicity analysis complete, it remains to minimize the nonmonotonic objective function of one variable i constrained by the single positive lower bound [4, (3, 2)], for convenience rearranged here as $i \ge 2(F/\pi P)^{1/2}$. Only two cases arise: The constraint is either active or it is not. In the former situation the minimizing diameter is constraint-bound; in the latter, it is unconstrained with one degree of freedom. Since finding the unconstrained minimum could require many uses of the finite element stress computation, the constraint-bound minimum internal diameter case will be considered first. In fortunate circumstances a subsequent bounding calculation will rule out the need for further exploration of larger internal diameters.

MINIMUM INTERNAL DIAMETER CASE

The inside diameter (i.d.), although not necessarily the outside one, is minimum when it equals $2(F/\pi P)^{1/2}$ and consequently the pressure p is at its maximum P. The corresponding wall thickness is $t(2(F/\pi P)^{1/2})$, found by implicit numerical solution using the FEA stress code as often as needed until the maximum stress is just at the allowed value S. This thickness value, abbreviated t_0, will be tested for local minimality as part of the relaxed case procedure outlined next. The resulting value d_0 of the objective outside diameter (o.d.) is computed for future reference as $d_0 = 2(F/\pi P)^{1/2} + 2t_0$.

UNCONSTRAINED MINIMUM (RELAXED) CASE

Before embarking on a search for any unconstrained minimum, the minimum i.d. design just described is checked for local minimality. This is done by increasing i by the smallest amount ε numerically significant to a value abbreviated as $i'(=2(F/\pi P)^{1/2} + \varepsilon)$. The corresponding pressure, computed from eliminated constraint (3, 2), is designated p' for use in the FEA computation. For the same purpose the thickness t' is taken as $(d_0 - i')/2$, for which the o.d. is the same as d_0. Then without iteration the single maximum stress value $s' = s(i', p', t')$ is computed by FEA. If this stress is excessive ($s' > S$), then the new design is infeasible, proving by the Relaxation Theorem that the minimum i.d. design is best, at least locally. However, any decrease in stress would definitely disqualify the minimum i.d. design and signal the need for an examination of larger i.d.s, at least those smaller than d_0, using a line search method from Chapters 4 and 7. This preliminary qualification or disqualification of an entire set of designs – before carrying out a possibly futile search – is an example of what is called a *branch-and-bound* procedure. The next section will use the same tactic to cope with design space discontinuities produced when the set of standard sizes is limited.

This completes the discussion of monotonicity analysis in generating parametric design procedures. The two hydraulic cylinder examples illustrated how algebraic solvability cuts down the need for numerical iteration, an important consideration in an age of computer simulation. Although opening up new design possibilities, such simulation-based optimization can be very expensive if not guided by careful analysis.

6.4 Discrete Variables

The final model reduction situation to be considered occurs when, as in most practical problems, some or all the variables are not free to assume arbitrary positive values. Instead, they are confined to a limited number of values determined by standard sizes available.

Suppose, for example, that in the air tank problem of Chapter 3 the shell length l has to be a standard size, say, a multiple of 10 cm, so that the value 602.5 cm found would be unacceptably nonstandard. Imagine similarly that head and shell thicknesses have to be multiples of 0.5 cm, causing the rejection of the values 13.75 and 1.02. To make things unanimous, suppose that the vessel must be rolled to standard inside radii that are multiples of 10 cm, so that 105.8 cm cannot be accepted. Overall length t and volume v are not, however, limited to standard sizes. Then l, h, s, and r are now *discrete* variables, whereas t and v remain *continuous* variables.

When an active constraint depends on at least one continuous variable, it is possible to satisfy it as a strict equality, as with the solution already found. But when all variables in a constraint are discrete, in which case the constraint will be called *discrete*, the optimum will, except by coincidence, only satisfy the constraint with strict inequality. Having every active constraint expressed as a directed equation, if it was not originally an inequality, is therefore especially important for discrete constraints.

In deciding how to generate a discrete solution, one must deal with the critical constraints, being aware which variable is bounded in each. Thus, the three critical constraints, written with the variables bounded by the constraint written on the left, are:

$$r^2 l \geqq 6.748(10^6), \quad h \geqq 130(10^{-3})r, \quad s \geqq 9.59(10^{-3})r. \tag{6.11}$$

The continuous variables v and t have been eliminated to simplify the analysis. Notice that since the first constraint can bound both r and l, both variables have been written on the left.

In the last two constraints, one cannot adjust the bounded variables because the right members contain another variable not yet determined. Only the first constraint has its right member known as a constant. But it is multiply critical in that it bounds both r and l. Thus, selecting a discrete value for one variable will determine the appropriate discrete value for the other.

If r is rounded up, l must be rounded down, and vice versa. Rounding one variable automatically fixes the discrete value of the other. Thus, rounding r from its continuous value of 105.8 up to the next allowable discrete value of 110 produces the inequality

$$l \geqq 6.748(10^6)/(110)^2 = 557.7. \tag{6.12}$$

Hence, l must be rounded up to 560. Rounding r down to 100 would require l to be 680. If l were rounded first from 602.5 to 610, then r would have to be rounded up to 110 from 105.2. Finally, if l were rounded down to 600, r would still have to be rounded up to 110 from 106.1.

Rather than enumerate all these possibilities, let us simply find the lowest value of one of the variables that will give a feasible discrete solution for all variables. In deciding whether to round r or l, it is wise to consider how sensitive the left member $r^2 l$ is to changes in each variable. Thus,

$$\partial(r^2 l)/\partial r = 2rl = 2(105.8)(602.5) = 127.5(10^3), \tag{6.13}$$

whereas

$$\partial(r^2 l)/\partial l = r^2 = (105.8)^2 = 11.19(10^3), \tag{6.14}$$

meaning that the constraint is much more sensitive to a change in r than to an equal adjustment of l. Hence, r should be rounded first, since the change it causes in l will be significant.

Then let r be rounded up to equal 110, for which $l = 560$ as already noted. The remaining critical constraints involve only the appropriate critical variable:

$$h \geqq 130(10^{-3})(110) = 14.3, \quad s \geqq 9.59(10^{-3})(110) = 1.1, \tag{6.15}$$

whence $h = 14.5$ and $s = 1.5$. The relaxed constraints must also be checked:

$$\begin{aligned} &l + 2h = 589 < 630, \quad 10 < l = 560 < 610, \\ &r + s = 111.5 < 150. \end{aligned} \tag{6.16}$$

Thus this discrete solution is feasible.

To prove that the value $r = 110$ is the smallest possible, consider the next smaller discrete value $r = 100$, for which $l = 680$. This is not feasible, violating $l \leq 610$ as well as $l \leq 630 - 2h < 630$. Hence $r = 110$ is the smallest discrete value for which a feasible design is possible. The metal volume corresponding is

$$\pi[2(110)(1.5)(560) + (1.5)^2(560) + 2(110)^2(14.5)] = 5.370\pi(10^5). \tag{6.17}$$

This last design has not, however, been proven optimal yet. One could imagine that larger values of r might allow a closer approach to the critical constraints that could produce a better design. But of course such a design could be no better than the continuous optimum with $r \geq 120$, the next allowable discrete value. Monotonicity analysis would indicate that $r = 120$ for this restricted problem and the three remaining critical constraints would give

$$\begin{aligned} &l = 6.748(10^6)/(120)^2 = 468.6, \quad h = 130(10^{-3})(120) = 15.6, \\ &s = 9.59(10^{-3})(120) = 1.151, \end{aligned} \tag{6.18}$$

whence

$$f(15.6, 468.6, 620, 1.151) = 5.793\pi(10^5). \tag{6.19}$$

But this is already greater than the discrete solution $(5.370\pi(10)^5)$ already found. Hence, no better solution, discrete or continuous, exists for $r \geq 120$. This completes the proof of discrete optimality.

The tactic of using the restricted continuous minimum as a lower bound on a restricted discrete minimum is an example of *implicit enumeration* or *branch-and-bound*. It is a powerful method for solving discrete optimization problems, especially when combined with monotonicity analysis to simplify computation of the lower bound.

6.5 Discrete Design Activity and Optimality

At this point we must reconsider the concepts of constraint activity and local optimality in the presence of discrete variables. In this section, the formal definitions of activity introduced first in Chapter 3 for optimization of continuous problems are clarified and extended to address discrete ones. Moreover, two distinct types of discrete local optima, one arising from the discreteness of the problem and the other from the underlying continuous problem, are noted.

Constraint Activity Extended

Constraint activity definitions, applicable to continuous global optimization problems, are presented first. These definitions include *inactivity, tightness, weak* and *strong activity*, and *weak* and *strong semiactivity*. Two brief examples are used to demonstrate the definitions. A "degenerate" set of the extended definitions, applicable to discrete optimization problems, is presented next. The applicability of the extended definitions to local optimization methods is also noted. A summary example, considered successively as a local, continuous global, and discrete optimization problem, is used to solidify the various definitions.

The extended definitions for global optimization problems with continuous variables and with discrete variables are mutually exclusive, eliminating any possible ambiguity. Local optimization is then reexamined in light of the extended definitions.

CONTINUOUS GLOBAL OPTIMIZATION

Given a continuous global optimization problem, two simple tests are applied to determine the activity status of an arbitrary inequality constraint, g_i. The first test considers the effect of deleting the constraint from the problem. In particular, such a deletion may alter both the global solution set and its objective function value, may alter the global solution set without affecting its objective function value, or may have no effect at all. The second test investigates whether or not constraint g_i is satisfied as an equality or as a strict inequality at the global optimum. The six mutually exclusive combinations of test outcomes, along with the definitions assigned to each combination, are detailed in Table 6.13. For continuous global optimization problems, the nonspecific term "active" is now taken to refer to *any* of the four types of activity in Table 6.13. That is, removal of an active constraint has some effect on the global solution set.

Example 6.1 Consider the model

$$\text{minimize} f(x_1, x_2) = x_2$$

subject to

$$g_1\colon (x_1 - 1)(x_1 - 3)^2(x_1 - 5) + x_1 + 4 - x_2 \leq 0,$$
$$g_2\colon -x_1 + 3 \leq 0,$$
$$g_3\colon -x_2 + 1 \leq 0.$$

Figure 6.5 shows the constraints and minimization direction for this problem. The global optimum is $x_* = (4.347, 4.381)^T$ with $f(x_*) = 4.381$. Removal of either g_1 or g_2 alters both the global solution set and its objective function value, so each is active. Since g_1 is satisfied as an equality at the optimum it is strongly active, while g_2 is weakly active. Constraint g_3 is satisfied as a strict inequality at the optimum, and its removal does not affect the global solution set, so g_3 is inactive. ■

Table 6.12. Activity Definitions: Continuous Global Optimization

	g_i satisfied as a strict inequality at the global optimum	**g_i satisfied as an equality at the global optimum**
Removal of g_i does not affect the set of globally optimal solutions	*inactive*	*tight*
Removal of g_i alters the global solution set but does not affect its objective function value	*weakly semiactive*	*strongly semiactive*
Removal of g_i alters both the global solution set and its objective function value	*weakly active*	*strongly active*

Example 6.2 Consider the model

$$\text{minimize} f(x_1, x_2) = x_2$$

subject to

$$g_1\text{: } (x_1 - 1)(x_1 - 3)^2(x_1 - 5) + 6 - x_2 \le 0,$$

$$g_2\text{: } -x_1 + 3 \le 0,$$

$$g_3\text{: } -x_2 + 2 \le 0.$$

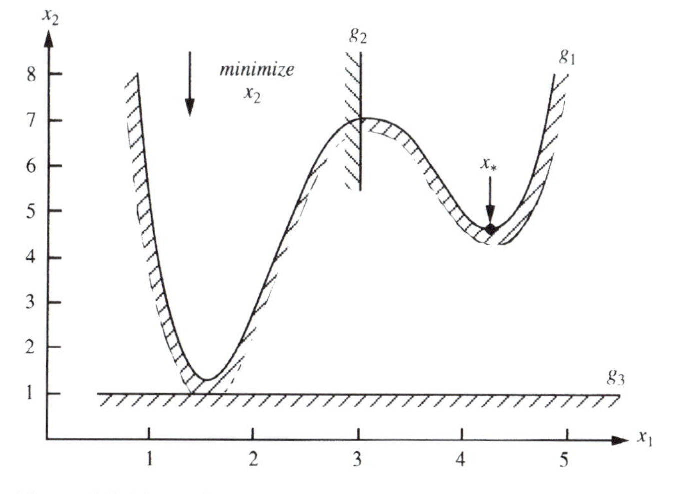

Figure 6.5. Plots of g_1, g_2, and g_3 for Example 6.1.

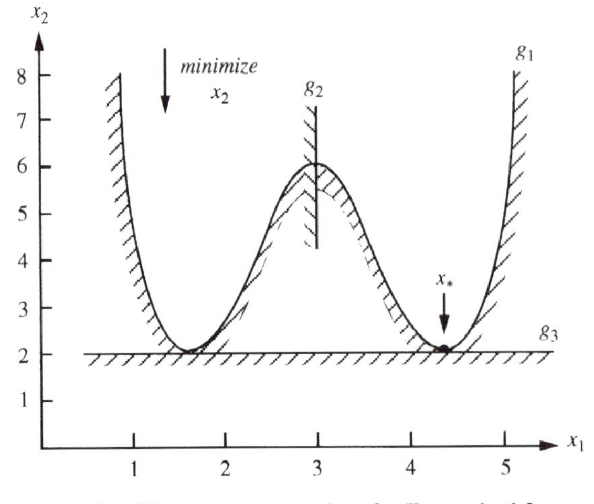

Figure 6.6. Plots of g_1, g_2, and g_3 for Example 6.2.

Figure 6.6 shows the constraints and minimization direction. The global optimum is $x_* = (4.414, 2.0)^T$ with $f(x_*) = 2.0$. Removal of either g_1 or g_2 alters the global solution set but not its objective function value, so g_1 and g_2 are semiactive. As in the previous example, g_1 is satisfied as an equality at the optimum, while g_2 is not. Thus, g_1 is strongly semiactive, while g_2 is weakly semiactive. Constraint g_3 is an equality at the optimum, and its removal does not affect the global solution set. Therefore, g_3 is tight. ■

DISCRETE OPTIMIZATION

Lacking a universally accepted definition for local optima within discrete design problems, only global solutions to discrete problems may be considered. Therefore, *discrete optimization* should be interpreted as *discrete global optimization* unless otherwise noted. The topic of discrete "local" optima is discussed further below. Unless specifically stated otherwise, the term *local optimization* will always refer only to continuous problems.

When considering discrete optimization problems, the test focusing on the effect of deleting a particular constraint remains applicable. However, the question of whether an arbitrary inequality constraint, containing only discrete variables, is satisfied as an equality at the optimum is of little use. Such satisfaction is purely coincidental and coincidences cannot be predicted ahead of time. Consequently active inequality constraints cannot be exploited to solve problems, as was done in the continuous case. In light of the above observations, Table 6.13 should be modified to show the limitations of activity definitions for discrete optimization. In particular, the second column should be deleted and the term "weakly" eliminated from the first column, as in Table 6.14.

LOCAL OPTIMIZATION

When dealing with local optima, many of the refined activity definitions for continuous global optimization become pointless. Focusing on local effects precludes consideration of optima in other regions of the solution space, and so weak activity

Table 6.13. Activity Definitions Restricted to Discrete and Local Optimization

Discrete Optimization		
	g_i satisfied as a strict inequality at the global optimum	
Removal of g_i does not affect the set of globally optimal solutions	*inactive*	
Removal of g_i alters the global solution set but does not affect its objective function value	*semiactive*	
Removal of g_i alters both the global solution set and its objective function value	*active*	
Local Optimization		
	g_i satisfied as a strict inequality at the local optimum	**g_i satisfied as an equality at the local optimum**
Removal of g_i does not affect the set of locally optimal solutions	*inactive*	*tight*
Removal of g_i alters both the local solution set and its objective function value		*(strongly) active*

and both forms of semiactivity become meaningless. Eliminating the *weakly active*, *weakly semiactive*, and *strongly semiactive* entries from Table 6.13, and replacing all occurrences of *global* with *local*, produces the definitions shown in Table 6.14.

Table 6.14 leads to two insights. First, as only one type of activity remains, the *strongly* label is superfluous. The single type of activity defined in Table 6.14 is the type referred to throughout a majority of the current optimization literature. Second, the empty cell indicates that no circumstances exist where the removal of an inequality constraint satisfied as strict inequality at a local optimum can affect the value of that local optimum. The constraint will have a zero multiplier in the KKT conditions (a degenerate case).

The following Example 6.3 demonstrates the application of the above definitions to a single optimization problem, considered in turn as a local, continuous global, and discrete problem. In the second and third instances, Example 6.3 is seen to have more active inequality constraints than problem variables!

Example 6.3 Consider the model

minimize $f(x_1, x_2) = x_1 + 10x_2$

subject to

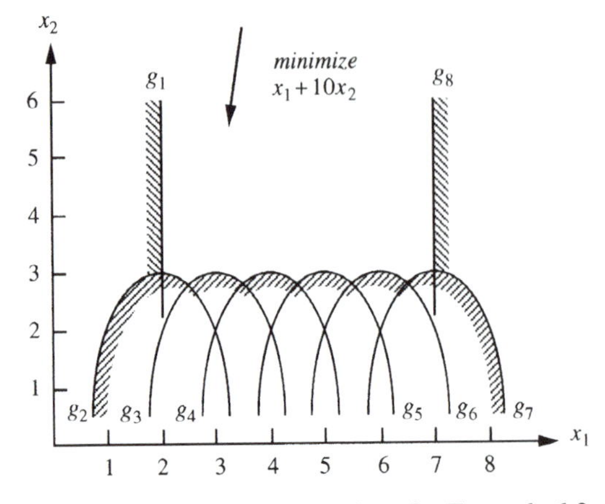

Figure 6.7. Plots of g_1 through g_8 for Example 6.3.

$$
\begin{aligned}
&g_1\colon -x_1 + 2 \le 0,\\
&g_2\colon -(x_1 - 2)^2 - x_2 + 3 \le 0,\\
&g_3\colon -(x_1 - 3)^2 - x_2 + 3 \le 0,\\
&g_4\colon -(x_1 - 4)^2 - x_2 + 3 \le 0,\\
&g_5\colon -(x_1 - 5)^2 - x_2 + 3 \le 0,\\
&g_6\colon -(x_1 - 6)^2 - x_2 + 3 \le 0,\\
&g_7\colon -(x_1 - 7)^2 - x_2 + 3 \le 0,\\
&g_8\colon x_1 - 7 \le 0.
\end{aligned}
$$

Figure 6.7 shows plots of g_1 through g_8, as well as the minimization direction. Local optima occur at $(2.0, 3.0)^T, (2.5, 2.75)^T, (3.5, 2.75)^T, (4.5, 2.75)^T, (5.5, 2.75)^T$, and $(6.5, 2.75)^T$, with objective values of 32, 30, 31, 32, 33, and 34, respectively. The continuous global optimum occurs at $x^*_{\text{cont}} = (2.5, 2.75)^T$ with $f(x^*_{\text{cont}}) = 30$, corresponding to the second local optimum listed. The discrete optimum, with both variables required to be integer, occurs at $x^*_{\text{disc}} = (2.0, 3.0)^T$ with $f(x^*_{\text{disc}}) = 32$.

Local optimization Considering the problem as a search for a local optimum, the two variables and the functional independence of the constraints ensure that at most two constraints will be active at an optimum. All remaining constraints are inactive. Finding the simultaneous solution of each pair of active constraints, checking for feasibility, and verifying the local optimality conditions yields the six solutions detailed above.

Continuous global optimization As a continuous global problem, g_2 and g_3 are satisfied as equalities at the optimum, making them strongly active. In contrast to local optimization, all remaining constraints in the continuous global problem are weakly active, as elimination of any one of them alters the value of the global optimum!

Discrete optimization Considered as a discrete problem, Example 6.3 shows that such a problem may have an arbitrarily large number of active constraints. As with the continuous global problem, elimination of any constraint alters the value of the optimum. Notice that g_1 and g_2 are satisfied as equalities at the optimum only coincidentally. Altering g_1 to $x_1 \leq 1.9$ and g_2 to $x_2 \leq 2.9 - (x_1 - 2)^2$ converts both active constraints into strict inequalities at the optimum, while not affecting the location of the optimum. ■

Discrete Local Optima

Local optima as studied in continuous optimization create significant difficulties in discrete problems. As demonstrated below, multiple discrete local optima arise from two entirely different causes. One type of discrete local optimum arises from the mathematical model under consideration, while the other type is largely a function of how "local" is defined. Alternate ways of defining discrete local minima are presented below, followed by an illustrative example employing these definitions. Note also that although multiple global optima arise in both continuous and discrete problems, occurrences of such optima are rare and are easily detectable in discrete problems.

DISCRETE LOCAL MINIMA DEFINITIONS

Four definitions are now presented to aid in the example and discussion that follows. Two assumptions are implicit in the definitions. First, the definitions apply only to two-dimensional solution spaces, and second, the definitions assume that all variables are integer. These restrictions are not severe and extension to n dimensions and to general discrete variables is straightforward.

Definitions (Discrete Variables)

1. A *neighborhood* of $(x_1, x_2)^T$ is the set of four points $\{(x_1, x_2 + 1)^T, (x_1, x_2 - 1)^T, (x_1 + 1, x_2)^T, (x_1 - 1, x_2)^T\}$.
2. A point $(x_1, x_2)^T$ is a *local minimum* if every point $(a, b)^T$, in the neighborhood of $(x_1, x_2)^T$, is either infeasible or has an objective function value $f(a, b) > f(x_1, x_2)$.
3. An *extended neighborhood* of $(x_1, x_2)^T$ is the set of eight points $\{(x_1, x_2 + 1)^T, (x_1, x_2 - 1)^T, (x_1 + 1, x_2)^T, (x_1 - 1, x_2)^T, (x_1 + 1, x_2 + 1)^T, (x_1 + 1, x_2 - 1)^T, (x_1 - 1, x_2 + 1)^T, (x_1 - 1, x_2 - 1)^T\}$.
4. A point $(x_1, x_2)^T$ is an *extended local minimum* if every point $(a, b)^T$, in the extended neighborhood of $(x_1, x_2)^T$, is either infeasible or has an objective function value $f(a, b) > f(x_1, x_2)$.

The above definitions are not universally recognized, and so they do not contradict the earlier claim that no such universal definitions exist. If the definition of a neighborhood is extended to n dimensions, two points along each dimension are investigated, so that the number of points in a neighborhood grows as $2n$. To define an extended neighborhood, three points along each dimension must be considered, with all possible combinations of such points belonging to the set of interest. For an n-dimensional problem, the number of points in an extended neighborhood is $3^n - 1$, where the one accounts for the center point. The cost of exponential growth over linear growth increases rapidly with the dimension of the problem, rendering useful implementations of "exponential" definitions unlikely.

The nature of discrete local minima is now considered from two different points of view. The first viewpoint concerns how the number of minima appearing in a given problem is related to the definition used to generate them. The second viewpoint deals with the importance of identifying various local minima within a particular problem.

Example 6.4 (Discrete multiple optima.) Consider the model

$$\begin{aligned}
&\text{minimize } f(x_1, x_2) = x_2 - 1.1x_1\\
&\text{subject to}\\
&\quad g_1\colon x_1 - x_2 + 1 \le 0,\\
&\quad g_2\colon x_1 - 5 \le 0,\\
&\quad g_3\colon -4x_1^2 + 28x_1 - x_2 - 40 \le 0,\\
&\quad x_1, x_2 \in \mathcal{Z}_p,
\end{aligned}$$

where $\mathcal{Z}_p$ is the set of positive integers. A relaxed version of Example 6.4 may be obtained by eliminating constraint g_3. The constraints, minimization direction, and several important discrete points for this example are shown in Figure 6.8. For each case, unrelaxed and relaxed, the global minimum occurs at $(5, 6)^T$. For the relaxed case, additional local minima occur at $(1, 2)^T$, $(2, 3)^T$, $(3, 4)^T$, and $(4, 5)^T$, with the later two points being infeasible in the unrelaxed case due to constraint g_3.

Noting that none of the additional local minima for the relaxed case are also extended local minima, it becomes obvious that a relatively minor change in the definition of a neighborhood may have a major effect on the number of local minima present in the problem. In the unrelaxed example, points $(1, 2)^T$, $(2, 3^T)$, and $(5, 6^T)$ are local minima, while only $(2, 3)^T$ and $(5, 6)^T$ are extended local minima. In general, the number of local minima identified by any particular definition of "local" decreases as the number of points in the definition of neighborhood employed by that definition increases.

Because the local minima of the relaxed case arise solely from the discreteness of the problem, identifying point $(2, 3)^T$ in the relaxation is of little concern. However, because g_3 effectively divides the solution space, identification of point $(2, 3)^T$ as a local minimum becomes more important to an engineer or designer when g_3 is included in the model.

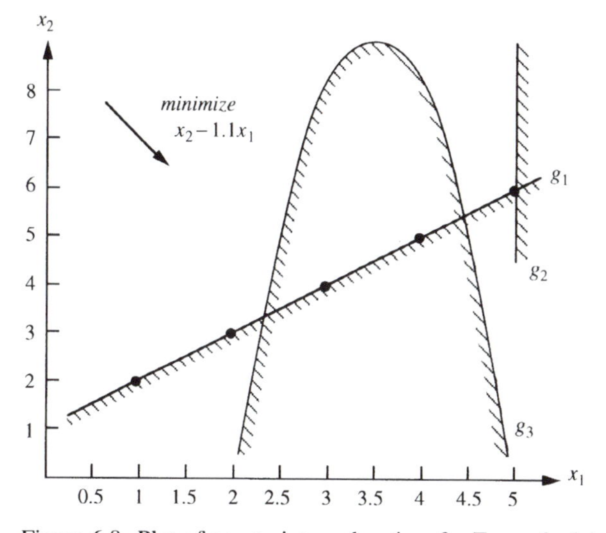

Figure 6.8. Plot of constraints and optima for Example 6.4.

Another way to view this situation is to consider the two cases as continuous optimization problems, where the variables are restricted to be positive real numbers rather than positive integers. In this design domain, the relaxed case has a unique minimum, while the unrelaxed case has two local minima occurring in different portions of the solution space. Identifying both of these minima and/or both portions of the solution space would be the goal of a thorough engineering study. ■

Although for Example 6.4 the use of extended local minima definitions allows detection of the important discrete local minima arising from problem characteristics, while it ignores minima resulting from problem discreteness, this advantageous result is not guaranteed. Moreover, recall that the cost of investigating extended local minima grows exponentially with the dimension of the problem. In light of these concerns about uncertainty and cost, the issue of defining and identifying discrete local minima remains an open research topic. Throughout the remainder of this chapter, discrete optimization implies that a global optimum is desired.

6.6 Transformer Design

This section introduces the problem of designing low-frequency transformers for minimum cost. This design problem will be used in subsequent sections to illustrate how model reduction techniques can help in solving discrete problems.

An extensive analysis of transformer performance was presented by Hamaker and Hehenkamp (1950, 1951). The problem was revisited by Schinzinger (1965) and has been used as a global optimization test problem by Ballard, Jelinek, and Schinzinger (1974), Bartholomew-Biggs (1976), and Hock and Schittkowski (1981). In the present section a brief discussion of the original problem statement by Hamaker

Table 6.14. Transformer Design Variables

x_1	core width [cm]
x_2	winding window width [cm]
x_3	winding window height [cm]
x_4	stacking depth [cm]
x_5	peak magnetic-flux density [Wb]
x_6	effective electric-current density [A mm^2]

and Hehenkamp leads to a mathematical model with six variables and two inequality constraints. In later sections we see that analysis of the model forces the variables representing magnetic-flux density and electric-current density to assume fixed values, resulting in a reduced, four-variable model. Rigorous reduction of the model eliminates infeasible regions of the solution space, so that from the original number of 8.9×10^{10} points only 4,356 remain. These points are then investigated exhaustively, identifying the global optimum for the discrete problem.

Model Development

As Hamaker and Hehenkamp suggest:

> Three equations play a part in the theoretical design of a transformer: (1) the 'price equation,' specifying the price as a function of the geometrical dimensions, (2) the 'power equation,' expressing the apparent power in terms of the dimensions and of the magnetic-flux density and electric-current density, and (3) the 'loss equation,' giving the losses in terms of the same set of variables.

Table 6.15 lists the six problem variables, including their units. The four geometric problem variables are used in Figure 6.9, which details the geometry of the transformer. The three equations described above are now discussed.

PRICE EQUATION

Summing the iron core (cross-hatched) regions depicted in Figure 6.9 and reducing yields

$$2x_1x_4(x_1 + x_2 + x_3), \tag{6.20}$$

while summing the copper-winding (double cross-hatched) regions gives

$$2x_2x_3(x_1 + \pi x_2/2 + x_4). \tag{6.21}$$

Introducing $\alpha/2$ and $\beta/2$ as "price densities" for iron and copper, respectively, leads to the equation for the price P,

$$P = \alpha x_1x_4(x_1 + x_2 + x_3) + \beta x_2x_3(x_1 + \pi x_2/2 + x_4). \tag{6.22}$$

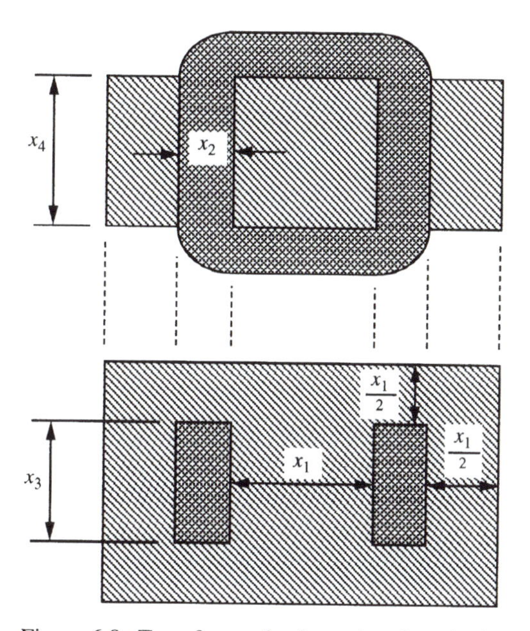

Figure 6.9. Transformer horizontal and vertical cross sections.

Schinzinger introduces the loss equation (developed below) into the objective function as a means of incorporating operating costs into the analysis. With weighting factors ϕ and κ on the respective terms, the price equation becomes

$$\begin{aligned} P = &\alpha x_1 x_4(x_1 + x_2 + x_3) + \beta x_2 x_3(x_1 + \pi x_2/2 + x_4) \\ &+ \phi x_5^2 x_1 x_4(x_1 + x_2 + x_3) + \kappa x_6^2 x_2 x_3(x_1 + \pi x_2/2 + x_4). \end{aligned} \tag{6.23}$$

POWER EQUATION

From Hamaker and Hehenkamp the power equation is given as

$$VA = \varepsilon x_1 x_2 x_3 x_4 x_5 x_6, \tag{6.24}$$

where VA represents the apparent power, and ε is a frequency and material specific constant.

LOSS EQUATION

Again from Hamaker and Hehenkamp, the loss equation is given as

$$W = \gamma x_5^2 x_1 x_4(x_1 + x_2 + x_3) + \delta x_6^2 x_2 x_3(x_1 + \pi x_2/2 + x_4), \tag{6.25}$$

where W is the loss, and $\gamma/2$ and $\delta/2$ are the specific losses per unit volume for the iron core and copper winding, respectively. Note that γ and δ are incorporated into ϕ and κ, respectively, in Equation (6.23).

TRANSFORMER DESIGN MODEL SUMMARY

$$\begin{aligned}\min P = {} & 0.0204x_1x_4(x_1+x_2+x_3)+0.0187x_2x_3(x_1+1.57x_2+x_4)\\ & +0.0607x_1x_4x_5^2(x_1+x_2+x_3)+0.0437x_2x_3x_6^2(x_1+1.57x_2+x_4)\end{aligned}$$

subject to (6.26)

$$\begin{aligned}& g_1\colon -0.001x_1x_2x_3x_4x_5x_6+2.07\le 0,\\ & g_2\colon 0.00062x_1x_4x_5^2(x_1+x_2+x_3)\\ & \quad +0.00058x_2x_3x_6^2(x_1+1.57x_2+x_4)-1.2\le 0,\\ & x_i\in\{0,1,2,3,\ldots,\infty\},\quad i=1 \text{ to } 6.\end{aligned}$$

Inserting appropriate values for α through κ (from Schinzinger) into Equations (6.23), (6.24), and (6.25) yields the objective and constraint functions given in (6.26). Note that the power equation has been converted into a lower bound on the required power, while the loss equation has been converted into an upper bound on the allowable loss.

The problem is now modified in several ways to serve the needs of the present discussion. First, all six problem variables are required to be integer. For the geometric problem variables, this restriction could result from standard material or tooling sizes. For the flux and current densities, the integrality requirement is somewhat artificial, as these are really continuous phenomena. Second, adjustment of the constant term in constraint g_2 from 1 to 1.2 makes the constraint easier to satisfy. This second modification is necessary to create feasible integer solutions to the problem. Finally, nonnegativity constraints on all variables in the continuous model are incorporated into the set constraints of the discrete model. Solution of this modified model motivates the remainder of this chapter.

Preliminary Set Constraint Tightening

Two of the six problem variables, x_5 and x_6, can be eliminated by proving that each must have a value of one in any feasible solution. Inspection of constraint g_1 indicates that no variable may be assigned a value of zero. The set constraints on all variables are therefore tightened to

$$x_i\in\{1,2,3,\ldots,\infty\},\quad i=1 \text{ to } 6. \tag{6.27}$$

Noting that constraint g_2 increases with respect to each of the problem variables, the following technique is applied. First, five of the variables are set equal to their minimum values of one. Second, the remaining single-valued monotonic function is maximized while continuing to satisfy the constraint. The variable value generating the maximum is then used to update the set constraint of the variable under consideration.

Applying the above process to each of the six variables leads to the updated set constraints

$$\begin{aligned}& x_1\in\{1,2,3,\ldots,42\}, && x_2\in\{1,2,3,\ldots,35\},\\ & x_3\in\{1,2,3,\ldots,445\}, && x_4\in\{1,2,3,\ldots,491\},\\ & x_5\in\{1,2,3,\ldots,25\}, && x_6\in\{1,2,3,\ldots,24\}.\end{aligned} \tag{6.28}$$

With x_5 and x_6 having the smallest set constraints, effort is aimed at reducing these ranges even further. Eventually, x_5 and x_6 are each forced to take on a value of one, and the degrees of freedom of the problem are reduced by two.

6.7 Constraint Derivation

This section develops a result that allows constraints of a desired form to be derived from constraints of other forms. The new constraints are helpful in the efficient reduction of the feasible domain.

Discriminant Constraints

Discriminant constraints are so named for reasons that will soon become obvious. Consider an inequality constraint, $g_j(\mathbf{x})$, parsed into the form

$$A(\mathbf{x})T^2(\mathbf{x}) + B(\mathbf{x})T(\mathbf{x}) + C(\mathbf{x}) \leq 0, \tag{6.29}$$

where A, B, C, and T are arbitrary functions of $\mathbf{x}$. Relaxing the dependence of A, B, C, and T on $\mathbf{x}$ leads immediately to

$$AT^2 + BT + C \leq 0. \tag{6.30}$$

Note that satisfaction of Equation (6.30) implies satisfaction of Equation (6.29), a more specific result. Assuming that $A > 0$, and multiplying through by A, yields

$$A^2T^2 + ABT + AC \leq 0. \tag{6.31}$$

Completing the square and simplifying leads to

$$(AT + B/2)^2 - B^2/4 + AC \leq 0. \tag{6.32}$$

Rearranging Equation (6.32) leads to

$$0 \leq 4(AT + B/2)^2 \leq B^2 - 4AC, \tag{6.33}$$

which after further rearrangement and reintroduction of $\mathbf{x}$ yields the desired result:

$$4A(\mathbf{x})C(\mathbf{x}) - B^2(\mathbf{x}) \leq 0. \tag{6.34}$$

Constraint Addition

The discriminant constraint of Equation (6.34) is now used as a problem reduction technique. A derivation similar to the one above produces the discriminant constraint for $A(\mathbf{x}) < 0$.

UPPER BOUND ON x_5x_6

An upper bound on the product of x_5 and x_6 is determined first. This upper bound is used below to form a lower bound on the sum $(x_1 + x_2 + x_3)$, which is then used to tighten the upper bound, and so on. Rewriting constraint g_1 as

$$x_1x_4x_5 \geq 2{,}070/x_2x_3x_6 \tag{6.35}$$

and substituting into constraint g_2 yields, with some rearrangement,

$$\begin{aligned}&128{,}340x_5(x_1 + x_2 + x_3)/x_2x_3x_6\\&\quad + 58x_2x_3x_6^2(x_1 + 1.57x_2 + x_4) \leq 120{,}000.\end{aligned} \tag{6.36}$$

With $(x_1 + x_2 + x_3) \geq 3$ and $(x_1 + 1.57x_2 + x_4) \geq 3.57$, Equation (6.36) is rewritten as

$$385{,}020x_5/x_2x_3x_6 + 207.06x_2x_3x_6^2 \leq 120{,}000. \tag{6.37}$$

Multiplying through by $x_2x_3x_6$ yields

$$385{,}020x_5 + 207.06x_6(x_2x_3x_6)^2 - 120{,}000x_2x_3x_6 \leq 0. \tag{6.38}$$

Noting the form of the left-hand side of Equation (6.38), we can form a discriminant constraint. With

$$\begin{aligned}&A(\mathbf{x}) = 207.06x_6, \quad B(\mathbf{x}) = -120{,}000,\\&C(\mathbf{x}) = 385{,}020x_5, \quad \text{and} \quad T(\mathbf{x}) = x_2x_3x_6,\end{aligned} \tag{6.39}$$

the discriminant constraint is found to be

$$4(207.06x_6)(385{,}020x_5) \leq 120{,}000^2. \tag{6.40}$$

Solving

$$318{,}888{,}964.8x_5x_6 \leq 120{,}000^2 \tag{6.41}$$

for the product x_5x_6 yields

$$x_5x_6 \leq 45.2, \tag{6.42}$$

which is tightened to

$$x_5x_6 \leq 45. \tag{6.43}$$

Equation (6.43) represents a global upper bound, valid for the entire problem.

LOWER BOUND ON $x_1 + x_2 + x_3$

To tighten the lower bound on $(x_1 + x_2 + x_3)$, an assumption is made and contradicted. Assuming $x_1 + x_2 + x_3 = 3$ leads to

$$x_1 = x_2 = x_3 = 1. \tag{6.44}$$

Substituting Equations (6.43) and (6.44) into constraint g_1 and reducing yields

$$x_4 \geq 46. \tag{6.45}$$

Substituting Equations (6.44) and (6.45) into constraint g_2 and reducing yields

$$8{,}556x_5^2 + 2{,}817.1x_6^2 \leq 120{,}000, \tag{6.46}$$

from which

$$x_5 \le 3, \ x_6 \le 6 \quad \text{and} \quad x_5 x_6 \le 18 \tag{6.47}$$

can be deduced. Note that the upper bounds of Equation (6.47) are valid only within the assumption above. Equation (6.47) now replaces Equation (6.43) and the process is repeated. A second round yields

$$x_5 \le 1, \ x_6 \le 1, \quad \text{and} \quad x_5 x_6 \le 1. \tag{6.48}$$

With all variables except x_4 now equal to one, x_4 must equal 2,070 to satisfy constraint g_1. However, this value fails to satisfy the set constraint on x_4, providing the needed contradiction to the assumption above. With $x_1 + x_2 + x_3 \neq 3$, the sum is increased to its next achievable discrete value, yielding

$$x_1 + x_2 + x_3 \ge 4. \tag{6.49}$$

Linear and Hyperbolic Constraints

This subsection continues the constraint derivation activities began above. In particular, linear constraints are derived from hyperbolic constraints, and vice versa. Examples demonstrate the application of these two types of constraint derivation. Note that the results here are applicable to integer variables only, with extension to general discrete variables left as future investigation.

DERIVATION OF LINEAR CONSTRAINTS

Linear constraints of the form

$$K_s \le x_1 + x_2 + x_3 + \cdots + x_n \tag{6.50}$$

can be derived from hyperbolic constraints of the form

$$K_p \le x_1 x_2 x_3 \cdots x_n, \tag{6.51}$$

where $K_p > 0$. Note that the s and p subscripts of Equations (6.50) and (6.51) refer to *sum* and *product*, respectively, denoting the relationship of the variables within the inequality. To determine the maximum value of K_s for a given value of K_p, the following example is solved as a continuous optimization problem.

Example 6.5 Consider the model with continuous variables

$$\text{minimize}\; x_1 + x_2 + x_3 + \cdots + x_n$$

subject to

$$g_1\colon K_p - x_1 x_2 x_3 \cdots x_n \le 0.$$

From the Lagrangian,

$$L(\mathbf{x}, \mu) = x_1 + x_2 + x_3 + \cdots + x_n + \mu(K_p - x_1 x_2 x_3 \cdots x_n), \tag{6.52}$$

the Karush–Kuhn–Tucker (KKT) conditions are found to be

$$\nabla L = (1 + \mu(-x_2 x_3 \cdots x_n),\quad 1 + \mu(-x_1 x_3 \cdots x_n), \ldots,\\ 1 + \mu(-x_1 x_2 \cdots x_{n-1})) = \mathbf{0}, \tag{6.53}$$

$$K_p - x_1 x_2 x_3 \cdots x_n \leq 0, \tag{6.54}$$

$$\mu \geq 0, \tag{6.55}$$

$$\mu(K_p - x_1 x_2 x_3 \cdots x_n) = 0. \tag{6.56}$$

Satisfaction of Equation (6.53) ensures that

$$\mu \neq 0, \tag{6.57}$$

allowing Equation (6.56) to force

$$K_p = x_1 x_2 x_3 \cdots x_n. \tag{6.58}$$

Substitution of Equation (6.58) into Equation (6.53) yields

$$\nabla L = (1 - \mu K_p/x_1,\ 1 - \mu K_p/x_2, \ldots, 1 - \mu K_p/x_n) = \mathbf{0}, \tag{6.59}$$

leading to

$$x_1 = x_2 = x_3 = \cdots = x_n = \mu K_p = K_p^{1/n}, \tag{6.60}$$

where the final equality in Equation (6.60) is obtained by reducing constraint g_1 to a function of a single variable and solving. The minimizer of the continuous optimization problem is

$$\mathbf{x}^c = \left(x_1^c, x_2^c, x_3^c, \ldots, x_n^c\right)^T = \left(K_p^{1/n}, K_p^{1/n}, \ldots, K_p^{1/n}\right)^T \tag{6.61}$$

with

$$f\left(x_1^c, x_2^c, x_3^c, \ldots, x_n^c\right) = n K_p^{1/n}. \tag{6.62}$$

■

ROUNDING OFF THE CONTINUOUS SOLUTION

Having solved Example 6.5 as a continuous optimization problem, we desire a discrete solution based on the continuous minimizer x^c. Using the rounding down and up operators $\lfloor\ \rfloor$ and $\lceil\ \rceil$, respectively, we define

$$K_{pl} = \left\lfloor K_p^{1/n} \right\rfloor, \quad \text{and} \quad K_{pu} = \left\lceil K_p^{1/n} \right\rceil. \tag{6.63}$$

A claim can be made that the discrete minimizer $\mathbf{x}^d$ of Example 6.5,

$$\mathbf{x}^d = \left(x_1^d, x_2^d, x_3^d, \ldots, x_n^d\right)^T, \tag{6.64}$$

satisfies

$$x_i^d \in \{K_{pl}, K_{pu}\}, \quad \text{for } i = 1 \text{ to } n. \tag{6.65}$$

That is, each component of $\mathbf{x}^c$ will be rounded either up or down to the nearest integer only. Rounding more than one increment is never required and often leads to infeasibility or nonoptimality. This claim is now justified.

The proof of the claim is broken into three steps. First, cases where x_i^c happen to be integer are addressed. The second and third steps demonstrate a procedure for converting solutions with

$$x_i^d < K_{pl} \quad \text{or} \quad x_i^d > K_{pu}, \qquad \text{for at least one } i, \tag{6.66}$$

into equal or better solutions satisfying Equation (6.65). The overall approach is to decrease components of the discrete minimizer having values in excess of K_{pu} while simultaneously increasing components less than K_{pl}. Three outcomes are possible. The solution may be shown to be infeasible or nonoptimal, or a new solution satisfying Equation (6.65) may be derived from the given solution.

Step 1 If $K_p^{1/n}$ happens to be an integer, then

$$x_i^c = x_i^d = K_{pl} = K_{pu} \tag{6.67}$$

and the claim is justified. Also,

$$K_s = f\left(x_1^d, x_2^d, x_3^d, \ldots, x_n^d\right) = nK_p^{1/n} \tag{6.68}$$

is the maximum bound for the linear constraint.

Step 2 Assume that x_i^c is noninteger and that

$$x_i^d < K_{pl}, \qquad \text{for some } i, \tag{6.69}$$

which leads immediately to

$$x_i^d \leq K_{pl} - 1, \tag{6.70}$$

exploiting the integrality of the variables. Now, note that

$$x_j^d > K_{pl}, \qquad \text{for some } j, \tag{6.71}$$

in order to satisfy inequality constraint g_1, leading to

$$x_j^d \geq K_{pu}, \tag{6.72}$$

again based on integrality arguments. Subtracting Equation (6.70) from (6.72) yields

$$x_j^d - x_i^d \geq K_{pu} - (K_{pl} - 1) = 2, \tag{6.73}$$

a result used below. Defining

$$y_i^d = x_i^d + 1, \tag{6.74}$$

$$y_j^d = x_j^d - 1, \tag{6.75}$$

we consider the effects of replacing x_i^d and x_j^d with y_i^d and y_j^d, respectively, in the discrete minimizer. Summing Equations (6.74) and (6.75) leads to

$$y_i^d + y_j^d = x_i^d + x_j^d, \tag{6.76}$$

ensuring that the proposed replacements do not affect the optimality of the minimizer. Multiplying the same equations yields

$$\begin{aligned} y_i^d y_j^d &= (x_i^d + 1)(x_j^d - 1) \\ &= x_i^d x_j^d + x_j^d - x_i^d - 1 \geq x_i^d x_j^d + 1, \end{aligned} \tag{6.77}$$

where the lower bound from Equation (6.73) has been substituted for $x_j^d - x_i^d$. Equation (6.77) ensures that the proposed replacements do not affect the feasibility of the minimizer. Thus, assuming a component of the discrete minimizer exists below K_{pl} requires another component to equal or exceed K_{pu}. Incrementing and decrementing (by one) the lower and upper components, respectively, leads to a new solution, which is both feasible and optimal, and whose components are two steps closer to the set of Equation (6.65). Repeating the analysis of Equations (6.70) through (6.77), until the assumption of Equation (6.69) is violated, leads to

$$x_i^d \geq K_{pl}, \qquad \text{for all } i. \tag{6.78}$$

Step 3 Following the analysis of Step 2, assume

$$x_i^d > K_{pu}, \qquad \text{for some } i, \tag{6.79}$$

leading immediately to

$$x_i^d \geq K_{pu} + 1. \tag{6.80}$$

Noting that

$$x_j^d = K_{pu}, \qquad \text{for all } j \neq i, \tag{6.81}$$

leads to a nonoptimal solution forces

$$x_j^d < K_{pu}, \qquad \text{for some } j \neq i. \tag{6.82}$$

Equation (6.82) implies

$$x_j^d = K_{pl}. \tag{6.83}$$

Subtracting Equation (6.83) from (6.80) yields

$$x_i^d - x_j^d \geq K_{pu} + 1 - K_{pl} = 2. \tag{6.84}$$

Defining

$$y_i^d = x_i^d - 1 \tag{6.85}$$

and

$$y_j^d = x_j^d + 1 \tag{6.86}$$

leads to both Equation (6.76), ensuring optimality, and to

$$\begin{aligned} y_i^d y_j^d &= (x_i^d - 1)(x_j^d + 1) \\ &= x_i^d x_j^d + x_i^d - x_j^d - 1 \geq x_i^d x_j^d + 1, \end{aligned} \tag{6.87}$$

where Equation (6.84) has been substituted. Again feasibility has been retained. Repeating the analysis of Equations (6.80) through (6.87), until the assumption of Equation (6.79) is violated, leads to

$$x_i^d \leq K_{pu}, \qquad \text{for all } i. \tag{6.88}$$

Taken together, along with the integrality of the variables, Equations (6.78) and (6.88) imply Equation (6.65), validating the claim. Determination of the discrete minimizer from within the set in Equation (6.65) is now considered.

DETERMINATION OF K

Exploiting the symmetry of objective and constraint functions of Example 6.5, a set of n-tuples

$$\begin{aligned} \mathcal{T} = \{&(K_{pl}, K_{pl}, K_{pl}, \ldots, K_{pl}), (K_{pu}, K_{pl}, K_{pl}, \ldots, K_{pl}), \\ &(K_{pu}, K_{pu}, K_{pl}, \ldots, K_{pl}), \ldots, (K_{pu}, K_{pu}, K_{pu}, \ldots, K_{pu})\} \end{aligned} \tag{6.89}$$

is formed. The set contains only $n + 1$ elements. Contrast this with the $2n$ elements requiring investigation were symmetry not present. Let T_i represent the n-tuple from Equation (6.89) containing i occurrences of K_{pu} and $n - i$ occurrences of K_{pl}, and let P_i and S_i represent the product and sum generated by T_i, respectively. Equation (6.89) is now rewritten as

$$\mathcal{T} = \{T_0, T_1, T_2, T_3, \ldots, T_n\}. \tag{6.90}$$

With the n-tuples ordered as in Equation (6.89), each n-tuple generates successively larger values of both P_i and S_i. We now desire the smallest value of i such that

$$K_p \leq P_i. \tag{6.91}$$

The increasing nature of both the P_is and the S_is now ensures that

$$S_i \leq x_1 + x_2 + x_3 + \cdots + x_n \tag{6.92}$$

for all T_i satisfying Equation (6.91). Further, increasing i to $i + 1$ creates a condition where a term, namely T_i, both satisfies constraint g_1 and violates

$$S_{i+1} \leq x_1 + x_2 + x_3 + \cdots + x_n. \tag{6.93}$$

Thus, S_i is the largest integer value satisfying Equation (6.92).

These ideas are demonstrated in the following example.

Example 6.6 Given $2{,}070 \leq x_1x_2x_3x_4$, find the maximum value K such that $K \leq x_1 + x_2 + x_3 + x_4$.

With

$$K_{pl} = \lfloor 2{,}070^{1/4} \rfloor = 6 \text{ and } K_{pu} = \lceil 2{,}070^{1/4} \rceil = 7, \tag{6.94}$$

the set

$$(x_1, x_2, x_3, x_4) \in \{(6,6,6,6), (7,6,6,6), (7,7,6,6), (7,7,7,6), (7,7,7,7)\} \tag{6.95}$$

is investigated. Finding

$$P_3 = 2{,}058 \leq 2{,}070 \leq 2{,}401 = P_4 \tag{6.96}$$

shows that $i = 4$ is required to satisfy the given hyperbolic constraint. Thus, $K = S_4 = 28$, leading to

$$28 \leq x_1 + x_2 + x_3 + x_4. \tag{6.97}$$

■

OTHER CONSTRAINT DERIVATIONS

Derivation of hyperbolic constraints of the form

$$x_1x_2x_3 \cdots x_n \leq K_p \tag{6.98}$$

from linear constraints of the form

$$x_1 + x_2 + x_3 + \cdots + x_n \leq K_s \tag{6.99}$$

is similar to the previous procedure and is left as an exercise.

Two other conversions are obtained by reversing the direction of the inequalities in the above analyses. The results are listed here for completeness. The constraint

$$x_1x_2x_3 \cdots x_n \leq K_p \tag{6.100}$$

implies

$$x_1 + x_2 + x_3 + \cdots + x_n \leq K_p + (n-1), \tag{6.101}$$

while

$$x_1 + x_2 + x_3 + \cdots + x_n \geq K_s \tag{6.102}$$

implies

$$x_1x_2x_3 \cdots x_n \geq K_s - (n-1). \tag{6.103}$$

All of the above results are extendable to handle weighted products and sums.

Further Upper and Lower Bound Generation

The following analysis picks up the upper and lower bound generation demonstrated above, adding linear and hyperbolic constraint derivations. The results are presented in an abbreviated form, with use of the new techniques more fully explained. Repeating the upper bound generation steps using the result of Equation (6.49) yields

$$x_5 x_6 \leq 33. \tag{6.104}$$

Assuming $x_1 + x_2 + x_3 = 4$, the hyperbolic constraint $x_1 x_2 x_3 \leq 2$ can be derived. The lower bound generation steps (6.45) through (6.47) are now repeated five times, leading to the same result as Equation (6.48). A contradiction is again formed as x_4 is forced to 2,070. The sum constraint is updated to

$$x_1 + x_2 + x_3 \geq 5. \tag{6.105}$$

The upper bound generation steps are now repeated in an abbreviated form:

$$641{,}700 x_5 + 207.06 x_6 (x_2 x_3 x_6)^2 - 120{,}000 x_2 x_3 x_6 \leq 0, \tag{6.106}$$

which leads to

$$4(207.06 x_6)(641{,}700 x_5) \leq 120{,}000^2, \tag{6.107}$$

which in turn yields

$$x_5 x_6 \leq 27. \tag{6.108}$$

For a new lower bound we assume $x_1 + x_2 + x_3 = 5$, which leads to $x_1 x_2 x_3 \leq 4$. We will generate the new bound by contradiction. The method used previously will not work in this case as it largely ignores the interaction between x_5 and x_6. Rather, all feasible (x_5, x_6) pairs must be considered, from which the maximum product may be determined. For instance,

$$x_1 x_2 x_3 \leq 4 \text{ implies } x_4 \geq 19, \tag{6.109}$$

which leads to

$$5{,}890\, x_5^2 + 1{,}251.1 x_6^2 \leq 120{,}000. \tag{6.110}$$

The result of Equation (6.110) admits the following maximal feasible pairs:

$$(x_5, x_6) \in \{(1, 9), (2, 8), (3, 7), (4, 4)\}. \tag{6.111}$$

The maximum product, generated by the third pair, is found to be 21. This product is used to reapply the process. After eight applications of the above steps, with intermediate products of 15, 10, 8, 6, 4, 3, and 2, the product is limited to a value of one. This result forces $x_4 \geq 518$, generating the contradiction. The updated sum constraint is

$$x_1 + x_2 + x_3 \geq 6. \tag{6.112}$$

Repetition of the upper bound generation steps now yields

$$x_5x_6 \leq 22. \tag{6.113}$$

The lower bound generation steps may also be applied to the sum $(x_1 + 1.57x_2 + x_4)$. With Equation (6.113) providing a bound on x_5x_6, it is possible to show that

$$x_1 + 1.57x_2 + x_4 \geq 6.57. \tag{6.114}$$

Making use of Equation (6.114), the upper bound generation steps yield

$$x_5x_6 \leq 12. \tag{6.115}$$

Case Analysis

There are 35 possible assignments of x_5 and x_6 satisfying Equation (6.115). These 35 cases are analyzed individually after assigning values to x_5 and x_6.

Consider the case where $(x_5, x_6) = (1, 12)$. Employing Equations (6.112) and (6.114), we rewrite constraint g_2 as

$$372x_1x_4 + 54{,}872.64x_2x_3 \leq 120{,}000, \tag{6.116}$$

while constraint g_1 reduces to

$$x_1x_2x_3x_4 \geq 173. \tag{6.117}$$

Eliminating x_1 and x_4 from Equation (6.116) using Equation (6.117) yields

$$64{,}356/x_2x_3 + 54{,}872.64x_2x_3 \leq 120{,}000, \tag{6.118}$$

which is feasible only when $x_2x_3 \leq 1$. With $x_2x_3 \leq 1$,

$$x_1x_4 \geq 173, \tag{6.119}$$

which implies, using the linear constraint derivation technique,

$$x_1 + x_4 \geq 27. \tag{6.120}$$

Equation (6.120) leads to

$$x_1 + 1.57\,x_2 + x_4 \geq 28.57, \tag{6.121}$$

allowing Equation (6.118) to be rewritten as

$$64{,}356/x_2x_3 + 238{,}616.64x_2x_3 \leq 120{,}000, \tag{6.122}$$

which has no feasible solution. Thus, the case $(x_5, x_6) = (1, 12)$ is eliminated from further consideration.

Twenty-eight other cases are eliminated using identical analysis. However, the analysis must often be repeated a number of times to prove infeasibility. To demonstrate this repetition, the case $(x_5, x_6) = (1, 4)$ is analyzed in an abbreviated form: $(x_5, x_6) = (1, 4)$ implies $x_1x_2x_3x_4 \geq 518$, leading to

$$192{,}696/x_2x_3 + 6{,}096.96x_2x_3 \leq 120{,}000 \text{ or } x_2x_3 \leq 17, \tag{6.123}$$

leading to $x_1x_4 \geq 31$, which implies $x_1 + x_4 \geq 12$, leading to

$$192{,}696/x_2x_3 + 12{,}592.96x_2x_3 \leq 120{,}000 \text{ or } x_2x_3 \leq 7, \quad (6.124)$$

leading to $x_1x_4 \geq 74$, which implies $x_1 + x_4 \geq 18$, leading to

$$192{,}696/x_2x_3 + 18{,}160.96x_2x_3 \leq 120{,}000 \text{ or } x_2x_3 \leq 3, \quad (6.125)$$

leading to $x_1x_4 \geq 173$, which implies $x_1 + x_4 \geq 27$, leading to

$$192{,}696/x_2x_3 + 26{,}512.96x_2x_3 \leq 120{,}000, \quad (6.126)$$

which proves to be infeasible.

Cases eliminated using the above analysis technique include

$$(x_5, x_6) \in \{(1,4), \ldots, (1,12), (4,1), \ldots, (12,1), (2,3), \ldots, (2,6), (3,2), \ldots, (6,2), (3,3), (3,4), (4,3)\}. \quad (6.127)$$

The remaining six cases are

$$(x_5, x_6) \in \{(1,1), (1,2), (1,3), (2,1), (2,2), (3,1)\}. \quad (6.128)$$

Constraint Substitution: Remaining Cases

With $(x_5, x_6) = (1, 1)$ known to lead to feasible solutions, the five remaining cases are now considered. The case $(x_5, x_6) = (2, 2)$ is described here, with the other cases following the same analysis procedure. Substituting $(x_5, x_6) = (2, 2)$ into constraint g_2 yields

$$248x_1x_4(x_1 + x_2 + x_3) + 232x_2x_3(x_1 + 1.57x_2 + x_4) \leq 120{,}000, \quad (6.129)$$

while constraint g_1 leads to

$$x_1x_2x_3x_4 \geq 518. \quad (6.130)$$

Eliminating x_1 and x_4 from Equation (6.129) using Equation (6.130) yields

$$128{,}464(x_1+x_2+x_3)/x_2x_3 + 232x_2x_3(x_1+1.57x_2+x_4) \leq 120{,}000. \quad (6.131)$$

Using discriminant constraint derivation, Equation (6.131) is transformed, after some rearrangement, into

$$(x_1 + x_2 + x_3)(x_1 + 1.57x_2 + x_4) \leq 120.8. \quad (6.132)$$

Equations having left-hand sides identical to Equation (6.132) may be derived for the other four (x_5, x_6) pairs. For $(x_5, x_6) = (1, 3)$ or $(3, 1)$, the right-hand side of Equation (6.132) equals 161.21, while for $(x_5, x_6) = (1, 2)$ or $(2, 1)$, the value is 241.81. None of the five equations resulting from the above analysis admit any feasible solutions, as we will see in the next section.

6.8 Relaxation and Exhaustive Enumeration

The final two techniques employed to solve the transformer design problem are already familiar. In the first technique we *relax* the discreteness requirement on the variables and solve the problem with continuous variables. The continuous solution provides a global lower bound on the optimal value of the nonrelaxed (discrete) objective. The second technique is a "brute force" solution: Compute all feasible points and compare them. This *exhaustive enumeration* makes sense when the number of points is relatively small, and so it is the last step after using all possible feasible domain reductions.

Continuous Relaxation: Global Lower Bounds

With the left-hand side of Equation (6.132) in mind, the following continuous problem is formed:

$$\begin{aligned} &\text{minimize } (x_1 + x_2 + x_3)(x_1 + 1.57x_2 + x_4) \\ &\text{subject to} \\ &\quad x_1x_2x_3x_4 \geq C^2, \end{aligned} \tag{6.133}$$

where C^2 is a constant. If $C^2 = 518$ leads to a global solution of problem (6.133) greater than the right-hand side of Equation (6.132), there is no feasible solution for the case $(x_5, x_6) = (2, 2)$. Similar results are obtained for the other four cases.

Problem (6.133) is solved as follows. First, because a lower bound is being generated, the 1.57 multiplier on x_2 is dropped. The resulting problem is symmetric with respect to x_1 and x_2, and also with respect to x_3 and x_4. Replacing the later symmetric variables with the former yields

$$\begin{aligned} &\text{minimize } (2x_1 + x_3)^2 \\ &\text{subject to} \\ &\quad x_1^2x_3^2 \geq C^2. \end{aligned} \tag{6.134}$$

The active constraint is now used to eliminate x_3 from the problem. The resulting unconstrained problem is

$$\text{minimize } (2x_1 + C/x_1)^2. \tag{6.135}$$

The minimum objective value and corresponding minimizer are

$$(2(2C)^{1/2})^2 = 8C, (x_1, x_3) = ((C/2)^{1/2}, (2C)^{1/2}). \tag{6.136}$$

For $(x_5, x_6) = (2, 2)$,

$$8C = 182.08 \geq 120.79 \tag{6.137}$$

so that Equation (6.132) can never be satisfied. The other four cases exhibit similar results.

Setting x_5 and x_6 equal to one and with the set constraints retained from Equation (6.28), the reduced model for the transformer design is as follows.

TRANSFORMER DESIGN REDUCED MODEL

$$\text{minimize } 0.0811 x_1 x_4(x_1 + x_2 + x_3) + 0.0624\, x_2 x_3(x_1 + 1.57x_2 + x_4)$$

subject to

$$\begin{aligned}
&g_1\colon -0.001\, x_1x_2x_3x_4 + 2.07 \le 0,\\
&g_2\colon 0.00062\, x_1x_4(x_1 + x_2 + x_3)\\
&\qquad + 0.00058\, x_2x_3(x_1 + 1.57x_2 + x_4) - 1.2 \le 0,\\
&x_1 \in \{1, 2, 3, \ldots, 42\}, \quad x_2 \in \{1, 2, 3, \ldots, 35\},\\
&x_3 \in \{1, 2, 3, \ldots, 445\}, \quad x_4 \in \{1, 2, 3, \ldots, 491\}.
\end{aligned} \tag{6.138}$$

Substituting constraint g_1 into constraint g_2 and rearranging yields

$$\begin{aligned}
&128{,}340(x_1 + x_2 + x_3) + 58(x_1 + 1.57x_2 + x_4)(x_2x_3)^2\\
&\quad - 120{,}000\, x_2x_3 \le 0.
\end{aligned} \tag{6.139}$$

Using the discriminant constraint procedure generates

$$(x_1 + x_2 + x_3)(x_1 + 1.57\, x_2 + x_4) \le 483.63. \tag{6.140}$$

Noting the similarity of (6.140) to (6.132), we employ relaxation methods once again. Eliminating x_3 and x_1 respectively yields the following two unconstrained optimization problems:

$$\min\,(2x_1 + (2070)^{1/2}/x_1)^2 \quad \text{and} \quad \min\,(x_3 + 2(2070)^{1/2}/x_3)^2. \tag{6.141}$$

The objective function values of the minimization problems (6.141) represent lower bounds on the left-hand side of (6.130) subject to constraint g_1.

These functions are used to tighten the set constraints on x_1 and x_3 by evaluating them for each possible variable value. Variable values generating results larger than 483.63 represent infeasible assignments and are eliminated from their set constraint. Symmetry provides similar set constraint tightening for x_2 and x_4. The results of this tightening ensure that

$$\begin{aligned}
&x_1 \in \{3, 4, 5, 6, 7, 8\}, \quad x_2 \in \{3, 4, 5, 6, 7, 8\},\\
&x_3 \in \{6, 7, \ldots, 16\}, \quad x_4 \in \{6, 7, \ldots, 16\}.
\end{aligned} \tag{6.142}$$

Problem Completion: Exhaustive Enumeration

At this point, the solution space has been reduced to $4{,}356\,(= 6^2 \times 11^2)$ discrete points. Comparing this number to the 1.9×10^{11} $(= 42 \times 35 \times 445 \times 491 \times 25 \times 24)$ points, existing after preliminary set constraint tightening, shows that the solution space has been reduced by a factor of over 20 million. While further application of reduction techniques and incumbent value updating could reduce this space even further, a decision is made to investigate the remaining points using exhaustive enumeration.

The global optimum to the original problem is found to be

$$(x_1, x_2, x_3, x_4, x_5, x_6)^T = (4, 4, 10, 13, 1, 1)^T, \tag{6.143}$$

having an objective function value of 134.0. This integer-valued optimum is near the continuous global optimum

$$(x_1, x_2, x_3, x_4, x_5, x_6)^T = (5.33, 4.66, 10.44, 12.08, 0.7525, 0.8781)^T, \tag{6.144}$$

as reported in the cited references. However, it is unlikely that the discrete optimum could have been discovered by rounding off the continuous optimum.

Substituting the discrete solution into constraint g_2 yields

$$g_2(x_1, x_2, x_3, x_4, x_5, x_6) = -0.080, \tag{6.145}$$

so that increasing the constant term was indeed necessary to ensure feasibility.

6.9 Summary

This chapter resumed the topic of monotonicity analysis by discussing its application to model reduction and case decomposition. Special emphasis fell on its systematic employment to construct parametric routines that generate optimal designs, with a minimum of iterative searching, directly from given values of the parameters defining the model. To illustrate these ideas, the two design problems of Chapter 3 – the air tank and the hydraulic cylinder – were expanded and carried all the way to completion of their optimized parametric design procedures. A major aid to this process was the Monotonicity Table, which guides not only the case decomposition but also the systematic generation of back substitution solution functions needed to recover optimal values of variables eliminated during the model reduction.

Originally developed for fully algebraic models, monotonicity analysis can now be applied when some or all of the model involves numerical simulation, rather than neat algebraic expressions. Distinction was made between *explicit algebraic elimination* employed to solve algebraic equations and *implicit numerical solution* needed to find numerical solutions of simulations, approximations, or large equations too complicated for closed-form solution.

Optimized parametric design procedures can be used to decompose large complicated optimization problems into components of manageable size. As in classical Dynamic Programming, this can be done when some component input parameters are outputs from other components, and the overall system objective is simply the sum of component objectives. To build large optimized systems from suboptimized components, simply reverse this approach by linking locally optimized parametric procedures.

When discrete variables are present things get complicated very quickly since the solution must really be a global one and the number of possible solutions increases exponentially with the number of design variables and their discrete choices. The thinking that guided the model analysis of continuous problems can come very handy in the discrete case as well. The main idea is to reduce the number of points that must be examined for optimality by proving points are either infeasible or lead to no better answer than the one already available. This is the essence of branch-and-bound techniques.

In the transformer design problem, introduced in this chapter as a discrete problem, systematic reduction involved repeated manipulation of the constraints to coax ever tighter bounds and a smaller solution set. The approach could be seen as heuristic but one should really perceive it as *opportunistic*: Every step taken is rigorous if the opportunity to use it arises.

Notes

In the notes of Chapter 3 we mentioned codes that have automated much of the bookkeeping and manipulation of Monotonicity Analysis (Agogino and Almgren 1987, Hansen, Jaumard, and Lu 1989c, Rao and Papalambros 1991, Hsu 1993). They do that primarily by manipulating the monotonicity tables. This chapter is a useful explanatory companion to such software. Conversely, using the software is a good companion to learning the theory, for it guides students through the unavoidable complexity and new terminology.

The original purpose of material in the first edition of this textbook that has been eliminated here was to start at a good case and bypass nonoptimal cases quickly with little computation. But this kind of cleverness is not needed when powerful computers and branch-and-bound are on the job. Scholars of optimization theory may still find useful the first edition concepts, techniques, and proofs concerning degrees of freedom, overconstrained models, monotonicity transformation, symmetry, and constraint activity mapping – useful, that is, for special applications and further research.

The new material added on discrete variables comes in its entirety from the doctoral thesis of Len Pomrehn at Michigan (Pomrehn 1993). Pomrehn actually automated a general process for performing all the reduction steps and manipulations using symbolic mathematics. Thus the chore of algebraic manipulations and bookkeeping was drastically reduced. Solving discrete design optimization problems is a difficult subject and nondeterministic methods, such as genetic algorithms, are important tools to address such problems. Examples of interesting applications are in topology and configuration design [see, e.g., Chapman, Saitou, and Jakiela (1994) and Hajela (1990)].

Exercises

6.1 Verify the air tank design of Chapter 3 using the parametric procedure of Figure 6.3. List each parametric test in the order in which it is performed.

6.2 Use the parametric procedure of Section 6.2 to design a hydraulic cylinder

(a) for $F = 98\,\text{N}$, $P = 2.45(10^4)\,\text{Pa}$, $S = 6(10^5)\,\text{Pa}$, $T = 0.3\text{cm}$;

(b) for a cheaper material having half the strength of the material in (a).

For each problem, list the parametric tests in the order they are performed.

6.3 Design an air tank with the same specifications as in Section 6.4 except that the integer intervals on all variables are (a) doubled; (b) halved.

6.4 Find all solutions to the problem

$$\begin{aligned}\min f &= 0.01(x_1 - 1)^2 + (x_2 - x_1^2)^2\\ \text{subject to } h_1 &= x_1 + x_3^2 + 1 = 0,\\ g_1 &= 2x_2 - x_1^2 + 1 \le 0,\\ g_2 &= -x_2 \le 0.\end{aligned}$$

6.5 Find and prove the global solution to

$$\begin{aligned}\min f &= (x_1 - x_2)^2 + (x_1 + x_2 - 10)^2 + (x_3 - 5)^2\\ \text{subject to } & x_1^2 + x_2^2 + 2x_3 - 48 \le 0,\\ & 0 \le x_1 \le 4.5, 0 \le x_2 \le 4.5, 0 \le x_3 \le 0.5.\end{aligned}$$

6.6 Solve the problem of Exercise 5.12 using monotonicity analysis.

6.7 Consider the problem

$$\begin{aligned}\min f &= 0.5(x_1^2 + x_2^2 + x_3^2) - (x_1 + x_2 + x_3)\\ \text{subject to } g_1 &= x_1 + x_2 + x_3 - 1 \le 0,\\ g_2 &= 4x_1 + 2x_2 - \tfrac{7}{3} \le 0,\end{aligned}$$

where $x_i \ge 0, i = 1, 2, 3$.

(a) For an identified active constraint, examine monotonicities in the reduced space using a constrained derivative. Analyze various cases and prove global solution. Do not forget to check feasibility.

(b) Write the KKT conditions and identify the number of possible cases implied by them without monotonicity arguments.

6.8 Consider the explosively activated cylinder model of Exercise 3.10. Clarify the ambiguity of rule (b) by dominance.
To do this:

(a) Rewrite constraints g_2 and g_3 in the form $x_4 \le b_2(x_5^-)$, $x_4 \le b_3(x_3^+, x_5^-)$, respectively, where b_2 and b_3 are the upper bounds for x_4. As required by the Monotonicity Principles, one of them will be critical.

(b) Show that

$$b_2(x_5^-) - b_3(x_3^+, x_5^-) \ge \frac{h(x_5)}{((2ND_{\max}^2)/S_{yt})x_5^2},$$

where

$$\begin{aligned}h(x_5) &\triangleq x_5^4 - D_{\max}^2 x_5^2 + K,\\ K &\triangleq (8ND_{\max}^2 F_{\max}/1000\pi S_{yt}).\end{aligned}$$

(c) Prove by contradiction that $h(x_5) < 0$ implies $b_2 - b_3 \le 0$.

(d) State the general (parametric) dominance conditions.

(e) Examine the possible cases for the particular set of parameters given in Exercise 3.10. Identify the optimal case.

(f) Based on the preceding, find the solution when

$$\text{(i) } x_1 \geq 0.5; \text{(ii)} W_{\min} = 400 \text{ lb-in;}$$

$$\text{(iii)} F_{\min} = 5{,}000 \text{ lbs.}$$

6.9 Use regional monotonicity and case decomposition to find the global solution to the problem

$$\min f = (x_1 - 2)^2 + (x_2 - 1)^2$$

$$\text{subject to } -x_1^2 + x_2 \geq 0,$$

$$-x_1 - x_2 + 2 \geq 0.$$

6.10 Find the global solution to the problem

$$\min f = 2x_1^2 - 6x_1x_2 + 9x_2^2 - 18x_1 + 9x_2$$

$$\text{subject to } g_1 = x_1 + 2x_2 - 12 \leq 0,$$

$$g_2 = 4x_1 - 3x_2 - 20 \leq 0,$$

$$x_1 \geq 0, x_2 \geq 0.$$

Hint: Prove that this is a quadratic problem with a positive-definite Hessian. Enumerate the cases using the KKT conditions and eliminate most cases as nonoptimal. Analyze only the few remaining cases.

6.11 A rectangular bar for supporting a vertical force on one end is to be welded at the other end to a metal column of known distance from the force (see figure, following page). The four variables whose values are to be selected are (in cm): weld length l, weld rod thickness h, bar width b, and bar depth t. The objective to be minimized is the sum of weld, stub, and bar costs

$$f = C_1h^2l + C_2lbt + C_3bt,$$

subject to

weld shear stress:	$K_1 - hl \leq 0,$	(1)
bar bending stress:	$K_2 - bt^2 \leq 0,$	(2)
weld cannot be wider than bar:	$h - b \leq 0,$	(3)
end deflection:	$K_3 - bt^3 \leq 0,$	(4)
bar buckling:	$(K_4t + K_5)/b^3t - 1 \leq 0,$	(5)
minimum size welding rod:	$K_6 - h \leq 0.$	(6)

(a) Use monotonicity tables to construct an optimized parametric procedure for sizing the bar weldment. Ragsdell and Phillips (1976) first formulated this algebraic model; it was analyzed extensively by other methods in the first edition of this book. (b) Test it for the following parameter values used by Ragsdell and Phillips: $C_1 = 1.105$, $C_2 = 0.6735$, $C_3 = 0.0481$, $K_1 = 1.52$,

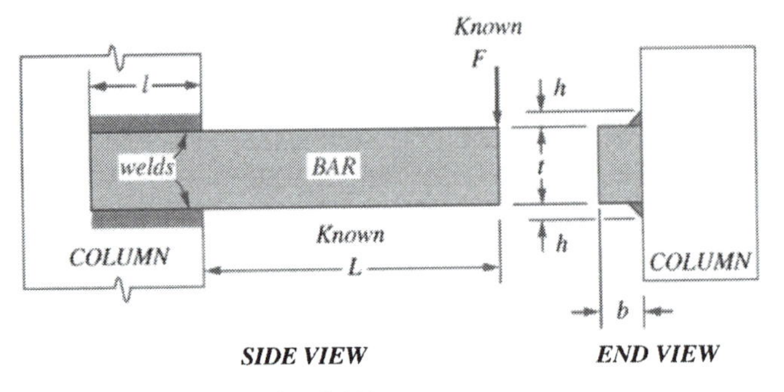

Bar weldment (Exercise 6.11).

$K_2 = 16.8$, $K_3 = 9.08$, $K_4 = 0.0278$, $K_5 = 0.0943$, $K_6 = 0.125$. They give the minimizing design as $l = 6.72$, $h = 0.226$, $b = 0.245$, $t = 8.27$.

6.12 Redo the preceding Exercise 6.11 after making the change of variable $a = bt$, where a is the vertical cross-sectional area of the bar. Compare the resulting procedures; which would be easier to implement? Which has fewer cases?

6.13 Construct a finite element model for evaluating the stresses in Exercise 6.11. Check the design given there for excessive stresses.

6.14 Redo Exercise 6.12 using the finite element model developed in Exercise 6.13. Delete the original constraints superseded by the FEA model.

6.15 Round the corners where appropriate to reduce the high stress concentrations in the design of Exercise 6.14. Generate the new optimal design and compare it with those produced in previous problems.

6.16 In the literature find an optimal design problem for which the full model is given and a numerical solution presented. Construct an optimized parametric procedure for resizing the design given when parameter values change. Be on the lookout – monotonicity analysis sometimes finds errors in published solutions obtained numerically.

6.17 In the text the optimized parametric procedure for designing air tanks was constructed informally to introduce the subject. Generate the corresponding monotonicity tables.

6.18 Program and test the parametric procedure given in Section 6.1 for designing optimized air tanks.

6.19 Program and test the parametric procedure given in Section 6.3 for designing optimized hydraulic cylinders using thin-wall theory.

6.20 Continue the derivation started in Equations (6.44) through (6.47), proving that $x_5 \leq 1$, $x_6 \leq 1$, and $x_5 x_6 \leq 1$.

6.21 Repeat the analysis detailed in the derivation of linear constraints subsection for hyperbolic constraints.

6.22 Given $x_1 + x_2 + x_3 + x_4 + x_5 \leq 17$, find the minimum value, K, such that $x_1x_2x_3x_4x_5 \leq K$.

6.23 Assuming $x_1+x_2+x_3 = 4$, start with the hyperbolic constraint $x_1x_2x_3 \leq 2$ and use the techniques of Equations (6.45) through (6.47) to force a contradiction.

6.24 Prove infeasibility for some of the cases in Equation (6.127).

6.25 Prove infeasibility for some of the cases in Equation (6.128).

7

Local Computation

> We knew that the islands were beautiful, around here somewhere, feeling our way a little lower or a little higher, a least distance.
> *George Seferis (1900–1971)*

Model analysis by itself can lead to the optimum only in limited and rather opportune circumstances. Numerical iterative methods must be employed for problems of larger size and increased complexity. At the same time, the numerical methods available for solving NLP problems can fail for a variety of reasons. Some of the reasons are not well understood or are not easy to remedy without changes in the model. It is safe to say that no single method exists for solving the general NLP problem with complete reliability. This is why it is important to see the design optimization process as an interplay between analysis and computation. Identifying model characteristics such as monotonicity, redundancy, constraint criticality, and decomposition can assist the computational effort substantially and increase the likelihood of finding and verifying the correct optimum. The literature has many examples of wrong solutions found by overconfident numerical treatment.

Our goal in this chapter is to give an appreciation of what is involved in numerical optimization and to describe a small number of methods that are generally accepted as preferable within our present context. So many methods and variations have been proposed that describing them all would closely resemble an encyclopedia. Workers in the field tend to have their own preferences. Probably the overriding practical criterion for selecting among the better methods is extensive experience with a particular method. This usually gives enough insight for the fine tuning and coaxing that is often necessary for solving practical design problems. Simply switching methods (if good ones are used) when failure occurs is not generally effective unless the cause of failure is clearly understood. There are many excellent texts that treat numerical optimization in detail, and we will frequently refer to them to encourage the reader to seek further information.

An in-depth study of numerical optimization requires use of numerical analysis, in particular, numerical linear algebra. Theoretical treatment of algorithms allows proof of convergence results. Such proofs generally do not guarantee good performance in practice, primarily because the mathematical assumptions about the nature of the functions, such as convexity, may not hold. However, because convergence analysis helps in understanding the underlying mathematical structure and aids in improving algorithmic reliability, it should not be ignored. Convergence proofs can be found in cited references and will not be included in the descriptions offered in this chapter. Development of good optimization algorithms requires performance of numerical experiments and inclusion of heuristics based on those experiments. Researchers develop special test problems that could expose the weaknesses and strengths of an algorithm. An ideal algorithm is "a good selection of experimental testing backed up by convergence and rate of convergence proofs" (Fletcher 1980). A general theory of algorithms as point-to-set mappings was developed by Zangwill (1969) and led to studying families of algorithms in a unified way. However, classification of algorithms, particularly in the constrained case, tends to be rather confusing. It is often possible to show that a particular implementation of an algorithm can be interpreted as a special case of another class of algorithms. Although many insights may be gained this way, it is not of particular importance to our interests here. So any apparent classifications we use should be viewed as a convenience in communication rather than as an attempt for a particular interpretation of algorithmic families.

In the algorithms described in this chapter no assumptions are made about positivity of the variables. This is consistent with the differential optimization theory of Chapters 4 and 5, on which the algorithms are based.

7.1 Numerical Algorithms

A numerical problem is a precise description of a functional relation between input and output data. These data are each represented by a vector of finite dimension. The functional relation may be expressed explicitly or implicitly, that is, it could be an algebraic expression or a procedure (subroutine). An *algorithm* for a problem is a precise description of well-defined operations through which an acceptable input data vector is transformed into an output data vector. Since here we are concerned only with numerical problems, the operations involved in the construction of an algorithm should be limited to arithmetic and logical ones, so that they can be performed by a computer. An *iteration* is a repetitive execution of a set of actions, for example, of numerical evaluations. An iteration may itself contain other iterations.

Local and Global Convergence

An iterative algorithm generates a sequence of points by repeated use of an iteration, so it can be thought of as a generalized function, or mapping. Theoretically, the mapping can be a point-to-set one, so that an algorithmic class may be defined. In practice of course, only a point-to-point mapping will be useful. Thus, to apply an algorithm A we need a *starting* or *initial point* $\mathbf{x}_0$ in a set $\mathcal{X}$, and an iteration formula

$$\mathbf{x}_{k+1} = A(\mathbf{x}_k) \tag{7.1}$$

that generates the sequence $\{\mathbf{x}_k\}$. For example, the typical iteration formula in optimization algorithms is

$$\mathbf{x}_{k+1} = \mathbf{x}_k + \alpha_k \mathbf{s}_k, \tag{7.2}$$

which we discovered in Chapter 4. Theoretically, we would like the sequence $\{\mathbf{x}_k\}$ to converge to at least a (local) minimizer $\mathbf{x}_*$. If $\mathbf{x}_*$ cannot be proven to be the limit of $\{\mathbf{x}_k\}$, it may be possible to prove that it is at least a cluster point of $\{\mathbf{x}_k\}$, which would mean that there is a subsequence of $\{\mathbf{x}_k\}$ converging to $\mathbf{x}_*$. In any case, the algorithm should generate the sequence of iterant $\mathbf{x}_k$s so that away from $\mathbf{x}_*$ a steady progress toward $\mathbf{x}_*$ is achieved and once near $\mathbf{x}_*$ a rapid convergence to $\mathbf{x}_*$ itself occurs. When we discuss the convergence of an algorithm, we can then identify two distinct characteristics of convergence behavior.

> *Global convergence refers to the ability of the algorithm to reach the neighborhood of $\mathbf{x}_*$ from an arbitrary initial point $\mathbf{x}_0$, which is not close to $\mathbf{x}_*$. The convergence of a globally convergent algorithm should not be affected by the choice of initial point.*
>
> *Local convergence refers to the ability of the algorithm to approach $\mathbf{x}_*$ rapidly from a starting point (or iterant $\mathbf{x}_k$) in the neighborhood of $\mathbf{x}_*$.*

Despite the definition above, global convergence can be proven typically for a stationary point, or a KKT point, rather than a minimizer.

An iterative algorithm generates an infinite sequence $\{\mathbf{x}_k\}_{k=0}^{\infty}$. If the elements $\mathbf{x}_k$ are distinct, so that $\mathbf{x}_k \neq \mathbf{x}_*$ for any k, we can define an error $\mathbf{e}_k \triangleq \mathbf{x}_k - \mathbf{x}_*$. Convergence of the sequence $\{\mathbf{x}_k\}$ to $\mathbf{x}_*$ means that the limit of $\{\mathbf{e}_k\}$ is zero. Given some iteration k_0, a local *rate of convergence* can be defined by measuring the decrease in the error in subsequent iterations (i.e., for all $k > k_0$). The usual rates of convergence are expressed in terms of an asymptotic *convergence ratio* γ as follows:

$$\begin{aligned}
&\|\mathbf{x}_{k+1} - \mathbf{x}_*\| \leq \gamma \|\mathbf{x}_k - \mathbf{x}_*\|, \quad 0 < \gamma < 1 \text{ (linear)}, \\
&\|\mathbf{x}_{k+1} - \mathbf{x}_*\| \leq \gamma_k \|\mathbf{x}_k - \mathbf{x}_*\|, \quad \gamma_k \to 0 \text{ (superlinear)}, \\
&\|\mathbf{x}_{k+1} - \mathbf{x}_*\| \leq \gamma \|\mathbf{x}_k - \mathbf{x}_*\|^2, \quad \gamma \in \Re \text{ (quadratic)}.
\end{aligned} \tag{7.3}$$

For a more general definition, see Ortega and Rheinboldt (1970). The best optimization algorithms have local convergence with quadratic rate, at least theoretically. For algorithms with linear rate, the value of the convergence ratio will determine the performance substantially.

An algorithmic property often mentioned for nonlinear programming methods is *quadratic termination*. An algorithm has this property if it can reach the minimizer of a quadratic function in a known finite number of iterations. It is usually true that algorithms that work well for quadratic functions will also do the same for general functions. But because this is not always true, quadratic termination is a useful but not absolute criterion of expected practical performance.

It should be mentioned here that global convergence corresponds to reliability or robustness of the algorithm, while local convergence corresponds to efficiency.

It is unfortunate that the efficient algorithms with quadratic rates (such as Newton's method) are not globally convergent in their pure form, while the reliable methods (such as the gradient one) are inefficient, with linear rates, and in some cases arbitrarily slow. When modifications are introduced in either case the result is a compromise between reliability and efficiency. It is somewhat ironical that quite often in practice the choice of algorithm is itself a form of optimization that attempts to compromise between computing resources and knowledge about the problem.

Example 7.1 The scalar sequence $x_k = k^{-k}$ has a superlinear rate of convergence, because

$$\lim_{k\to\infty} \frac{x_{k+1}}{x_k} = \lim_{k\to\infty} \frac{1}{k(1+1/k)^{k+1}} = 0.$$

The meaning of such a rate should be evident if we perform a few iterations. After two iterations, one correct figure is added at each step. ■

Example 7.2 The gradient method with exact line search has a worst case convergence ratio of $\gamma = (\lambda_{\max} - \lambda_{\min})/(\lambda_{\max} + \lambda_{\min})$, where $\lambda_{\max}$ and $\lambda_{\min}$ are respectively the largest and smallest eigenvalues of the Hessian of f at $\mathbf{x}_*$. (For a proof see, for example, Luenberger, 1973.) The ratio $R = \lambda_{\max}/\lambda_{\min}$ is the *condition number* of a matrix, which is termed *ill-conditioned* if R is large. The above convergence ratio will approach the value of one for large values of R. Then the gradient method will show slow progress.

Consider the problem of minimizing $f = 20x_1^2 + x_2^2$. We have

$$\nabla f = (40x_1, 2x_2), \quad \mathbf{H} = \begin{pmatrix} 40 & 0 \\ 0 & 2 \end{pmatrix}.$$

Thus the condition number is $R = 20$ and the asymptotic convergence ratio $\gamma = \frac{19}{21} = 0.905$. The exact line search is performed by calculating

$$\begin{aligned} f(\mathbf{x} - \alpha\mathbf{g}) &= 20(x_1 - 40\alpha x_1)^2 + (x_2 - 2\alpha x_2)^2 \\ &= 20x_1^2(1 - 40\alpha)^2 + x_2^2(1 - 2\alpha)^2 \end{aligned}$$

and setting $\partial f/\partial\alpha = 0$, which gives

$$\alpha = \left(800x_1^2 + 2x_2^2\right)/\left(32000x_1^2 + 4x_2^2\right).$$

Note that this is exactly what Equation (4.47) would use since there we had

$$\alpha = \frac{\mathbf{g}^T\mathbf{g}}{\mathbf{g}^T\mathbf{H}\mathbf{g}} = \frac{(40x_1, 2x_2)(40x_1, 2x_2)^T}{(40x_1, 2x_2)\begin{pmatrix} 40 & 0 \\ 0 & 2 \end{pmatrix}\begin{pmatrix} 40x_1 \\ 2x_2 \end{pmatrix}} = \frac{1600x_1^2 + 4x_2^2}{64000x_1^2 + 8x_2^2}.$$

Taking the starting point $\mathbf{x}_0 = (1, 1)^T$ with $f_0 = 21$, we calculate the following:

$$\mathbf{x}_1 = \begin{pmatrix} 1 \\ 1 \end{pmatrix} - 0.02506\begin{pmatrix} 40 \\ 2 \end{pmatrix} = \begin{pmatrix} -0.0024 \\ 0.9499 \end{pmatrix}, \quad f_1 = 0.9024,$$

$$\mathbf{x}_2 = \begin{pmatrix} -0.0024 \\ 0.9499 \end{pmatrix} - 0.4769\begin{pmatrix} -0.096 \\ 1.900 \end{pmatrix} = \begin{pmatrix} 0.0434 \\ 0.0438 \end{pmatrix}, \quad f_2 = 0.0396,$$

$$\mathbf{x}_3 = \begin{pmatrix} 0.0434 \\ 0.0438 \end{pmatrix} - 0.02506\begin{pmatrix} 1.736 \\ 0.088 \end{pmatrix} = \begin{pmatrix} -0.0001 \\ 0.0416 \end{pmatrix}, \quad f_3 = 0.0017,$$

$$\mathbf{x}_4 = \begin{pmatrix} -0.0001 \\ 0.0416 \end{pmatrix} - 0.4790\begin{pmatrix} -0.0040 \\ 0.0832 \end{pmatrix} = \begin{pmatrix} 0.0018 \\ 0.0017 \end{pmatrix}, \quad f_4 = 6.8(10^{-5}).$$

Since the solution here is $\mathbf{x}_* = (0, 0)T$, we calculate

$$\|\mathbf{x}_1 - \mathbf{x}_*\| / \|\mathbf{x}_0 - \mathbf{x}_*\| = 0.9499/1.4142 = 0.6717,$$
$$\|\mathbf{x}_2 - \mathbf{x}_*\| / \|\mathbf{x}_1 - \mathbf{x}_*\| = 0.0617/0.9499 = 0.0649,$$
$$\|\mathbf{x}_3 - \mathbf{x}_*\| / \|\mathbf{x}_2 - \mathbf{x}_*\| = 0.0416/0.0617 = 0.6742,$$
$$\|\mathbf{x}_4 - \mathbf{x}_*\| / \|\mathbf{x}_3 - \mathbf{x}_*\| = 0.0025/0.0416 = 0.0601.$$

Accounting for round-off errors, the observed worst case ratio appears of about the same magnitude as predicted by the theory, but only every other iteration. Note also that the step size is the same at every other iteration.

Next consider the problem of minimizing $f = (10^3)x_1^2 + x_2^2$. Here we have

$$\nabla f = (2000x_1, 2x_2), \quad \mathbf{H} = \begin{pmatrix} 2000 & 0 \\ 0 & 2 \end{pmatrix}, \quad R = 1000, \quad \gamma = 0.998.$$

The step size will be

$$\alpha = \frac{4(10^6)x_1^2 + 4x_2^2}{8(10^9)x_1^2 + 8x_2^2} \cong 0.5(10^{-3}) \text{ for small } \mathbf{x}.$$

A first iteration from $\mathbf{x}_0 = (1, 1)^T$ gives

$$\mathbf{x}_1 = \begin{pmatrix} 1 \\ 1 \end{pmatrix} - 0.5(10^{-3})\begin{pmatrix} 2000 \\ 2 \end{pmatrix} = \begin{pmatrix} 0.000 \\ 0.999 \end{pmatrix}.$$

The correct value of the second component of $\mathbf{x}_*$ will be reached in the next iteration. If, however, $\mathbf{x}$ is not small, then the value of α cannot be taken as approximately $0.5(10^{-3})$ over all iterations. For example, if a new starting point $\mathbf{x}'_0 = (1, 1000)^T$ is used, then

$$\alpha'_1 = \frac{4(10^6) + 4(10^6)}{8(10^9) + 8(10^6)} = 9.99(10^{-4}),$$

$$\mathbf{x}'_1 = \begin{pmatrix} 1 \\ 1000 \end{pmatrix} - 9.99(10^{-4})\begin{pmatrix} 2000 \\ 2000 \end{pmatrix} = \begin{pmatrix} -0.998 \\ 998.0 \end{pmatrix}.$$

For the next iteration α will be essentially unchanged, and so

$$\mathbf{x}_2' = \begin{pmatrix} -0.998 \\ 998.0 \end{pmatrix} - 9{,}99(10^{-4})\begin{pmatrix} -1996.0 \\ 1996.0 \end{pmatrix} = \begin{pmatrix} 0.9960 \\ 996.0 \end{pmatrix}.$$

The value of α will slowly change as more points are generated, but the numerical convergence process will be very slow. ■

The successive search directions of the gradient method are *orthogonal.* To see this, consider the iteration $\mathbf{x}_{k+1} = \mathbf{x}_k - \alpha_k \mathbf{g}_k$ and find α_k as the minimizer of $f(\mathbf{x}_{k+1}) = f(\mathbf{x}_k - \alpha \mathbf{g}_k)$. The stationary point $\partial f/\partial \alpha = (\nabla f_{k+1})(-\mathbf{g}_k) = 0$ gives $\mathbf{g}_{k+1}\mathbf{g}_k = 0$. For a quadratic function we can get a geometric insight into the convergence situation because the condition number of the Hessian is a measure of the eccentricity of the objective contours. A quadratic represents an n-dimensional ellipsoid with axes coinciding with the n orthogonal eigenvectors of the Hessian. The axes lengths are proportional to λ^{-1} where λ is the eigenvalue corresponding to the particular eigenvector. Thus, large eigenvalues will correspond to short axes while if all eigenvalues are about equal, the ellipsoid will approximate a hypersphere. In the case of an exact sphere, the gradient method will converge in one step. As ellipticity increases, the number of required steps will increase and the step size will also become smaller. This situation is shown in Figure 7.1. The eigenvectors corresponding to $\lambda_{\min}$ and $\lambda_{\max}$ are directions of minimum and maximum curvature, respectively. The spacing of the contours could make the steps be longer in one direction and shorter in another.

Computing the eigensystem of a Hessian is computationally expensive and rarely necessary. Knowledge of ill-conditioning, however, is important. In algorithms that utilize a Cholesky factorization $\mathbf{LDL}^T$ for the Hessian, as in the stabilized Newton's method of Section 4.7, a lower bound on the condition number of the Hessian is given by

$$r = d_{\max}/d_{\min}, \tag{7.4}$$

where $d_{\max}$ and $d_{\min}$ are respectively the largest and smallest elements of $\mathbf{D}$. Moreover, r and R have the same magnitude (see Gill, Murray, and Wright, 1981, for more details).

Termination Criteria

Let us now examine the question of how to terminate the iterations in a minimization algorithm. A typical optimization algorithm reaches the solution in the limit, and so *termination criteria* are required. This is true even when it is known that the algorithm terminates in a finite number of steps. The criteria should address the following situations:

1. An acceptable estimate of the solution has been found.
2. Slow or no progress is encountered.

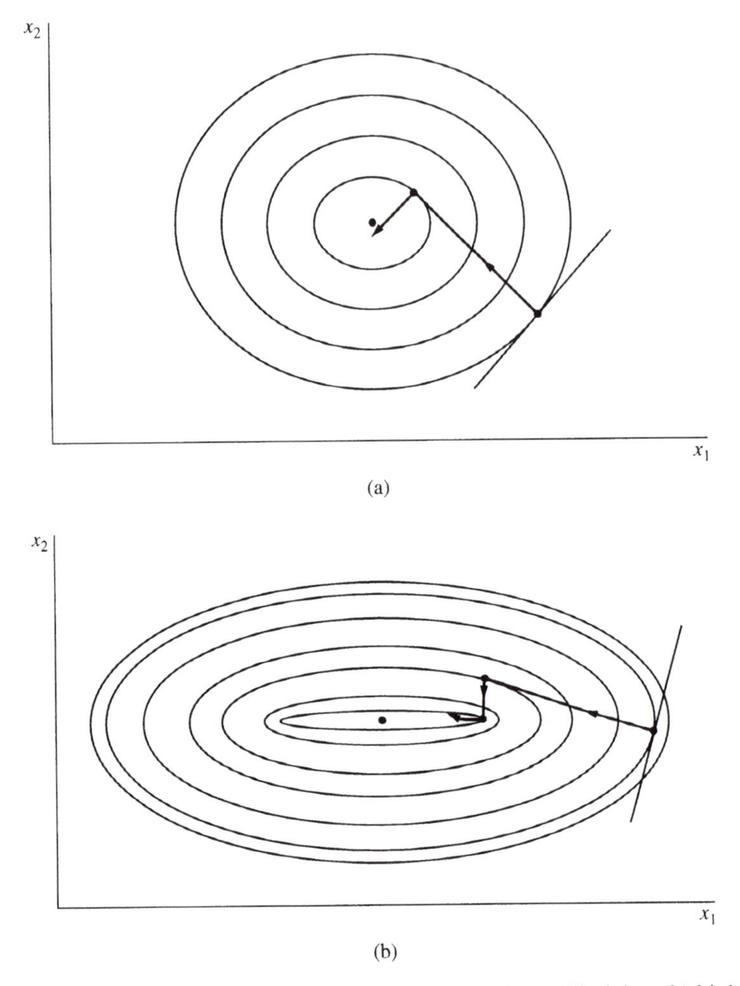

Figure 7.1. Progress of the gradient method. (a) Low ellipticity; (b) high ellipticity.

3. An acceptable number of iterations has been exceeded.
4. An acceptable solution does not exist.
5. The algorithm cycles.

Practical means for addressing such questions are tests that measure how close the optimality conditions are satisfied and examine if the sequence $\{\mathbf{x}_k\}$ is convergent.

Ideally, a test such as $|f_k - f_*| \leq \varepsilon$, or $|x_{i,k} - x_{i,*}| \leq \varepsilon_i$, where ε and ε_i are "small" constants determined by the user, would be desirable. But since the optimal values are not known, this is not useful. Instead we can use the computable tests

$$|f_k - f_{k+1}| \leq \varepsilon, \tag{7.5}$$

$$|x_{i,k} - x_{i,k+1}| \leq \varepsilon_i, \quad i = 1, \ldots, n. \tag{7.6}$$

If selecting a vector ε is not convenient, Equation (7.6) can be replaced by

$$\|\mathbf{x}_k - \mathbf{x}_{k+1}\| \leq \varepsilon. \tag{7.7}$$

The above criteria essentially test convergence. Many algorithms generate sequences $\{\mathbf{x}_k\}$ that converge in a manner such that $\{\nabla f_k\} \rightarrow \mathbf{0}^T$. This indicates only that a stationary point is reached. However, if the algorithm is constructed to guarantee descent at each step, the stationary point will be a minimizer, except in rare cases of hidden saddles. Therefore, another termination criterion is

$$\|\nabla f_k\| \leq \varepsilon \tag{7.8}$$

together with a satisfactory positive-definiteness of the Hessian at $\mathbf{x}_k$. For problems with large numbers of variables, (7.8) may be too restrictive and the condition $\max_i |(\partial f/\partial x_i)_k| \leq \varepsilon$ may be used instead.

For algorithms solving constrained problems, termination criteria are more involved and tend to vary depending on the particular algorithm. In general, apart from convergence tested by (7.7) and possibly by (7.5), we need to give some tolerance in the acceptable violation of the active constraints, inequalities and equalities. The constraints may be tested individually or collectively by setting

$$\|\mathbf{g}_k\| \leq \varepsilon, \tag{7.9}$$

where $\mathbf{g}_k$ is the vector of all active equalities and inequalities. Moreover, the optimality criterion is now the KKT conditions. So a test such as (7.8) can be used with the *reduced* gradient instead of the usual one. Equivalently, the test may be imposed on the *KKT norm*:

$$\left\|\nabla f_k + \boldsymbol{\lambda}_k^T \nabla \mathbf{h}_k + \boldsymbol{\mu}_k^T \nabla \mathbf{g}_k\right\| \leq \varepsilon. \tag{7.10}$$

In active set strategies (Section 7.4), acceptable limits on the values of Lagrange multiplier estimates may be also required (see Gill, Murray, and Wright, 1981). In summary, in constrained cases the tests aim at verifying that, in addition to convergence, a KKT point has been found.

Determining the proper termination criteria is a matter of experience and expertise. It is an issue related to the question of assessing the performance of optimization algorithms in general, as well as to how much knowledge prior model analysis has yielded about the specific problem. The subject is treated more extensively in the cited specialized texts.

7.2 Single Variable Minimization

The simplest numerical optimization problem is the one-dimensional minimization. In Chapter 4 we saw that to perform the n-dimensional iteration $\mathbf{x}_{k+1} = \mathbf{x}_k + \alpha_k \mathbf{s}_k$ we need to minimize the function $f(\mathbf{x}_{k+1})$ with respect to the step size parameter

α and thus determine the value of α_k. Since this minimization is viewed as a "search" along the line that passes through $\mathbf{x}_k$ and has direction $\mathbf{s}_k$ we termed it a *line search*.

Single variable minimization can be encountered also in situations where exactly one degree of freedom remains after identification (and possible elimination) of all active constraints. In this case, solving the one-dimensional minimization problem leads immediately to the solution. This is in contrast to the line search where one-dimensional minimization is part of one iteration and will be repeated in a subsequent iteration. Thus, in the line search the goal is not to find an exact value of the minimizer but to ensure that acceptable descent progress is being achieved during the iteration. In fact, usually it is difficult and computationally expensive to perform a line search with perfect accuracy, so we prefer to settle for an *inexact line search*, which finds a suitable step size α_k rather than the optimal α_*.

The choice of a suitable line search algorithm depends on the constrained algorithm in which it is used. For example, for an augmented Lagrangian method (Section 7.6) an exact line search is necessary, while for a sequential quadratic programming method (Section 7.7) a crude line search is sufficient.

Bracketing, Sectioning, and Interpolation

It is useful to recognize from the start that finding the minimum of $f(x)$ is equivalent to solving $f'(x) = g(x) = 0$, if it is known that a unique stationary point exists that is also a minimum. Then the one-dimensional minimization methods would simply be the numerical methods used for finding roots of functions of one variable.

A large number of strategies have been developed for single variable minimization. Here we will concentrate on describing the basic elements of any good strategy, which can also be applied effectively to the line search. Such a strategy will have three parts:

1. Find an interval $[a, b]$ that contains the minimum. This is usually called a *bracket* or an *interval of uncertainty*.
2. Develop a *sectioning* procedure that can reduce the size of the bracket with successive iterations at a uniform rate.
3. Find an approximation to the bracketed minimum by calculating the minimum of a simple function (usually quadratic or cubic polynomial) that approximates the true one in the bracket. This is referred to as *interpolation*, because the polynomial is constructed to curvefit function data at points in the interval.

The sectioning part will be effective only if the function has a special property called *unimodality* (Figure 7.2). A unimodal function does not have to be continuous or differentiable, but it has a unique minimum and contains no intervals of finite length where its slope is zero. This means that $f(x)$ will decrease monotonically to the minimum and then increase again monotonically. In Figure 7.3, the original bracket is $[a, b]$. We evaluate the function at two new interior points c and d. If the

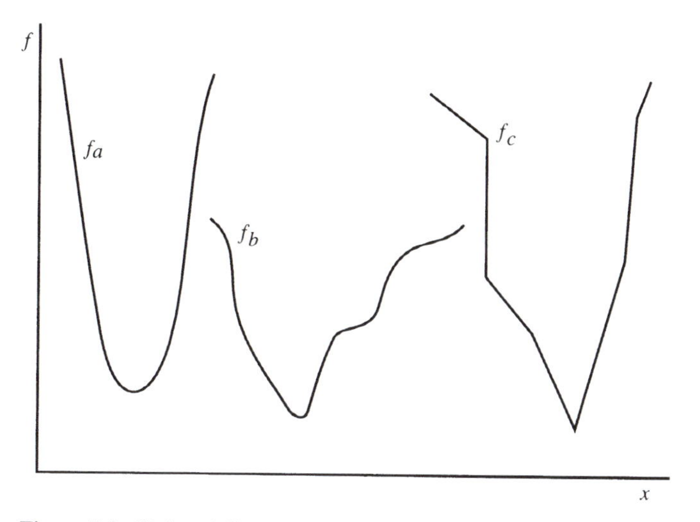

Figure 7.2. Unimodality.

function is unimodal, the minimum must be in the interval $[c, b]$, which is then the second bracket.

A simple sectioning procedure is the *bisection* method for root finding applied to the derivative of the function $g(x) = f'(x)$ (Figure 7.4). Starting with an interval $[a, b]$ where $g(a) > 0$, $g(b) < 0$, we find the point c that bisects $[a, b]$, that is, $c = (a+b)/2$. From the two new intervals we select the one with opposite signs for g at its ends, that is, $[c, b]$, and so the $g = 0$ value is bracketed again. Bisection will converge to any accuracy we want as long as the evaluation of g is precise enough to determine its sign. In the minimization version we must bracket with intervals that have opposite *slope* signs at their ends. For example, in Figure 7.3 we must have $f'(a) < 0$, $f'(b) > 0$ and subsequently $f'(c) < 0$, $f'(b) > 0$ for the new interval. As a rough verification, we may observe that $(f_c - f_a)/(c - a) < 0$ and $(f_b - f_d)/(b - d) > 0$.

The obvious question now is how to select the new points inside a given bracket to make sectioning more efficient. In the bisection method the relative values of f

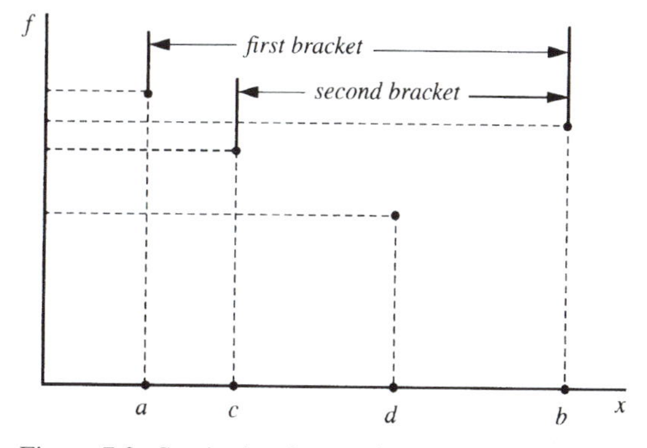

Figure 7.3. Sectioning for a unimodal function.

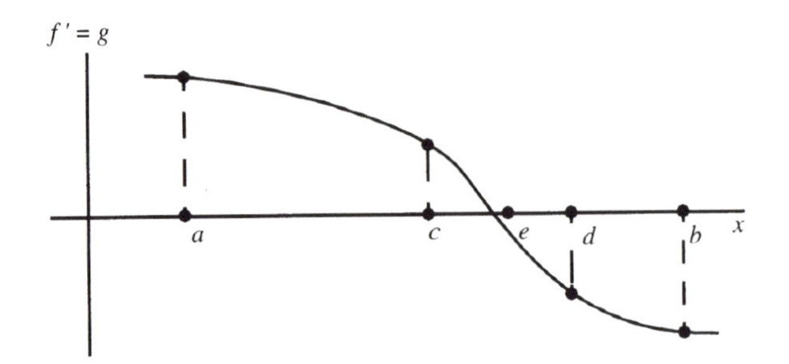

Figure 7.4. Bisection.

at the measurement points are ignored. If we expect that the function is reasonably well behaved in the interval, we may approximate the original function by another one whose minimum is easy to calculate and then take this minimizer as a new point for bracketing. This is the interpolation part mentioned above.

A *quadratic interpolation* procedure starts with an interval of uncertainty (bracket) $[x_1, x_3]$ and an additional point x_2 such that $x_1 < x_2 < x_3$ and $f(x_2) < f(x_1)$, $f(x_2) < f(x_3)$ (Figure 7.5). A quadratic approximation $q(x) = ax^2 + bx + c$ is constructed so that it coincides with $f(x)$ at the three points, that is, $q(x_i) = f(x_i), i = 1, 2, 3$. Thus, a uniquely defined parabola is fitted in the bracket $[x_1, x_3]$ that is guaranteed to have a minimum at the point

$$\hat{x} = \left(\frac{1}{2}\right) \frac{(x_2^2 - x_3^2) f(x_1) + (x_3^2 - x_1^2) f(x_2) + (x_1^2 - x_2^2) f(x_3)}{(x_2 - x_3) f(x_1) + (x_3 - x_1) f(x_2) + (x_1 - x_2) f(x_3)}. \tag{7.11}$$

This can be verified using the theory in Chapter 4. The point $\hat{x}$ will be used instead of x_1 or x_3 for a new bracket and a new interpolation. In the situation of Figure 7.5, the new bracket will be $[x_2, x_3]$ and the new quadratic will pass through the points $x_2 < \hat{x} < x_3$. One difficulty that may prematurely terminate this process is that the distance $|x_2 - \hat{x}|$ may be very small so that no new point can be identified numerically. In this case instead of $\hat{x}$ another point $\hat{x} \pm \delta$ is selected, with δ a small positive number.

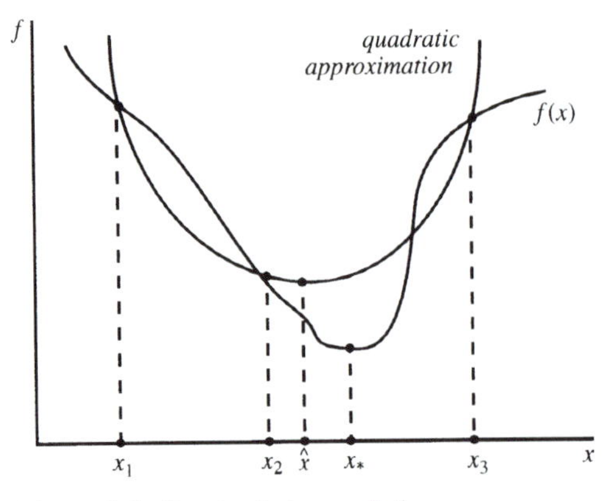

Figure 7.5. Quadratic interpolation.

Whether $\hat{x}$ will be shifted to the right or to the left depends on which side will result in a smaller bracket.

Another form of a quadratic interpolation can be employed if we have derivative information available. Using only the end points of a bracket $[x_1, x_2]$ we construct a quadratic $q(x)$ with $q(x_1) = f(x_1) \triangleq f_1$, $q(x_2) = f(x_2) \triangleq f_2$, and $q'(x_1) = f'(x_1) \triangleq f_1'$. The resulting minimizer for q is given by (see Exercise 7.6)

$$\hat{x} = x_1 + \left(\frac{1}{2}\right) \frac{f_1'(x_2 - x_1)^2}{f_1'(x_2 - x_1) + (f_1 - f_2)}. \tag{7.12}$$

If we wanted to use only one point x but include second derivative information, the minimizer of the resulting quadratic would be $\hat{x} = x_1 - f'(x_1)/f''(x_1)$. Iterative use of this equation is nothing else but Newton's method applied to the one-dimensional problem.

A more reliable interpolation is a *cubic* one that uses the polynomial $p(x) = ax^3 + bx^2 + cx + d$ and fits it in the bracket $[x_1, x_2]$ by setting $p(x_1) = f(x_1)$, $p(x_2) = f(x_2)$, $p'(x_1) = f'(x_1)$, and $p'(x_2) = f'(x_2)$ where we *must* have $f'(x_1) < 0$ and $f'(x_2) > 0$. The sign restriction on these derivatives is necessary for excluding non-minima stationary points of $p(x)$ in the interval $[x_1, x_2]$ (Figure 7.6). In case (b) of this figure, the interpolation will include also a maximum. In practice, this situation can happen in line search applications of the cubic fit and then special measures must be taken, usually heuristic.

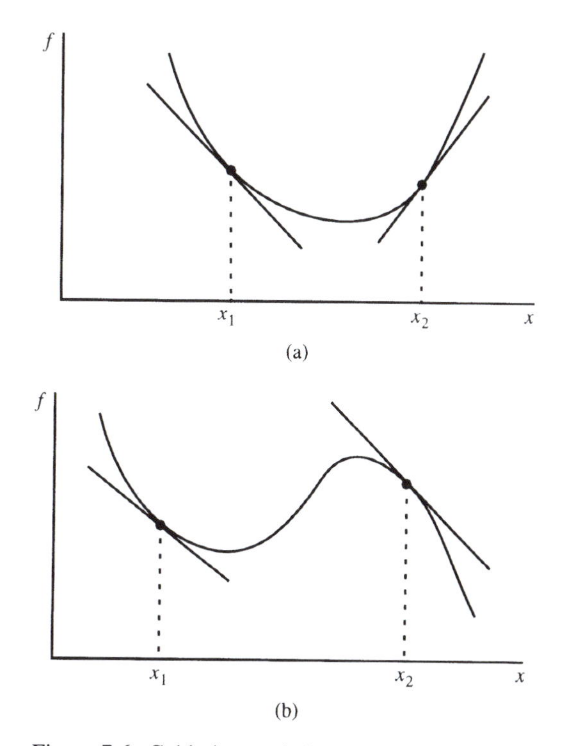

Figure 7.6. Cubic interpolation derivative sign restriction.

The minimizer for the cubic interpolating polynomial as defined above is given by

$$\hat{x} = x_1 + (x_2 - x_1)\left\{1 - \frac{f'(x_2) + u_2 - u_1}{f'(x_2) - f'(x_1) + 2u_2}\right\}, \tag{7.13}$$

where

$$\begin{aligned} u_1 &= \frac{3[f(x_1) - f(x_2)]}{(x_2 - x_1)} + f'(x_1) + f'(x_2), \\ u_2 &= \left[u_1^2 - f'(x_1)f'(x_2)\right]^{1/2}. \end{aligned} \tag{7.14}$$

Though they appear complicated, these expressions are easy to evaluate in the computer.

Example 7.3 Use the quadratic interpolation with derivatives, Equation (7.12), to minimize the function $f = 3x^3 - 18x + 27$, for $x \geq 0$.

We will assume that an initial interval of uncertainty is $[0, 5]$, that is, $x_1 = 0$, $x_2 = 5$ to start the iteration (Figure 7.7). Usually this information may not be available and a bracketing procedure must be used to arrive at an initial interval. The derivative of f is $df/dx = 9x^2 - 18$. Therefore, the minimizer in the first interpolation is found from $\hat{x} = 0 + (\frac{1}{2})[(-18)(5-0)^2/((-18)(5-0) + (27 - 312))] = 0.60$. Taking the midpoint $x_3 = 2.5$, we evaluate $f_3 = 28.90$. From Figure 7.7 we see that the new bracket is $[0, 2.5]$. Note that the minimum is $x_* = \sqrt{2}$ with $f_* = 10.029$, so the point $\hat{x} = 0.6$ with $\hat{f} = 16.85$ is not yet a good approximation. The new approximation to the minimizer is found from Equation (7.12) to be $\hat{x} = 1.2$ with $f(\hat{x}) = 10.58$. To bracket again we

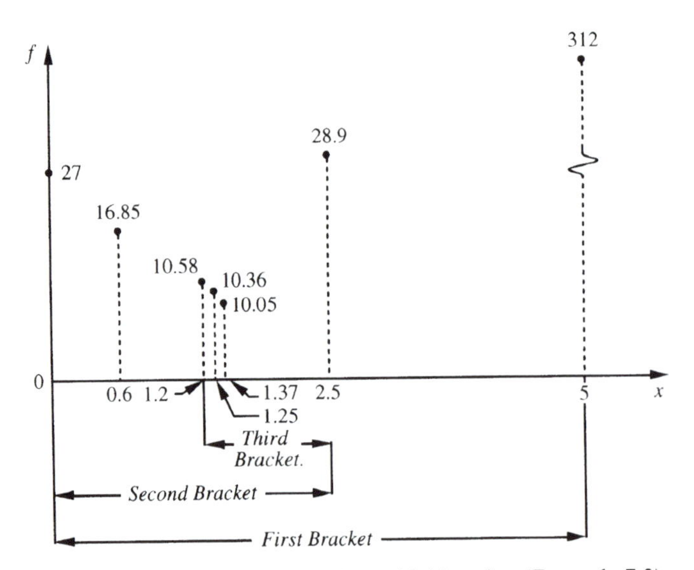

Figure 7.7. Quadratic interpolation with bisection (Example 7.3).

take the midpoint $x_3 = 1.25$ with $f_3 = 10.36$. From the figure it is clear that the new bracket is [1.20, 2.50]. Another interpolation gives $\hat{x} = 1.37$ with $f(\hat{x}) = 10.05$. We see that we are very close to the solution and termination will depend upon the desired accuracy in f_* or x_*. ■

The Davies, Swann, and Campey Method

To illustrate an example of constructing one-dimensional algorithms we will describe an old but still useful algorithm generally known as the Davies, Swann, and Campey (DSC) method (Box, Davies, and Swann 1969). The method combines the elements described earlier in a simple way.

Starting at a point x_0, the function value $f(x_0) = f_0$ is calculated and a test is made at a point $x + \Delta x$ or $x - \Delta x$ to determine which way the function is decreasing (Figure 7.8). Once such a direction is established, successive steps $x_1, x_2, x_3, x_4, \ldots,$ at doubling increments $\Delta x, 2\Delta x, 4\Delta x, 8\Delta x, \ldots,$ are taken while calculating the function values $f_1, f_2, f_3, f_4, \ldots,$ where $f_i \equiv f(x_i)$, $i = 1, 2, 3, 4, \ldots,$ for simplicity. Note that this numbering commences to the right or to the left of x_0 according to the above test. When a function value shows an increase over the previous one, it signals that the minimum has been bracketed. It should lie in the interval that contains the last three points, that is, $[x_2, x_4]$ in Figure 7.8. Another point x_a is taken at the middle of the last interval. With four points available, a sectioning, such as we did in Figure 7.3, is performed and one of the points is discarded. In Figure 7.8, point x_2 is discarded and the remaining three points $x_3 < x_a < x_4$ are equally spaced at $4\Delta x$ apart. These points are now used to fit a quadratic. The minimizer of the interpolating quadratic is the new starting point and a DSC iteration is complete. Note that because of the equal spacing of the points, the quadratic minimizer, Equation (7.11), is given

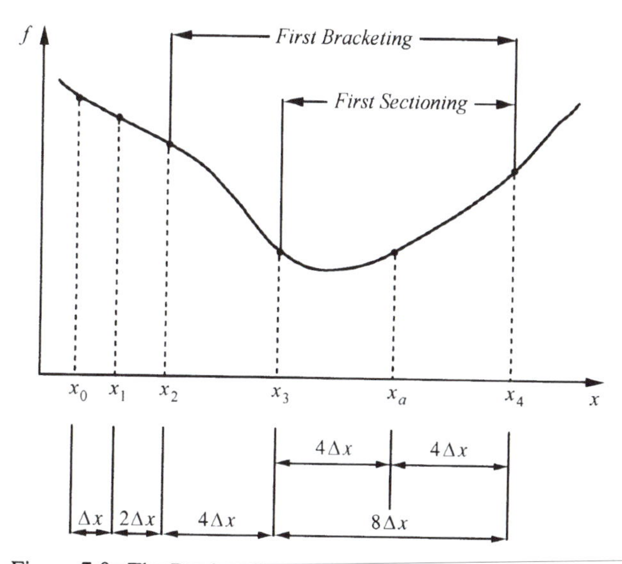

Figure 7.8. The Davies, Swann, and Campey method.

now by the simpler expression

$$\hat{x} = x_2 + \frac{(f_1 - f_3)\Delta x}{2(f_3 - 2f_2 + f_1)}, \tag{7.15}$$

where the quadratic coincides with the function at points x_1, x_2, and x_3 in Figure 7.5. In the DSC search these will be the last three points retained after sectioning, namely, x_3, x_a, and x_4.

In iterative format, the DSC search generates the points

$$x_{n,k} = x_{n-1,k} + 2^{n-1}\Delta x_k,$$

where k is the iteration number and n is the number of function evaluations within an iteration, so that $f_{n,k} > f_{n-1,k}$. After $x_{n,k}$, the additional function evaluation is taken at the point $x_{a,k} = x_{n,k} - 2^{n-2}\Delta x_k$. During sectioning, we must discard either of the points $x_{n,k}$ and $x_{n-2,k}$. Thus, the quadratic minimizer is found from (see Exercise 7.8):

$$x_{0,k+1} = x_{n-1,k} + \frac{2^{n-2}\Delta x_k(f_{n-2,k} - f_{a,k})}{2(f_{a,k} - 2f_{n-1,k} + f_{n-2,k})}, \quad \text{if } f_{a,k} > f_{n-1,k}, \tag{7.16a}$$

$$x_{0,k+1} = x_{a,k} + \frac{2^{n-2}\Delta x_k(f_{n-1,k} - f_{n,k})}{2(f_{n,k} - 2f_{a,k} + f_{n-1,k})}, \quad \text{if } f_{a,k} < f_{n-1,k}. \tag{7.16b}$$

Note that in Figure 7.8, Equation (7.16b) would apply. The new iteration $k + 1$ is performed exactly the same way, except that the step size Δx is *reduced* by a factor K. If we set $x_{-1,k} = x_{0,k} - \Delta x_k$ the algorithmic steps will be as follows (Adby and Dempster 1974):

1. $k = 0$; input $x_{0,1}$, Δx_1, ε, K.
2. $k = k + 1$; evaluate $f_{0,k}$, $f_{1,k}$.
3. If $f_{1,k} < f_{0,k}$, set $\sigma = 1$ and go to 5; otherwise go to 4.
4. Evaluate $f_{-1,k}$. If $f_{-1,k} < f_{0,k}$, set $\sigma = -1$ and go to 5; otherwise go to 9.
5. Find n such that $x_{n,k} = x_{n-1,k} + 2^{n-1}\sigma\Delta x_k$ and $f_{n,k} > f_{n-1,k}$.
6. Evaluate $f_{a,k}$ at $x_{a,k} = x_{n,k} - 2^{n-2}\sigma\Delta x_k$.
7. If $f_{a,k} > f_{n-1,k}$, calculate $x_{0,k+1}$ from Equation (7.16a); otherwise calculate $x_{0,k+1}$ from Equation (7.16b).
8. If $2^{n-2}\Delta x_k \leq \varepsilon$, go to 11; otherwise set $\Delta x_{k+1} = K\Delta x_k$ and go to 2.
9. Set

$$x_{0,k+1} = x_{0,k} + \frac{\Delta x_k(f_{-1,k} - f_{1,k})}{2(f_{1,k} - 2f_{0,k} + f_{-1,k})}. \tag{7.17}$$

10. If $\Delta x_k \leq \varepsilon$, go to 11; otherwise set $\Delta x_{k+1} = K\Delta x_k$ and go to 2.
11. Output $x_{\min} = x_{0,k+1}$, $f_{\min} = f_{0,k+1}$.

Example 7.4 Apply the DSC algorithm to find the minimum of $f = x^2 + 4x^{-1}$ for $x > 0$. The reader is encouraged to construct graphically the progress of the following iterations.

Initialization

$$x_{0,1} = 4, \quad \Delta x_1 = 0.5, \quad K = 0.5, \quad \varepsilon = 0.3.$$

First Iteration

$$f_{0,1} = f(4) = 17, \quad f_{1,1} = f(4.5) = 21.14, \quad f_{-1,1} = f(3.5) = 13.39.$$

Set $\sigma = -1$ and find $x_{1,1} = 3.5$, $x_{2,1} = 2.5$, $x_{3,1} = 0.5$ with $f_{1,1} = 13.39$, $f_{2,1} = 7.85$, $f_{3,1} = 8.25$. So the minimum is bracketed between $x_{1,1}$ and $x_{3,1}$. Evaluate $x_{a,1} = 0.5 + 2(0.5) = 1.5$ with $f_{a,1} = 4.92$. Since $f_{a,1} < f_{2,1}$, we use Equation (7.16b) to calculate $x_{0,2} = 1.5 + [2(0.5)(7.85 - 8.25)/2(8.25 - 2(4.92) + 7.85)] = 1.47$. The new iterant $x_{0,2}$ is the starting point of the new bracketing procedure.

Second Iteration To demonstrate how the algorithm terminates without performing further iterations here, and since $x_{0,2}$ is close enough to $x_{a,1}$, let us set $x_{0,2} = x_{a,1} = 1.5$. Then,

$$x_{0,2} = 1.5, \quad f_{0,2} = 4.92, \quad \Delta x_2 = K\Delta x_1 = (0.5)(0.5) = 0.25.$$

Thus we have $f_{1,2} = f(1.75) = 5.35$, $f_{-1,2} = f(1.25) = 4.76$. Set $\sigma = -1$ and find $x_{1,2} = 1.25$ with $f_{1,2} = 4.76$, and $x_{2,2} = 0.75$ with $f_{2,2} = 5.89$. We then evaluate $x_{a,2} = 1.00$ and $f_{a,2} = 5.00$. Since $f_{a,2} > f_{1,2}$, we find the minimizer of the interpolated quadratic from Equation (7.16a), that is, $x_{0,3} = 1.23$ and $f_{0,3} = 4.77$. Since $2^{2-2}\Delta x_2 = 0.25 < \varepsilon = 0.3$, we stop and declare $x_* = 1.23$ with $f_* = 4.77$. In fact, $df/dx = 2x - 4x^{-2} = 0$ gives $x_* = 1.26$ with $f_* = 4.76$, if we maintain the same significant digits as above. ■

Inexact Line Search

An n-dimensional algorithm iterating according to $\mathbf{x}_{k+1} = \mathbf{x}_k + \alpha_k \mathbf{s}_k$ requires the descent property $\mathbf{g}_k^T \mathbf{s}_k < 0$ for every k, while the gradient $\mathbf{g}_k$ is not zero. The purpose of the line search that determines the value α_k is to ensure an acceptable decrease from f_k to f_{k+1}, that is, $f_k - f_{k+1} > 0$. When the changes in f are very small, while the sequence $\{f_k\}$ may be monotonically decreasing, the actual convergence could be *asymptotic*, that is $\mathbf{x}_* = \lim \mathbf{x}_k (k \to \infty)$. This poses a practical stability problem when the line search does not locate the minimizer α_* exactly. Therefore, an important special requirement for inexact line searches is that an *acceptable sufficient* decrease in f_k is realized. This will guarantee a *finite convergence*, that is, $\|\mathbf{x}_k - \mathbf{x}_*\| < \varepsilon$ in a finite number of iterations.

The situation is depicted in Figure 7.9. The function $\phi(\alpha) \triangleq f(\mathbf{x}_k + \alpha \mathbf{s}_k)$ is plotted in the interval $(0, a)$ for which $f_k - f_{k+1} > 0$ is satisfied. A sufficient decrease in

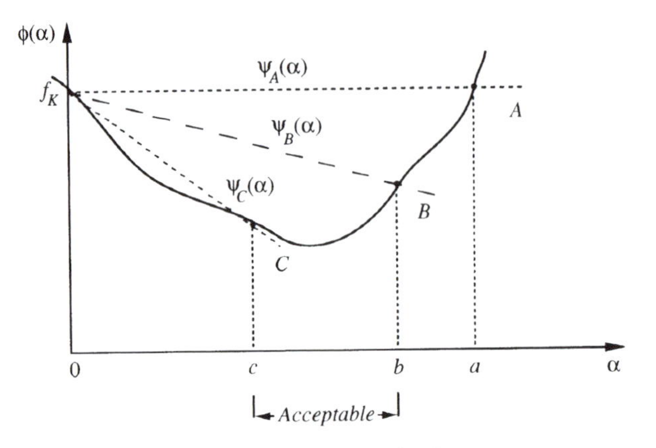

Figure 7.9. The Armijo–Goldstein criteria.

f_k is guaranteed if we choose α_k away from the end points of $(0, a)$. Graphically, we show this by drawing the lines B and C, which analytically are expressed by $\psi(\alpha) = \phi(0) + \varepsilon\phi'(0)\alpha$ with $0 < \varepsilon < 1$, a fixed parameter. The reduction in f_k will be bounded away from zero when α is chosen only from the interval $[c, b]$ (Figure 7.9). To do this, $\phi(\alpha)$ must be below line B for $\alpha < b$ and above line C for $\alpha > c$. Noting that $\psi(\alpha) = f_k + \varepsilon\alpha f'_k = f_k + \varepsilon\alpha\nabla f_k \mathbf{s}_k = f_k + \varepsilon\mathbf{g}_k^T\partial\mathbf{x}_k$, where the directional derivative $f'_k = \nabla f_k \mathbf{s}_k$ by definition, we find that these two conditions are satisfied by the inequalities (see Exercise 7.9)

$$f_k - f_{k+1} \geq -\varepsilon_1 \mathbf{g}_k^T \partial\mathbf{x}_k \quad (\text{i.e., } \alpha \notin [b, a]), \tag{7.18}$$

$$f_k - f_{k+1} \leq -(1 - \varepsilon_1)\mathbf{g}_k^T \partial\mathbf{x}_k \quad (\text{i.e., } \alpha \notin [0, c]), \tag{7.19}$$

where $0 < \varepsilon_1 < \frac{1}{2}$. Practical values for ε_1 are 10^{-1} (Fletcher 1980) to 10^{-4} (Dixon, Spedicato, and Szegö 1980). The line search will terminate for any α found in the interval $[c, b]$. The conditions (7.18) and (7.19) will be referred to as the *Armijo–Goldstein criteria*, being due to original ideas by Armijo (1966) and Goldstein (1965).

There is another danger to convergence. The search vector $\mathbf{s}_k$ may not have a significant component in the negative gradient direction $-\mathbf{g}_k$, that is, the two vectors may be nearly orthogonal so that $\mathbf{g}_k^T\mathbf{s}_k \cong 0$ and the descent property is jeopardized. To prevent this situation, another criterion must be included that will uniformly bound the angle between $-\mathbf{g}_k^T$ and $\mathbf{s}_k$ away from 90°. Recall that the cosine of the angle θ between two vectors $\mathbf{x}$ and $\mathbf{y}$ is given by

$$\cos\theta = \mathbf{x}^T\mathbf{y}/\|\mathbf{x}\| \cdot \|\mathbf{y}\|. \tag{7.20}$$

Thus, the new condition imposed is

$$-\mathbf{g}_k^T\mathbf{s}_k \geq \varepsilon_2\|\mathbf{g}_k\| \cdot \|\mathbf{s}_k\| \tag{7.21}$$

where $\varepsilon_2 > 0$. An often chosen value is $\varepsilon_2 = 10^{-3}$ (Dixon, Spedicato, and Szego 1980).

The three criteria (7.18), (7.19), and (7.21) will guarantee the stability of the inexact line search and convergence proofs may be derived. Other variations of these

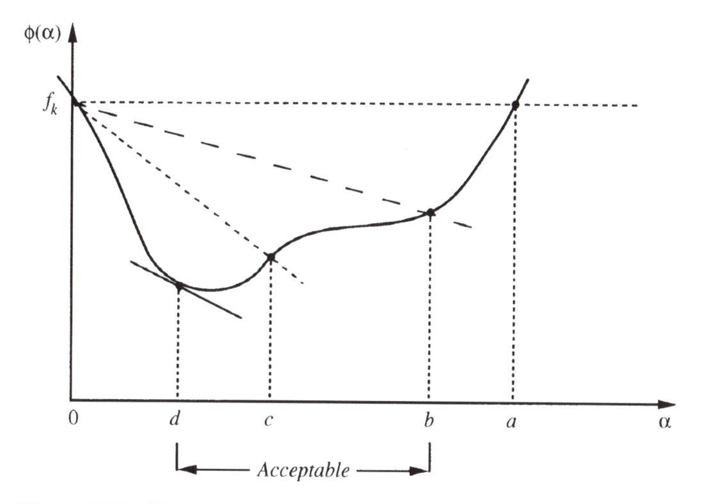

Figure 7.10. Slope reduction criterion as a lower bound on the step size.

criteria may also be useful. For example, the Armijo–Goldstein criterion (7.19) that bounds the step size away from zero may be replaced by an equivalent criterion requiring that the final slope along the line decreases sufficiently to a value that is a fraction ε_3 of the initial slope. Analytically,

$$\left|\mathbf{g}_{k+1}^T \mathbf{s}_k\right| \leq -\varepsilon_3 \mathbf{g}_k^T \mathbf{s}_k, \tag{7.22}$$

where the absolute value is used to allow the search to become an exact one as $\varepsilon_3 \to 0$. In Figure 7.10, a situation is shown where the criterion (7.19) would have excluded the exact minimum, but the use of (7.22) would instead increase the acceptable interval to $[d, b]$ and include the minimizer of $f_{k+1}(\alpha)$. The smaller the value of ε_3, the more restrictive (accurate) the line search will be. Again, an often chosen value is $\varepsilon_3 = 0.5$ (Dixon et al. 1980).

Example 7.5 An example of a very simple but effective inexact line search that does not employ interpolation is the *Armijo Line Search* (Armijo 1966). In this method the conditions (7.18) and (7.19) are replaced by the conditions

$$\phi(2\alpha) > \psi_1(\alpha) = \phi(0) + 2\varepsilon\alpha\phi'(0), \tag{a}$$

$$\phi(\alpha) < \psi_2(\alpha) = \phi(0) + \varepsilon\alpha\phi'(0), \tag{b}$$

where $\phi(\alpha) = f(\mathbf{x}_k + \alpha\mathbf{s}_k)$ as defined earlier in this section. The algorithm will start with an arbitrary α and continue by doubling or halving the value of α until conditions (a) and (b) are satisfied. Usually ε is selected in the range of $0.1 \leq \varepsilon \leq 0.2$.

Let us apply this idea to the function $f = x_1^2 + x_2^2$ for the gradient method so that $\mathbf{s}_k = -\mathbf{g}_k$. We have $\mathbf{g}_k^T = \nabla f = (2x_1, 2x_2)$. Assume that we have arrived at an iteration k, where $\mathbf{x}_k = (1, 1)^T$. At this point we have

$$\phi(\alpha) = f(\mathbf{x}_k - \alpha\mathbf{g}_k) = (1 - 2\alpha)^2 + (1 - 2\alpha)^2 = 2(1 - 2\alpha)^2.$$

Similarly, $\phi(2\alpha) = 2(1-4\alpha)^2$. Since $\phi'(\alpha) = -8(1-2\alpha)$, we have $\phi(0) = 2$ and $\phi'(0) = -8$. For $\varepsilon = 0.1$, conditions (a) and (b) become

$$2(1-4\alpha)^2 > 2 - 1.6\alpha, \quad 2(1-2\alpha)^2 < 2 - 0.8\alpha. \tag{c}$$

It can be quickly seen that these are equivalent to $0.45 < \alpha < 0.9$. Numerically, the algorithm will select an α, say, $\alpha = 0.2$, and double it until conditions (c) are satisfied. This will give $\alpha_k = 0.8$ and the line search ends. ■

7.3 Quasi-Newton Methods

The modified Newton's method examined in Chapter 4 uses the iteration

$$\mathbf{x}_{k+1} = \mathbf{x}_k - \alpha_k \mathbf{H}_k^{-1} \mathbf{g}_k, \tag{7.23}$$

where $\mathbf{H}_k$ and $\mathbf{g}_k$ are the Hessian and gradient at iteration k. The method has quadratic convergence provided that $\mathbf{H}_k$ remains positive-definite. Thus, a stabilized Newton's method is very attractive, except for the need to calculate second derivatives. This could be a serious drawback in practice, except when these are easy to calculate.

In Section 8.2 we will discuss finite difference approximations to derivatives in general. For the specific case of iteration (7.23), the key question is how to include curvature information, as the one carried by the Hessian, while not computing the Hessian at each iteration. A class of methods called *quasi-Newton* methods build up curvature information by observing the behavior of f_k and $\mathbf{g}_k$ during a sequence of descent iterations. A sequence of points is used to generate an approximation to the Hessian, rather than evaluating the Hessian at a single point. We should mention right away that quasi-Newton methods today are regarded as the best general-purpose methods for solving unconstrained problems.

Hessian Matrix Updates

A typical quasi-Newton method operates exactly as the modified Newton's method, Equation (7.23), except for the calculation of the matrix $\mathbf{H}_k$. Instead of using the Hessian, an initial positive-definite matrix $\mathbf{H}_0$ is chosen (usually $\mathbf{H}_0 = \mathbf{I}$, the identity matrix) and is subsequently updated by an *update* formula

$$\mathbf{H}_{k+1} = \mathbf{H}_k + \hat{\mathbf{H}}_k, \tag{7.24}$$

where $\hat{\mathbf{H}}_k$ is the update matrix. It should be emphasized that $\mathbf{H}_k$ and $\mathbf{H}_{k+1}$ in (7.24) are *not* the Hessians at $\mathbf{x}_{k+1}$ and $\mathbf{x}_k$ but approximations to them. We avoid using different symbols for simplicity. Since (7.23) requires the inverse $\mathbf{H}_k^{-1}$, many update formulas are applied directly to the inverse so that the solution of a linear system of equations at each iteration is avoided and the order of number of multiplications per iteration is reduced from n^3 (for Newton's) to n^2. If we define

$$\mathbf{B}_k \triangleq \mathbf{H}_k^{-1} \tag{7.25}$$

then the updating formula (7.24) for the inverse is also of the form

$$\mathbf{B}_{k+1} = \mathbf{B}_k + \hat{\mathbf{B}}_k \tag{7.26}$$

(provided the update is of "rank one" or "rank two"; see end of this subsection).

The specific choice for the updating matrices is the distinguishing characteristic among the large variety of quasi-Newton methods that have been proposed. The desirable properties of the updating scheme are that $\mathbf{H}_{k+1}$ remains positive-definite to assure descent and that only first derivative information is needed in the update. In addition, successive updates should transform the initial (possibly arbitrary) matrix $\mathbf{H}_0$ to a successively better approximation of the true Hessian. A practical way to achieve this is to observe that the Taylor expansion for the *gradient function* to first order gives

$$\mathbf{g}_{k+1} - \mathbf{g}_k = \mathbf{H}_k(\mathbf{x}_{k+1} - \mathbf{x}_k). \tag{7.27}$$

In shorthand, if $\partial\mathbf{g}_k = \mathbf{g}_{k+1} - \mathbf{g}_k$ and $\partial\mathbf{x}_k = \mathbf{x}_{k+1} - \mathbf{x}_k$, this expression is simply $\partial\mathbf{g}_k = \mathbf{H}_k\partial\mathbf{x}_k$. Solving for $\partial\mathbf{x}_k = \mathbf{H}_k^{-1}\,\partial\mathbf{g}_k$, we see that a matrix $\mathbf{H}_k$ in general will not satisfy this condition for the Hessian, so we *correct* it in order to relate the perturbations $\partial\mathbf{x}_k$ and $\partial\mathbf{g}_k$ properly. A good update correction then should satisfy the requirement

$$\partial\mathbf{x}_k = \mathbf{H}_{k+1}^{-1}\partial\mathbf{g}_k, \quad \text{or} \quad \partial\mathbf{x}_k = \mathbf{B}_{k+1}\,\partial\mathbf{g}_k. \tag{7.28}$$

This is called the *quasi-Newton condition.* Substituting (7.26) in (7.28), we get an equation that must be satisfied by the update matrix $\hat{\mathbf{B}}_k$:

$$\partial\mathbf{x}_k = \mathbf{B}_k\partial\mathbf{g}_k + \hat{\mathbf{B}}_k\partial\mathbf{g}_k. \tag{7.29}$$

Finding a $\hat{\mathbf{B}}_k$ that satisfies (7.29) is not a procedure leading to a unique solution. A general form of updating matrix is

$$\hat{\mathbf{B}}_k = a\mathbf{u}\mathbf{u}^T + b\mathbf{v}\mathbf{v}^T, \tag{7.30}$$

where a and b are scalars and $\mathbf{u}$ and $\mathbf{v}$ are vectors, all appropriately selected to satisfy (7.29) and other requirements of $\mathbf{B}_{k+1}$ such as symmetry and positive-definiteness. The quantities $a\mathbf{u}\mathbf{u}^T$ and $b\mathbf{v}\mathbf{v}^T$ are symmetric matrices of rank one. Quasi-Newton methods that take $b=0$ are using *rank one* updates. For $b \neq 0$ more flexibility is possible and the resulting methods are said to use *rank two* updates. Rank two updates are currently the most widely used schemes. The rationale for choices in (7.30) can be quite complicated and it will be suppressed here by referring the reader to specialized texts. We will only give two update formulas that have received wide acceptance.

The DFP and BFGS Formulas

If we substitute (7.30) for $\hat{\mathbf{B}}_k$ in (7.29), we have

$$\partial\mathbf{x}_k = \mathbf{B}_k\partial\mathbf{g}_k + a\mathbf{u}\mathbf{u}^T\partial\mathbf{g}_k + b\mathbf{v}\mathbf{v}^T\partial\mathbf{g}_k. \tag{7.31}$$

We choose to set $\mathbf{u} = \partial \mathbf{x}_k$, $\mathbf{v} = \mathbf{B}_k \partial \mathbf{g}_k$ and let $a\mathbf{u}^T \partial \mathbf{g}_k = 1$, $b\mathbf{v}^T \partial \mathbf{g}_k = -1$ to determine a and b. The resulting update formula is

$$\mathbf{B}_{k+1}^{(\mathrm{DFP})} = \mathbf{B}_k + \left[\frac{\partial \mathbf{x} \partial \mathbf{x}^T}{\partial \mathbf{x}^T \partial \mathbf{g}}\right]_k - \left[\frac{(\mathbf{B}\partial \mathbf{g})(\mathbf{B}\partial \mathbf{g})^T}{\partial \mathbf{g}^T \mathbf{B} \partial \mathbf{g}}\right]_k. \tag{7.32}$$

This is popularly known as the DFP formula from Davidon (1959) and Fletcher and Powell (1963) who are credited with its development. The superscript (DFP) for $\mathbf{B}_{k+1}$ in (7.32) is used as a mnemonic shorthand for the expression on the right-hand side.

Another important rank two update formula was independently suggested by Broyden (1970), Fletcher (1970), Goldfarb (1970), and Shanno (1970) and is known as the *BFGS formula* (and method). The update formula is

$$\mathbf{B}_{k+1}^{(\mathrm{BFGS})} = \mathbf{B}_k + \left[1 + \frac{\partial \mathbf{g}^T \mathbf{B} \partial \mathbf{g}}{\partial \mathbf{x}^T \partial \mathbf{g}}\right]_k \left[\frac{\partial \mathbf{x} \partial \mathbf{x}^T}{\partial \mathbf{x}^T \partial \mathbf{g}}\right]_k - \left[\frac{\partial \mathbf{x} \partial \mathbf{g}^T \mathbf{B} + \mathbf{B} \partial \mathbf{g} \partial \mathbf{x}^T}{\partial \mathbf{x}^T \partial \mathbf{g}}\right]_k. \tag{7.33}$$

This expression is complicated but straightforward to apply.

In summary, a quasi-Newton algorithm will have the following structure:

1. Input $\mathbf{x}_0$, $\mathbf{B}_0$, termination criteria.
2. For any k, set $\mathbf{s}_k = -\mathbf{B}_k \mathbf{g}_k$.
3. Compute a step size and set $\mathbf{x}_{k+1} = \mathbf{x}_k + \alpha_k \mathbf{s}_k$.
4. Compute the update matrix $\hat{\mathbf{B}}_k$ according to a given formula (say, DFP or BFGS) using the values $\partial \mathbf{g}_k = \mathbf{g}_{k+1} - \mathbf{g}_k$, $\partial \mathbf{x}_k = \mathbf{x}_{k+1} - \mathbf{x}_k$, and $\mathbf{B}_k$.
5. Set $\mathbf{B}_{k+1} = \mathbf{B}_k + \hat{\mathbf{B}}_k$.
6. Continue with next k until termination criteria are satisfied.

An interesting result is obtained if we prefer to work with the Hessian matrix $\mathbf{H}_k$ rather than its inverse $\mathbf{B}_k$. If we take the expression for $\mathbf{B}_{k+1}^{(\mathrm{BFGS})}$ and make the interchanges

$$\mathbf{B}_k \to \mathbf{H}_k, \quad \mathbf{B}_{k+1} \to \mathbf{H}_{k+1}, \quad \partial \mathbf{x}_k \to \partial \mathbf{g}_k, \quad \partial \mathbf{g}_k \to \partial \mathbf{x}_k \tag{7.34}$$

we arrive at the expression

$$\mathbf{H}_{k+1}^{(\mathrm{DFP})} = \mathbf{H}_k + \left[1 + \frac{\partial \mathbf{x}^T \mathbf{H} \partial \mathbf{x}}{\partial \mathbf{g}^T \partial \mathbf{x}}\right]_k \left[\frac{\partial \mathbf{g} \partial \mathbf{g}^T}{\partial \mathbf{g}^T \partial \mathbf{x}}\right]_k - \left[\frac{\partial \mathbf{g} \partial \mathbf{x}^T \mathbf{H} + \mathbf{H} \partial \mathbf{x} \partial \mathbf{g}^T}{\partial \mathbf{g}^T \partial \mathbf{x}}\right]_k. \tag{7.35}$$

Similarly, if we make the same interchanges in the $\mathbf{B}_{k+1}^{(\mathrm{DFP})}$ formula, we find

$$\mathbf{H}_{k+1}^{(\mathrm{BFGS})} = \mathbf{H}_k + \left[\frac{\partial \mathbf{g} \partial \mathbf{g}^T}{\partial \mathbf{g}^T \partial \mathbf{x}}\right]_k - \left[\frac{(\mathbf{H}\partial \mathbf{x})(\mathbf{H}\partial \mathbf{x})^T}{\partial \mathbf{x}^T \mathbf{H} \partial \mathbf{x}}\right]_k. \tag{7.36}$$

These last two expressions are indeed the correct ones and are symmetrical to the inverse expressions (7.33) and (7.32), respectively. This relation among the formulas is referred to as *complementarity.*

Example 7.6 Consider the function $f = 4x_1^2 + 3x_1x_2 + x_2^2$. We will apply the BFGS formula (7.36), starting at $\mathbf{x}_0 = (1, 1)^T$. The gradient is given by $\mathbf{g}(\mathbf{x}) = (8x_1+3x_2, 3x_1+2x_2)^T$.

Initialization

$$\mathbf{x}_0 = \begin{pmatrix} 1 \\ 1 \end{pmatrix}, \quad \mathbf{H}_0 = \begin{pmatrix} 1 & 0 \\ 0 & 1 \end{pmatrix}, \quad \mathbf{s}_0 = -\begin{pmatrix} 1 & 0 \\ 0 & 1 \end{pmatrix}\begin{pmatrix} 11 \\ 5 \end{pmatrix} = -\begin{pmatrix} 11 \\ 5 \end{pmatrix}, \quad f_0 = 8.$$

First Iteration

$$\mathbf{x}_1 = \mathbf{x}_0 + \alpha_0 \mathbf{s}_0 = (1 - 11\alpha_0, 1 - 5\alpha_0)^T.$$

A line search must be performed to determine α_0 according to the procedures described in Section 7.2. For simplicity, let us select $\alpha_0 = 0.1$, which produces descent (see Exercise 7.18). In fact, then $\mathbf{x}_1 = (-0.1, 0.5)^T$ with $f_1 = 0.14$. At $\mathbf{x}_1$ we have

$$\partial \mathbf{x}_0 = (-1.1, -0.5)^T,$$

$$\partial \mathbf{g}_0 = (0.7, 0.7)^T - (11, 5)^T = (-10.3, -4.3)^T.$$

The updating formula (7.36) gives

$$\mathbf{H}_1 = \begin{pmatrix} 1 & 0 \\ 0 & 1 \end{pmatrix} + \frac{\begin{pmatrix} -10.3 \\ -4.3 \end{pmatrix}\begin{pmatrix} -10.3 \\ -4.3 \end{pmatrix}^T}{\begin{pmatrix} -10.3 \\ -4.3 \end{pmatrix}^T\begin{pmatrix} -1.1 \\ -0.5 \end{pmatrix}} - \frac{\begin{pmatrix} 1 & 0 \\ 0 & 1 \end{pmatrix}\begin{pmatrix} -1.1 \\ -0.5 \end{pmatrix}\begin{pmatrix} -1.1 \\ -0.5 \end{pmatrix}^T\begin{pmatrix} 1 & 0 \\ 0 & 1 \end{pmatrix}}{\begin{pmatrix} -1.1 \\ -0.5 \end{pmatrix}^T\begin{pmatrix} 1 & 0 \\ 0 & 1 \end{pmatrix}\begin{pmatrix} -1.1 \\ -0.5 \end{pmatrix}}$$

$$= \begin{pmatrix} 1 & 0 \\ 0 & 1 \end{pmatrix} + (13.48)^{-1}\begin{pmatrix} 106.09 & 44.29 \\ 44.29 & 18.49 \end{pmatrix} - (1.46)^{-1}\begin{pmatrix} 1.21 & 0.55 \\ 0.55 & 0.25 \end{pmatrix}$$

$$= \begin{pmatrix} 1 & 0 \\ 0 & 1 \end{pmatrix} + \begin{pmatrix} 7.87 & 3.29 \\ 3.29 & 1.37 \end{pmatrix} - \begin{pmatrix} 0.83 & 0.38 \\ 0.38 & 0.17 \end{pmatrix} = \begin{pmatrix} 8.04 & 2.91 \\ 2.91 & 2.20 \end{pmatrix}.$$

Of course, the correct Hessian anywhere is

$$\mathbf{H} = \begin{pmatrix} 8 & 3 \\ 3 & 2 \end{pmatrix}$$

since the function is quadratic. Note that the first iteration did give a positive-definite matrix and an approximation to **H** that is much better than the identity matrix. Also, no exact line search was performed. In fact, the BFGS formula performs remarkably well with low accuracy line searches. ■

Some general comments should be made about the DFP and BFGS methods. Both methods have theoretical properties that guarantee superlinear convergence rate and global convergence, under certain conditions. Theoretically, this global convergence for DFP requires exact line searches, while for BFGS Armijo–Goldstein criteria will suffice. However, both methods could fail for general nonlinear models. A particular problem is possible convergence to a saddlepoint. In Newton's method this can be detected during modifications of the Hessian and measures can be taken as pointed out in Chapter 4. But in quasi-Newton methods the Hessian is unknown and its approximations are forced to be positive-definite. So a saddlepoint may be reached without any warning. It is therefore advisable to search around the final point in case a descent direction can be found. This may be done rather easily if a Cholesky $\mathbf{LDL}^T$ factorization is used in the representation of the approximating matrices (Gill and Murray 1972). It appears, however, that such factorization schemes for quasi-Newton methods are not presenting practical advantages (Fletcher 1980).

7.4 Active Set Strategies

So far in this chapter we have examined iterative methods for unconstrained models. In the remaining sections we discuss iterative methods for constrained models. The reduced gradient and gradient projection methods were described in Section 5.5 for equality constraints. In Chapter 6 global active set strategies were studied, while local active set strategies were introduced in linear programming (Section 5.8). In the present section we discuss a generic NLP local active set strategy.

Given that a method for equality constraints is available, how do we extend such a method to handle inequalities iteratively?

The difficulty with inequalities is that we do not know which ones are active at the optimum. Such information is in itself very important for design models and a relatively large part of this book is devoted to strategies for exploring constraint activity. If we knew the active constraints a priori, for example, through monotonicity arguments, then a method for equality constraints could be applied directly. Methods for discovering global constraint activity were examined in Chapters 3 and 6. Unfortunately, the complexity of a model may prevent attainment of such global conclusions. Then an iterative procedure must be employed exploiting only *local* information computable at a given point $\mathbf{x}_k$. Apart from exceptional cases, inequality constraints satisfied as exact equalities at $\mathbf{x}_k$ may not be active at $\mathbf{x}_*$ and, conversely, active constraints at $\mathbf{x}_*$ may be strict inequalities at $\mathbf{x}_k$. We need a strategy that activates and deactivates constraints during the course of iterations while progress toward $\mathbf{x}_*$ is achieved. Such a procedure we called an *active set strategy*. This is exactly the approach we used in linear programming for moving from one face to another while decreasing the objective function.

We will assume throughout the present discussion that all equality constraints in the model are active and that all inequality constraints can be satisfied as strict equalities, meaning that an active inequality constraint will be automatically tight.

Recall that the *active set* contains all equalities and active inequalities and is often expressed as an index set, that is, containing the indices of the equalities and the active inequalities. A *working* (or *currently active*) *set* is the set of equalities and inequalities taken as active *at a given point* $\mathbf{x}_k$, not necessarily the optimal one. The ideal working set would be the active set (at $\mathbf{x}_*$). That not being the case, updates of the working set must be made by deleting or adding new constraints. A *candidate set* is the set of constraints, one (or more) of which must be selected as a new member of the working set. Typically, one would like the candidate set to be a small subset of the constraints not present in the working set. Finally, an *active set strategy* is a set of rules for changing the working set at a point $\mathbf{x}_k$ to another working set through constraint addition or deletion. Geometrically, this would mean decisions for moving from one constraint surface to another while progressing toward the optimum. In LP the selection rules for moving from one face to another constitute an active set strategy.

Adding and Deleting Constraints

A general local active set strategy works as follows. Starting at an initial feasible point and an initial working set, we minimize the objective function subject to the equalities in the working set, as shown in Figure 7.11. In this figure we start at $\mathbf{x}_0$ with g_1 in the working set. The objective function is minimized subject to $g_1 = 0$ and the points $\mathbf{x}_1$, $\mathbf{x}_2$, $\mathbf{x}_3$, $\mathbf{x}_4$ are generated by a descent algorithm. The direction

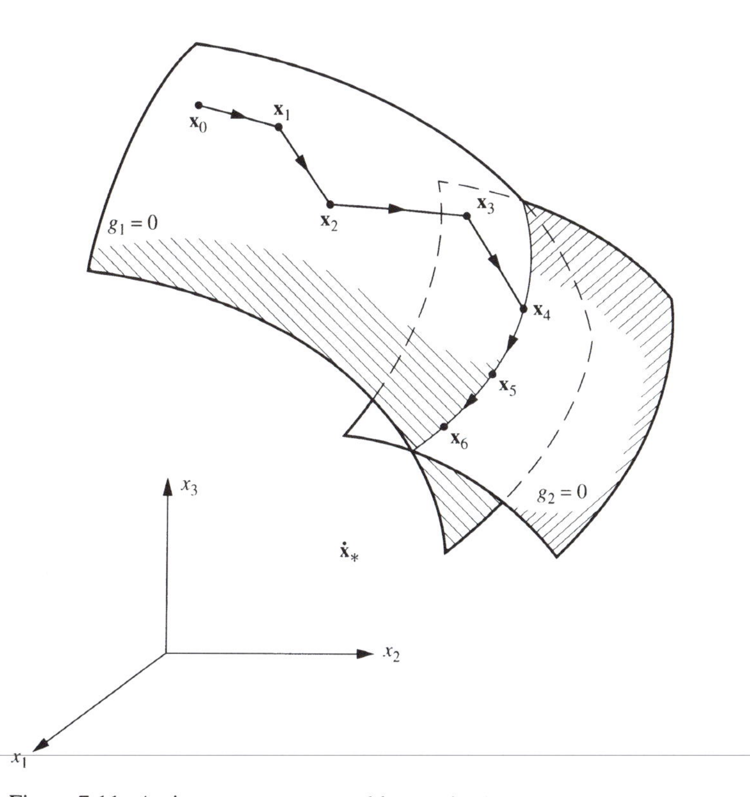

Figure 7.11. Active set strategy; working set is changed by adding a constraint.

$\mathbf{x}_4 - \mathbf{x}_3$ is a descent one and further reduction would be possible along it, except that $g_2 \leq 0$ would be violated. Thus, $\mathbf{x}_4$ is the minimum of f on $\mathbf{x}_4 - \mathbf{x}_3$ without violating feasibility. Further progress is initiated by adding the constraint g_2 to the working set at $\mathbf{x}_4$. The new working set contains g_1 and g_2. Minimization of f proceeds subject to $g_1 = 0$, $g_2 = 0$, that is, along the intersection of the two surfaces, points $\mathbf{x}_5$, $\mathbf{x}_6$. We continue in this way until no further reduction in f is possible by adding constraints in the working set. During this entire process, no constraints have been deleted. If at any point several constraints were candidates for the working set, we would select only one, say, the one promising steeper descent. Having arrived at a point $\mathbf{x}_F$ where no progress is possible by adding constraints, we may examine if further progress can be achieved by *deleting* constraints. To this end recall that the KKT condition at a feasible point $\mathbf{x}_F$ is

$$\nabla f(\mathbf{x}_F) + \sum_i \mu_i \nabla g_i(\mathbf{x}_F) = \mathbf{0}^T, \quad i \in \mathcal{A}, \tag{7.37}$$

where the equalities have been ignored for convenience and $\mathcal{A}$ is the index working set of constraints currently active at $\mathbf{x}_F$. If the point is feasible and all the multipliers are nonnegative, the point $\mathbf{x}_F$ is a KKT point and also the minimizer for the original problem, so the process terminates. If, however, a multiplier for a constraint g_i is negative, then the objective can be made smaller by moving away from $\mathbf{x}_F$ and deleting the constraint g_i from the working set. This should be evident from the view of multipliers as sensitivity coefficients. Recalling Equation (5.82), we will have for a constraint $g_i \leq 0$,

$$(\partial z/\partial g_i)_F = -(\mu_i)_F > 0, \quad \text{if } \mu_i < 0, \tag{7.38}$$

where z is the reduced objective. Moving in the negative $(\partial z/\partial g_i)_F$ direction will give a lower value for f. Geometrically, in the original $\mathbf{x}$-space, this means abandoning the surface $g_i = 0$ and moving in the interior. In Figure 7.12 such a situation is shown

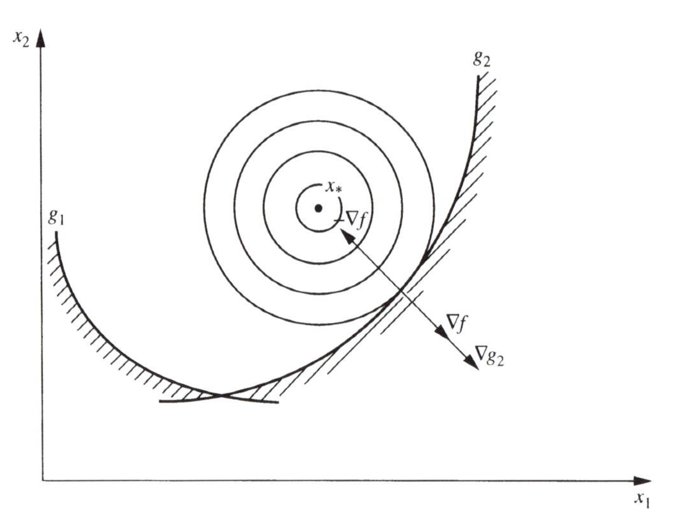

Figure 7.12. Active set strategy; working set is changed by deleting a constraint.

for two variables. In summary, constraints are added to the working set (one at a time) when they are violated, and constraints are deleted (one at a time) when corresponding multipliers are found negative at working set minimizers.

Several important remarks must be made now:

1. Convergence of such a method requires that an exact, unique, nondegenerate, global minimum is found for each working set. This is unlikely to happen in practice, but measures can be taken to assure that the working set does not change infinitely many times.
2. At nonoptimal points the values of $\boldsymbol{\lambda}$ and $\boldsymbol{\mu}$ are not the Lagrange multipliers but estimates of them. If these estimates are not good, the wrong signs may be assumed, which could lead the deleting strategy astray.
3. When several constraints have negative multiplier estimates, it is tempting to delete all of them, but this tends to invalidate global convergence properties. So usually only one is chosen. An often used rule is to delete the constraint with the most negative multiplier estimate. This is not always a good choice. However, all such strategies are heuristic and may fail occasionally.
4. If more than one constraint is violated, a heuristic choice is made to add the "most violated" one to the active set. Again, such rules do not always work.

In spite of the above difficulties, active set strategies have been implemented successfully with a variety of heuristic rules. In many special cases, the difficulties described will just not happen, as is the case with linear programming.

In an algorithm employing an active set strategy, the solution of the equality constrained problem on the working set is often called the *equality subproblem* and it is an example of an *inner iteration*. An *outer iteration* in this case will be concluded each time a KKT test is performed. A typical algorithm is outlined as follows.

ACTIVE SET ALGORITHM

1. Input initial feasible point and working set.
2. Termination test (including KKT test). If the point is not optimal, either continue with same working set or go to 7.
3. Compute a feasible search vector $\mathbf{s}_k$.
4. Compute a step length α_k along $\mathbf{s}_k$, such that $f(\mathbf{x}_k + \alpha_k\mathbf{s}_k) < f(\mathbf{x}_k)$. If α_k violates a constraint, continue; otherwise go to 6.
5. Add a violated constraint to the constraint set and reduce α_k to the maximum possible value that retains feasibility.
6. Set $\mathbf{x}_{k+1} = \mathbf{x}_k + \alpha_k\mathbf{s}_k$.
7. Change the working set (if necessary) by deleting a constraint, update all quantities, and go to step 2.

One should note that there is considerable flexibility implied in the actual algorithmic implementation of such a strategy.

Lagrange Multiplier Estimates

Let us now discuss briefly how the Lagrange multiplier estimates are calculated at an arbitrary point $\mathbf{x}$ other than $\mathbf{x}_*$. This was in fact discussed in Section 5.9 on sensitivity, but we summarize it again here in the active set strategy context.

From the KKT conditions we have (ignoring again the equality constraints for simplicity)

$$(\partial f/\partial \mathbf{x})_* = -\boldsymbol{\mu}^T(\partial \mathbf{g}/\partial \mathbf{x})_*, \tag{7.39}$$

where $\mathbf{g}$ is the vector of active inequality constraints $\mathbf{g} \leq \mathbf{0}$ (it should not be confused with the convenience symbol $\mathbf{g}^T = \nabla f$ employed in the description of unconstrained algorithms). Letting $\mathbf{x}_* = \mathbf{x} + \partial\mathbf{x}$ with a point $\mathbf{x}$ near $\mathbf{x}_*$, we perform a Taylor expansion of $(\partial f/\partial \mathbf{x})_*$ to obtain

$$(\partial f/\partial \mathbf{x})_*^T = (\partial f/\partial \mathbf{x})^T + (\partial^2 f/\partial \mathbf{x}^2)\partial \mathbf{x} \tag{7.40}$$

to first order. Similarly, we have for the constraints

$$(\partial \mathbf{g}/\partial \mathbf{x})_* = (\partial \mathbf{g}/\partial \mathbf{x}) + (\partial^2 \mathbf{g}/\partial \mathbf{x}^2)^T\, \partial \mathbf{x}, \tag{7.41}$$

where $(\partial^2\mathbf{g}/\partial\mathbf{x}^2)$ is defined as in Equation (5.24). We may ignore the second-order information about f and $\mathbf{g}$ included in (7.40) and (7.41). When $\mathbf{x}$ is far from $\mathbf{x}_*$, the equations equivalent to (7.39) at $\mathbf{x}$, namely,

$$(\partial f/\partial \mathbf{x}) + \boldsymbol{\mu}^T(\partial \mathbf{g}/\partial \mathbf{x}) = \mathbf{0}^T, \tag{7.42}$$

will not have a solution, that is, in general, no $\boldsymbol{\mu}$ will satisfy them exactly. Instead, we may get a *least squares* solution to (7.42) by solving the problem

$$\underset{\mu_j}{\text{minimize}} \sum_{i=1}^{n} \left[(\partial f/\partial x_i) + \sum_j \mu_j(\partial g_j/\partial x_i) \right]^2 . \tag{7.43}$$

The minimizing vector $\boldsymbol{\mu}$ will be the multiplier *estimate* at $\mathbf{x}$, more specifically, a *first-order estimate* of the multipliers. The values $\boldsymbol{\mu}$ thus obtained will be estimates even when the residual is zero in (7.43). In some algorithms, compatible equations (7.42) are automatically generated giving zero residuals in (7.43) but this should not be misunderstood as exact calculation of multipliers, which would require that both (7.42) and $\mathbf{g} \leq \mathbf{0}$ be satisfied exactly.

Sometimes more reliable estimates of the multipliers may be calculated using the second-order information in (7.40) and (7.41). A least squares problem such as (7.43) can be set up again. The procedure is more complicated and not always guaranteed to give better results. It may be rather efficiently employed when the constraints are all linear so that $\partial\mathbf{g}/\partial\mathbf{x}$ is fixed. For a summary of this approach, see Scales (1985) and Gill, Murray, and Wright (1981).

A final comment should be made about computational implementation of active set strategies. When changes in the working set occur, usually the sets will differ only by one constraint. The various matrices required in the computations, particularly for the multiplier estimates, do not change drastically. We saw the same situation in the LP computations, Equation (5.64). In practice, the matrices are factorized and the updated factorizations can be found with more efficient schemes than complete recomputation. For further details see Scales (1985) and Gill and Murray (1974).

7.5 Moving along the Boundary

In Section 5.5 we introduced two methods – generalized reduced gradient (GRG) and gradient projection – that are essentially strategies for moving along the boundary of the feasible space. The movement between two consecutive iterates is along a feasible arc; hence we have the name *feasible directions* for an entire class of methods.

Since the methods were described in Chapter 5, in the present short section we will make only some comments about computational implementation. Recall that the common strategy behind these methods is to make a (first-order) move along the (linearized) boundary and then return to the actual nonlinear surface by some projection that is equivalent to solving the nonlinear system of equations in an implicit elimination method (such as GRG).

First note that since these algorithms track the boundary, an initial feasible point must be found. One way to do this is to minimize a norm of constraint violations, for example, the l_1 ("taxicab") norm

$$\|\mathbf{g}\|_1 \triangleq \sum_{j=1}^{m} |g_i|.$$

For descent in the decision space, methods other than the gradient one of Section 5.5 can be used. Current GRG implementations favor safeguarded quasi-Newton iterations for evaluating the new decision vector $\mathbf{d}_{k+1}$. In principle, any unconstrained method can be used for iterating in the decision space.

The inner iterations required for return to feasibility are usually Newton–Raphson (N–R) iterations [recall Equation (5.41)]. Since the basic N–R method may diverge, the algorithm may fail at that point. This is one of the drawbacks of the feasible direction-type methods. One remedy is to shorten the step size until N–R convergence is realized. Unfortunately, this lowers the efficiency of the algorithm. Sometimes the step size reductions lead to very small values for $\partial \mathbf{x}_k$ and/or ∂f_k, and the algorithm fails to progress. In these cases, a restart from another point may be the best remedy. One should also recall that specifically for GRG algorithms, failure due to poor variable partitioning is not uncommon.

In spite of these difficulties, computer code implementations of GRG methods for general NLP problems have proven remarkably competitive in tests and in practice

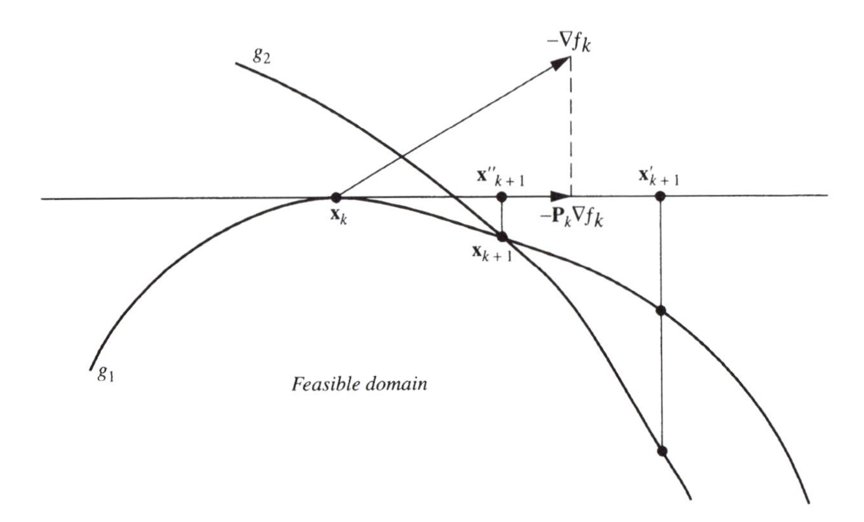

Figure 7.13. Step size reduction to maintain feasibility, or constraint addition to the working (active) set.

(see, e.g., Lasdon et al. 1978). Several popular codes for engineering applications are of the GRG type. Also, gradient projection algorithms have been used extensively and successfully. Current thinking favors GRG versions modified to handle simple bounds separately and to take advantage of sparsity in large problems. Convergence rates of these algorithms using BFGS methods for the unconstrained subproblem can be superlinear.

When an active set strategy is used to deal with inequalities, all methods may suffer a reduction in theoretical efficiency owing to extra step length calculations required as new nonlinear constraints are encountered. Convergence rates then tend to be only linear. An example of required step size reduction is given in Figure 7.13.

Yet active set strategies can be very advantageous in design problems with large numbers of constraints – far exceeding the number of variables – particularly when the computation of the constraint functions is expensive. The actual strategies tend to have a heuristic character in the way they alter the working set. This may be unappealing to the theoretically inclined but it offers efficiency of computation for practical problems, at the expense of a less elegant algorithm.

7.6 Penalties and Barriers

In contrast to the boundary tracking methods, we may consider minimizing the objective function while maintaining explicit control over the violations of constraints by *penalizing* the objective at points that violate or tend to violate the constraints. Such methods are generally called *penalty transformation* methods and represent one of the oldest classes of optimization techniques. This section gives only a review of the main ideas.

Consider the problem

$$\min f(\mathbf{x}) \quad (7.44)$$
$$\text{subject to } \mathbf{g}(\mathbf{x}) \leq \mathbf{0}.$$

A general transformation may be

$$\min T(\mathbf{x}, \mathbf{r}) \triangleq f(\mathbf{x}) + F(\mathbf{g}(\mathbf{x}), \mathbf{r}), \quad (7.45)$$

where $\mathbf{r}$ is a general vector of *controlling parameters* and F is a real-valued function, which imposes the penalty in a way controlled by $\mathbf{r}$. A local minimum for T, if it exists, is denoted by $\mathbf{x}(\mathbf{r})$. For an appropriate choice of F and $\mathbf{r}$, any good unconstrained method should solve the transformed problem without any further effort.

Two approaches have been developed in the choices of transformation algorithms. The first requires choosing an F and a sequence $\{\mathbf{r}_k\}$ such that a sequence $\{\mathbf{x}(\mathbf{r}_k)\}$ is determined to have as a limit the true $\mathbf{x}_*$ of (7.44), as $k \to \infty$. Methods with this approach are called *sequential penalty transformations*. The second approach requires an F, $\mathbf{r}_*$, and δ such that the condition $\|\mathbf{r} - \mathbf{r}_*\| < \delta$ implies $\mathbf{x}(\mathbf{r}) = \mathbf{x}_*$. This means that no limiting procedure is needed, but rather a threshold value $\mathbf{r}_*$, is required to reach $\mathbf{x}_*$. Methods using this approach are called, for this reason, *exact penalty transformations*. Thus, the unconstrained minimizer of an exact penalty transformation for a certain penalty parameter can be the constrained minimizer of the original problem. Typically, exact penalty transformations are either nondifferentiable or need first derivatives, with second derivative information required for efficiency of solution procedure. The sequential methods were developed first and are described in detail in an early work by Fiacco and McCormick (1968) as Sequential Unconstrained Minimization Techniques (SUMT). The exact penalty methods are relatively more recent. Here we will demonstrate the basic penalty transformation idea for sequential penalty methods.

There are two major classes within the sequential methods. The first is characterized by the property of preserving constraint feasibility at all times. They are referred to as *barrier function* methods, or *interior-point penalty function* methods. From the optimal design viewpoint, such methods are generally preferable, if one decides to use the penalty approach, because even when they do not converge, they still terminate at a useful feasible design. The second class uses a sequence of infeasible points and feasibility is obtained only at the optimum (although some methods can generate feasible points). They are referred to as *penalty function* methods, or *exterior-point penalty function* methods. Their primary use has been in solving equality constrained problems.

Barrier Functions

A barrier transformation has the general form

$$T(\mathbf{x}, r) \triangleq f(\mathbf{x}) + rB(\mathbf{x}), \quad r > 0, \quad (7.46)$$

where $B(\mathbf{x})$ is the *barrier function* defined to be positive in the interior of the constraint set $\mathbf{g}(\mathbf{x}) < \mathbf{0}$ and to approach infinity as $\mathbf{x}$ approaches the boundary. Note that a barrier transformation cannot be used for equality constrained problems. Typical barrier functions are the logarithmic

$$B(\mathbf{x}) \triangleq -\sum_{j=1}^{m} \ln[-g_j(\mathbf{x})] \tag{7.47}$$

and the inverse

$$B(\mathbf{x}) \triangleq -\sum_{j=1}^{m} g_j^{-1}(\mathbf{x}). \tag{7.48}$$

The desired behavior of the function is obtained if the sequence $\{r_k\}$ converges to zero. We can see that if g_j is active, then $g_j(\mathbf{x}) \to 0$ as $\mathbf{x} \to \mathbf{x}_*$ and this implies $B(\mathbf{x}) \to \infty$ for both the above functions. But, if the sequence $\{r_k\} \to 0$, then the growth of $B(\mathbf{x})$ is arrested and the value of T will approach the value of $f(\mathbf{x})$, as $\mathbf{x}$ approaches $\mathbf{x}_*$.

In any design situation we need to know which constraints are active. Barrier methods will produce estimates of the Lagrange multipliers, for example,

$$\mu_j(r_k) = -r_k/g_j, \tag{7.49a}$$

$$\mu_j(r_k) = -r_k/g_j^2 \tag{7.49b}$$

when (7.47) or (7.48) are used, respectively. The actual multiplier can be obtained at the limit

$$\mu_{j*} = \lim_{k\to\infty} \mu_j(r_k). \tag{7.50}$$

A barrier function algorithm will then have the following steps:

1. Find an interior point $\mathbf{x}_0$. Select a monotonically decreasing sequence $\{r_k\} \to 0$ for $k \to \infty$. Set $k = 0$.
2. At iteration k, minimize the function $T(\mathbf{x}, r_k)$ using an unconstrained method and $\mathbf{x}_k$ as the starting point. The solution $\mathbf{x}_*(r_k)$ is set equal to $\mathbf{x}_{k+1}$.
3. Perform a convergence test. If the test is not satisfied, set $k = k + 1$ and return to step 2.

Example 7.7 Consider the problem

$$\begin{aligned} &\text{minimize} f = 3x_1^2 + x_2^2 \\ &\text{subject to } g = 2 - x_1 - x_2 \le 0, \quad x_1 > 0, x_2 > 0. \end{aligned}$$

Use the logarithmic barrier function Equation (7.47) to define the equivalent problem

$$\min T(\mathbf{x}, r) = 3x_1^2 + x_2^2 - r \ln(-2 + x_1 + x_2).$$

We may solve this problem by taking a sequence $\{r_k\}$ with limit zero. However, since the problem is simple, let us state the stationarity conditions for $T(\mathbf{x}, r)$:

$$6x_1 + r(2 - x_1 - x_2)^{-1} = 0,$$
$$2x_2 + r(2 - x_1 - x_2)^{-1} = 0.$$

These can be solved exactly to give

$$3x_1 = x_2, \quad 2x_1^2 - x_1 - (r/12) = 0.$$

Solving further for x_1 as a function of r,

$$x_1(r) = 0.25 + 0.25(1 + 0.666r)^{1/2},$$

we get $\lim x_1(r) = 0.5$ for $r \to 0$. Then $x_{1*} = 0.5$ and $x_{2*} = 1.5$. An estimate of the multiplier as a function of r is found from Equation (7.49a):

$$\mu(r) = -r[2 - x_1(r) - x_2(r)]^{-1} = -r[2 - 4x_1(r)]^{-1},$$

$$\lim_{r\to 0} \mu(r) = \lim_{r\to 0} \frac{r}{4x_1(r) - 2} = \lim_{r\to 0} \frac{1}{4(\partial x_1/\partial r)} = 3$$

since $\partial x_1/\partial r = (\frac{1}{4})(\frac{1}{2})(\frac{2}{3})(1 + 0.666r)^{-1/2}$. These values can be easily verified using the KKT conditions for the original problem. ■

The method appears very attractive for a general-purpose NLP program. Unfortunately, major computational difficulties are inherent to the basic method. As r_k becomes smaller, a severe eccentricity is introduced in $T(\mathbf{x}, r)$ and an attendant ill-conditioning of the Hessian makes the minimization of $T(\mathbf{x}, r)$ near the solution very difficult (Figure 7.14).

The difficulty of slow convergence near the solution can be met in various ways. One is to control the diminishing of r within a fixed percentage of its previous value. Another is to switch to a Newton-type method to accelerate convergence. A third idea

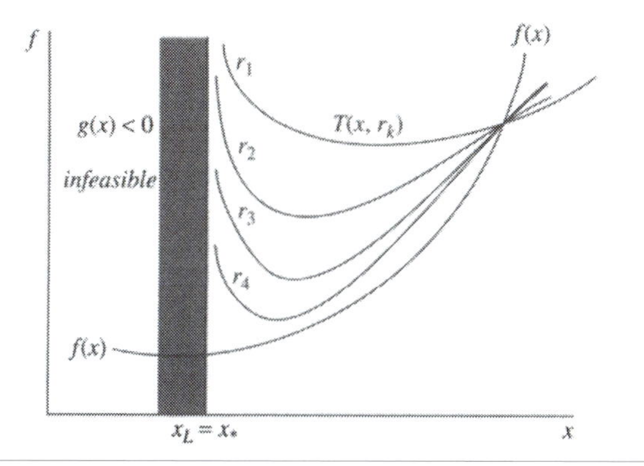

Figure 7.14. Increasing eccentricity in the transformation $T(x, r_k)$ that utilizes a barrier function, as $r_k \to 0$.

is to extrapolate the barrier function in the infeasible region, after a certain transition point, and use the modified barrier instead of the original one. This gives rise to so-called linear or quadratic *extended* barrier functions.

Penalty Functions

Typical penalty function transformations are of the form

$$T(\mathbf{x}, r) \triangleq f(\mathbf{x}) + r^{-1} P(\mathbf{x}), \quad r > 0, \tag{7.51}$$

where the *penalty function* $P(\mathbf{x})$ is nonnegative for all $\mathbf{x}$ in $\Re^n$ and $P(\mathbf{x}) = 0$ if and only if $\mathbf{x}$ is an interior point. For example, we have the *quadratic loss function* for inequality constraints:

$$P(\mathbf{x}) \triangleq \sum_{j=1}^{m} [\max\{0, g_j(\mathbf{x})\}]^2. \tag{7.52}$$

For equality constraints, an appropriate penalty function would be

$$P(\mathbf{x}) \triangleq \sum_{j=1}^{m} [h_j(\mathbf{x})]^2. \tag{7.53}$$

For decreasing sequences $\{r_k\}$, the behavior of these transformations is similar to that for barrier functions. The same eccentricity occurs near the solution (Figure 7.15). Lagrange multiplier estimates can be obtained as well. For example, if the quadratic loss function is used, then

$$[\mu_j]_k = (2/r_k) \max\{0, g_j(\mathbf{x}_k)\}. \tag{7.54}$$

Example 7.8 Consider again the problem in Example 7.7. The penalty transformation with quadratic loss penalty function is

$$\min T(\mathbf{x}, r) = 3x_1^2 + x_2^2 + r^{-1}(2 - x_1 - x_2)^2.$$

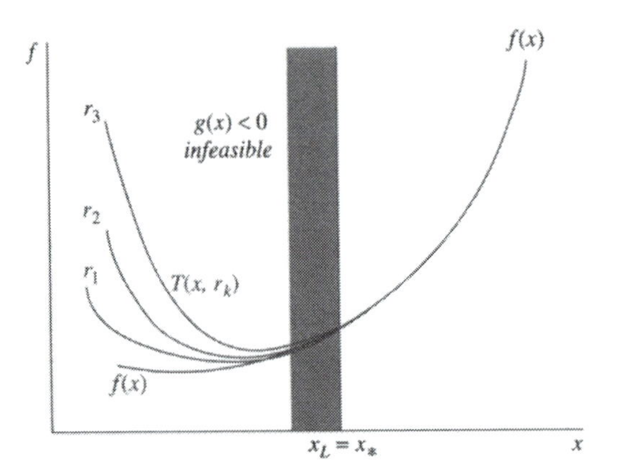

Figure 7.15. Ill-conditioning of transformation using penalty functions.

The stationary points are found from

$$6x_1 - 2r^{-1}(2 - x_1 - x_2) = 0,$$
$$2x_2 - 2r^{-1}(2 - x_1 - x_2) = 0,$$

giving $3x_1 = x_2$ and $x_1 = 2/(3r + 4)$. At $r \to 0$ we find $x_{1*} = 0.5$ and $x_{2*} = 1.5$. The multiplier is found from Equation (7.54),

$$\begin{aligned}\mu_* &= \lim_{r\to 0}[(2/r)\max\{0, g(r)\}]\\ &= \lim_{r\to 0}\left[\left(\frac{2}{r}\right)\max\left\{0, \frac{6r}{3r+4}\right\}\right] = \lim_{r\to 0}\left[\frac{12}{3r+4}\right] = 3.\end{aligned}$$

Again, in an iterative scheme, the values for x_1, x_2, and μ found above would be estimates of the optimal ones, since $r \neq 0$. ■

Augmented Lagrangian (Multiplier) Methods

The ill-conditioning of the original penalty transformation methods may be avoided by generating a new class of methods in which the penalty term is added not to the objective function but to the Lagrangian function of the problem. These methods operate by updating the estimates of the multipliers and so they are known both as *augmented Lagrangian* and as *multiplier methods*.

We will describe the rationale behind these methods for the case of equality constraints. Recall from Chapter 5 that the optimality conditions for the problem $\{\min f(\mathbf{x}), \text{ subject to } \mathbf{h}(\mathbf{x}) = \mathbf{0}\}$ are equivalent to having a stationary point for the Lagrangian at $\mathbf{x}_*$ and a positive curvature of the Lagrangian at $\mathbf{x}_*$ on the subspace $\nabla\mathbf{h}^T\partial\mathbf{x} = \mathbf{0}$ (tangent to the constraint surface). Thus, a function that could augment the Lagrangian should enforce the stationarity of $\mathbf{x}_*$ by altering only the Hessian of the Lagrangian projected on the tangent subspace. This implies that the function should have the following properties at $\mathbf{x}_*$: (a) a stationary point; (b) positive curvature in the tangent subspace; and (c) zero curvature in the normal subspace.

Such a function, multiplied by an appropriately large parameter, can be added to the Lagrangian and the "augmented" Lagrangian will have a minimum at $\mathbf{x}_*$. Another way to view this construction is that the augmentation *shifts* the origin of the penalty term so that an optimum value for the transformed function can be found without the penalty parameter going to the limit.

A function that can be used for augmentation is

$$P = \mathbf{h}^T\mathbf{h} = \|\mathbf{h}\|^2.$$

We can see easily that

$$[\partial P(\mathbf{x})/\partial\mathbf{x}]_* = 2[(\nabla\mathbf{h})\mathbf{h}]_* = \mathbf{0}^T$$

and

$$\left(\frac{1}{2}\right)(\partial^2 P/\partial\mathbf{x}^2)_* = \nabla\mathbf{h}_*\nabla\mathbf{h}_*^T,$$

which on the tangent subspace becomes

$$\nabla \mathbf{h}_*^T(\nabla \mathbf{h}_* \nabla \mathbf{h}_*^T)\nabla \mathbf{h}_* = (\nabla \mathbf{h}_*^T \nabla \mathbf{h}_*)^2$$

and is positive-definite. On the normal space we have $\partial \mathbf{x}^T(\nabla \mathbf{h}_* \nabla \mathbf{h}_*^T)\partial \mathbf{x} = (\nabla \mathbf{h}_*^T \partial \mathbf{x})(\nabla \mathbf{h}_*^T \partial \mathbf{x}) = 0$. The desired augmented Lagrangian function is thus

$$T(\mathbf{x}, \boldsymbol{\lambda}, r) = f(\mathbf{x}) + \boldsymbol{\lambda}^T \mathbf{h}(\mathbf{x}) + r^{-1} \sum_{j=1}^{m} [h_j(\mathbf{x})]^2 \tag{7.55}$$

for $r > 0$. Note that for $\lambda_j = 0,\ j = 1, \ldots, m$, this reduces to a usual penalty transformation, while for $\boldsymbol{\lambda} = \boldsymbol{\lambda}_*$ a minimization of T with respect to $\mathbf{x}$ will give the solution for *any* value of r (provided the minimizer is feasible). Hence, we have the motivation for the following iterative scheme:

1. Given an initial choice for $\boldsymbol{\lambda}$ and r minimize T wrt $\mathbf{x}$ and find a point $\mathbf{x}(\boldsymbol{\lambda}, r)_\dagger$.
2. Update $\boldsymbol{\lambda}$ and r so that $\boldsymbol{\lambda}_k \to \boldsymbol{\lambda}_*$ and repeat the minimization wrt $\mathbf{x}$, finding a new $\mathbf{x}_\dagger$.
3. Stop when termination criteria are satisfied.

The initial choice for $\boldsymbol{\lambda}$ may be an estimate of the multipliers, but the proper choice of r can be significant. As with other penalty methods, proper choice of penalty parameters is often problem dependent and a result of experience. A typical updating strategy for r must guarantee feasibility of the limit point; for example, one can set $r_{k+1} = \gamma r_k,\ 0 < \gamma < 1$, when $\|\mathbf{h}(\mathbf{x}_{k+1})\| > \delta \|\mathbf{h}(\mathbf{x}_k)\|$ with $\delta > 1$. The reader is further referred to more specialized texts. The important point is that convergence will occur for small values of r but not so small that ill-conditioning will result.

The updating of the multipliers may be derived as follows. From the stationarity condition of the augmented Lagrangian we have, for an *estimated* λ_j,

$$(\partial f/\partial x_i) + \sum_j (\lambda_j + 2r^{-1}h_j)(\partial h_j/\partial x_i) = 0, \tag{7.56}$$

while for an *optimal* λ_{j^*}, we have

$$(\partial f/\partial x_i) + \sum_j \lambda_{j^*}(\partial h_j/\partial x_i) = 0. \tag{7.57}$$

Then we expect that as $\mathbf{x}(\boldsymbol{\lambda}, r)_\dagger$ approaches $\mathbf{x}_*$, so $\lambda_j + 2r^{-1}h_j$ must approach λ_{j^*}. We then set the updating as

$$[\lambda_j]_{k+1} = [\lambda_j + 2r^{-1}h_j]_k. \tag{7.58}$$

More efficient updating schemes for λ_j may be devised by employing Newton-like or quasi-Newton methods to solve an associated *dual* problem in each iteration, but discussion of general duality is not included in this book.

The method may be extended to inequalities with the aid of an active set strategy. Details of implementation may be found again in specialized texts (e.g., Pierre and Lowe 1975, Bertsekas 1982).

Example 7.9 Consider the problem

$$\begin{aligned} &\min f = x_1^2 - x_2^2 \\ &\text{subject to } h = x_2 + 2 = 0. \end{aligned} \tag{a}$$

The augmented Lagrangian is given by

$$T(\mathbf{x}, \lambda, r) = x_1^2 - x_2^2 + \lambda(x_2 + 2) + r^{-1}(x_2 + 2)^2. \tag{b}$$

Then we have $\partial T/\partial \mathbf{x} = \mathbf{0}^T$, giving

$$2x_1 = 0, \quad -2x_2 + \lambda + 2r^{-1}(x_2 + 2) = 0 \tag{c}$$

or

$$x_1 = 0, \quad x_2 = (4r^{-1} + \lambda)/(2 - 2r^{-1}). \tag{d}$$

Let us select $r_0 = 8$, $\lambda_0 = -2$, which gives $\mathbf{x}_0 = (0, -0.86)^T$, $h_0 = 1.14$. Iterating on the multiplier estimate according to Equation (7.58), we get $\lambda_1 = (-2) + (\frac{2}{8})(1.14) = -1.715$. Keeping $r_1 = 8$, we get, from (d), $\mathbf{x}_1 = (0, -0.694)^T$ and $h_1 = 1.306$. Note that for $r = 1$, x_2 would be undefined, even if (c) were used. Note also that the Hessian of the augmented function (b) is

$$\partial^2 T/\partial \mathbf{x}^2 = \begin{pmatrix} 2 & 0 \\ 0 & -2 + 2r^{-1} \end{pmatrix}.$$

So if $-2 + 2r^{-1} < 0$, there will be no minimizer for T. This happens if $r > 1$, so the iteration above is inappropriate. The value of r should be kept small and be sufficiently reduced to maintain adequate reduction of infeasibility at each iteration. ■

It should be emphasized that the performance of augmented Lagrangian methods depends critically on the accuracy of the multiplier estimates. The update given in Equation (7.58) will tend to make the convergence rate of the overall algorithm only linear. Faster rates can be obtained by using quasi-Newton methods to solve the unconstrained minimization of $T(\mathbf{x}, \boldsymbol{\lambda}, r)$, while second-order updates for $\boldsymbol{\lambda}$ are enforced.

7.7 Sequential Quadratic Programming

This section deals with a class of methods that became prominent in the 1970s and contains what are considered the most efficient general purpose NLP algorithms today. We will first give the motivation for the approach and then describe an implementation.

The Lagrange–Newton Equations

Consider the equality constrained problem

$$\begin{aligned}&\min f(\mathbf{x})\\&\text{subject to } \mathbf{h}(\mathbf{x}) = \mathbf{0}.\end{aligned} \tag{7.59}$$

The stationarity condition for the Lagrangian of this problem with respect to both $\mathbf{x}$ and $\boldsymbol{\lambda}$ is $\nabla L(\mathbf{x}_*, \boldsymbol{\lambda}_*) = \mathbf{0}^T$. We may choose to solve this equation using Newton's method to update $\mathbf{x}$ and $\boldsymbol{\lambda}$. This is done by using the Taylor expansion to first order

$$[\nabla L(\mathbf{x}_k + \partial\mathbf{x}_k, \boldsymbol{\lambda}_k + \partial\boldsymbol{\lambda}_k)]^T = \nabla L_k^T + \nabla^2 L_k(\partial\mathbf{x}_k, \partial\boldsymbol{\lambda}_k)^T \tag{7.60}$$

and setting the left-hand side $\nabla L_{k+1}^T = 0$. Then Equation (7.60) gives

$$\nabla^2 L_k \begin{pmatrix} \partial\mathbf{x}_k \\ \partial\boldsymbol{\lambda}_k \end{pmatrix} = -\nabla L_k^T. \tag{7.61}$$

We compute easily that

$$\nabla^2 L_k = \begin{pmatrix} \nabla^2{}_{\mathbf{x}} L & \nabla^2_{\mathbf{x}\boldsymbol{\lambda}} L \\ \nabla^2_{\boldsymbol{\lambda}\mathbf{x}} L & \nabla^2_{\boldsymbol{\lambda}} L \end{pmatrix}_k = \begin{pmatrix} \nabla^2 f + \boldsymbol{\lambda}^T \nabla^2 \mathbf{h} & \nabla \mathbf{h}^T \\ \nabla \mathbf{h} & \mathbf{0} \end{pmatrix}_k, \tag{7.62}$$

$$\nabla L_k = (\nabla f^T + \nabla \mathbf{h}^T \boldsymbol{\lambda}, \mathbf{h})_k^T \quad \text{(transposed elements)}. \tag{7.63}$$

Define further for convenience the shorthand

$$\mathbf{W} \triangleq \nabla^2 f + \boldsymbol{\lambda}^T \nabla^2 \mathbf{h} \quad \text{and} \quad \mathbf{A} \triangleq \nabla \mathbf{h}. \tag{7.64}$$

Then Equation (7.61) is written as

$$\begin{pmatrix} \mathbf{W}_k & \mathbf{A}_k^T \\ \mathbf{A}_k & \mathbf{0} \end{pmatrix} \begin{pmatrix} \partial\mathbf{x}_k \\ \partial\boldsymbol{\lambda}_k \end{pmatrix} = \begin{pmatrix} -\nabla f_k^T - \mathbf{A}_k^T \boldsymbol{\lambda}_k \\ -\mathbf{h}_k \end{pmatrix}. \tag{7.65}$$

Setting $\partial\mathbf{x}_k = \mathbf{s}_k$ and $\partial\boldsymbol{\lambda}_k = \boldsymbol{\lambda}_{k+1} - \boldsymbol{\lambda}_k$, we can rewrite Equation (7.65) as

$$\begin{pmatrix} \mathbf{W}_k & \mathbf{A}_k^T \\ \mathbf{A}_k & \mathbf{0} \end{pmatrix} \begin{pmatrix} \mathbf{s}_k \\ \boldsymbol{\lambda}_{k+1} \end{pmatrix} = \begin{pmatrix} -\nabla f_k^T \\ -\mathbf{h}_k \end{pmatrix}. \tag{7.66}$$

Solving Equation (7.66) iteratively, we obtain the iterates $\mathbf{x}_{k+1} = \mathbf{x}_k + \mathbf{s}_k$ and $\boldsymbol{\lambda}_{k+1}$, which eventually should approach $\mathbf{x}_*$ and $\boldsymbol{\lambda}_*$. Thus, any method solving Equation (7.66) can be referred to as a *Lagrange–Newton* method for solving the constrained problem (7.59). The solution will be unique if $\mathbf{A}_*$ has full rank (regularity assumption) and $\mathbf{W}_*$ is positive-definite on the tangent subspace.

We may write Equation (7.66) explicitly as

$$\begin{aligned}&\mathbf{W}_k \mathbf{s}_k + \mathbf{A}_k^T \boldsymbol{\lambda}_{k+1} + \nabla f_k^T = \mathbf{0},\\&\mathbf{A}_k \mathbf{s}_k + \mathbf{h}_k = \mathbf{0}\end{aligned} \tag{7.67}$$

and observe that they may be viewed as the KKT conditions for the quadratic model

$$\begin{aligned} &\min_{\mathbf{s}_k} q(\mathbf{s}_k) = f_k + \nabla_{\mathbf{x}} L_k \mathbf{s}_k + \tfrac{1}{2}\mathbf{s}_k^T \mathbf{W}_k \mathbf{s}_k \\ &\text{subject to } \mathbf{A}_k \mathbf{s}_k + \mathbf{h}_k = \mathbf{0}, \end{aligned} \tag{7.68}$$

where $\nabla_{\mathbf{x}} L_k = \nabla f_k + \boldsymbol{\lambda}_k^T \nabla \mathbf{h}_k$ and the multipliers of problem (7.68) are $\partial\boldsymbol{\lambda}_k$. In fact, the Lagrangian stationarity condition of (7.68) is

$$\nabla_{\mathbf{x}} L_k + \mathbf{s}_k^T \mathbf{W}_k + (\partial\boldsymbol{\lambda}_k)^T \mathbf{A}_k = \mathbf{0}^T$$

or, transposing,

$$\nabla f_k^T + \mathbf{A}_k^T \boldsymbol{\lambda}_k + \mathbf{W}_k \mathbf{s}_k + \mathbf{A}_k^T(\boldsymbol{\lambda}_{k+1} - \boldsymbol{\lambda}_k) = \mathbf{0}, \tag{7.69}$$

which is easily reduced to the first of Equations (7.67). Thus, solving the *quadratic programming subproblem* (7.68) gives $\mathbf{s}_k$ and $\boldsymbol{\lambda}_{k+1}$ exactly as solving Equation (7.67), and the two formulations are equivalent. If the second formulation is selected for solving the Lagrange–Newton equations, the values of $\mathbf{x}_*$ and $\boldsymbol{\lambda}_*$ will be obtained from solving a sequence of quadratic programming (QP) subproblems; hence, the relevant algorithms are known as *sequential quadratic programming* (SQP) methods. The QP subproblem may not be exactly as in (7.68). For example, the subproblem

$$\begin{aligned} &\min_{\mathbf{s}_k} q(\mathbf{s}_k) = f_k + \nabla f_k \mathbf{s}_k + \tfrac{1}{2}\mathbf{s}_k^T \mathbf{W}_k \mathbf{s}_k \\ &\text{subject to } \mathbf{A}_k \mathbf{s}_k + \mathbf{h}_k = \mathbf{0} \end{aligned} \tag{7.70}$$

gives also a solution $\mathbf{s}_k$ and multipliers $\boldsymbol{\lambda}_{k+1}$ directly, rather than $\partial\boldsymbol{\lambda}_k$ as (7.68) does.

A simple SQP algorithm has the following general structure.

SQP ALGORITHM (WITHOUT LINE SEARCH)

1. Select initial point $\mathbf{x}_0, \boldsymbol{\lambda}_0$; let $k = 0$.
2. For $k = k + 1$, solve the QP subproblem and determine $\mathbf{s}_k$ and $\boldsymbol{\lambda}_{k+1}$.
3. Set $\mathbf{x}_{k+1} = \mathbf{x}_k + \mathbf{s}_k$.
4. If termination criteria are not satisfied, return to 2.

Enhancements of the Basic Algorithm

The QP subproblem (7.68) is minimizing the quadratic approximation to the Lagrangian subject to the linearized constraints. A solution exists if $\mathbf{W}_k$ is positive-definite, a condition implicit in Equation (7.67) for stability of Newton's method. Local convergence of the algorithm can be proven to be quadratic if $\mathbf{W}_k$ behaves properly. The difficulty, however, is with global convergence. For points far from $\mathbf{x}_*$, the QP subproblem may have an unbounded or infeasible solution. A stabilization procedure must be devised. One possibility is to view $\mathbf{s}_k$ as a search direction and define the iteration as $\mathbf{x}_{k+1} = \mathbf{x}_k + \alpha_k \mathbf{s}_k$, where the step size α_k results from minimizing an appropriate function along the search direction $\mathbf{s}_k$. This function (called a *merit function*) should measure how good $\mathbf{x}_k$ and $\boldsymbol{\lambda}_k$ are (including feasibility), and it should be locally minimized at the solution. This points to some form of penalty function that properly weighs objective function decrease and constraint violations. Penalizing constraint violations only could lead to stationary points that are not minima.

For example, the merit function

$$\phi(\mathbf{x}, \boldsymbol{\lambda}) = \|\nabla_{\mathbf{x}} L(\mathbf{x}, \boldsymbol{\lambda})\|^2 + \|\mathbf{h}(\mathbf{x})\|^2 \tag{7.71}$$

representing the KKT conditions' violation would not be very satisfactory. The choice for merit function varies in different SQP implementations. We will give a suggested one later in this subsection.

Another point of practical interest is the evaluation of $\mathbf{W}_k$ in the QP subproblem. The obvious enhancement is to avoid computing second derivatives and employ an updating formula that approximates $\mathbf{W}_k$ as in the quasi-Newton methods. One possibility is to use the DFP formula Equation (7.35) with $\partial \mathbf{g}_k$ defined by (Han 1976)

$$\partial \mathbf{g}_k = \nabla L(\mathbf{x}_{k+1}, \boldsymbol{\lambda}_{k+1})^T - \nabla L(\mathbf{x}_k, \boldsymbol{\lambda}_{k+1})^T. \tag{7.72}$$

Another possibility (Powell 1978a,b) is to use the BFGS formula Equation (7.36) and keep the approximating matrix positive-definite by using the linear combination

$$\partial \mathbf{g}_k = \theta_k \mathbf{y}_k + (1 - \theta_k)\mathbf{H}_k \partial \mathbf{x}_k, \quad 0 \le \theta \le 1, \tag{7.73}$$

where

$$\mathbf{y}_k = \nabla L(\mathbf{x}_{k+1}, \boldsymbol{\lambda}_{k+1})^T - \nabla L(\mathbf{x}_k, \boldsymbol{\lambda}_{k+1})^T, \tag{7.74}$$

$$\theta_k = \begin{cases} 1 & \text{if } \partial \mathbf{x}_k^T \mathbf{y}_k \ge (0.2)\partial \mathbf{x}_k^T \mathbf{H}_k \partial \mathbf{x}_k, \\ \dfrac{(0.8)\partial \mathbf{x}_k^T \mathbf{H}_k \partial \mathbf{x}_k}{\partial \mathbf{x}_k^T \mathbf{H}_k \partial \mathbf{x}_k - \partial \mathbf{x}_k^T \mathbf{y}_k} & \text{if } \partial \mathbf{x}_k^T \mathbf{y}_k < (0.2)\partial \mathbf{x}_k^T \mathbf{H}_k \partial \mathbf{x}_k, \end{cases}$$

and $\partial \mathbf{x}_k = \mathbf{x}_{k+1} - \mathbf{x}_k$; $\mathbf{H}_k$ is the current BFGS approximation to the Hessian of the Lagrangian. The values (0.2) and (0.8) result from numerical experiments. The choice of θ_k is made to assure that $\partial \mathbf{g}_k$ remains as close to $\mathbf{y}_k$ as possible, while $\partial \mathbf{x}_k^T \mathbf{y}_k \ge (0.2)\partial \mathbf{x}_k^T \mathbf{H}_k \partial \mathbf{x}_k$ is always satisfied. That way, $\partial \mathbf{x}^T \partial \mathbf{g}_k > 0$ is preserved and $\mathbf{H}_{k+1}$ will be always positive-definite if $\mathbf{H}_k$ is as well.

In the construction above, $\partial \mathbf{x}_k$ is the search direction $\mathbf{s}_k$ if no line search is performed. However, when a merit function is used for line search, $\partial \mathbf{x}_k = \alpha_k \mathbf{s}_k$ and the updating of $\mathbf{H}_k$ is done after the line search is completed. The new $\mathbf{H}_{k+1}$ approximates $\mathbf{W}_{k+1}$ in the next QP subproblem (7.68) or (7.70).

An alternative to solving the QP subproblem is to solve a linear least squares subproblem based on a Cholesky $\mathbf{LDL}^T$ factorization of the approximating Hessian (Schittkowski 1981).

The presentation of the algorithm so far was for equality constraints. For the general NLP problem

$$\begin{aligned} &\min f(\mathbf{x}) \\ &\text{subject to } \mathbf{h}(\mathbf{x}) = \mathbf{0}, \\ &\qquad\qquad \mathbf{g}(\mathbf{x}) \leq \mathbf{0}, \end{aligned} \tag{7.75}$$

the inequality constraints may be handled in two different ways. First, an active set strategy may be employed on the original problem (7.75) so that the QP subproblem of the inner iteration will be always equality constrained. This is sometimes called a *preassigned active set strategy* in SQP. The merit function must then include *all* constraints, active and inactive, to guard against failure when the wrong active set is used to determine $\mathbf{s}_k$. The second way to handle inequalities is to pose the QP subproblem with the linearized inequalities included, that is, $\mathbf{A}_k\mathbf{s}_k + \mathbf{g}_k \leq \mathbf{0}$, and use an active set strategy on the subproblem. The resulting active set for the subproblem may be employed as a prediction of the active set for the original problem, but it is not clear if this is in fact an advantage over the first method.

When the QP subproblem contains inequalities, a merit function using the exact penalty idea has been proposed (Powell 1978a) and successfully implemented in a widely used algorithm (Crane, Hillstrom, and Minkoff 1980). This function is as follows:

$$\phi(\mathbf{x}, \boldsymbol{\lambda}, \boldsymbol{\mu}) = f(\mathbf{x}) + \sum_{j=1}^{m_1} w_j|h_j| + \sum_{j=1}^{m_2} w_j|\min\{0, -g_j\}|, \tag{7.76}$$

where m_1 and m_2 are the numbers of equality and inequality constraints and w_j are *weights* used to balance the infeasibilities. The suggested values are

$$\begin{aligned} &w_j = |\lambda_j| \quad \text{for } k = 0 \text{ (first iteration)}, \\ &w_{j,k} = \max\{|\lambda_{j,k}|, 0.5(w_{j,k-1} + |\lambda_{j,k}|)\} \quad \text{for } k \geq 1, \end{aligned} \tag{7.77}$$

where μ_j would be used for the inequalities ($j = 1, \ldots, m_2$). This rather obscure choice was reported as a result of numerical experiments (Powell 1978a). A merit function using an augmented Lagrangian formulation has also been proposed and implemented (Schittkowski 1981, 1983, 1984).

In summary, a complete SQP algorithm will have the following steps.

SQP ALGORITHM (WITH LINE SEARCH)

1. Initialize.
2. Solve the QP subproblem to determine a search direction $\mathbf{s}_k$.
3. Minimize a merit function along $\mathbf{s}_k$ to determine a step length α_k.
4. Set $\mathbf{x}_{k+1} = \mathbf{x}_k + \alpha_k\mathbf{s}_k$.
5. Check for termination. Go to 2 if not satisfied.

Solving the Quadratic Subproblem

A final question is how to solve efficiently the quadratic programming subproblem. The methods we described for general NLP in previous sections (e.g., reduced space or projection methods or augmented Lagrangian methods) can be specialized to QP. As an example of this, consider the QP problem

$$\begin{aligned} &\min q(\mathbf{x}) = \tfrac{1}{2}\mathbf{x}^T\mathbf{Q}\mathbf{x} + \mathbf{c}^T\mathbf{x} \\ &\text{subject to } \mathbf{h}(\mathbf{x}) = \mathbf{A}\mathbf{x} - \mathbf{b} = \mathbf{0}. \end{aligned} \tag{7.78}$$

The Lagrange–Newton equations for this problem are simply

$$\begin{pmatrix} \mathbf{Q} & \mathbf{A}^T \\ \mathbf{A} & \mathbf{0} \end{pmatrix}\begin{pmatrix} \mathbf{x} \\ \boldsymbol{\lambda} \end{pmatrix} = \begin{pmatrix} -\mathbf{c} \\ \mathbf{b} \end{pmatrix} \tag{7.79}$$

[compare with Equation (7.65)]. The *Lagrangian matrix* of coefficients to the left of Equation (7.79) is symmetric but may not be positive-definite. If it can be inverted, the solution to the QP problem is

$$\begin{pmatrix} \mathbf{x} \\ \boldsymbol{\lambda} \end{pmatrix} = \begin{pmatrix} \mathbf{Q} & \mathbf{A}^T \\ \mathbf{A} & \mathbf{0} \end{pmatrix}^{-1}\begin{pmatrix} -\mathbf{c} \\ \mathbf{b} \end{pmatrix} \triangleq \begin{pmatrix} \mathbf{H} & \mathbf{T}^T \\ \mathbf{T} & \mathbf{U} \end{pmatrix}\begin{pmatrix} -\mathbf{c} \\ \mathbf{b} \end{pmatrix}, \tag{7.80}$$

where the element matrices in the inverse can be explicitly calculated as

$$\begin{aligned} \mathbf{H} &= \mathbf{Q}^{-1} - \mathbf{Q}^{-1}\mathbf{A}^T(\mathbf{A}\mathbf{Q}^{-1}\mathbf{A}^T)^{-1}\mathbf{A}\mathbf{Q}^{-1}, \\ \mathbf{T}^T &= \mathbf{Q}^{-1}\mathbf{A}^T(\mathbf{A}\mathbf{Q}^{-1}\mathbf{A}^T)^{-1}, \\ \mathbf{U} &= -(\mathbf{A}\mathbf{Q}^{-1}\mathbf{A}^T)^{-1}. \end{aligned} \tag{7.81}$$

These expressions assume of course that $\mathbf{Q}^{-1}$ exists. This should be generally satisfied by the construction of the QP subproblem in an SQP algorithm. The optimizer for the QP is explicitly calculated from Equation (7.80):

$$\begin{aligned} \mathbf{x}_* &= -\mathbf{H}\mathbf{c} + \mathbf{T}^T\mathbf{b}, \\ \boldsymbol{\lambda}_* &= -\mathbf{T}\mathbf{c} + \mathbf{U}\mathbf{b}. \end{aligned} \tag{7.82}$$

These expressions can be used to solve the QP problem directly (Fletcher 1971), but in practice the solution of Equation (7.79) would be done by an efficient factorization, such as a modified $\mathbf{LDL}^T$ one. A dual method particularly suitable for SQP-generated subproblems is one proposed by Goldfarb and Idnani (1983).

Example 7.10 Let us consider the following problem:

$$\begin{aligned} \min f &= x_1x_2^{-2} + x_2x_1^{-1} \\ \text{subject to } h &= 1 - x_1x_2 = 0, \\ g &= 1 - x_1 - x_2 \le 0, \\ & x_1, x_2 > 0. \end{aligned} \tag{a}$$

We can find its solution quickly by setting $x_2 = 1/x_1$ from $h = 0$ and eliminating x_2 to get

$$\begin{aligned} &\min f = x_1^3 + x_1^{-2} \\ &\text{subject to } g = -x_1^2 + x_1 - 1 \le 0. \end{aligned} \tag{b}$$

The constraint is a quadratic with negative discriminant so it can be active only for imaginary x_1; moreover, it would be always negative for real x_1, so it is inactive. The solution to (b) is then the stationary point $x_{1*} = 0.922$, and the solution to (a) is $\mathbf{x}_* = (0.922, 1.08)^T$.

We will now perform an SQP iteration with line search starting at the feasible point $\mathbf{x}_0 = (2, 0.5)^T$. To formulate the QP subproblem, we evaluate

$$f_0 = 8.25, \quad \nabla f_0 = \left(x_2^{-2} - x_2 x_1^{-2}, -2x_1 x_2^{-3} + x_1^{-1}\right)_0 = (3.87, -31.5),$$
$$\nabla \mathbf{g} = (\nabla h, \nabla g)^T = [(-x_2, -x_1), (-1, -1)]^T.$$

To evaluate $\nabla_{\mathbf{x}} L_0 = \nabla f_0 + \lambda_0 \nabla h_0 + \mu_0 \nabla g_0$, we need estimates for the multipliers λ_0 and μ_0. To make the calculations easy let us take $\lambda_0 = \mu_0 = 0$. Then $\nabla_{\mathbf{x}} L_0 = \nabla f_0 = (3.87, -31.5)$. The Hessian of the Lagrangian $\mathbf{W}_0$ will be equal to the Hessian of f, for zero multipliers, so

$$\mathbf{W}_0 = \nabla^2 f_0 = \begin{pmatrix} 2x_2 x_1^{-3} & -2x_2^{-3} - x_1^{-2} \\ -2x_2^{-3} - x_1^{-2} & 6x_1 x_2^{-4} \end{pmatrix}_0$$
$$= \begin{pmatrix} 0.125 & -16.25 \\ -16.25 & 192.0 \end{pmatrix}.$$

Finally,

$$\mathbf{A}_0 = \nabla \mathbf{g}_0 = [(-0.5, -2), (-1, -1)]^T,$$
$$h_0 = 0, \quad g_0 = -1.5,$$

so that the QP subproblem is

$$\begin{aligned} &\min q = f_0 + \nabla_{\mathbf{x}} L_0 \mathbf{s} + \frac{1}{2}\mathbf{s}^T \mathbf{W}_0 \mathbf{s} \\ &\text{subject to } \nabla h_0 \mathbf{s} + h_0 = 0, \\ &\qquad \nabla g_0 \mathbf{s} + g_0 \le 0, \end{aligned}$$

or

$$\min q = 8.25 + (3.87, -31.5)\begin{pmatrix} s_1 \\ s_2 \end{pmatrix}$$
$$+ \frac{1}{2}\begin{pmatrix} s_1 \\ s_2 \end{pmatrix}^T \begin{pmatrix} 0.125 & -16.25 \\ -16.25 & 192.0 \end{pmatrix} \begin{pmatrix} s_1 \\ s_2 \end{pmatrix}$$

$$\begin{aligned} &\text{subject to } (-0.5, -2)(s_1, s_2)^T + 0 = 0, \\ &\qquad (-1, -1)(s_1, s_2)^T - 1.5 \le 0, \end{aligned}$$

or

$$\min q = 8.25 + 3.87s_1 - 31.5s_2 + 0.062s_1^2 - 16.25s_1s_2 + 96s_2^2$$
$$\text{subject to } -0.5s_1 - 2s_2 = 0,$$
$$-s_1 - s_2 - 1.5 \leq 0.$$

This is solved by any method to give $\mathbf{s}_0 = (-0.580, 0.145)^T$. The associated multipliers can be easily found to be $\lambda_0 = 2.88$, $\mu_0 = 0$.

To determine the step size we can use the merit function, Equation (7.76),

$$\phi_0(\mathbf{x}, \lambda, \mu) = f_0(\mathbf{x}) + w_1|h_0| + w_2|\min\{0, -g_0\}|,$$

where $w_1 = |\lambda_0| = 2.88$, $w_2 = |\mu_0| = 0$. Since at $\mathbf{x}_0$ we have $h_0 = 0$, the merit function reduces to just the objective function. This is as it should be since no infeasibilities are present at $\mathbf{x}_0$. Then the step size is found from

$$\min_{\alpha \geq 0} f(\mathbf{x}_0 + \alpha\mathbf{s}_0) = \frac{(2 - 0.580\alpha)}{(0.5 + 0.145\alpha)^2} + \frac{(0.5 + 0.145\alpha)}{(2 - 0.580\alpha)}.$$

This minimization can be done with one of the inexact line search procedures described in Section 7.2 with the additional restriction of possible upper bounds on α to avoid violating a new constraint. ■

7.8 Trust Regions with Constraints

The restricted step or trust region concept introduced in Section 4.8 can be used for constrained problems as well. When constraints are present simply adding the constraint $\|\mathbf{s}\| \leq \Delta$ will not suffice. If $\mathbf{x}_k$ is infeasible and the trust region radius Δ is sufficiently small, then the resulting quadratic program will be infeasible. We will look at two ways to deal with infeasibility in trust region algorithms.

Relaxing Constraints

Here the constraints are relaxed by introducing a parameter α. The resulting quadratic program appears as follows:

$$\min f \approx \mathbf{gs} + \frac{1}{2}\mathbf{s}^T\mathbf{Hs} \tag{7.83}$$

subject to

$$\nabla\mathbf{gs} + \alpha\mathbf{g} \leq \mathbf{0}, \tag{7.84}$$
$$\nabla\mathbf{hs} + \alpha\mathbf{h} = \mathbf{0},$$
$$\|\mathbf{s}\| \leq \Delta.$$

The parameter α always has a value between 0 and 1. It is chosen so that the intersection of the trust region and the relaxed constraints is nonempty. But α must not be so small that the resulting step does not move toward the feasible region.

Theoretically, every step can be uniquely divided into two components: one component in the space spanned by the gradients of the active set of constraints (the "normal" component $\mathbf{s}_{\text{normal}}$) and the other lying in the null space. Then

$$\mathbf{s} = \mathbf{s}_{\text{normal}} + \mathbf{s}_{\text{null}}. \tag{7.85}$$

The normal component of $\mathbf{s}$ must advance toward the feasible region, so Byrd, Schnabel, and Shultz (1987) introduced a requirement that every step have a finite contribution, that is,

$$\|\mathbf{s}_{\text{normal}}\| \geq \theta\Delta, \tag{7.86}$$

where θ is a strictly positive number. Similar requirements are used in trust region methods such as by Alexandrov (1993) and El-Alem (1988). Requirement (7.85) is often used to produce a "two-step" algorithm in which the normal and null space steps are calculated separately, as in Maciel (1992).

Using Exact Penalty Functions

Another approach to dealing with infeasibility introduced by Fletcher (1981) is to replace the quadratic program with an approximation of an exact penalty function. Using a maximum-violation penalty function with a penalty parameter σ and the infinity norm $\|\mathbf{x}\|_\infty \triangleq \max_i\{|x_i|\}$, steps could be calculated by

$$\begin{aligned} &\min \phi(\mathbf{s}) \approx \mathbf{gs} + \frac{1}{2}\mathbf{s}^T\mathbf{Hs} + \sigma \max(0, \|(\nabla\mathbf{gs} + \mathbf{g})^+\|_\infty, \|\nabla\mathbf{hs} + \mathbf{h}\|_\infty) \\ &\text{subject to } \|\mathbf{s}\| \leq \Delta. \end{aligned} \tag{7.87}$$

The superscript "+" means that only the violated inequality constraints will be considered. Thus every point within the trust region is feasible, and if the penalty parameter is large enough, the constraint violation will automatically decrease.

Solving (7.86) is about as computationally expensive as solving a quadratic program. One can show that near the optimum the solution to (7.86) and the solution to a quadratic program are identical; thus the local convergence behavior is the same.

Modifying the Trust Region and Accepting Steps

Regardless of how the step is calculated, a penalty function is always used when deciding if a step is acceptable and in altering the size of the trust region. A step is usually accepted if there is any improvement in a particular penalty function (i.e., $\Phi(\mathbf{x}_k) - \Phi(\mathbf{x}_k + \mathbf{s}_k) > 0$, where Φ is some penalty function). However, to modify the size of the trust region, we must discern how good the step calculation is at producing descent for the penalty function.

An approximation of the penalty function is built from the objective and constraint functions used in the step calculation. For the step calculation given in (7.86) the actual penalty function may be

$$\Phi(\mathbf{x}) = f(\mathbf{x}) + \sum_j \mu_j \max(0, g_j(\mathbf{x})) + \sum_j \lambda_j |h_j(\mathbf{x})|, \tag{7.87}$$

where the μ_j and λ_j are weights. The approximate penalty function would then be

$$\phi(\mathbf{s}) = \mathbf{gs} + \frac{1}{2}\mathbf{s}^T\mathbf{Hs} + \sum_j \mu_j \max(0, \nabla g_j\mathbf{s} + g_j) + \sum_j \lambda_j |\nabla h_j\mathbf{s} + h_j|. \quad (7.88)$$

By construction, the quantity $\phi(\mathbf{0}) - \phi(\mathbf{s}_k) > 0$, so we compare how the two penalty functions have changed using the ratio r_k, where r_k = actual/predicted, as it was used in Section 4.8, but now,

$$\text{actual reduction} = \Phi(\mathbf{x}_k) - \Phi(\mathbf{x}_k + \mathbf{s}_k) \quad (7.89)$$

and

$$\text{predicted reduction} = \phi(\mathbf{0}) - \phi(\mathbf{s}_k). \quad (7.90)$$

The approximation is considered good for producing descent if r_k is about one or larger and the trust region is increased. If the ratio r_k is small or negative, then the approximation is considered inaccurate and the trust region is reduced.

Yuan's Trust Region Algorithm

An algorithm proposed by Yuan (1995) exemplifies the general approach taken in modern trust region algorithms. The underlying approximate problem is a nonsmooth minimization problem of the approximate function $\varphi(\mathbf{s})$ as follows:

$$\begin{aligned} &\min_{\mathbf{s}\in\Re^n} \varphi(\mathbf{s}) = \nabla f(\mathbf{x}_k)\mathbf{s} + \frac{1}{2}\mathbf{s}^T\mathbf{B}_k\mathbf{s} + \sigma_k\|(\nabla\mathbf{c}(\mathbf{x}_k)\mathbf{s} + \mathbf{c}(\mathbf{x}_k))^+\|_\infty \\ &\text{subject to } \|\mathbf{s}\|_\infty \le \Delta_k. \end{aligned} \quad (7.91)$$

Here $\mathbf{c}$ is the vector of all equality and inequality constraints, σ_k is the penalty parameter, Δ_k is the trust region radius, $\mathbf{x}_k$ is the current iterate, and $\mathbf{B}_k$ is the current approximation for the Hessian of the Lagrangian.

To evaluate the accuracy of the approximate model, the predicted change in the approximate model $(\varphi(\mathbf{0}) - \varphi(\mathbf{s}_k))$ is compared to the actual change $(\Phi(\mathbf{x}_k) - \Phi(\mathbf{x}_k + \mathbf{s}_k))$, as mentioned above:

$$r_k = \frac{\Phi(\mathbf{x}_k) - \Phi(\mathbf{x}_k + \mathbf{s}_k)}{\varphi(\mathbf{0}) - \varphi(\mathbf{s}_k)}. \quad (7.92)$$

If the ratio r_k is near unity, the approximate model is deemed accurate, and the trust region is increased; if the ratio r_k is small or negative, the approximate model is inaccurate and the trust region is reduced according to the following rule:

$$\Delta_{k+1} = \begin{cases} \max\{2\Delta_k, 4\|\mathbf{s}_k\|_\infty\}, & 0.9 < r_k, \\ \Delta_k, & 0.1 \le r_k \le 0.9, \\ \max\left\{\dfrac{\Delta_k}{4}, \dfrac{\|\mathbf{s}_k\|_\infty}{2}\right\}, & r_k < 0.1. \end{cases} \quad (7.93)$$

The flow chart in Figure 7.16 portrays the basic steps in Yuan's algorithm as given below.

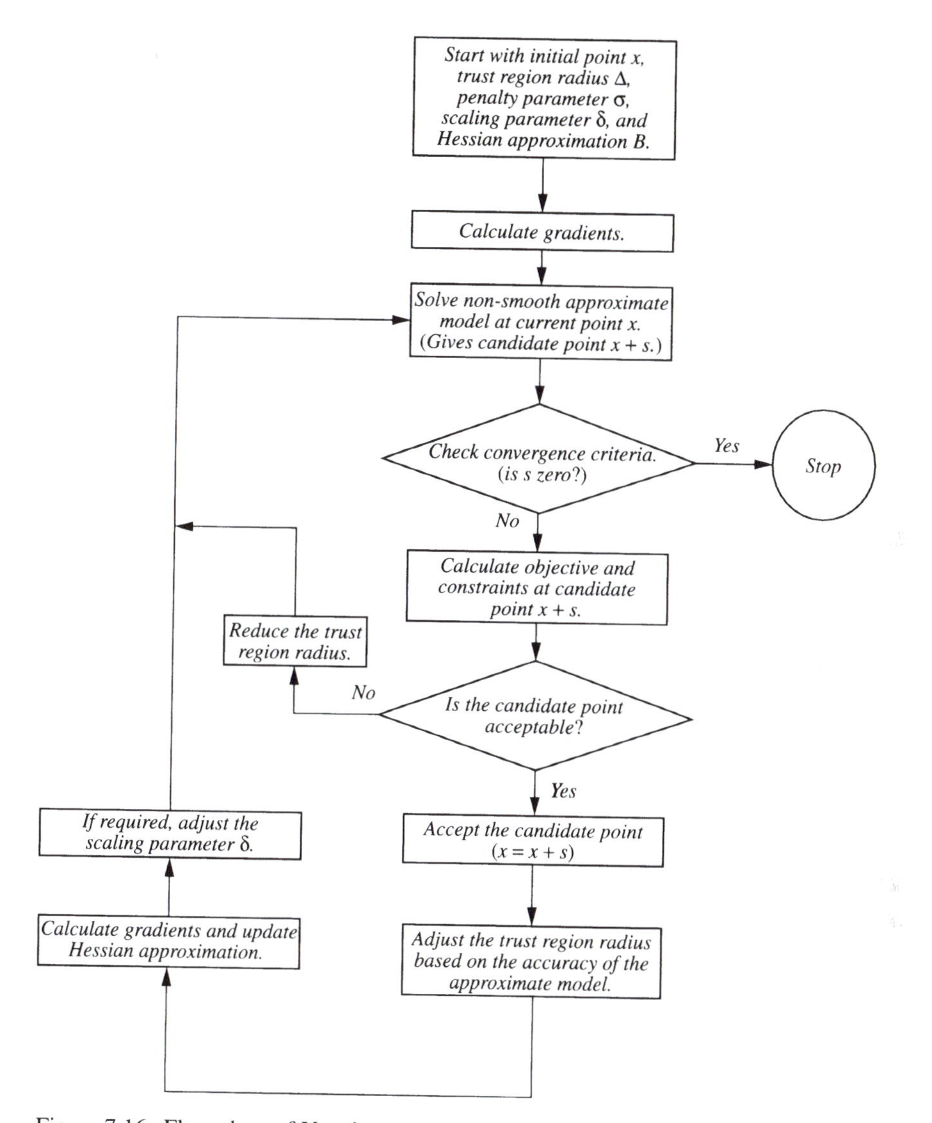

Figure 7.16. Flow chart of Yuan's trust region algorithm.

CONSTRAINED TRUST REGION ALGORITHM (AFTER YUAN)

1. Set the iteration counter $k = 1$. Choose some $\mathbf{x}_1$ as an initial point and a symmetric $\mathbf{B}_1$ as an initial Hessian estimate. Also define a scaling parameter $\delta_1 > 0$, a penalty parameter $\sigma_1 > 0$, and a trust region radius $\Delta_1 > 0$.
2. At $\mathbf{x}_k$ solve the approximate problem in Equation (7.91) and denote the solution $\mathbf{s}_k$. If $\mathbf{s}_k = \mathbf{0}$ then stop.
3. Calculate the penalty function $\Phi_k(\mathbf{x})$, the approximate penalty function $\varphi_k(\mathbf{s})$, and the ratio of the two, r_k. If $r_k > 0$ accept the step (go to Step 4); otherwise reduce the trust region and try again (set $\Delta_{k+1} = \|\mathbf{s}_k\|_\infty/4$, $\mathbf{x}_{k+1} = \mathbf{x}_k$, $k = k + 1$, and go to Step 2.)

4. Alter the trust region radius Δ_k according to Equation (7.93).
5. Define the next point as $\mathbf{x}_{k+1} = \mathbf{x}_k + \mathbf{s}_k$.
6. Generate $\mathbf{B}_{k+1}$ with a quasi-Newton update as in Section 7.3.
7. If $\varphi_k(\mathbf{0}) - \varphi_k(\mathbf{s}_k) < \delta_k \sigma_k \min\{\Delta_k, \|\mathbf{c}(\mathbf{x}_k)^+\|_\infty\}$ then set $\sigma_{k+1} = 2\sigma_k$ and $\delta_{k+1} = \delta_k/4$; otherwise, set $\sigma_{k+1} = \sigma_k$ and $\delta_{k+1} = \delta_k$.
8. Set $k = k + 1$ and go to Step 2.

Several variants can be created based on this algorithm to take advantage of special problem structures, such as hierarchical decomposition (Nelson 1997).

7.9 Convex Approximation Algorithms

Convex approximation algorithms is an important class of trust region algorithms that was developed independently over a period of many years, motivated by problems in optimal structural design. Early structural optimization problems involved sizing of truss, beam, plate, and other structures using a minimum weight objective function and constraints on stresses and deflections. Such problems can have quite complicated analysis models, typically using finite element formulations. However, the underlying mathematical functions tend to be monotonic with respect to the design variables, and the optimal solutions are constraint-bound or nearly so. Therefore, algorithms that efficiently identify active constraints are very effective on such problems.

One way to develop such algorithms is to solve a sequence of approximation problems that preserve the monotonicity of the original functions. For example, a general NLP model can be approximated by an LP model where the objective and all the constraints are linearized at an initial point using a Taylor expansion. The solution of the LP problem is checked against the KKT conditions, and if it does not satisfy them a new LP approximation is computed at that point. This algorithm is known as *sequential linear programming* (SLP) and generally is *not* convergent unless a trust region is utilized to control the quality of the approximation. The trust region is imposed in the form of simple bounds on the variables; these traditionally have been called *move limits*.

To account for curvature information more involved algorithms have been created using a variety of convex approximations to the original NLP (Fleury 1982). These *convex approximation* (CA) *algorithms* do not use quadratic Taylor approximations to capture curvature behavior. Rather, they employ special nonlinear approximations that preserve the monotonic behavior of the original functions and are easy to compute based only on first-order information.

Convex Linearization

A common CA algorithm is the convex linearization algorithm CONLIN (Fleury and Braibant 1986). An optimal structural design problem is given in negative null form as

$$\begin{aligned}&\text{minimize } f_0(\mathbf{x})\\&\text{subject to } f_j(\mathbf{x}) \le 0, \quad j = 1, \ldots, m,\\&\qquad\qquad LB_i \le x_i \le UB_i \quad i = 1, \ldots, n,\end{aligned} \tag{7.94}$$

where LB_i and UB_i are simple bounds on the design variables.

The objective and constraint functions are approximated depending on the sign of the derivative with respect to the design variables. If the derivative is positive, the function is approximated with respect to the design variable whereas if it is negative the function is approximated with respect to the *reciprocal* of the design variable. Thus the following convex approximations are obtained:

$$f_j^k(\mathbf{x}) = r_j^k + \sum_{i=1}^{n} \left(P_{ji}^k x_i + \frac{Q_{ji}^k}{x_i} \right), \tag{7.95}$$

where

$$P_{ji}^k = \begin{cases} \frac{\partial f_j(\mathbf{x}^k)}{\partial x_i}, & \text{if } \frac{\partial f_j(\mathbf{x}^k)}{\partial x_i} \ge 0 \\ 0, & \text{if } \frac{\partial f_j(\mathbf{x}^k)}{\partial x_i} < 0 \end{cases}, \tag{7.96a}$$

$$Q_{ji}^k = \begin{cases} 0, & \text{if } \frac{\partial f_j(\mathbf{x}^k)}{\partial x_i} \ge 0 \\ -\left(x_i^k\right)^2 \frac{\partial f_j(\mathbf{x}^k)}{\partial x_i}, & \text{if } \frac{\partial f_j(\mathbf{x}^k)}{\partial x_i} < 0 \end{cases}, \tag{7.96b}$$

$$\begin{aligned}&r_j^k = f_j(\mathbf{x}^k) - \sum_{i=1}^{n} \left(P_{ji}^k x_i^k + \frac{Q_{ji}^k}{x_i^k} \right),\\&\alpha_i^k \le x_i \le \beta_i^k, \quad i = 1, \ldots, n \quad \text{and} \quad j = 0, \ldots, m.\end{aligned} \tag{7.96c}$$

Here we are modifying the notation to avoid too many multiple subscripts. So the iteration counter k is placed as a superscript. Then $\mathbf{x}^k$ is the current design variable, $f_j^k(\mathbf{x})$ represent the approximated objective and constraint functions, $f_j(\mathbf{x}^k)$ and $\partial f_j(\mathbf{x}^k)/\partial \mathbf{x}$ are the actual functions and their gradients evaluated at $\mathbf{x}^k$, and α_i^k and β_i^k are the *move limits*. The problem in (7.96) can be solved using special convex optimization methods.

Moving Asymptotes

The basic CONLIN will not work well when there are nonmonotonic constraints or when the optimum is not truly constraint-bound. Oscillations near the optimum occur when there is a change in the sign of the partial derivative of one of the design variables. Also, convergence is slow if the approximation is too conservative.

To account for some of these difficulties Svanberg (1987) proposed the *method of moving asymptotes* (MMA). The problem functions are approximated with respect to $1/(U_i - x_i)$ or $1/(x_i - L_i)$, where U_i and L_i are the "moving" asymptotes that can be adjusted during iterations analogously to the move limits. The resulting approximation functions are

$$f_j^k(\mathbf{x}) = r_j^k + \sum_{i=1}^{n} \left(\frac{P_{ji}^k}{U_i^k - x_i} + \frac{Q_{ji}^k}{x_i - L_i^k} \right), \tag{7.97}$$

where

$$P_{ji}^k = \begin{cases} \left(U_i^k - x_i^k\right)^2 \frac{\partial f_j(\mathbf{x}^k)}{\partial x_i}, & \text{if } \frac{\partial f_j(\mathbf{x}^k)}{\partial x_i} \geq 0 \\ 0, & \text{if } \frac{\partial f_j(\mathbf{x}^k)}{\partial x_i} < 0 \end{cases}, \tag{7.98a}$$

$$Q_{ji}^k = \begin{cases} 0, & \text{if } \frac{\partial f_j(\mathbf{x}^k)}{\partial x_i} \geq 0 \\ -\left(x_i^k - L_i^k\right)^2 \frac{\partial f_j(\mathbf{x}^k)}{\partial x_i}, & \text{if } \frac{\partial f_j(\mathbf{x}^k)}{\partial x_i} < 0 \end{cases}, \tag{7.98b}$$

$$r_j^k = f_j(\mathbf{x}^k) - \sum_{i=1}^{n} \left(\frac{P_{ji}^k}{U_i^k - x_i^k} + \frac{Q_{ji}^k}{x_i^k - L_i^k} \right),$$
$$L_i^k < \alpha_i^k \leq x_i \leq \beta_i^k < U_i^k, \quad i = 1, \ldots, n \quad \text{and} \quad j = 0, \ldots, m. \tag{7.98c}$$

Here U_i^k and L_i^k are the current *upper and lower moving asymptotes*, respectively, while α_i^k and β_i^k are the move limits.

The upper and lower moving asymptotes use the curvature information of the approximate functions to prevent oscillation. If the asymptotes are very close to the design variables convergence is slow, and thus at the next iteration, the asymptotes are relaxed to coax faster convergence.

The resulting approximate problem now becomes

$$\begin{aligned} &\min f_0^k(\mathbf{x}) \\ &\text{subject to } f_j^k(\mathbf{x}) \leq 0, \quad j = 1, \ldots, m, \\ &\qquad L_i^k < \alpha_i^k \leq x_i \leq \beta_i^k < U_i^k, \quad i = 1, \ldots, n, \end{aligned} \tag{7.99}$$

where the functions are all convex. The special techniques employed to solve this problem are based on *duality theory*, a topic not discussed in this book. The main idea in duality is that the so-called primal problem (7.99) has an associated *dual* problem, where *the variables are the Lagrange multipliers* $\boldsymbol{\mu}$ *of the primal constraints*. Under convexity, the primal and the dual problems have the same optimal value for their corresponding objectives. Moreover, the optimal primal variables can be computed from the optimal dual variables. An advantage is gained when the dual is a much simpler problem to solve.

The dual of the convex primal (7.99) is the maximization problem below, where superscripts k are dropped for simplicity.

$$\max W(\boldsymbol{\mu}) = r_0 + \boldsymbol{\mu}^T \mathbf{r} + \sum_{i=1}^{n} W_i(\mathbf{x}, \boldsymbol{\mu}), \quad \boldsymbol{\mu} \geq \mathbf{0} \tag{7.100}$$

$$W_i(\mathbf{x}, \boldsymbol{\mu}) = \min_{x_i} \left(\frac{P_{0i} + \boldsymbol{\mu}^T \mathbf{P}_i}{U_i - x_i} + \frac{\mathbf{Q}_{0i} + \boldsymbol{\mu}^T \mathbf{Q}_i}{x_i - L_i} \right)$$

subject to $\max\{\alpha_i, LB_i\} \leq x_i \leq \min\{\beta_i, UB_i\}$,

where

$$\mathbf{r} = (r_1, \ldots, r_m)^T, \quad \mathbf{P}_i = (P_{1i}, \ldots, P_{mi})^T,$$
$$\mathbf{Q}_i = (Q_{1i}, \ldots, Q_{mi})^T, \quad \boldsymbol{\mu} = (\mu_1, \ldots, \mu_m)^T.$$

The dual function W is dependent on two variables, $\mathbf{x}$ and $\boldsymbol{\mu}$, but $\mathbf{x}$ is optimized out by solving the minimization problem first. An expression for $W_i(\mathbf{x}, \boldsymbol{\mu})$ is found if the derivatives with respect to x_i are set to zero:

$$\frac{P_{0i} + \boldsymbol{\mu}^T \mathbf{P}_i}{(U_i - x_i)^2} - \frac{Q_{0i} + \boldsymbol{\mu}^T \mathbf{Q}_i}{(x_i - L_i)^2} = 0, \tag{7.101}$$

giving the closed-form solution

$$x_i(\boldsymbol{\mu}) = \frac{(P_{0i} + \boldsymbol{\mu}^T \mathbf{P}_i)^{1/2} L_i + (Q_{0i} + \boldsymbol{\mu}^T \mathbf{Q}_i)^{1/2} U_i}{(P_{0i} + \boldsymbol{\mu}^T \mathbf{P}_i)^{1/2} + (Q_{0i} + \boldsymbol{\mu}^T \mathbf{Q}_i)^{1/2}}. \tag{7.102}$$

The resulting "pure" dual function (expressed only in terms of $\boldsymbol{\mu}$) is convex and is subject only to nonnegativity constraints:

$$\max W(\boldsymbol{\mu}) = r_0 + \boldsymbol{\mu}^T \mathbf{r} + \sum_{i=1}^{n} \left(\frac{P_{0i} + \boldsymbol{\mu}^T \mathbf{P}_i}{U_i - x_i(\boldsymbol{\mu})} + \frac{Q_{0i} + \boldsymbol{\mu}^T \mathbf{Q}_i}{x_i(\boldsymbol{\mu}) - L_i} \right) \tag{7.103}$$

subject to $\boldsymbol{\mu} \geq 0$.

This dual problem is solved for the optimal dual variables, and the optimal primal design variables are obtained using (7.102). This is done iteratively until convergence is achieved. Note that the number of dual variables is equal to the number of constraints in the primal.

The method of moving asymptotes is a generalization of the other convex approximation methods. For instance, if both moving asymptotes are chosen to be $-\infty$ and $+\infty$, respectively, then MMA becomes SLP. If the lower moving asymptote (L_i) is chosen to be 0 and the upper moving asymptote (U_i) to be $+\infty$, then MMA becomes CONLIN.

Choosing Moving Asymptotes and Move Limits

The success of MMA depends on how the moving asymptotes are chosen. Several methods have been proposed on choosing moving asymptotes. The simplest form proposed by Svanberg is given below.

For $k = 0$ and $k = 1$,

$$L_i^k = x_i^k - (UB_i - LB_i), \qquad U_i^k = x_i^k + (UB_i - LB_i).$$

For $k \geq 2$,

1. If $(x_i^k - x_i^{k-1})$ and $(x_i^{k-1} - x_i^{k-2})$ have opposite signs, implying oscillation, the asymptotes are chosen as follows:

$$L_i^k = x_i^k - s\left(x_i^{k-1} - L_i^{k-1}\right), \quad U_i^k = x_i^k + s\left(U_i^{k-1} - x_i^{k-1}\right).$$

2. If $(x_i^k - x_i^{k-1})$ and $(x_i^{k-1} - x_i^{k-2})$ have equal signs, implying that the asymptotes are slowing down convergence, then

$$L_i^k = x_i^k - \left(x_i^{k-1} - L_i^{k-1}\right)/s, \quad U_i^k = x_i^k + \left(U_i^{k-1} - x_i^{k-1}\right)/s,$$

where s is a fixed real number less than 1.

Another method proposed by Jiang (1996) consists of a two-method strategy. The first method ensures that the first derivative of the approximate function is equal to that of the original function at the previous and current iteration points. This results in the following asymptotes:

$$L_{ji}^k = x_i^k + \frac{x_i^k - x_i^{k-1}}{C_{ji}^{1/2} - 1}, \quad U_{ji}^k = x_i^k + \frac{x_i^k - x_i^{k-1}}{C_{ji}^{1/2} - 1},$$

where

$$C_{ji} = \frac{\frac{\partial f_j(\mathbf{x}^k)}{\partial x_i}}{\frac{\partial f_j(\mathbf{x}^{k-1})}{\partial x_i}}. \tag{7.104}$$

The C_{ji} are the coefficients of the moving asymptotes. These coefficients determine how aggressive or conservative the approximations are.

The second method uses the approximate second derivative information of the functions. This results in the following asymptotes:

$$L_{ji}^k = x_i^k + \frac{2\frac{\partial f_j(\mathbf{x}^k)}{\partial x_i}}{H_{ji}}, \quad \text{if} \quad \frac{\partial f_j(\mathbf{x}^k)}{\partial x_i} \le 0,$$

$$U_{ji}^k = x_i^k + \frac{2\frac{\partial f_j(\mathbf{x}^k)}{\partial x_i}}{H_{ji}}, \quad \text{if} \quad \frac{\partial f_j(\mathbf{x}^k)}{\partial x_i} > 0, \tag{7.105}$$

where

$$H_{ji} = \frac{\left(\frac{\partial f_j(\mathbf{x}^k)}{\partial x_i} - \frac{\partial f_j(\mathbf{x}^{k-1})}{\partial x_i}\right)\left(x_i^k - x_i^{k-1}\right)}{\left\| x_i^k - x_i^{k-1} \right\|^2}.$$

Here H_{ji} is the approximate second derivative and can be updated by any convenient update method such as BFGS. The first method can be used initially to avoid the additional computational cost of computing the second derivative while the second can be used near the optimum.

Choosing the move limits is critical to avoid unexpected division by zero and to help prevent oscillation. Jiang (1996) suggests that adding move limits between the current iteration point and the previous point will reduce oscillation. The move limits are therefore chosen as follows:

$$\alpha_i^k = \max\{LB_i, A_i^k, 0.99L_{ji}^k + 0.01x_i^k\}, \tag{7.106}$$

$$\beta_i^k = \min\{UB_i, B_i^k, 0.99U_{ji}^k + 0.01x_i^k\},$$

$$A_i^k = \frac{x_i^k + x_i^{k-1}}{2}, \quad \text{if } x_i^k > x_i^{k-1},$$

$$B_i^k = \frac{x_i^k + x_i^{k-1}}{2}, \quad \text{if } x_i^k < x_i^{k-1}.$$

The major drawback of MMA methods is that global convergence is generally not guaranteed and any remedies to that end tend to make the method arbitrarily slow. So one must use it heuristically and only when extensive monotonicities are expected to exist.

7.10 Summary

This chapter showed some of the variety of techniques and ideas employed in local iterative methods that use derivative information. It also pointed out many of the limitations, both theoretical and practical. The exposition was cautionary, to allow the proper perspective to be achieved. The global model analysis described in previous chapters becomes more attractive and meaningful when the complexity and uncertainty of local computation are fully understood.

At the same time, many practical situations exist where local computation is the only recourse. Complicated or implicit models may resist any other solution approach. In spite of their weaknesses, local computation methods can be very powerful and can produce valuable results, otherwise impossible to achieve. Theoretical developments and software improvements make the efforts of informed users worthwhile. Hence one should not feel discouraged after becoming knowledgeable about difficulties and limitations.

Another interesting point was the use of active set strategies in local computation. This is in direct analogy with the case decomposition and global computation described in Chapter 6.

The intricacies of local computation and the difficulties described in this chapter demonstrate once again the main thesis in this book, namely, that modeling is inextricably connected with computation. In this sense, the present chapter is a closure to the book's general theme. Chapter 8 serves to consolidate the ideas explored in all previous chapters and summarizes the tactics in conducting practical design studies.

Notes

The techniques of numerical linear algebra have played a major role in the development of efficient optimization algorithms and code implementations. To gain some perspective, the reader may consult Rabinowitz (1970) and Rice (1981).

Several texts on nonlinear programming guided the descriptions in this chapter and the reader is encouraged to consult them for more information, both on theory and implementation of methods: see Fletcher (1980, 1981), Gill, Murray, and Wright (1981), Luenberger (1984), and Scales (1985). Some texts that resulted from conference proceedings after careful editorial work are good sources and have become classics: Gill and Murray (1974), Dixon, Spedicato, and Szegö (1980), Powell (1982), and Schittkowski (1985). An interesting perspective is also given by Evtushenko (1985). The material on trust regions was based on the dissertation by Nelson (1997), where SQP and trust region algorithms are extended to handle models with special structures amenable to decomposition. The material on convex approximations is based on the theses by Jiang (1996) and Nwosu (1998) that addressed structural design applications, including topology and multicomponent problems.

Several methods are not described in this chapter. The important class of *conjugate direction* methods is useful for large, sparse problems. A thorough study of these methods can be found in Hestenes (1980). Older but still very useful studies of large-scale systems can be found in Lasdon (1970), Wismer (1971), and Himmelblau (1973). A more modern comprehensive book on this subject is still lacking. Recent efforts in systems design optimization are motivated by *multidisciplinary design optimization* (MDO) problems, involving several complicated analysis models from diverse disciplines within a single optimal design model (Olhoff and Rozvany 1995, Alexandrov and Hussaini 1997). The book by Conn, Gould, and Toint (1992) on LANCELOT, a software package for large-scale nonlinear problems, provides a good background on the issues involved.

A description of *direct search* or *derivative-free* methods that utilize only function values can be found in Reklaitis, Ravindran, and Ragsdell (1983). A careful and comprehensive study of algorithms that do not use derivative information is the one by Brent (1973). Several very recent research articles motivated by expensive and noisy simulation models have renewed interest in this area; see, for example Lewis and Torczon (1996, 1998), Torczon (1997), and Torczon and Trosset (1998).

One should also look for methods described under different names. We noted in the text that quasi-Newton methods are also called *variable-metric* ones. Another name for SQP methods is *projected Lagrangian* or *constrained variable-metric* methods.

Finally, we have completely left out discussion on stochastic or random methods, such as simulated annealing, genetic algorithms, or the so-called Lipschitzian algorithms. Some sources for study in these topics are Ansari and Hou (1997), Horst, Pardalos, and Thoai (1995), Kouvelis and Yu (1997), Lootsma (1997), and Pflug (1996). These methods are becoming increasingly important and should be considered alongside the gradient methods when facing practical design situations.

Exercises

7.1 Recall that the inverse $\mathbf{A}^{-1}$ of $\mathbf{A}$ can be found by solving the system $\mathbf{A}(\mathbf{x}_1, \ldots, \mathbf{x}_n) = (\mathbf{e}_1, \ldots, \mathbf{e}_n)$ where $\mathbf{x}_i$ are the column vectors of $\mathbf{A}^{-1}$ and $\mathbf{e}_i$ are the column vectors of the identity matrix. Using this result, find the inverse of

$$\mathbf{A} = \begin{pmatrix} 2 & -3 & 2 & 5 \\ 1 & -1 & 1 & 2 \\ 3 & 2 & 2 & 1 \\ 1 & 1 & -3 & -1 \end{pmatrix}.$$

Solve using Gauss elimination and back-substitution.

7.2 Find the rank of the matrices below by reducing them to row-echelon form:

$$\mathbf{A} = \begin{pmatrix} -1 & 0 & 1 & 2 \\ -1 & 1 & 0 & -1 \\ 0 & -1 & 1 & 3 \\ 1 & -2 & 1 & 4 \end{pmatrix}, \quad \mathbf{B} = \begin{pmatrix} 1 & 3 & -2 & 1 \\ 2 & -1 & 3 & 4 \\ 3 & -5 & 8 & 7 \end{pmatrix}.$$

7.3 Check if the following vectors are linearly independent:

$$\mathbf{x}_1 = (1, -1, -1, 2)^T, \quad \mathbf{x}_3 = (2, -3, -3, 2)^T,$$

$$\mathbf{x}_2 = (-1, 2, 3, 1)^T, \quad \mathbf{x}_4 = (1, 1, 1, 6)^T.$$

Hint: Do this by writing the vectors as columns of a matrix and reducing this matrix to row-echelon form.

7.4 Consider the function

$$f = x_1^2 - 3x_1x_2 + 4x_2^2 + x_1 - x_2.$$

(a) Find a minimizer $\mathbf{x}_*$ analytically.

(b) For the iterative scheme $\mathbf{x}_{k+1} = \mathbf{x}_k + \alpha_k \mathbf{s}_k$, take $\mathbf{s}_k$ to be the negative gradient vector and $\alpha_k = \arg\min f(\mathbf{x}_k + \alpha \mathbf{s}_k)$. Find an *explicit* expression for α_k.

(c) For $\mathbf{x}_0 = (2, 2)^T$, calculate at least three iterations giving the values for $\mathbf{x}_k$, $\mathbf{s}_k$, α_k, and f_k.

(From Vanderplaats 1984.)

7.5 The *secant* (or *linear interpolation*) method for one-dimensional minimization combines Newton's method with a sectioning method for finding the root of $f'(x) = 0$ in $[a, b]$ if it exists. The algorithm starts with two points where $f'(x)$ has opposite signs (see figure). It then approximates $f'(x)$ by a straight line (the so-called secant line) between these two points. The intersection point x_3 is the new approximation to x_*.

(a) Derive an expression for evaluating x_3.

(b) Give three complete step-by-step iterations of the secant method applied to solving the problem $\min f(x) = 0.5x^2 + (27/x)$ over $1 \leq x \leq 8$.

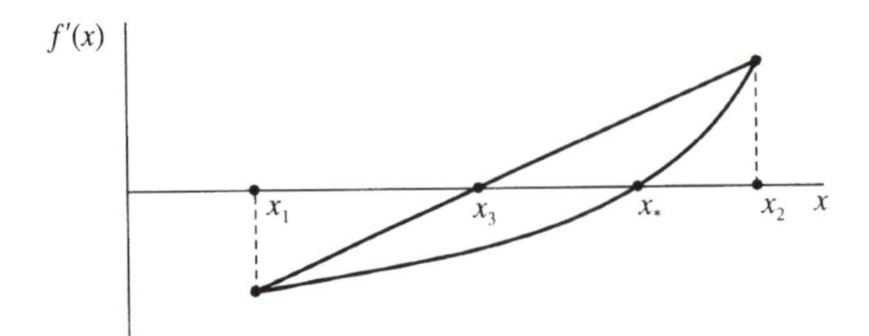

Secant method.

(c) Compare the secant with the bisection and Newton's methods. Itemize differences, advantages, and disadvantages.

7.6 Prove the quadratic interpolation formula Equation (7.12) by setting $q = ax^2 + bx + c$ and calculating $x_* = -(b/2a)$. Then add and subtract x_1 to arrive at the desired expression.

7.7 Perform at least three iterations of the quadratic interpolation Equation (7.12) applied to the function of Example 7.4. Start with points $x_1 = 0$, $x_2 = 2.5$, $x_3 = 5$.

7.8 Derive the DSC formulas, Equations (7.16a,b), from Equation (7.11). Draw two sketches similar to Figure 7.8 with points x_2, x_3, x_a, and x_4 replaced by x_{n-2}, x_{n-1}, x_a, and x_n. The first sketch should show $f_a > f_{n-1}$, and the second $f_a < f_{n-1}$.

7.9 Prove the Armijo–Goldstein criteria (7.18) and (7.19) using the geometry of Figure 7.9.

7.10 A popular purely sectioning procedure is the *golden section search*. It aims at creating bracket lengths that are in a sense "optimal," that is, they minimize

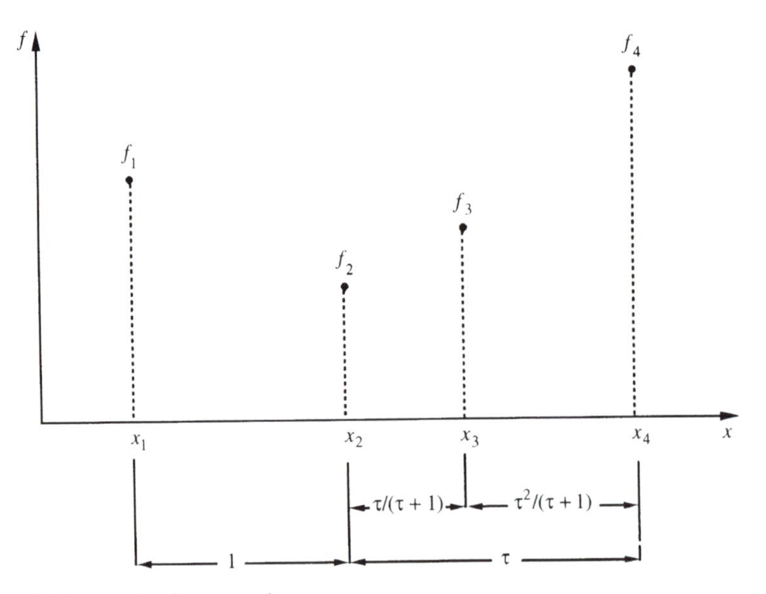

Golden section line search.

the worst possible reduction of the uncertainty interval. Consider a function as shown in the figure with three points x_1, x_2, x_3 already placed with the ratio $(x_3 - x_2)/(x_2 - x_1)$ fixed at τ, where $\tau > 1$ is a constant. Insert the fourth trial point x_4 so that both *potential* new brackets $[x_1, x_4]$ and $[x_2, x_3]$ have intervals in the same ratio τ. Based on this construction, evaluate the lengths of the intervals $[x_2, x_4]$ and $[x_4, x_3]$. Prove that $\tau \cong 1.618$ (the so-called golden section ratio of the ancient Greeks). Write out the steps of an algorithm based on repeating this sectioning, noting that in the figure the interval $[x_2, x_3]$ is discarded. (The method has linear convergence with respect to the uncertainty interval reduction.)

7.11 Find the minimum of $f = (10 - x)^2 + x$ with the golden section method starting with the interval [5, 20].

7.12 Apply the DSC and the golden section methods to the function of Exercise 7.5(b). Perform at least five iterations.

7.13 Consider the function $f = 1 - x\exp(-x)$.

(a) Find the minimum using the golden section method, terminating when $|x_{k+1} - x_k| < 0.1$ and starting from [0, 2].

(b) Find the value(s) for x that satisfy the Armijo–Goldstein criteria with $\varepsilon_1 = 0.1$.

(c) Find a value for x using the Armijo Line Search of Example 7.5.

(From Fletcher 1980.)

7.14 Repeat Exercise 7.13 for the function $f = 1 - (5x^2 - 6x + 5)^{-1}$ with the same starting interval and parameter values (from Fletcher 1980.)

7.15 The secant method given in Exercise 7.5 may diverge if the linear approximation locates the solution estimate by *extrapolation*. Sketch an example when this happens, that is, when $g_k g_{k+1} > 0$. To obtain global convergence, a modification is suggested by the *method of false position* (*regula falsi*). In this case, x_{k+1} replaces either x_{k-1} or x_k depending on which corresponding derivative value gives a sign opposite to g_{k+1}. Sketch an example where the regula falsi method converges very slowly, for example, when the initial point x_0 in the first bracket $[x_0, x_1]$ can never be discarded. (Note that this method has linear rate of convergence but the asymptotic error constant can be arbitrarily close to one.)

7.16 A "safeguarded" linear interpolation method considered more efficient than the golden section method is as follows (Dixon, Spedicato, and Szego 1980):

Bracketing Given x_1 such that $g(x_1) < 0$, find x_2 such that $g(x_2) > 0$. For example:

1. Select h, set $k = 0$.
2. Calculate $g(x_1 + h)$ and set $k = k + 1$.
3. If $g(x_1 + h) \geq 0$, set $x_2 = x_1 + h$ and go to interpolation; otherwise, set $x_1 = x_1 + 3h$ and return to Step 2.

Interpolation Given $x_1 < x_2$ with $g(x_1) < 0$, $g(x_2) > 0$, find the linear prediction of $g(x_*) = 0$ [see Exercise 7.5(a)], and call it x_p. Evaluate $g(x_p)$. If $g(x_p) < \varepsilon_1$ and/or $|x_1 - x_2| < \varepsilon_2$, stop; otherwise go to sectioning.

Sectioning If $g(x_p) < 0$, set $h = (x_p - x_1)/13$ and $x_1 = x_p$; otherwise $[g(x_p) > 0]$, set $h = -(x_p - x_1)/13$ and $x_1 = x_p$. Return to bracketing. (The method has global convergence and at least linear rate of convergence.) Write a computer code that implements this method.

7.17 Prove that the choices for $\mathbf{u}$, $\mathbf{v}$, a, and b in Equation (7.30) leading to the DFP update Equation (7.32) are correct, that is, the update satisfies the quasi-Newton condition.

7.18 In Example 7.6, show that the choice $\alpha_0 = 0.1$ is correct if an Armijo Line Search (Example 7.5) is used with $\varepsilon = 0.1$ or $\varepsilon = 0.2$. Repeat with $\alpha_0 = 0.4$ in Example 5.9.

7.19 Find the minimum of the function $f = x_1^2 - 3x_1x_2 + 4x_2^2 + x_1 - x_2$ using the DFP formula (7.32). Repeat with the BFGS formula (7.33). Compare with Example 4.2.

7.20 Consider the function $f = (3 - x_1)^2 - (4 - x_2)^2$. Find and characterize the nature of its stationary point(s). How do you expect the BFGS method to behave when applied to this function? Select at least two different starting points and demonstrate this expected behavior by carrying out three iterations from each starting point.

7.21 Consider the function $f = x_1^2 + 2x_2^2 + 3x_3^2 + 3x_1x_2 + 4x_1x_3 - 3x_2x_3$. Repeat Exercise 7.20 using the DFP method.

7.22 Solve the problem $\min f = x_1^4 - 2x_1^2x_2 + x_2^2$ using the BFGS method and starting at $\mathbf{x}_0 = (1.1, 1)^T$. Recall Example 4.17.

7.23 (*Memoryless BFGS Updates*) A simplification of the BFGS update is obtained if the matrix $\mathbf{H}_{k+1}$ (or $\mathbf{B}_{k+1}$) is defined by a BFGS update applied to $\mathbf{H} = \mathbf{I}$ rather than to $\mathbf{H}_k$; hence, it is named memoryless BFGS. Derive the update formulas (a) for an inexact, and (b) for an exact line search. Then give the steps required to implement such an algorithm.

7.24 Scale the variables of the function $f = 4x_1^2 + 3x_1x_2 + 100x_2^2$ so that a gradient-type method can converge fast, for example, starting from the point $(4, 4)^T$. See also Section 8.3.

7.25 Use the KKT conditions to prove the *local* validity of the monotonicity principles given in Chapter 3. Examine how a (local) active set strategy could use these principles.

7.26 To avoid violating the nonlinear constraint surface by moving along the tangent subspace, a feasible directions method has been proposed that calculates the direction $\mathbf{s}$ (at each iteration from a point $\mathbf{x}$) by solving the problem

$$
\begin{aligned}
&\max_{\mathbf{s}} \beta \\
&\text{subject to } [\nabla f(\mathbf{x})]^T\mathbf{s} + \beta \leq 0, \\
&\qquad [\nabla g_j(\mathbf{x})]^T\mathbf{s} + \theta_j\beta \leq 0, \quad \text{all } j \text{ (active)}, \\
&\qquad -1 \leq s_i \leq 1, \quad \text{all } i,
\end{aligned}
$$

where θ_j are nonnegative constants called *push-off factors*. Sketch graphically for a 2-D problem the effect of θ_j and show that the solution to the above problem is a feasible descent direction. (For a description of algorithms based on this idea, see Vanderplaats 1984.)

7.27 Using the penalty transformation $T = f + \frac{1}{2}r\mathbf{h}^T\mathbf{h}$, evaluate and sketch the progress of the penalty method (sequential unconstrained minimization) for the problem $\{\min f = x, \text{subject to } h = x - 1 = 0\}$, with $r = 1, 10, 100, 1000$. Repeat, using the augmented Lagrangian transformation $T = f + {}^T\mathbf{h} + \frac{1}{2}r\mathbf{h}\,{}^T\mathbf{h}$. (From Fletcher 1981.)

7.28 In Exercise 7.27 compare the two sketches and examine the augmented Lagrangian formulation as a mechanism for shifting the penalty origin. What is the role of the multiplier?

7.29 It can be shown rigorously (e.g., Luenberger 1984) that the BFGS update preserves positive-definiteness, if $\partial x_k^T \partial \mathbf{g}_k > 0$ – even with inexact line search. Prove that Powell's modification of the BFGS formula, (7.73) and (7.74), for the Hessian of the Lagrangian, preserves positive-definiteness, that is, prove that $\partial x_k^T \partial \mathbf{g}_k > 0$ as defined in this case.

7.30 Consider the problem of Example 5.9. Apply an SQP algorithm with line search, starting from $\mathbf{x}_0 = (1, 1)^T$. Solve the QP subproblem using (7.82) and BFGS approximation (7.73), (7.74) for the Hessian of the Lagrangian. Use the merit function (7.76) and the Armijo Line Search to find step sizes. Perform at least three iterations. Discuss the results.

7.31 Repeat Exercise 7.30 for the problem

$$
\begin{aligned}
&\min f = (x_1 - x_2)^2 + (x_2 - x_3)^2 + (x_3 - x_4)^2 + (x_4 - x_5)^2 \\
&\text{subject to } h_1 = x_1 + x_2^2 + x_3^3 - 3 = 0, \\
&\qquad h_2 = x_2 - x_3^2 + x_4 - 1 = 0, \\
&\qquad h_3 = x_1 x_5 - 1 = 0,
\end{aligned}
$$

with starting point $(2, \sqrt{2}, -1, 2 - \sqrt{2}, 0.5)^T$. (From Hock and Schittkowski 1981, Problem No. 47.)

7.32 Repeat as above for the problem

$$\min f = x_1^2 + x_2^2 + 2x_3^2 + x_4^2 - 5x_1 - 5x_2 - 21x_3 + 7x_4$$
$$\text{subject to } 8 - x_1^2 - x_2^2 - x_3^2 - x_4^2 - x_1 + x_2 - x_3 + x_4 \geq 0,$$
$$10 - x_1^2 - 2x_2^2 - x_3^2 - 2x_4^2 + x_1 + x_4 \geq 0,$$
$$5 - 2x_1^2 - x_2^2 - x_3^2 - 2x_1 + x_2 + x_4 \geq 0,$$

with starting point $\mathbf{x} = \mathbf{0}^T$. (From Hock and Schittkowski 1981, Problem No. 43.)

7.33 Consider the problem

$$\min f = 6x_1x_2^{-1} + x_1^{-2}x_2$$
$$\text{subject to } h = x_1x_2 - 2 = 0,$$
$$g = 1 - x_1 - x_2 \leq 0.$$

(a) Obtain the QP subproblem at point $(1, 2)^T$; at this point the multiplier estimates $(\lambda, \mu) = (0.5, 0)$.

(b) Show that $(0, 0)^T$ is the solution to this QP subproblem. What can you conclude about the optimality of $(1, 2)^T$ for the original problem?

7.34 Consider the problem

$$\min f = x_1^2 + x_2^2 - 3x_1x_2$$

subject to

$$g_1 = \frac{1}{6}x_1^2 + \frac{1}{6}x_2^2 - 1 \leq 0, \quad g_2 = -x_1 \leq 0, \quad g_3 = -x_2 \leq 0.$$

(a) Solve the problem analytically and provide a graphical sketch to aid visualization.

(b) Linearize *f* and g_1 about the point $\mathbf{x}_0 = (1, 1)^T$.

(c) Solve the resulting linear programming problem (including g_2 and g_3) using monotonicity analysis.

(d) Confirm that point $\mathbf{x}_1 = (2, 2)^T$ is a solution of the problem defined in (c). Linearize the original problem again at $\mathbf{x}_1$ and solve again.

(e) Steps (b)–(d) can form an algorithm that solves a nonlinear problem with a sequence of approximating subproblems. Describe formally what the steps for such an algorithm may be.

(f) List advantages and disadvantages (briefly) for such a *Sequential Linear Programming* (SLP) algorithm.

(g) Discuss (briefly) how the basic algorithm in (e) can be modified for better performance.

8

Principles and Practice

Im Anfang war die Tat. (In the beginning was the Act.)
J. W. von Goethe (1749–1832)

In designing, as in other endeavors, one learns by doing. In this sense the present chapter, although at the end of the book, is the beginning of the action. The principles and techniques of the previous chapters will be summarized and organized into a problem-solving strategy that can provide guidance in practical design applications. Students in a design optimization course should fix these ideas by applying them to a term project. For the practicing designer, actual problems at the workplace can serve as first trials for this new knowledge, particularly if sufficient experience exists for verifying the first results.

The chapter begins with a review of some modeling implications derived from the discussion in previous chapters about how numerical algorithms work. Although the subject is quite extensive, our goal here is to highlight again the intimacy between modeling and computation that was explored first in Chapters 1 and 2. The reader should be convinced by now of the validity of this approach and experience a sense of closure on the subject.

The next two sections deal with two extremely important practical issues: the computation of derivatives and model scaling. Local computation requires knowledge of derivatives. The accuracy by which derivatives are computed can have a profound influence on the performance of the algorithm. A closed-form computation would be best, and this has become dramatically easier with the advent of *symbolic computation* programs. When this is not possible, numerical approximation of the derivatives can be achieved using *finite differences*. The newer methods of *automatic differentiation* are a promising way to compute derivatives of functions defined implicitly by computer programs.

Scaling the model's functions and variables is probably the single most effective "trick" that one can do to tame an initially misbehaving problem. Although there are strong mathematical reasons why scaling works (and sometimes even why it *should not* work), effective scaling in practice requires a judicious trial and error approach.

Most optimization tools present the solution to the user as a set of numerical results, such as optimal values of functions, variables, and multipliers. The designer must understand what these numbers really say as an output of a numerical process. Although there is an increased effort in commercial software to provide diagnostic or even "learning" capabilities, these are still rather limited. A section is included here with a few basic comments on the somewhat detective-like work that is often needed to reach an understanding about what the numerical results really tell you.

One of the most common questions asked at this point, typically at the end of an optimization course, is what are the "best" algorithms and corresponding software to use. As with most everything in this book, answering this question requires making a trade-off decision. This issue is addressed briefly in Section 8.5, along with a few comments on commercial software.

Next, a design optimization checklist is provided, which the reader may find useful in the early stages of accumulating design optimization experience. The checklist itemizes issues that one should address, or at least think about, during an optimal design study. It should be regarded as a prompt sheet rather than a rigid step-by-step procedure. The list can be used effectively for pacing term projects in a course, or just for guiding one's thoughts during a study. As experience and practice accumulates, many items will automatically become part of the designer's approach to any new problem.

Finally, some of the most important concepts, rules, conditions, and principles developed in this book are listed without regard to the order or frequency in which they might find application. This recapitulation is intended to seed the designer's mind with ideas for cutting through to the core of an optimal design problem.

8.1 Preparing Models for Numerical Computation

The detailed form of the model used for numerical processing evidently has strong influence on the success or failure of the numerical algorithm. Numerical algorithms have an uncanny ability to exploit any modeling weaknesses, leading to unwanted answers. In the present section we review a few ideas that should help make a model more amenable to successful numerical treatment.

Modeling the Constraint Set

The overriding concern in preparing the constraint set model is that *it should remain equivalent to the original model of the physical problem addressed.* Any transformations that appear convenient for solution purposes should not alter the feasible set, nor change the nature of the optima, nor introduce new or delete existing optima, unless this is done consciously and purposefully. Careless model transformations will have such adverse effects.

Modern NLP algorithms have provisions for handling linear and simple variable bound constraints separately with increased efficiency. In such cases, it is often not desirable to eliminate linear constraints explicitly. Some popular old transformations are also not recommended for modern algorithms. For example, it is almost never beneficial to use the transformation $x = y^2$ for replacing $x \geq 0$.

Putting explicit simple bounds on each model variable, $l_i \leq x_i \leq u_i$, is always useful for numerical algorithms. The feasible set is compact and an optimal solution will exist. The smaller the interval $l_i \leq x_i \leq u_i$, the smaller the search space will be, which is generally an advantage. The larger the interval, the less likely is that the simple bound constraints will become active. From a physical modeling viewpoint, simple bounds often are evidence of simplistic modeling, replacing physically meaningful but more complicated bounding constraints. The inclusion of restrictive simple bounds may force these bounds to dominate other more meaningful constraints that are then rendered inactive. When numerical algorithms identify simple bounds as active, one should always attempt to understand the physical meaning of this constraint activity. Moreover, it would be a good idea to relax the bounds to less restrictive values and solve the problem again.

Inequalities usually should be handled with active set strategies. This is done automatically by existing algorithms, such as GRG. However, an *extended* active set strategy may be necessary for some problems. In this type of active set strategy, a small subset of constraints that are *likely* to include the active ones is included in each iteration. This substantially reduces computational and memory requirements. Changing this set at each iteration is usually done heuristically and is programmed by the user. Such a strategy may be used if there is a very large number of similar constraints only one of which may be active. For example, a maximum stress constraint in a structural problem solved using the finite element method may be represented by setting an inequality constraint for the stress at each element. Once the likely set of active constraints is determined a regular active set strategy may still be applied to it, as described in Section 7.4.

Another method sometimes advocated for handling inequalities is the use of *slack variables*. Here a constraint $g(\mathbf{x}) \leq 0$ is transformed into the equivalent set $g(\mathbf{x}) + y = 0$, $y \geq 0$, where y is a new (slack) variable. As mentioned before, this is much better than setting $g(\mathbf{x}) + y^2 = 0$. The main argument against slack variables is that computations include all the constraints in the model. In cases where there are many more constraints than variables, the dimensionality is increased substantially, leading to very expensive individual iterations. There is no evidence that the number of iterations is reduced when slack variables are used. Therefore, no advantages seem to exist, at least for general engineering design problems. Slack variables can be used effectively in certain special situations, when a uniform way of handling all the constraints is advantageous. One case is in LP algorithms. Another case is in large sparse problems, where taking advantage of sparsity improves the computational burden dramatically, as in the LANCELOT package (Conn, Gould, and Toint 1992).

Recall that two standard assumptions for numerical algorithms are functional independence of the constraints and linear independence of the gradients of the constraints (regularity). The second assumption is rarely invalid in practice, but failure of the first is not uncommon because of careless modeling. Mostly in larger models, constraints may be accidentally repeated or combined with each other, possibly with scaling factors. The usual result of this situation will be poor or no convergence. The difficulty must be remedied by careful reexamination of the model.

Sometimes constraint satisfaction may be considered acceptable within a *tolerance*, say, t. An equality constraint $h(\mathbf{x}) = 0$ is replaced by $-t \leq h(\mathbf{x}) \leq t$ where t is a small, positive constant but larger than a termination tolerance ε. This technique may serve in modeling situations with uncertainty, but it can be a source of numerical difficulties because the feasibility interval for $h(\mathbf{x})$ is very small (see Gill, Murray, and Wright 1981, for an example). It is preferable to treat uncertainty directly using the methodology of probabilistic design (see, e.g., Siddall 1972, 1983, Hazelrigg 1996).

Equality constraints are occasionally treated by replacing $h(\mathbf{x}) = 0$ by $h(\mathbf{x}) \geq 0$, $h(\mathbf{x}) \leq 0$. This is only helpful if the equality can be directed (see Chapters 5 and 6) and replaced by either $h(\mathbf{x}) \equiv >0$ or $h(\mathbf{x}) \equiv <0$ but not both. If an a priori direction is not possible, this practice should be avoided.

When representing relatively large systems the model may be built up by assembling repeated elements or subsystems that are "hooked together." Examples include electronic and chemical process plant design problems. In the mathematical model this will appear as a collection of functions that have the same form but different variables. The resulting model will exhibit sparsity, for example, in the Jacobian matrix of the constraints. NLP algorithms can be modified to handle very large problems by exploiting sparsity. So if extensive sparsity exists in the model, it is worth looking into special algorithms for sparse problems.

Modeling the Functions

It should be evident by now that three major concerns exist in modeling objective or constraint functions: (a) ease of evaluation, (b) continuity, and (c) differentiability. Our recommendation is to avoid defining functions in the model through tables or series. Instead, perform a curve-fitting or metamodeling procedure and use a model function that reasonably approximates the original data. One may even advocate this idea for functions calculated from complicated analysis models that solve, for example, systems of differential equations. It may be very advantageous to use the analysis model for computational experiments, sample the data generated, and use curve fitting to create function approximations for the optimization model. The solution of the optimization model thus constructed can always be refined by using it as a starting point for one or two iterations with the full model. This procedure will address all concerns mentioned above. Note, however, that the expense associated with creating metamodels must be justified.

Function models may be changed by transforming the variables in them. Examples include $x = y^2$, $x = \cos y$, and $x = e^{-y}$. In general, nonlinear and transcendental transformations require very careful and purposeful implementation to avoid introducing disastrous changes in the model. The point here is that variable transformations should not be done casually.

A very common practical source of failure for many algorithms is that in the course of an iteration, function evaluations must be performed at points where *the function is not defined.* If one is fortunate, the host computer's operating system will issue an error message and immediately terminate the algorithm. Although this was more likely with older compilers, modern high level programming environments may not issue such a message and just provide whatever (erroneous) results are computed. Searching for such errors could require special interrogation of the operating system and the computing language used. In engineering design problems, functions such as square roots and logarithms often cause failure for this reason. Including bounds may not help even with a tolerance. If $f(x)$ is defined for $a \leq x \leq b$, a bound of the form $a + t \leq x \leq b - t$, where t is a tolerance, can still be violated during an inner iteration, for example, in the QP part of an SQP algorithm.

A proposed remedy for this situation (Fletcher, in Powell, 1982) is given by an example: If $f(x) = (x_1^3 + x_2^3)^{1/2}$, then set $x_3 = x_1^3 + x_2^3$, $x_3 \geq 0$, and $f(x) = x_3^{1/2}$; then modify the algorithm so that simple bounds are never violated. Although this remedy may work sometimes, modifying an algorithm, particularly in the form of a commercial code, is not always a simple matter. A user should always check if the code at hand has this capability. Another possibility is to modify the subroutine with the model input so that the functions in question are never evaluated at invalid points; instead they are assigned a valid value close to the violated bound. This value may be fixed a priori or vary depending on the location of the most recent point in the iterations. When this function evaluation is used as an intermediate point within an iteration, the effect on the algorithmic behavior should be negligible. If this "correction" appears to be excessive then some form of penalty term would need to be added in the objective to force the algorithm to return to the desirable region.

Finally, as mentioned in Chapter 1, when models are cast into standard form, it is always preferable to avoid any divisions by quantities that may become zero, or near-zero, during the iterations. The reasons for that recommendation should now be obvious.

Modeling the Objective

Sometimes a problem of the form $\{\min f(\mathbf{x})\}$ is transformed to $\{\min y$, subject to $y = f(\mathbf{x})\}$. For example, this may be needed for transforming a model to a special form. Most good algorithms should not be affected by this change. However, efficiency can always be improved by keeping the number of equality constraints small. It is preferable to use equalities in the function definition rather than as constraints. For example, if a problem has the form

$$\begin{aligned} &\min f(\mathbf{x}, y_1, y_2) \\ &\text{subject to } y_1 = h_1(\mathbf{x}), \\ &\qquad y_2 = h_2(\mathbf{x}, y_3), \\ &\qquad y_3 = h_3(\mathbf{x}, y_1), \end{aligned} \tag{8.1}$$

it is better to model it as an unconstrained problem where f is directly evaluated in the sequence: $\mathbf{x} \rightarrow y_1 \rightarrow y_3 \rightarrow y_2 \rightarrow f$.

An exception to the above is in minimax formulations where an *upper bound formulation* may be employed for a variety of reasons, including lack of differentiability. For example, the problem

$$\underset{\mathbf{x}}{\text{minimize}} \ \underset{f_i}{\max} \ \{f_1(\mathbf{x}), \ldots, f_m(\mathbf{x})\} \tag{8.2}$$

is transformed to the problem

$$\begin{aligned} &\underset{\mathbf{x},\,\beta}{\text{minimize}} \ \beta \\ &\text{subject to} \ \ f_i(\mathbf{x}) \leq \beta, \quad i = 1, 2, \ldots, m, \end{aligned} \tag{8.3}$$

where β is the new upper bound *variable*.

Finally, the objective should be scaled for the reasons described in Section 8.3, in spite of the fact that the minimizer's value is not affected in theory.

8.2 Computing Derivatives

For functions that are simple or of special form, for example, power functions, derivatives may be easy to calculate analytically and the resulting expressions can be used for gradient or Hessian evaluations required by a numerical method. It is also possible to use symbolic manipulation programs to derive such expressions automatically. The difficulty is that many important engineering problems are based on implicit models for function evaluation. For example, the function evaluations may be obtained from numerical solution of a system of differential equations. Then explicit evaluation of derivative information is impossible or impractical. Fortunately, analytical or semianalytical derivative approximations can be derived in many specific cases. Finite element codes available for structural optimization often include the capability for computing partial derivatives or "sensitivities" in their jargon.

Techniques from the finite difference calculus in numerical analysis can be used when the derivatives of a function must be approximated using information only from function values. In many practical design problems this situation will be unavoidable. Use of finite differences is appealing in general purpose optimization algorithms. In the following discussion we will summarize a few basic ideas and refer the reader to the specialized texts for more information (see, e.g., Gill, Murray, and Wright, 1981).

Automatic differentiation refers to techniques that take the source code of a computer program and automatically determine how to calculate derivatives with respect to a set of program inputs. This approach, which is different from symbolic computation, is particularly attractive if we expect significant errors in finite difference approximations. Such errors occur when a function has a large amount of curvature or when it becomes somewhat noisy, for example, because it results from numerical simulation.

Finite Differences

A differentiable function may be approximated by a Taylor series as we discussed in Section 4.2. Consider a function of a single variable for simplicity

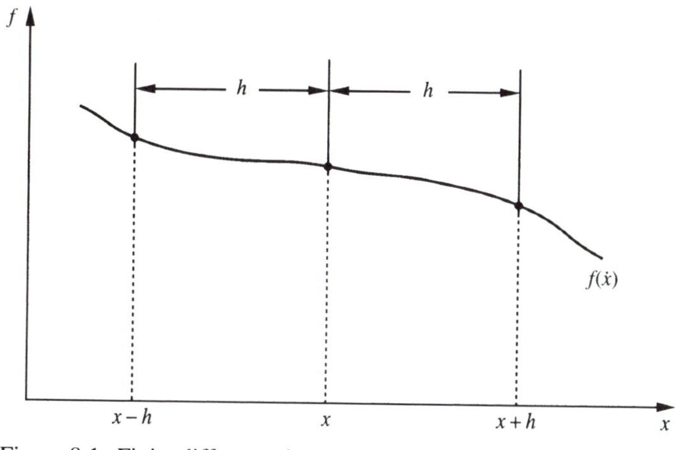

Figure 8.1. Finite difference intervals.

(Figure 8.1), and express $f(x+h)$ by a Taylor expansion about x. (Recall from Section 4.2 that the order definition $o(h^n)$ means terms of degree higher than n are negligible.) We have

$$f(x+h) = f(x) + hf'(x) + (h^2/2)f''(x) + o(h^2). \tag{8.4}$$

Solving for $f'(x)$ and neglecting the higher order terms, we find

$$f'(x) = \frac{f(x+h) - f(x)}{h} + o(h), \tag{8.5}$$

where the order term is just $(h/2)f''(\zeta), x \leq \zeta \leq x+h$. This is the *forward difference* approximation to the derivative and the quantity $(h/2)f''(\zeta)$ is the *truncation error* due to the neglected terms of the Taylor expansion.

If the expansion is taken backwards at point $x - h$, then we get

$$f(x-h) = f(x) - hf'(x) + (h^2/2)f''(x) + o(h^2). \tag{8.6}$$

Subtracting (8.6) from (8.4), simplifying, and solving for $f'(x)$, we get

$$f'(x) = \frac{f(x+h) - f(x-h)}{2h} + o(h^2), \tag{8.7}$$

where the main order term is $(h^2/6)f'''(\zeta)$. This is called the *central difference* approximation.

If we use a subscript notation $f(x) = f_j$, $f(x+h) = f_{j+1}$, $f(x-h) = f_{j-1}$, etc., we can easily find the following expressions for both first and second derivative approximations:

Forward		**Central**	
$f' = (f_{j+1} - f_j)/h,$	(a)	$f' = (f_{j+1} - f_{j-1})/(2h),$	(c)
$f'' = (f_{j+2} - 2f_{j+1} + f_j)/h^2,$	(b)	$f'' = (f_{j+1} - 2f_j + f_{j-1})/h^2.$	(d)

(8.8)

These expressions can be generalized for derivatives of any order (see Dahlquist and Björck 1974). Expressions such as (8.8) (a), (c) can be used to generate a finite-difference approximation of the gradient, for multidimensional cases. A finite-difference approximation of the Hessian can be generated by finite differences of the gradient, $\mathbf{g}$, since $\partial\mathbf{g} \cong \mathbf{H}\,\partial\mathbf{x}$. For example, the jth column of $\mathbf{H}$ at a point $\mathbf{x}$ may be given by the forward difference $[\mathbf{g}(\mathbf{x} + h_j\mathbf{e}_j) - \mathbf{g}(\mathbf{x})]/h_j$ where h_j and $\mathbf{e}_j$ are a finite-difference interval and unit vector along the jth component, respectively (see Gill, Murray, and Wright 1981, for more details). Note that because such approximations will not yield, in general, symmetric matrices, the transformation to a symmetric form (Section 4.2) should be used.

The truncation error that stems from neglecting the higher order terms is not the only one involved in the finite-difference calculation. There is an error in the computed value of the function itself, usually called the *condition* or *cancellation* error, which is proportional to $1/h$ (Gill, Murray, and Wright 1981). The *rounding error* due to arithmetic operations by the computer is negligible here. Thus, we see that the overall error in these approximations has two terms: truncation, proportional to h, and cancellation, proportional to $1/h$. In principle there should be an optimal differentiation step h_* that minimizes the total error.

Finding such an h_* is not readily evident, particularly if the function calculation has some numerical noise. In fact, one can say that the precise selection of a finite-difference step is not very important for functions that can be computed with high accuracy, while it is relatively important for functions with low accuracy, such as functions computed using simulations.

An error analysis, under some mild assumptions on the behavior of f and on probability distributions for its values, can lead to algorithms for evaluating "optimal" finite-difference intervals. The following, somewhat heuristic, procedure illustrates what is involved in the case of a first forward difference.

First calculate a bound on the *absolute error* ε_A of $f(x)$, which is the difference between the theoretical value of the function and the computed one; that is, ε_A is the sum of all errors. A "true" value $f(x)$ will be computed to have a (floating point) value $\overline{f}(\overline{x})$ at a computed point $\overline{x}$, so that $|\overline{f}(\overline{x}) - f(x)| < \varepsilon_A$. The error ε_A can be estimated by computing $f(x)$ at various points x_i properly sampled within the neighborhood of the point where the finite difference will be computed, and creating an envelope of the sampled points, as in Figure 8.2. The idea is to get some measure of the spread of values of f.

The optimal (minimum error) value for h can be found to be (Gill et al. 1981)

$$h = 2(\varepsilon_A/|f''|)^{1/2}. \tag{8.9}$$

Since f'' is not known, if an additional assumption is made that $|f''|$ is of the same order as $|f|$, the following expressions can be used:

$$h \cong (\varepsilon_A/|f|)^{1/2} \quad \text{(forward difference)}, \tag{8.10}$$

$$h \cong (\varepsilon_A/|f|)^{1/3} \quad \text{(central difference)}, \tag{8.11}$$

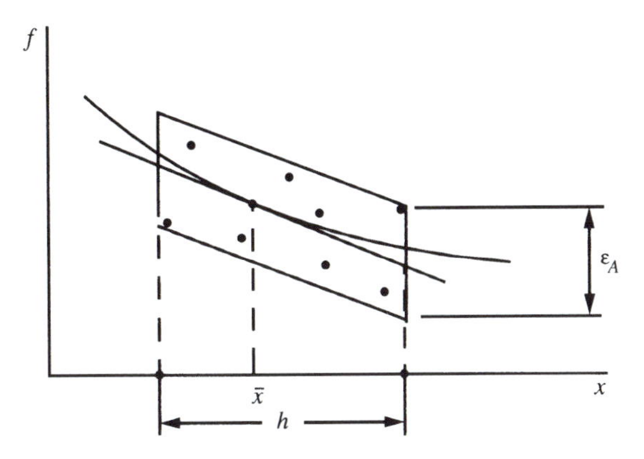

Figure 8.2. Estimating the error ε_A at various points x_i.

where $|f|$ is computed at a nonzero "typical" value of f in the range where the finite differencing will take place.

Selecting h with a procedure, such as the above one, can be tedious and is usually done only at x_0. Clearly, widely varying values of f will give different recommended intervals at different parts of the domain of f. Procedures for automatic estimation of finite-difference intervals have been developed (Gill, Murray, and Wright 1981).

A simple and appealing strategy for choosing h_* as the optimization iterations progress is described by Barton (1992) and is based on tracking changes in the significant digits of the function calculation. The cited article by Barton is comprehensive and provides sufficient detail for implementing the strategy.

Another common approach in general-purpose optimization codes is to set the forward finite-difference interval equal to $x_k(10^{-q})$ at each iteration k, where q is the number of digits of machine precision for real numbers. Although not "optimal," this selection works well in practice if the functions do not have too much noise (e.g., they can be computed with three or four significant digits of precision).

At this point, it should be emphasized that proper selection of the finite-difference interval h is not a theoretical curiosity but a practical concern. The expressions (8.8)(a) or (c) are used to approximate the components of the gradient vector. Quasi-Newton implementations of finite differences are very effective, if proper care is taken. Moreover, the estimation of an absolute error bound ε_A can help in selecting termination criteria rationally, avoiding costly unsuccessful trials.

Comparing the use of forward and central differences, we observe that forward differences will work efficiently until $\|\nabla f_k(\mathbf{x})\|$ becomes very small. Thus, near an optimum the approximation will become increasingly poor. Away from the optimum, forward differences may become inadequate when the step size α_k is smaller than h, or when $|f_{j+1} - f_j| \cong h$. Higher accuracy is then required and the central differencing may be used instead. There is unfortunately a trade-off since two additional function evaluations per iteration are required for central differences, as opposed to one for forward differences.

In engineering problems where the cost of function evaluations is high, this is a major disadvantage. Therefore, it is usually recommended that central-difference approximations be used only when it is absolutely necessary. Carrying this rationale a little further, we may observe that in a finite-difference implementation it usually makes sense to do the following:

1. Use a line search algorithm that does not need gradient information.
2. Perform a more accurate (but still inexact) line search since this will tend to reduce the total number of iterations (although sometimes $\alpha_k = 1$ may work better).
3. Prefer a quasi-Newton rather than a modified Newton method.

All of the above will tend to decrease the number of function evaluations required, *if the line search is not too expensive*. However, it is possible that a penalty function line search, such as in SQP, may require too many function evaluations, thus offsetting any gains from reducing the overall number of iterations.

As Barton (1992) points out, when you are forced to use finite differences with low accuracy function evaluations, the choice of finite-difference interval is very important, more so than the choice of the optimization algorithm itself. A "black box" simulation used to evaluate functions may often contain a numerical integration scheme. The integration step affects function accuracy or noise and thus interacts with the differentiation interval used in a finite-difference approximation. These are complicated issues to resolve in practical problems and require substantial experimentation and increased knowledge about the analysis model inside the simulation "black box." Under these circumstances, the recent developments in automatic differentiation, described below, become increasingly appealing.

Automatic Differentiation

The theory behind automatic differentiation relies on a few simple ideas. All mathematical formulas are built from a small set of basic mathematical functions, such as $\sin x$, $\cos x$, or $\exp x$, and the arithmetic operations of multiplication, addition, subtraction, and division. The derivatives of all of these functions are members of the same basic set. For example, $d\sin(\alpha t)/dt = \alpha\cos(\alpha t)$, $d\exp(\alpha t)/dt = \alpha\exp(\alpha t)$, and so on. Then, at least in theory, when a source code is compiled into an executable program, it is not difficult to add extra functions to calculate the derivatives automatically. To do that we can apply the chain rule.

Let u be the quantity of interest and t be the variable for which we would like to take derivatives. Consider the generic "pseudo code" statements

$$s = p(t), \tag{8.12}$$

$$u = q(s). \tag{8.13}$$

The derivative of s with respect to t is

$$\frac{ds}{dt} = \frac{d}{dt}p(t). \tag{8.14}$$

For each quantity that depends on t we must store its derivative with respect to t. If the code for dp/dt already exists, then when s is calculated, ds/dt can also be calculated and stored. When it comes time to calculate u, the chain rule gives

$$\frac{du}{dt} = \frac{d}{dt}q(s(t)) = \frac{d}{ds}q(s)\frac{ds}{dt}. \tag{8.15}$$

Note that ds/dt is available because it has been previously calculated and stored in the computer's memory. Therefore, if code for dq/dt exists, the derivative du/dt can be calculated at the same time that u is calculated.

Given the source code for a program and a list of design variables it is then possible for a compiler to produce not only an executable program that calculates the desired values but also the derivatives of those values with respect to the design variables. Several such programs are currently available (see the Notes section).

Example 8.1 To give a concrete idea of how these programs work, let x and y be the dimensions of a bar with rectangular cross section, and let F be the longitudinal force applied to the bar. The following few expressions are used to calculate the tensional stress σ in the bar:

$$\begin{aligned} &x = 3, \quad y = 5, \quad F = 100, \\ &A = xy, \quad \sigma = AF^{-1}. \end{aligned} \tag{8.16}$$

For each quantity that depends on one of the design variables, not only is the value of the quantity stored, but an array is used to store the derivatives with respect to each variable. With respect to Equations (8.16) there would be an array of length three.

The computer starts by assigning the variable x a value of 3, y a value of 5, and F a value of 100. Since x, y, and F are variables for which we are interested in computing their derivatives, when the computer runs into the code lines $x = 3$, $y = 5$ the computer would make the following assignments internally:

		Array for Derivatives		
Variable	**Value**	$\frac{\partial}{\partial x}$	$\frac{\partial}{\partial y}$	$\frac{\partial}{\partial F}$
x	3	1	0	0
y	5	0	1	0
F	100	0	0	1

When A is calculated, the operation of multiplying x and y, not only is the value of xy placed into A, but the derivatives of A with respect to each of the variables is also placed into an array associated with A. Examining the rule for differentiation with respect to multiplication

$$\frac{\partial}{\partial x}(xy) = y\frac{\partial x}{\partial x} = y, \qquad \frac{\partial}{\partial y}(xy) = x\frac{\partial y}{\partial y} = x,$$

the computer not only has the current value of x and y but also the values of $\partial x/\partial x$ and $\partial y/\partial y$ that are read from the arrays associated with the derivatives of x and y. Thus, the following assignment is automatically made within the computer:

		Array for Derivatives		
Variable	**Value**	$\frac{\partial}{\partial x}$	$\frac{\partial}{\partial y}$	$\frac{\partial}{\partial F}$
A	15	5	3	0

For the next calculation the chain rule is similarly applied:

$$\frac{\partial}{\partial x}\left(\frac{A}{F}\right)=\frac{F\frac{\partial A}{\partial x}-A\frac{\partial F}{\partial x}}{F^2}=\frac{(100\cdot 5)-(15\cdot 0)}{10000}=\frac{5}{100},$$

$$\frac{\partial}{\partial y}\left(\frac{A}{F}\right)=\frac{F\frac{\partial A}{\partial y}-A\frac{\partial F}{\partial y}}{F^2}=\frac{(100\cdot 3)-(15\cdot 0)}{10000}=\frac{3}{100},$$

$$\frac{\partial}{\partial F}\left(\frac{A}{F}\right)=\frac{F\frac{\partial A}{\partial F}-A\frac{\partial F}{\partial F}}{F^2}=\frac{(100\cdot 0)-(15\cdot 1)}{10000}=\frac{-15}{10000}.$$

Again, all of the values necessary for calculating the derivatives come either from the values in the operation or from the arrays, and so the following assignment is made:

		Array for Derivatives		
Variable	**Value**	$\frac{\partial}{\partial x}$	$\frac{\partial}{\partial y}$	$\frac{\partial}{\partial F}$
σ	0.15	0.05	0.03	−0.0015

As we can see, the computer is doing more work than just calculating σ. However, when compared to the use of finite differences, automatic differentiation will not only do less work (recall that finite differences require at least $n+1$ function calls for the gradient of a function of n variables) but the derivatives will be calculated exactly. ■

8.3 Scaling

Scaling is a term often used loosely to describe efforts aimed at resolving numerical difficulties associated with large differences in the values of computed quantities, usually of many orders of magnitude, but also associated with computing values that may be too small to distinguish from each other in the finite arithmetic of digital computers. Scaling can be applied to the design variables as well as the constraints and the objective function. We will discuss scaling of variables first, and then make a few comments on scaling the functions.

In optimization algorithms, proper scaling is critical when actual design models are used. Initially the design quantities are expressed in units natural for the problem at hand. For example, volume may be in units of m^3 with typical values of 10^{-2} m^3, yield strength may be in MPa with typical values of order 10^2 MPa, while length in mm may have values of order 10^3 mm. Thermal design problems will often tend to have units with variable values differing in orders of magnitude. In general, a model should be scaled with respect to its variables and functions, so that *all variables and functions have similar magnitudes in value in the feasible domain*. The order of magnitude selected is often unity or, for functions, the range $[-1, 1]$. Large differences in the order of magnitude of variables and functions will result in ill-conditioning matrices such as Jacobians and Hessians, making algorithmic calculations unstable or inefficient.

However, the advice above must be used with some caution. Occasionally, the current scaling of the problem may result in an objective function that becomes insensitive to changes in some variable or function near the optimum. The numerical algorithm can achieve no progress and terminates prematurely. Rescaling, for example, by simply multiplying the objective by some constant, can allow the algorithm to continue and terminate properly.

This becomes more evident when we consider that termination criteria include several tolerances to test for practical convergence. For example, if we use the criterion $|f_k - f_{k+1}| < \varepsilon$ to terminate, the value of ε used for functions with values of order 10^3 would not be appropriate for functions of order 10^{-3}. Scaling will affect any comparison of quantities with each other, when their values are really close, including the decision of when a variable is really zero. As we will see below, scaling affects gradients, Hessians, and Lagrange multipliers – in short, all the important quantities we use in a numerical optimization algorithm.

Scaling can be viewed as a linear transformation of the form

$$\mathbf{y} = \mathbf{A}\mathbf{x} + \mathbf{b}, \tag{8.17}$$

where $\mathbf{A}$ is a nonsingular matrix and $\mathbf{b}$ is a nonzero vector. Thus, the coordinate system $\mathbf{x}$ is replaced by the coordinate system $\mathbf{y}$. Since $\mathbf{A}$ is assumed nonsingular, it is invertible and the transformation is one to one. It is easy to prove that under a linear transformation $\mathbf{A}$ we get

$$\nabla_{\mathbf{x}} f = (\nabla_{\mathbf{y}} f)\mathbf{A}, \qquad \nabla_{\mathbf{x}}^2 f = A^T(\nabla_{\mathbf{y}}^2 f)\mathbf{A}. \tag{8.18}$$

An optimization algorithm is called *invariant* under the linear transformation, if $\mathbf{y}_{k+1} = \mathbf{A}\mathbf{x}_{k+1} + \mathbf{b}$. Note that any algorithm that takes the identity matrix $\mathbf{I}$ as an approximation to the Hessian will *not* be invariant. Thus, the gradient method, the modified Newton with stabilization, Equation (4.55), and quasi-Newton methods with $\mathbf{H}_0 = \mathbf{I}$ will not be invariant. But Newton's method with line search will be invariant. The invariance property is theoretically significant because invariant algorithms tend to be insensitive to ill-conditioning. In practice, no algorithm is invariant and the property is only marginally useful.

The most common and useful transformations are those where **A** is a diagonal matrix, **D**,

$$\mathbf{y} = \mathbf{D}\mathbf{x}, \quad \mathbf{y} = \mathbf{D}\mathbf{x} + \mathbf{b}. \tag{8.19}$$

The first of these simply indicates that typical values of x_i would be set equal to the inverse of the diagonal elements d_i.

Example 8.2 Let $\mathbf{x} = (x_1, x_2, x_3)^T$ with typical values $x_1 = 10$, $x_2 = 3(10^{-3})$, $x_3 = 400$. Suppose we define new variables $y_1 = x_1/10$, $y_2 = x_2/0.003$, $y_3 = x_3/400$. Note that then

$$\begin{pmatrix} y_1 \\ y_2 \\ y_3 \end{pmatrix} = \begin{pmatrix} 1/10 & 0 & 0 \\ 0 & 1/0.003 & 0 \\ 0 & 0 & 1/400 \end{pmatrix} \begin{pmatrix} x_1 \\ x_2 \\ x_3 \end{pmatrix};$$

hence the matrix **D** has elements d_i equal to the inverse of the typical values of x_i. ■

This scaling is point dependent and may be inappropriate if the values of **x** change over a range of orders of magnitude. A more sophisticated scaling is the second of (8.19), which can be used if good upper and lower bounds for the x_is are known. If we know that $l_i \leq x_i \leq u_i$, then the transformation $\mathbf{y} = \mathbf{D}\mathbf{x} + \mathbf{b}$ is given by

$$d_i = 2/(u_i - l_i), \quad b_i = -(u_i + l_i)/(u_i - l_i). \tag{8.20}$$

This will guarantee that $-1 \leq y_i \leq 1$ for all i. This scaling should be used if at least the order of magnitude of the bounds l_i, u_i is known.

Example 8.3 Consider the vector $\mathbf{x} = (x_1, x_2)^T$ with $10 \leq x_1 \leq 200, 0.5 \leq x_2 \leq 2.3$. We define new variables

$$y_1 = \frac{x_1}{(200 - 10)/2} - \frac{200 + 10}{200 - 10} = 0.01053x_1 - 1.105,$$

$$y_2 = \frac{x_2}{(2.3 - 0.5)/2} - \frac{2.3 + 0.5}{2.3 - 0.5} = 1.111x_2 - 1.556.$$

If we take $x_1 = 100$ and $x_2 = 1$, we find $y_1 = -0.052$ and $y_2 = -0.445$. ■

Scaling may be used to improve the conditioning of the Hessian. Several techniques have been proposed for "optimal" conditioning of the Hessian by minimizing the condition number of the Hessian at the solution using a linear transformation. These elaborate methods are often not as efficient as one would hope. The simplest attempt that could be beneficial is to use a diagonal scaling based on the Hessian at $\mathbf{x}_0$, the starting point of the iterations. The best-conditioned matrix is the identity, so if we selected a matrix equal to $\mathbf{H}^{1/2}$, the transformed problem will have a Hessian that is identity. Since we cannot do that at $\mathbf{x}_*$, we may do it at $\mathbf{x}_0$ or a later point. Then we have $\mathbf{y} = \mathbf{D}\mathbf{x}$ with $d_i = (h_{ii})^{1/2}$, where h_{ii} are the elements of the diagonalized

Hessian. This implies the need for a spectral decomposition to find the eigenvalues. However, if a Cholesky factorization is employed, this task may be somewhat simplified (Gill, Murray, and Wright 1981).

Example 8.4 Consider the function

$$f = 3x_1^2 + 100x_2^2 \quad \text{with} \quad \mathbf{H} = \begin{pmatrix} 6 & 0 \\ 0 & 200 \end{pmatrix}.$$

Since this is quadratic and the Hessian is fixed, scaling for conditioning is straightforward. Here the function is separable and the Hessian is diagonal. So the scaling would simply be

$$x_1 = (1/\sqrt{6})y_1, \quad x_2 = (1/\sqrt{200})y_2,$$

and substituting in f we get $f = (\frac{1}{2})(y_1^2 + y_2^2)$. ■

Scaling techniques may be also used in specific algorithms. For example, a simple scaling that has been suggested for the BFGS method and verified by computational experiments is as follows (Shanno and Phua 1978): In the BFGS procedure take $\mathbf{H}_0^{-1} = \mathbf{I}$ and determine $\mathbf{x}_1$ by a suitable choice of α_0 satisfying Armijo–Goldstein type criteria. *Before* $\mathbf{H}_1^{-1}$ is calculated, scale $\mathbf{H}_0$ by setting

$$\hat{\mathbf{H}}_0^{-1} = \alpha_0 \mathbf{H}_0^{-1}. \tag{8.21}$$

Then compute $\mathbf{H}_1^{-1}$ using $\hat{\mathbf{H}}_0^{-1}$ rather than $\mathbf{H}_0^{-1}$. Another approach in the same spirit is to perform the initial scaling

$$\hat{\mathbf{H}}_0^{-1} = (b/a)\mathbf{H}_0^{-1}, \tag{8.22}$$

where $a = \partial \mathbf{g}_0^T \mathbf{H}_0^{-1} \partial \mathbf{g}_0$, $b = \partial \mathbf{x}_0^T \partial \mathbf{g}_0$. Both these techniques are referred to as *self-scaling* and appear to be effective.

So far we have concentrated on scaling the model variables. However, the same rationale applies to the functions. Scaling the constraint functions will affect Jacobian matrices, Lagrange multipliers, Hessians, and so on. In general, a diagonal scaling of the form $\mathbf{g}_{\text{scaled}} = \mathbf{D}\mathbf{g}_{\text{unscaled}}$ can be used to make all constraint values have the same order of magnitude, say, in the interval $[-1, 1]$, at the initial point $\mathbf{x}_0$, as we did for the variables above.

It is important to note that because scaling affects Lagrange multiplier estimates it will indirectly affect the identification of active constraints, the progress of an active set strategy, and the eventual satisfaction of the KKT conditions.

Less obvious is the need for scaling the objective function. Again there seems no theoretical justification to scale the objective. However, one can quickly see that if the objective is too small throughout the domain, this will trigger termination criteria prematurely. Moreover, the "sensitivity" of the objective function to changes

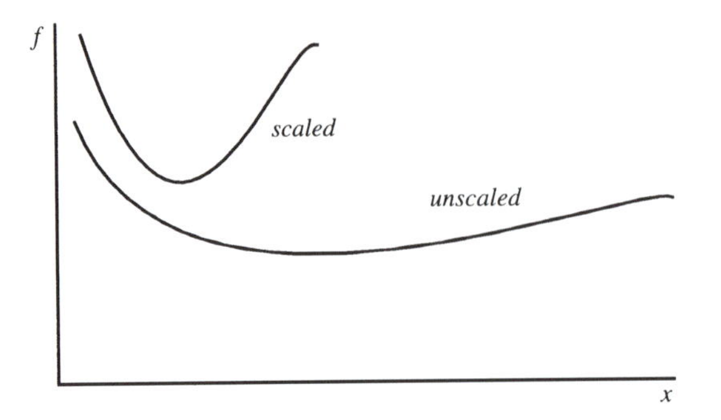

Figure 8.3. Scaling the objective function to eliminate flat spots.

in the variable or constraint values can have a profound influence on convergence. Independently of any other scaling that may have been done, it is still possible that the objective function becomes substantially "flat" at some iteration point, including the minimum. The iterations may terminate prematurely and a different optimum may be computed for different starting points, often giving the false impression of multiple optima. Scaling the objective by simply multiplying it by a constant will increase its "sensitivity" and eliminate the "flatness" (Figure 8.3). Using a multiplying constant works better than adding a constant, exactly because of the effect on reducing the "flattening" uniformly.

Because overscaling can also have undesirable effects some experimentation is required before consistently convergent results are achieved. In any case, one cannot overemphasize that *scaling is the single most important, but simplest, reason that can make the difference between success and failure of a design optimization algorithm.*

8.4 Interpreting Numerical Results

A typical optimization package will provide a set of output values produced by the NLP algorithm. These sets of numbers must be looked at as a whole to interpret what the results really mean. This would seem particularly important if the NLP algorithm terminated without apparently identifying an optimum. Indeed it is important to check what you have even when the code declares success. "False-positive" results are possible, namely, the termination test is passed successfully but the algorithm has really not identified a minimum.

As an example, recall that the most common and efficient unconstrained algorithms for minimization (say, BFGS) guard against reaching maxima but may stop at a saddle point. The constrained algorithms that use some unconstrained optimization algorithm within them will also suffer from the same difficulty. Termination criteria based on satisfying the KKT conditions will declare a normal successful termination.

When things go wrong after repeated optimization runs, one is often required to do some detective work: Collect evidence and interpret the facts to arrive at the truth.

Code Output Data

A typical NLP code will provide output containing values for the following quantities:

- design variables, objective, and constraint functions at the initial and final points;
- lower and upper bounds on the variables and distance of design variables from these bounds at the initial and final points;
- constraint violations at the initial and final points;
- Lagrange multipliers at the optimum for all constraints, including the inactive ones;
- termination criteria, such as KKT norm, number of iterations, and change in the objective and design point during the last iteration;
- algorithm-specific data, such as number of subproblems solved, number of line searches performed, sizes of various arrays, etc.; and
- all of the above, at all intermediate iterations.

The output information will not only depend on the specific algorithm but also on the code. Code programmers with a numerical analysis bent may include a variety of raw data helping the user to follow closely the numerical operations of the algorithm. Others may opt for diagnostic messages that provide possible reasons for an unsuccessful run and potential remedies. Some efforts have been made to use expert systems that assist users and also learn automatically from accumulated problem-solving experience so that they can provide increasingly better diagnostics, tagged to different problem classes. Some packages may also provide graphing capabilities that track the values listed above. Such graphs are useful in visualizing and interpreting results.

When looking at output data the overall philosophy should be to ascertain data consistency. For example, a constraint that has reached its bound should be considered as active, but only after it has been verified that the associated Lagrange multiplier is nonzero, or indeed positive for an inequality in the standard negative null form. If the model has been analyzed for boundedness, along the lines of Chapters 3 and 6, there may be some known activities or some conditional activities for constraints. The numerical output should be verified to satisfy these boundedness results, typically derived from monotonicity analysis.

Degeneracy

Degeneracy arises when zero or near-zero Lagrange multipliers are associated with active constraints. The multipliers are first-order sensitivity indicators, and so a zero value for them may prove optimality (unless the inequality is inactive). Usually, a zero multiplier of an active constraint will mean that the constraint is redundant and can be removed from the model (recall Example 5.14). This is not always the case. A zero multiplier will occur if the constraint surface contains a saddlepoint of the objective. The constraint is not redundant, since descent will occur if it is deleted from the working set. Thus, if the geometry of the constraint surface is not well understood, when a very small multiplier is estimated for an active constraint the best strategy is to delete the constraint from the working set anyway. While doing this, one should remember that (a) only multiplier estimates are found away from the solution and (b) multiplier estimates are affected by the scaling of the constraints.

8.5 Selecting Algorithms and Software

When one is newly exposed to nonlinear optimization techniques, the large number of proposed algorithms and attendant variants makes one wonder which is the best algorithm. Seasoned users typically will answer by saying that "the best algorithm is the one *you* understand best"! There is much truth to such an attitude, as the effectiveness of a particular code depends not only on the underlying theoretical algorithm but also on the specific code implementation and its individual algorithmic elements, such as linear or quadratic solvers.

Some codes *are* better than others in terms of programming elegance and specific numerical processes, including extensive empirical testing. Every nonlinear optimization code in the end includes some heuristics no matter how comprehensive its theoretical analysis may be. Heuristics are a result of experience, including your own experience with a particular code.

Different forms of models can be solved more effectively by different kinds of algorithms. For example, a model with extensive monotonicity in the functions is likely to have a constraint-bound solution or nearly one. Convex approximation algorithms, such as sequential linear programming or the method of moving asymptotes, use approximations that maintain the essential monotonicity of the functions often enabling them to identify quickly the active set. Close to the optimum, if there are remaining degrees of freedom (the problem has an interior optimum in one or more dimensions), these algorithms may become very slow as the trust region tries to "zero in" on the interior solution. The (usually ad hoc) strategy for the trust region modification becomes critical for the efficiency of the algorithm. Structural optimization problems involving sizing of components under stress and deflection constraints have extensive monotonicity and so convex approximation algorithms will work well for them.

In engineering design of systems with realistic complexity we rarely demand the identification of the mathematical optimum with precision, and we often will settle for a design that represents a substantial improvement over an existing one. The limiting concern in such problems is often the cost of model computation, which could run to several hours or days for one single function computation. There is a strong preference to use algorithms that guarantee intermediate feasible designs, such as generalized reduced gradient or feasibility-preserving SQP variants, for example, the code FSQP (Zhou and Tits 1996). The price to pay is the increased computational cost per iteration which, numerical purists find unacceptable. Engineers, however, find preservation of feasibility as being worth the extra cost.

In the past twenty years or so significant efforts have been expended trying to evaluate algorithms. Such efforts have rather decreased in recent years owing in part to general acceptance of the attitude we discussed above, and in part to the wide availability of several high quality, well-tested algorithms. One study of special interest for optimal structural design was published by Schittkowski, Zillober, and Zotemantel (1994).

Evaluation of NLP software usually involves keeping track of quantities such as computer central processing unit (CPU) time and number of function evaluations until termination, accuracy of termination, success rates over a given test bank of problems, and ease of implementation. There is no universal agreement on the criteria or on their priority or on the algorithm performance itself. Among the several studies that have been published an accessible discussion of performance evaluation can be found in Reklaitis, Ravindran, and Ragsdell (1983). A comprehensive study was conducted and reported by Hock and Schittkowski (1981). The Hock and Schittkowski collection of test problems is still used as a standard for trying out new algorithms. Early but still useful references on evaluating optimization software are Schittkowski (1980) and Mulvey (1981). The particularly attractive article by Lootsma (1985) is a follow-up of a previous report of his (Lootsma 1984) based on interviews with experts and users in the field.

There is probably universal agreement today that the class of SQP algorithms is the best general purpose one. They offer a good balance between efficiency and robustness, and they require no "fine tuning" of internal program parameters. Some proper individual scaling will usually allow them to work well for most practical problems.

Partial List of Software Packages

Accessibility of optimization software is increasing rapidly. An excellent reference is the *Optimization Software Guide* by Moré and Wright (1993) (found also on-line in the *NEOS Server*; see list below). There are a number of excellent Internet sites that compile optimization software information. Below we provide a partial list of software packages that may be most appropriate for the optimal design problems studied in this book. The list is not comprehensive and no attempt is made to make specific recommendations. We do not include optimizers that are part of a larger analysis package. Most of the major finite element method packages for structural analysis now include optimization algorithms or at least readily computed derivatives, which they refer to as sensitivities. Many of the new packages have strong graphical user interface (GUI) features. GUIs are attractive to new users but more experienced users will prefer access to direct editing of input files.

ACSL Optimize (*Advanced Continuous Simulation Language*) is a product suite available from MGA Software (www.mga.com) that provides support for mathematical simulation, parameter estimation, and sensitivity analysis. It is designed for researchers and engineers in chemistry, chemical engineering, and related fields.

DOT (*Design Optimization Tools*) from Vanderplaats R&D (www.vrand.com) contains general purpose nonlinear programming codes designed to interface with simulation analysis programs. Original development was motivated by structural optimization applications.

EASY-OPT is a full-fledged interactive optimization package under the Microsoft Windows platform from K. Schitkowski (www.uni-bayreuth.de/departments/math/~kschittkowski/easy_opt). It can solve general nonlinear programming, least squares, L_1, min-max, and multicriteria problems interactively.

The *Harwell Subroutine Library* is a suite of ANSI Fortran 77 subroutines and Fortran 90 modules for scientific computation, including modules for optimization, written by members of the Numerical Analysis Group and other experts (www.dci.clrc.ac.uk).

Several popular spreadsheet programs, such as *Microsoft Excel*, *Lotus 1-2-3*, and *QuatroPro*, can be used with a GRG algorithm provided by Frontline Systems (www.frontsys.com).

IMSL originally created as the International Mathematical and Statistical Libraries in the late 1960s for use in business, engineering, and sciences, and now marketed by Visual Numerics (www.vni.com), is fully modernized and contains many algorithms including Schittkowski's SQP.

iSIGHT offered by Engineous Software (www.engineous.com) is a GUI-based optimization package with its early motivation coming from artificial intelligence techniques.

LMS Optimus is a GUI-based optimization package from LMS International (www.lmsitl.com) that has powerful design of experiments and response surface modeling capabilities and offers graphical linking with external simulations.

The *MATLAB Optimization Toolbox* in the popular MATLAB package from Mathworks (www.mathworks.com) includes unconstrained and SQP algorithms.

Mathematica from Wolfram Research (www.wolfram.com) is a widely used symbolic mathematics package that includes unconstrained and SQP algorithms.

The *NEOS* (*Network-Enabled Optimization System*) *Server* for solving optimization problems remotely over the Internet is maintained by the Optimization Technology Center of Argonne National Laboratory (www.mcs.anl.gov/otc). The server contains software and an on-line version of the optimization software guide by Moré and Wright (1993) under www.mcs.anl.gov/otc/Guide/SoftwareGuide.

OPTDES-X distributed by Design Synthesis and based on work at Brigham Young University (www2.et.byu.edu/~optdes) has several optimizers and good graphic display capabilities.

OR-Objects maintained by DRA Systems (www.opsresearch.com) is a freeware library of more than 250 operations research (OR) objects for Java that can be used to develop OR applications.

Partial List of Internet Sites

The short list of web sites we provide below may serve as a starting point for an Internet search. A search with specific keywords is the best way to obtain updated information on software and algorithms. Addresses and links on the Internet change frequently so references here are likely to become obsolete quickly. Nevertheless it may be useful to list a few sites that have proven robust over time and that are associated with groups active in design optimization or provide links to other sites.

The *ASME* Design Automation Committee (www.me.washington.edu/~asmeda) provides links to research and events related to design optimization.

The International Society for Structural and Multidisciplinary Optimization (www.aero.ufl.edu/~issmo) also has similar information and links.

The Mathematical Programming Glossary maintained by H. J. Greenberg (carbon.cudenver.edu/~hgreenbe/glossary/glossary) contains a large set of definitions of terms specific to mathematical programming.

Michael Trick's Operations Research Page (mat.gsia.cmu.edu) is a well-maintained personal page with pointers to many topics in operations research.

A collection of computer sciences bibliographies useful for optimization is maintained at the University of Karlsruhe (liinwww.ira.uka.de/bibliography).

The Multidisciplinary Optimization Branch at NASA Langley Research Center (fmad-www.larc.nasa.gov/mdob) has a long history of interest in the design of complex engineering systems and subsystems.

The Operations Research site (OpsResearch.com) contains the *OR-Bookstore*, with books rated by level of difficulty; the *OR-Objects*, a freeware library of Java objects, mentioned above; and the *OR-Links*, a collection of OR Internet links.

A list of Internet sites of special interest for design optimization is maintained at the homepage of the Optimal Design Laboratory at the University of Michigan (http://ode.engin.umich.edu/links.html).

8.6 Optimization Checklist

The early stages of an optimal design project are critical for the success of the entire effort. Selecting the proper framework to define the project requires care and rigor. It is extremely helpful to express the optimization problem in words before describing it mathematically. With this in mind, we now suggest the following tersely listed items.

Problem Identification

1. Describe the design problem verbally. What goals do you want to achieve? Do they compete?
2. Identify quantities you could vary and others you should consider fixed. What specific trade-offs result from trying to satisfy competing goals?
3. Verbally express the trade-offs in optimization terms.
4. Search the literature for optimization studies related to your design problem.
5. Verify, in the final description of your problem, that the trade-offs are meaningful enough to justify further optimization study. Recognize that constraint-bound problems have obvious solutions.

Initial Problem Statement

1. Verbally define the objective function and constraints. If several objectives exist, decide how you will translate the multiobjective problem to a model with a single objective.
2. If you chose one objective and treat other objectives as constraints consider how you will set proper bounds. Expect to do a post-optimal parametric study on these bounds. If you chose a weighted scalar objective, consider how you will chose the weights. Expect to perform a post-optimal parametric study on these weights to generate a Pareto solution set.
3. List all modeling assumptions, updating the list as you proceed through the study. Remember that these will have to be checked at the end of your study to see whether they are satisfied by the final design.
4. List separately which design quantities have been classified as variables, parameters, and constants. You may decide to update or modify this classification as you proceed with the problem. Recall that only what *you* can choose can be a design variable in *your* problem statement.
5. Evaluate the availability of mathematical analysis model(s) for describing the objective and constraints. Will the models be explicitly (algebraically) or computationally defined?
6. When the analysis models are separate codes, "black box" simulations, or legacy codes, consider how the different software components will communicate with the numerical optimizer.

Analysis Models

1. Identify the system to be modeled, its boundary, and its links to the environment. Use appropriate system diagrams, for example, control volumes or free-body diagrams. Describe the system's function verbally.
2. Identify the system's hierarchical level and choose the limits of your decision-making capability. This will allow you to define more clearly the design variables and parameters in your model.
3. Verbally list inputs, outputs, and relating functions for each analysis model. If your model is a "black box" simulation you still need to understand what are the exact inputs/outputs of the simulation.
4. State the natural laws and principles of engineering science governing the system.
5. List all modeling assumptions, continuing from the initial problem statement.
6. Develop the analysis model(s) if needed. For readily available models, state the mathematical form and take results directly from the reference source. Recheck that you know explicitly the assumptions made in the reference.
7. Gather any empirical information from available data or generate empirical data if you have to.
8. Use curve-fitting techniques to develop appropriate analytical expressions from empirical data available in tables or graphs.
9. Distinguish carefully among variables, parameters, and constants for the analysis model(s). Compare these lists with the input/output list of (3).
10. Collect all results from analysis and curve fitting and assemble all equations relating the system quantities. Verify that you have a complete capability for computing outputs from inputs.
11. Check the model by calculating, for simple cases, desired quantities for selected values of parameters and constants.
12. Do a preliminary evaluation of model efficiency, accuracy, and operation under the stated assumptions.
13. For an implicit model or simulation, estimate typical computation time for each analysis run. Is computing time too long?
14. For computationally expensive models, consider creating a metamodel. Use such a metamodel instead of the full one in your optimization model. Recognize that building metamodels is also computationally expensive.

Optimal Design Model

1. State the design model, listing objective and constraint functions and deriving their mathematical forms.
2. Carefully define and list all design variables, parameters, and constants.

3. Distinguish clearly between equality and inequality constraints, keeping them in their original form without attempting to simplify them.
4. If some functional relations are still unavailable, try to state them in a general symbolic form, for example, $f(x_1, x_2, x_3) = x_4$, which at least shows which variables are involved. Attempt to find an approximate relation from an analysis model, experiments, or a consultant's opinion.
5. If the design model requires iteration of a computer-based model, consider using metamodels based on computational experiments, as mentioned earlier.
6. Review all modeling assumptions to see if any new ones have been added to the design modeling stage.
7. For typical values of the parameters, verify that the feasible domain is not empty. Check for inconsistency in the constraints. Finding at least one feasible point should be attempted using your engineering knowledge about the problem or information on a current design.

Model Transformation

1. Count degrees of freedom. Verify this invariant count after each further model transformation or simplification.
2. Identify monotonicities in the objective and constraint functions.
3. Use monotonicity analysis procedures wherever possible. Every monotonic variable in the objective must have an appropriate bound. If not, check for missing constraints, missing terms in the objective, or oversimplified approximations. Also check the hierarchical system level assumptions. Is the model well bounded?
4. Check validity of all transformations. For example, if you divided by something, it cannot be zero. Must this restriction be made explicit?
5. Look for hidden constraints such as asymptotes. For example, you may eliminate asymptotic expressions by case decomposition and additional constraints as appropriate.
6. Check for redundant constraints.
7. Perform a reduction of the feasible domain to tighten up the set constraint and perhaps find explicit bounds based only on natural constraints.
8. Drop all practical constraints not needed for well boundedness. Mark any remaining for further study.
9. For any revised model, (i) make a monotonicity table; (ii) apply monotonicity analysis to identify constraint activity; (iii) perform case decomposition based on activity status and solve each well-bounded case if their number is small.
10. Does the problem have discrete variables? If a discrete variable takes only a few (discrete) values treat it as a parameter and resolve the problem for each parameter value.

11. Does a discrete variable have an underlying continuous form? If so, consider relaxing the discreteness requirement. The continuous relaxation problem solution will be a lower bound on the discrete one. Set up a branch-and-bound strategy if necessary.
12. If several variables are truly discrete (e.g., 0,1) methods beyond those of this book are likely necessary.

Local Iterative Techniques

1. If you have equality constraints, see if you can put them into a function definition format. For example, if the problem is: min $f(x_1, x_2, x_3)$, subject to $h_1(x_1, x_2) - x_3 = 0$, use $x_3 = h_1(x_1, x_2)$ as a definition and compute $f[x_1, x_2, h_1(x_1, x_2)]$ instead of treating $h_1 - x_3 = 0$ as an equality constraint.
2. Watch out for function forms defined over intervals of the variables that may be violated during iterations. Recall that some algorithms will temporarily violate constraints. Make transformations to avoid such violations if they will lead to a computer "crash."
3. Select a starting base case that is feasible. Although most codes accept infeasible starting points, as a designer you should be able to find a feasible start by iterating the analysis models. If not, beware of constraint inconsistency.
4. Scale the problem using the base case values. Rescale as necessary, so that (i) all variables and functions have about the same order of magnitude and/or (ii) the objective function is properly "sensitive" to changes in variables and constraints, so that you avoid getting trapped in flat areas.
5. Carefully choose values of the termination criteria, taking scaling effects into account. Beware of noisy functions!
6. Are the model functions all smooth as required by local methods? If lack of smoothness is modest, Newton-like methods work better. If there is substantial noise, discontinuities, or nondifferentiabilities, consider your options: Look for nongradient-based methods, develop surrogate models, or try to eliminate the sources of trouble in the model.
7. If the model is explicitly algebraic, finite differencing can be avoided by computing gradients symbolically.
8. Do not assume a constraint is active just because its Lagrange multiplier does not vanish completely. Small values may indicate degeneracy or ill-conditioned computations. Monotonicity arguments may be needed for reliable results.
9. Numerical results should be verified analytically if possible, particularly for determining constraint activity. Apparently reliable results obtained by local methods will justify an attempt at global verification.
10. For parametric study, a special purpose program combining case decomposition and local iteration may be needed to generate reliable results efficiently.

Final Review

1. Is the solution sufficiently robust? Try different starting points. If you get different results, examine if this indicates sensitivity to the termination criteria or existence of multiple minima.
2. Consider using a multistart strategy, namely, select starting points that will tend to "cover" the feasible domain, in an effort to find all local minima.
3. Is the model accurate enough near the optimum?
4. Are the model validity constraints satisfied? Should some assumptions and simplifications be discarded or modified?
5. Should the classification of variables and parameters be changed?
6. Would a new model be at the same hierarchical level, but more complex and accurate? Should you move up in the hierarchy of a larger system thus increasing your decision-making options?
7. Define and explore a second-generation model.

8.7 Concepts and Principles

Finally, it is time to summarize the key ideas reviewed and developed in this book. The checklist of the previous section mixed them in with the many routine but important steps of a design optimization study. Here we list concepts and principles introduced throughout this book that should by now be familiar to the reader. The aim of the book is to create design optimization practitioners. The list can serve as a self-test to a designer's readiness to face the challenges of real-world problem solving.

Unlike the earlier checklist, this compilation emphasizes function rather than position in the project time schedule. We discern three major functions: building the model, analyzing the model, and searching for local optima. Concepts are listed within these categories without any particular order, along with pertinent section number.

Model Building

Model setup

- Negative null form (1.2)
- Hierachical levels (1.1, 1.2)
- Configuration vs. proportional design (1.2)
- Design variables, parameters, constants (1.1)
- Analysis vs. design models (1.1)
- Degrees of freedom (1.3)

Model validity constraints (1.4, 2.5)

Surrogate models

Curve fitting (2.1)

Families of curves (2.1)

Least squares (2.2)

Neural nets (2.3)

Kriging (2.4)

Multiple criteria (1.2)

Pareto set (1.2)

Natural and practical constraints (2.8)

Model properties

Feasibility (1.3)

Activity (1.3, 3.2, 6.5)

Well boundedness (1.3, 3.1, 3.2)

Monotonicity (3.3, 3.4)

Anticipated solution

Interior vs. boundary optima (1.4)

Constraint-bound optima (1.3)

Local vs. global optima (1.4)

Model Analysis

Monotonicity

Monotonicity theorem (3.3)

First monotonicity principle (3.3)

Second monotonicity principle (3.7)

Recognition of monotonicity (3.4)

Monotonicity table (6.2)

Regional monotonicity (3.6)

Directing equalities (1.3, 3.6)

Functional monotonicity analysis (6.3)

Activity

Activity theorem (3.2)

Dominance (1.4, 3.5)

Constraint relaxation (3.5)

- Semiactivity (3.2, 6.5)
- Weak vs. strong activity (6.5)
- Tight constraints (6.5)

Criticality (3.3)

- Conditional (3.5)
- Multiple (3.5)
- Uncriticality (3.5)

Parametric solution

- Parametric models (6.1)
- Parametric functions (6.1)
- Case decomposition (6.2)
- Branching (6.2)
- Parametric testing (6.1)

Partial minimization (3.2)

Model repair (3.3)

Model reduction and simplification

- Variable transformation (1.2, 2.5, 2.7, 2.8)
- Variable elimination (1.3, 3.2, 3.3)
- Asymptotic substitution (2.8)
- Redundant constraint (1.4)
- Feasible domain reduction (2.8, 6.7)

Discrete variables

- Discrete local optima (6.5)
- Multiple optima (6.5)
- Constraint tightening (6.5, 6.6)
- Constraint derivation (2.8, 6.7)
- Rounding off (6.7)
- Continuous relaxation (6.8)
- Branch-and-bound (6.8)
- Exhaustive enumeration (6.8)

Local Searching

Model forms

- Weierstrass Theorem (4.1)
- Compact set (4.1)

- Convexity (4.4)
- Unimodality (7.2)
- Taylor Series (4.2)
- Linear Approximation (4.2)
- Quadratic Approximation (4.2)
- Regularity–constraint qualification (5.2)

Slope
- Gradient (4.2, 5.3)
- First-order necessity (4.3, 5.6)
- Karush–Kuhn–Tucker conditions (5.6)
- Reduced gradient – constrained derivatives (5.3, 5.5)
- Projected gradient (5.5)
- Derivative approximations (8.2)
- Finite differences (8.2)
- Automatic differentiation (8.2)

Lagrange multipliers (5.3)
- Constraint sensitivity (5.9)
- Active set strategies (5.8, 7.4)
- Constraint activation (5.8, 7.4)
- Degeneracy (8.4)

Curvature
- Hessian (4.2)
- Positive-definite matrix (4.3)
- Positive-semidefinite matrix (4.3)
- Indefinite matrix (4.3)
- Second-order sufficiency (4.3, 5.4)
- Cholesky factorization (4.7)

Convergence
- Local (7.1)
- Global (7.1)
- Stabilization (4.7)
- Scaling (8.3)
- Termination criteria (7.1)

Algorithms
- Line search (4.6, 7.2)

Inexact, exact line search (7.2)

Armijo–Goldstein criteria (7.2)

Unconstrained

Gradient method (4.5, 4.6)

Newton's method (4.5, 4.6)

Quasi-Newton methods (7.3)

Constrained

Generalized reduced gradient (5.5, 7.5)

Projected gradient (5.5, 7.5)

Penalties and barriers (7.6, 7.8)

Augmented Lagrangian (7.6)

Trust region (4.8, 7.8)

Sequential linear programming (7.8)

Sequential quadratic programming (7.7)

Sequential convex approximations (7.8)

Method of moving asymptotes (7.8)

Scanning this list from time to time may jog the designer's memory of concepts and techniques (perhaps appearing farfetched on first reading) that are of potential value to a current study. Continual use can polish even the more abstract mathematical ideas into widely applied designer's expertise.

8.8 Summary

A complete and successful optimization study requires many skills and a bit of luck. One should be thoroughly familiar with the artifact being modeled and understand the technology behind it. This allows mathematical simplification of the model that would not be evident to the unfamiliar analyst. Often functional trends relating the variables are sufficient for finding meaningful preliminary design results. Monotonicity analysis can be particularly effective in this respect.

In larger complicated models, boundedness checking is not only initially useful but also helpful for focusing on the more difficult aspects of the problem. Local iterative methods may be the only way to reach a complete solution. Even then, model transformations and special procedures may be needed for convergence.

The two checklists are intended to guide the reader through new situations. One list gives an operating procedure; the other suggests useful principles for getting past the hard problems. A blend of rigorous global analysis, local exploration, and engineering heuristics is the only realistic way to solve design problems efficiently and reliably.

Notes

One can learn a lot by reviewing case studies on optimal design applications. The interested reader can find many of them usually as articles in technical journals, or in design optimization textbooks (see Notes for Chapters 1 and 2). The availability of powerful commercial software, as well as the inclusion of optimization algorithms in standard general purpose mathematical software, has dramatically increased the use of design optimization tools in industry. In fact, the most interesting and complex applications are performed internally by company design engineers and are rarely published in detail, if at all. Design optimization is now used extensively in the aerospace, automotive, electronic, and chemical process industries.

A term design project is indispensable in a design course that uses this book. An optimal design course has plenty of elegant mathematics to absorb instructors and students. Yet the concepts and methods come alive only when we apply ourselves to a real problem. In the University of Michigan course, projects are proposed by students in a great variety of domains, and the students must take responsibility for their models. Properly formulating a model, taking into account both design and numerical solution needs, is the greatest value added in the course.

Several texts discuss modeling implications for local computation. Among them, the book by Gill, Murray, and Wright (1981) remains a most useful companion. Examples of automatic differentiation codes available for processing Fortran, C, and C++ programs are the codes ADIFOR (Bischof et al. 1992), ADIC (Bischof, Roh, and Mauer 1997), AdolC (Griewank, Juedes, and Utke 1996), and PCOPT (Dobmann, Liepelt, and Schittkowski 1994). Recent versions are available for downloading from Internet sites, such as those listed in Section 8.5.

References

Abadie, J., and J. Carpentier, 1969. Generalization of the Wolfe reduced gradient method to the case of nonlinear constraints. In *Optimization* (R. Fletcher, ed.). Academic Press, London, Chapter 4.

Adby, P. R., and M. A. H. Dempster, 1974. *Introduction to Optimization Methods*. Chapman and Hall, London.

Agogino, A. M., and A. S. Almgren, 1987. Techniques for integrating qualitative reasoning and symbolic computation in engineering optimization. *Engineering Optimization*, Vol. 12, No. 2, pp. 117–135.

Alexander, R. McN., 1971. *Size and Shape*. Edward Arnold, London.

Alexandrov, N., 1993. *Multilevel Algorithms for Nonlinear Equations and Equality Constrained Optimization*. Doctoral dissertation, Dept. of Computational and Applied Mathematics, Rice University, Houston.

Alexandrov, N. M., and M. Y. Hussaini (eds.), 1997. *Multidisciplinary Design Optimization: State of the Art*. SIAM, Philadelphia.

Ansari, N., and E. Hou, 1997. *Computational Intelligence for Optimization*. Kluwer, Boston.

Aris, R., 1964. *The Optimal Design of Chemical Reactors*. Academic Press, New York.

Armijo, L., 1966. Minimization of functions having Lipschitz continuous first partial derivatives. *Pacific Journal of Mathematics*, Vol. 16, No. 1, pp. 1–3.

Arora, J. S., 1989. *Introduction to Optimum Design*. McGraw-Hill, New York.

Athan, T. W., 1994. *A Quasi-Monte Carlo Method for Multicriteria Optimization*. Doctoral dissertation, Univ. of Michigan, Ann Arbor.

Athan, T. W., and P. Y. Papalambros, 1996. A note on weighted criteria methods for compromise solutions in multi-objective optimization. *Engineering Optimization*, Vol. 27, pp. 155–176.

Avriel, M., 1976. *Nonlinear Programming – Analysis and Methods*. Prentice-Hall, Englewood Cliffs, New Jersey.

Avriel, M., and B. Golany, 1996. *Mathematical Programming for Industrial Engineers*. Marcel Dekker, New York.

Avriel, M., Rijckaert, M. J., and D. J. Wilde (eds.), 1973. *Optimization and Design*. Prentice-Hall, Englewood Cliffs, New Jersey.

Ballard, D. H., Jelinek, C. O., and R., Schinzinger, 1974. An algorithm for the solution of constrained generalised polynomial programming problems. *Computer Journal*, Vol. 17, No. 3, pp. 261–266.

Bartholomew-Biggs, M. C., 1976. *A Numerical Comparison between Two Approaches to Nonlinear Programming Problems*. Technical Report No. 77, Numerical Optimization Center, Hatfield, UK.

Barton, R. R., 1992. Computing forward difference derivatives in engineering optimization. *Engineering Optimization*, Vol. 20, No. 3, pp. 205–224.

Bazaraa, M. W., and C. M. Shetty, 1979. *Nonlinear Programming – Theory and Algorithms*. Wiley, New York.

Beachley, N. H., and H. L. Harrison, 1978. *Introduction to Dynamic System Analysis*. Harper & Row, New York.

Beale, M., and H. Demuth, 1994. *The Neural Net Toolbox User's Guide*. Math Works.

Beightler, C., and D. T. Phillips, 1976. *Applied Geometric Programming*. Wiley, New York.

Beightler, C., Phillips, D. T., and D. J. Wilde, 1979. *Foundations of Optimization*. Prentice-Hall, Englewood Cliffs, New Jersey. First edition: Wilde and Beightler (1967).

Belegundu, A. D., and T. R. Chandrupatla, 1999. *Optmization Concepts and Applications in Engineering*. Prentice-Hall, Upper Saddle River, New Jersey.

Bender, E. A., 1978. *An Introduction to Mathematical Modeling*. Wiley-Interscience, New York.

Bendsøe, M., and N., Kikuchi, 1988. Generating optimal topologies in structural design using a homogenization method. *Computer Methods in Applied Mechanics and Engineering*, Vol. 71, pp. 197–224.

Ben-Israel, A., Ben-Tal, A., and S. Zlobek, 1981. *Optimality in Nonlinear Programming: A Feasible Directions Approach*. Wiley-Interscience, New York.

Bertsekas, D. P., 1982. *Constrained Optimization and Lagrange Multiplier Methods*. Academic Press, New York.

Best, M. J., and K. Ritter, 1985. *Linear Programming: Active Set Analysis and Computer Programs*. Prentice-Hall, Englewood Cliffs, New Jersey.

Bischof, C., Carle, A., Corliss, G., Griewank, A., and P. Hovland, 1992. {ADIFOR}-Generating derivative codes from Fortran programs. *Scientific Programming*, Vol. 1, No. 1, pp. 11–29.

Bischof, C., Roh, L., and A. Mauer, 1997. ADIC: An extensible automatic *differentiation tool for ANSI-C*. Preprint ANL/MCS-P626-1196, Argonne National Laboratory.

Borel, E., 1921, La theorie du jeu et les equations integrals a noyau symetrique gauche, *Comptes Rendus de L'Académie des Sciences*, Paris, France, Vol. 173, pp. 1304–1308. This is considered the inception of game theory, cited in [Stadler, 1979].

Box, M. J., Davies, D., and W. H. Swann, 1969. *Nonlinear Optimization Techniques*. Oliver and Boyd, Edinburgh.

Bracken, J., and G. P. McCormick, 1967. *Selected Applications of Nonlinear Programming*. Wiley, New York.

Brent, R. P., 1973. *Algorithms for Minimization without Derivatives*. Prentice-Hall, Englewood Cliffs, New Jersey.

Broyden, C. G., 1970. The convergence of a class of double rank minimization algorithms: Parts I and II. *J. Inst. Math. Appl.*, Vol. 6, pp. 76–90 and 222–31.

Byrd, R. H., Schnabel, R. B., and G. A. Shultz, 1987. A trust region algorithm for nonlinearly constrained optimization. *SIAM Journal of Numerical Analysis*, Vol. 24, No. 5, pp. 1152–1170.

Carmichael, D. G., 1982. *Structural Modeling and Optimization*. Halsted Press, New York.

Carnahan, B., Luther, H. A., and J. O. Wilkes, 1969. *Applied Numerical Methods*. Wiley, New York.

Carrol, R. K., and G. E. Johnson, 1988. Approximate equations for the AGMA J-factor. *Mechanisms and Machine Theory*, Vol. 23, No. 6, pp. 449–450.

Chapman, C., Saitou, K., and M., Jakiela, 1994. Genetic algorithms as an approach to configuration and topology design. *ASME Journal of Mechanical Design*, Vol. 116, pp. 1005–1012.

Chirehdast, M., Gea, H. C., Kikuchi, N., and P. Y. Papalambros, 1994. Structural cofiguration examples of an integrated optimal design process. *ASME J. of Mechanical Design*, Vol. 116, No. 4, pp. 997–1004.

Conn, A. R., Gould, N. I. M., and Ph. L. Toint, 1992. *LANCELOT: A FORTRAN Package for Large-Scale Nonlinear Optimization (Release A)*. Springer-Verlag, Berlin.

Crane, R. L., Hillstrom, K. E., and M. Minkoff, 1980. *Solution of the General Nonlinear Programming Problem with Subroutine VMCON*. Report ANL-80-64, Argonne National Laboratory, Argonne, Illinois.

Cressie, N., 1988. Spatial prediction and ordinary kriging. *Mathematical Geology*, Vol. 20, No. 4, pp. 405–421.

Cressie, N., 1990. The Origins of Krigin. *Mathematical Geology*, Vol. 22, No. 3, pp. 239–252.

Dahlquist. G., and A. Björck, 1974. *Numerical Methods*. Prentice-Hall, Englewood Cliffs, New Jersey.

Dantzig, G. B., 1963. *Linear Programming and Extensions*. Princeton University Press, Princeton, New Jersey.

Davidon, W. C., 1959. *Variable Metric Method for Minimization*. U. S. Atomic Energy Commission Research and Development Report No. ANL-5990, Argonne National Laboratories.

Davis, L. (ed.), 1991. *Handbook of Genetic Algorithms*. Van Nostrand-Reinhold, New York.

Dimarogonas, A. D., 1989. *Computer Aided Machine Design*. Prentice-Hall International, Hemel Hempstead, UK.

Dixon, L. C. W. (ed.), 1976. *Optimization in Action*. Academic Press, London.

Dixon, L. C. W., Spedicato, E., and G. P. Szegö (eds.), 1980. *Nonlinear Optimization – Theory and Algorithms*. Birkhauser, Boston.

Dobmann, M., Liepelt, M., and K. Schittkowski, 1994. Algorithm 746: PCOMP: A FORTRAN code for automatic differentiation. *ACM Transactions on Mathematical Software*, Vol. 21, No. 3, pp. 233–266.

Draper, N. R., and H. Smith, 1981. *Applied Regression Analysis*. Wiley, New York.

Duffin, R. J., Peterson, E. L., and C. Zener, 1967. *Geometric Programming*. Wiley, New York.

Dym, C. L., and E. S. Ivey, 1980. *Principles of Mathematical Modeling*. Academic Press, New York.

Edgeworth, F. Y., 1881, *Mathematical Physics*. P. Keagan, London.

El-Alem, M. M., 1988. *A Global Convergence Theory for a Class of Trust Region Algorithms for Constrained Optimization*. Doctoral dissertation, Dept. of Mathematical Sciences, Rice University, Houston.

Eschenauer, H., Koski, J., and A. Osyczka (eds.), 1990. *Multicriteria Design Optimization*. Springer-Verlag, Berlin.

Evtushenko, Y. G., 1985. *Numerical Optimization Techniques*. Optimization Software, Inc., New York.

Fiacco, A. V., and G. P. McCormick, 1968. *Nonlinear Programming: Sequential Unconstrained Minimization Techniques*. Wiley, New York.

Fletcher, R., 1970. A new approach to variable metric algorithms. *Computer Journal*, Vol. 13, pp. 317–322.

Fletcher, R., 1971. A general quadratic programming algorithm. *J. Inst. Math. Appl.*, Vol. 7, pp. 76–91.

Fletcher, R., 1980, 1981. *Practical Methods of Optimization (Vol. 1: Unconstrained, Vol. 2: Constrained)*. Wiley, Chichester.

Fletcher, R., and M. J. D. Powell, 1963. A rapidly convergent descent method for minimization. *Computer Journal*, Vol. 6, pp. 163–168.

Fleury, C., 1982. Reconciliation of mathematical programming and optimality criteria methods. In *Foundations of Structural Optimization: A Unified Approach* (A. J. Morris, ed.). Wiley, Chichester, pp. 363–404.

Fleury, C., and V. Braibant, 1986. Structural optimization: a new dual method using mixed variables. *Int. J. Num. Meth. Eng.*, Vol. 23, pp. 409–428.

Floudas, C., 1995. *Nonlinear and Mixed-Integer Optimization*. Oxford University Press, New York.

Fowler, A. C., 1997. *Mathematical Models in the Applied Sciences*. Cambridge University Press, New York.

Fox, R. L., 1971. *Optimization Methods for Engineering Design*. Addison-Wesley, Reading, Massachusetts.

Friedman, L. W., 1996. *The Simulation Metamodel*. Kluwer Academic Publishers, Norwell, Massachusetts.

Gajda, W. J., and W. E. Biles, 1979. *Engineering: Modeling and Computations*. Houghton Mifflin, Boston.

Gill, P. E., and W. Murray, 1972. Quasi-Newton methods for unconstrained optimization. *J. Inst. Math. Appl.*, Vol. 9, pp. 91–108.

Gill, P. E., and W. Murray, 1974. *Numerical Methods for Constrained Optimization*. Academic Press, London.

Gill, P. E., Murray, W., and M. H. Wright, 1981. *Practical Optimization*. Academic Press, London.

Glover, F., and M. Laguna, 1997. *Tabu Search*. Kluwer Academic Publishers, Boston.

Goldberg, D., 1989. *Genetic Algorithms in Search, Optimization, and Machine Learning*. Addison-Wesley, Reading, Massachusetts.

Goldfarb, D., 1970. A family of variable metric methods derived by variational means. *Math. Comput.*, Vol. 24, pp. 23–26.

Goldfarb, D., and A. Idnani, 1983. A numerically stable dual method for solving strictly convex quadratic programs. *Math. Prog.*, Vol. 27, pp. 1–33.

Goldstein, A. A., 1965. On steepest descent. *SIAM J. on Control*, Vol. 3, pp. 147–151.

Griewank, A., Juedes, D., and J. Utke, 1996. ADOL-C: A package for the automatic differentiation of algorithms written in C/C++. *ACM Transactions on Mathematical Software*, Vol. 22, No. 2, pp. 131–167.

Haftka, R. T., and M. P. Kamat, 1985. *Elements of Structural Optimization*. Martinus Nijhoff, Dordrecht.

Hajela, P., 1990. Genetic search – an approach to the nonconvex optimization problem. *AIAA Journal*, Vol. 28, No. 7, pp. 1205–1210.

Hamaker, H. C., and G., Hehenkamp, 1950. Minimum-cost transformers and chokes, Part 1. *Philips Research Report 5*, pp. 357–394.

Hamaker, H. C., and G., Hehenkamp, 1951. Minimum-cost transformers and chokes, Part 2. *Philips Research Report 6*, pp. 105–134.

Han, S. P., 1976. Superlinearly convergent variable metric algorithms for general nonlinear programming problems. *Math. Progr.*, Vol. 11, pp. 263–282.

Hancock, H., 1917. *Theory of Maxima and Minima.* Reprinted in 1960 by Dover, New York.

Hansen, P., Jaumard, B., and S. H., Lu, 1989a. Some further results on monotonicity in globally optimal design. *ASME J. of Mechanisms, Transm. and Automation in Design*, Vol. 111, No. 3, pp. 345–352.

Hansen, P., Jaumard, B., and S. H., Lu, 1989b. A framework for algorithms in globally optimal design. *ASME J. of Mechanisms, Transm. and Automation in Design*, Vol. 111, No. 3, pp. 353–360.

Hansen, P., Jaumard, B., and S. H., Lu, 1989c. An automated procedure for globally optimal design. *ASME J. of Mechanisms, Transm. and Automation in Design*, Vol. 111, No. 3, pp. 360–366.

Haug, E. J., and J. S. Arora, 1979. *Applied Optimal Design.* Wiley-Interscience, New York.

Hazelrigg, G. A., 1996. *Systems Engineering: An Approach to Information-Based Design.* Prentice-Hall, Upper Saddle River, New Jersey.

Hestenes, M. R., 1980. *Conjugate Direction Methods in Optimization.* Springer-Verlag, Heidelberg.

Heywood, J. B., 1980. Engine combustion modeling – an overview. In *Combustion Modeling in Reciprocating Engines* (J. N. Mattavi, and Amann C. A. eds.). Plenum Press, New York.

Hillier, F. S., and G. J. Lieberman, 1967. *Introduction to Operations Research.* Holden-Day, San Francisco.

Himmelblau, D. M., 1972. *Applied Nonlinear Programming.* McGraw-Hill, New York.

Himmelblau, D. M., (ed.), 1973. *Decomposition of Large-Scale Problems.* North-Holland, Amsterdam.

Hock, W., and K. Schittkowski, 1981. *Test Examples for Nonlinear Programming Codes.* Lecture Notes in Economics and Mathematical Systems, No. 187. Springer-Verlag, New York.

Hornbeck, R. W., 1975. *Numerical Methods.* Quantum Publishers, New York.

Horst, R., Pardalos, P. M., and N. V. Thoai, 1995. *Introduction to Global Optimization.* Kluwer Academic Publishers, Dordrecht.

Horst, R., and H. Tuy, 1990. *Global Optimization-Deterministic Approaches.* Springer-Verlag, Berlin.

Hsu, Y. L., 1993, Notes on Interpreting Monotonicity Analysis Using Karush-Kuhn-Tucker Conditions and MONO: A Logic Program for Monotonicity Analysis. *Advances in Design Automation 1993* (B. J. Gilmore, D. A. Hoeltzel, S. Azarm, and H. A. Eschenauer eds.), Vol. 2, ASME, New York, pp. 243–252.

Ignizio, J. P., 1976. *Goal Programming and Extensions.* Heath, Boston.

Jelen, F. C., 1970. *Cost and Optimization Engineering.* McGraw-Hill, New York.

Jiang, T., 1996. *Topology Optimization of Structural Systems Using Convex Approximation Methods.* Doctoral dissertation, Dept. of Mechanical Engineering, University of Michigan, Ann Arbor.

Johnson, R. C., 1961, 1980. *Optimum Design of Mechanical Elements.* Wiley-Interscience, New York.

Johnson, R. C., 1971. *Mechanical Design Synthesis with Optimization Applications.* Van Nostrad-Reinhold, New York.

Jones, C. V., 1996. *Visualization and Optimization.* Kluwer Academic Publishers, Norwell, Massachusetts.

Juvinall, R. C., 1983. *Fundamentals of Machine Component Design*. Wiley, New York.

Kamat, M. P. (ed.), 1993. *Structural Optimization: Status and Promise*. AIAA, Washington, DC.

Karush, W., 1939. *Minima of Functions of Several Variables with Inequalities as Side Conditions*. MS thesis, Dept. of Mathematics, University of Chicago, Chicago, Illinois.

Kirsch, U., 1981. *Optimum Structural Design*. McGraw-Hill, New York.

Kirsh, U., 1993. *Structural Optimization: Fundamentals and Applications*. Springer-Verlag, Berlin.

Klamkin, M. S. (ed.), 1987. *Mathematical Modelling: Classroom Notes in Applied Mathematics*. SIAM, Philadelphia.

Koenigsberger, F., 1965. *Design Principles of Metal Cutting Machine Tools*. Macmillan, New York.

Koski, J., 1985. Defectiveness in weighting method in multicriteria optimization of structures. *Communications in Applied Numerical Methods*, Vol. 1, pp. 333–337.

Kouvelis, P., and G. Yu, 1997. *Robust Discrete Optimization and Its Applications*. Kluwer Academic Publishers, Dordrecht.

Kuhn, H. W., and A. W. Tucker, 1951. Nonlinear programming. In *Proceedings of the Second Berkeley Symposium on Mathematical Statistics and Probability* (J. Neyman, ed.). University of California Press, Berkeley, California.

Lasdon, L. S., 1970. *Optimization Theory for Large Systems*. MacMillan, New York.

Lasdon, L. S., Warren, A. D., Jain, A., and M. Ratner, 1978. Design and testing of a generalized reduced gradient code for nonlinear programming. *ACM Trans. on Mathematical Software*. March 1978, pp. 35–50.

Law, A. M., and W. D. Kelton, 1991. *Simulation Modeling and Analysis*. McGraw-Hill, New York.

Leitman, G., 1962. *Optimization Techniques with Applications to Aerospace Systems*. Academic Press, New York.

Lev, O. E. (ed.), 1981. *Structural Optimization – Recent Developments and Applications*. ASCE, New York.

Levenberg, K., 1944. A method for the solution of certain nonlinear problems in least squares. *Quart. J. Appl. Math.*, Vol. 2, pp. 164–168.

Lewis, R. M., and V. J. Torczon, 1996. *Pattern Search Algorithms for Bound-Constrained Minimization*. ICASE Technical Report 96-20, NASA Langley Research Center, Hampton, Virginia.

Lewis, R. M., and V. J. Torczon, 1998. *A Globally Convergent Augmented Lagrangian Pattern Search Algorithm for Optimization with General Constraints and Simple Bounds*. ICASE Technical Report 98-31, NASA Langley Research Center, Hampton, Virginia.

LMS Optimus, 1998. *Training Manual*. Rev. 2.0, LMS International, Louvain, Belgium.

Lootsma, F. A., 1984. *Performance Evaluation of Nonlinear Optimization Methods via Pairwise Comparison and Fuzzy Numbers*. Report No. 84-40, Dept. of Mathematics and Informatics, Delft University of Technology, Delft.

Lootsma, F. A., 1985. Comparative performance evaluation, experimental design, and generation of test problems in nonlinear optimization. In Schittkowski (1985), pp. 249–260.

Lootsma, F. A., 1997. *Fuzzy Logic for Planning and Decision Making*. Kluwer Academic Publishers, Dordrecht.

Luenberger, D. G., 1973, 1984. *Introduction to Linear and Nonlinear Programming*. Addison-Wesley, Reading, Massachusetts.

Maciel, M. C., 1992. *A Global Convergence Theory for a General Class of Trust Region Algorithms for Equality Constrained Optimization*. Doctoral dissertation, Dept. of Computational and Applied Mathematics, Rice University, Houston.

Mangasarian, O. L., 1969. *Nonlinear Programming*. Krieger, New York.

Marquardt, D. W., 1963. An algorithm for least-squares estimation of nonlinear parameters. *J. SIAM*, Vol. 11, pp. 431–441.

Matlab, 1997. *Matlab: The Neural Net Toolbox*. Math Works, Natick, MA.

Mickle, M. J., and T. W. Sze, 1972. *Optimization in Systems Engineering*. International Textbook, Philadelphia.

Miele, A. (ed.), 1965. *Theory of Optimum Aerodynamic Shapes*. Academic Press, New York.

Mistree, F., Smith, W. F., and B. A. Bras, 1993. A decision-based approach to concurrent engineering. In *Handbook of Concurrent Engineering*, Chapman & Hall, New York.

Moré, J., and S. J. Wright, 1993. *Optimization Software Guide*. SIAM, Philadelphia.

Morris, A. J. (ed.), 1982. *Foundations of Structural Optimization: A Unified Approach*. Wiley, Chichester.

Mulvey, J. M. (ed.), 1981. *Evaluating Mathematical Programming Techniques*. Lecture Notes in Economics and Mathematical Systems Vol. 199, Springer-Verlag, Berlin.

Murty, K. G., 1983. *Linear Programming*. Wiley, New York.

Murty, K. G., 1986. *Linear Complementarity, Linear and Nonlinear Programming*. Heldermann Verlag, West Berlin.

Nelson, S. A., II, 1997. *Optimal Hierarchical System Design via Sequentially Decomposed Programming*. Doctoral dissertation, Dept. of Mechanical Engineering, University of Michigan, Ann Arbor.

Nemhauser, G. L., and L. A. Wolsey, 1988. *Integer and Combinatorial Optimization*. Wiley-Interscience, New York.

Noble, B., 1969. *Applied Linear Algebra*. Prentice-Hall, Englewood Cliffs, New Jersey.

Nwosu, N., 1998. *Object-Oriented Optimization Using Convex Approximations*. MS thesis, Dept. of Mechanical Engineeering, University of Michigan, Ann Arbor.

Olhoff, N., and G. I. N. Rozvany (eds.), 1995. *Proceedings of the First World Congress on Structural and Multidisciplinary Optimization*. Pergamon-Elsevier, Oxford.

Ortega, J. M., and W. C. Rheinboldt, 1970. *Iterative Solution of Nonlinear Equations in Several Variables*. Academic Press, New York.

Osyczka, A., 1984, *Multicriteria Optimization in Engineering*. Wiley, New York.

Pakala, R., 1994. *A Study on Applications of Stackelberg Game Strategies in Concurrent Design Models*. Doctoral dissertation, University of Houston.

Papalambros, P., 1979. *Monotonicity Analysis in Engineering Design Optimization*. PhD dissertation, Design Division, Mechanical Engineering, Stanford University, Stanford, California.

Papalambros, P., 1988. Codification of semiheuristic global processing of optimal design models. *Engineering Optimization*, Vol. 13, pp. 464–471.

Papalambros, P. Y., 1994. Model reduction and verification techniques. In *Advances in Design Optimization* (H. Adeli, ed.). Chapman and Hall, London.

Pareto, V., 1971, *Manuale di Economia Politica*, Societa Editrice Libraria, Milano, Italy, 1906. Translated into English by A. S. Schwier as *Manual of Political Economy*. Macmillan, New York.

Pflug, G. Ch., 1996. *Optimization of Stochastic Models: The Interface between Simulation and Optimization.* Kluwer, Boston.

Pierre, D. A., and M. J. Lowe, 1975. *Mathematical Programming via Augmented Lagrangians.* Addison-Wesley, Reading, Massachusetts.

Pintér, J. D., 1996. *Global Optimization in Action-Continuous and Lipschitz Optimization: Algorithms, Implementations and Applications.* Kluwer Academic Publishers, Dordrecht.

Pomrehn, L., 1993. *A Recursive Opportunistic Optimization Tool for Discrete Optimal Design.* Doctoral dissertation, Dept. of Mechanical Engineering, University of Michigan, Ann Arbor.

Powell, M. J. D., 1978a. A fast algorithm for nonlinearly constrained optimization calculations. In *Numerical Analysis, Dundee 1977* (G. A. Watson, ed.), Lecture Notes in Mathematics 630, Springer-Verlag, Berlin.

Powell, M. J. D., 1978b. Algorithms for nonlinear constraints that use Lagrangian functions. *Mathematical Programming*, Vol. 14, pp. 224–248.

Powell, M. J. D. (ed.), 1982. *Nonlinear Optimization 1981.* Academic Press, New York.

Protter, M. H., and C. B. Morey, Jr., 1964. *Modern Mathematical Analysis.* Addison-Wesley, Reading, Massachusetts.

Rabinowitz, P. (ed.), 1970. *Numerical Methods for Nonlinear Algebraic Equations.* Gordon and Breach, London.

Ragsdell, K. M., and D. T. Phillips, 1976. Optimal design of a class of welded structures using geometric programming. *Trans. ASME J. of Engi. for Industry*, Vol. 98, Series B, No. 3, pp. 1021–1025.

Rao, S. S., 1978. *Optimization – Theory and Applications.* Wiley Eastern, New Delhi.

Rao, S. S., and S. K. Hati, 1979. Game theory approach in multicriteria optimization of function generating mechanisms. *ASME Journal of Mechanical Design*, Vol. 101, pp. 398–406.

Rao, J. J. R., and P. Papalambros, 1991. PRIMA: a production-based implicit elimination system for monotonicity analysis of optimal design models. *Trans. ASME J. of Mechanical Design*, Vol. 113, No. 4, pp. 408–415.

Ratschek, H., and J. Rokne, 1988. *New Computer Methods for Global Optimization.* Ellis Horwood, Chichester.

Reklaitis, G. V., Ravindran, A., and K. M. Ragsdell, 1983. *Engineering Optimization – Methods and Applications.* Wiley-Interscience, New York.

Rice, J. R., 1981. *Matrix Computations and Mathematical Software.* McGraw-Hill, New York.

Rosen, J. B., 1960. The gradient projection method for nonlinear programming, Part I: Linear constraints. *J. SIAM*, Vol. 8, pp. 181–217.

Rosen, J. B., 1961. The gradient projection method for nonlinear programming, Part II: Nonlinear constraints. *J. SIAM*, Vol. 9, 514–532.

Roubens, M., and P. Vincke, 1985. *Preference Modeling.* Springer-Verlag, Berlin.

Roy, B., 1996. *Multicriteria Methodology for Decision Aiding.* Kluwer Academic Publishers, Dordrecht.

Rubinstein, M. F., 1975. *Patterns of Problem Solving.* Prentice-Hall, Englewood Cliffs, New Jersey.

Rudd, D. F., Powers, G. J., and J. J. Siirola, 1973. *Process Synthesis.* Prentice-Hall, Englewood Cliffs, New Jersey.

Russell, D., 1970. *Optimization Theory.* Benjamin, New York.

Saaty, T. L., 1980. *The Analytic Hierarchy Process – Planning, Priority Setting, Resource Allocation*. McGraw-Hill, New York.

Saaty, T. L., and J. M. Alexander, 1981. *Thinking with Models*. Pergamon Press, Oxford.

Sacks, J., Welch, W., Mitchell, T. J., and H. P. Wynn, 1989. Design and Analysis of Computer Experiments. *Statistical Science*, Vol. 4, No. 4, pp. 409–435.

Sasena, M. J., 1998. *Optimization of Computer Simulations via Smoothing Splines and Kriging Metamodels*. MS thesis, Dept. of Mechanical Engineering, University of Michigan, Ann Arbor.

Scales, L. E., 1985. *Introduction to Nonlinear Optimization*. Springer-Verlag, New York.

Schinzinger, R., 1965. Optimization in electromagnetic system design. In *Recent Advances in Optimization Techniques* (Lavi, A., and T. P. Vogl eds.). Wiley, New York, pp. 163–213.

Schittkowski, K., 1980. *Nonlinear Programming Codes*. Lecture Notes in Economics and Mathematical Systems, No. 180. Springer-Verlag, Berlin.

Schittkowski, K., 1981. The nonlinear programming method of Wilson, Han and Powell with an augmented Lagrangian type line search function. Part 1: Convergency analysis. *Numerische Mathematic*, Vol. 38, pp. 83–114; Part 2: An efficient implementation with linear least squares subproblems, op. cit., pp. 115–127.

Schittkowski, K., 1983. On the convergence of a sequential quadratic programming method with an augmented Lagrangian line search function. *Optimization, Math. Operationsforschung und Statistik*, Vol. 14, pp. 197–216.

Schittkowski, K., 1984. *User's Guide for the Nonlinear Programming Code NLPQL*. Institut fur Informatik, University of Stuttgart, Stuttgart.

Schittkowski, K. (ed.), 1985. *Computational Mathematical Programming*. NATO ASI Series F, Vol. 15. Springer-Verlag, Berlin.

Schittkowski, K., Zillober, C., and R. Zotemantel, 1994. Numerical comparison of nonlinear programming algorithms for structural optimization. *Strutural Optimization*, Vol. 7, No. 1, pp. 1–28.

Schmidt, L. C., and J. Cagan, 1998. Optimal configuration design: an integrated approach using grammars. *ASME Journal of Mechanical Design*, Vol. 120, No. 1, pp. 2–9.

Shanno, D. F., 1970. Conditioning of quasi-Newton methods for function minimization. *Math. Comput.*, Vol. 24, pp. 647–656.

Shanno, D. F., and K-H Phua, 1978. Matrix conditioning and nonlinear optimization. *Math. Prog.*, Vol. 14, pp. 149–160.

Shannon, R. E., 1975. *Systems Simulation: The Art and Science*. Prentice-Hall, Englewood Cliffs, New Jersey.

Shigley, J. E., 1977. *Mechanical Engineering Design*, 3d ed. McGraw-Hill, New York.

Siddall, J. N., 1972. *Analytical Decision-Making in Engineering Design*. Prentice-Hall, Englewood Cliffs, New Jersey.

Siddall, J. N., 1982. *Optimal Engineering Design*. Marcel Dekker, New York.

Siddall, J. N., 1983. *Probabilistic Engineering Design*. Marcel Dekker, New York.

Smith, M., 1993. *Neural Nets for Statistical Modeling*. Van Nostrand Reinbolt, New York.

Spiegel, M. R., 1968. *Mathematical Handbook of Formulas and Tables*. Schaum's Outline Series, McGraw-Hill, New York.

Spunt, L., 1971. *Optimum Structural Design*. Prentice-Hall, Englewood Cliffs, New Jersey.

Stadler, W., 1979. A survey of multicriteria optimization or the vector maximum problem, part I: 1776-1960. *J. of Optimization Theory and Application*, Vol. 29, No. 1.

Stark, R. M., and R. L. Nichols, 1972. *Mathematical Foundations for Design: Civil Engineering Systems*. McGraw-Hill, New York.

Statnikov, R. B., and J. B. Matusov, 1995. *Multicriteria Optimization and Engineering*. Chapman and Hall, New York.

Stoecker, W. F., 1971, 1989. *Design of Thermal Systems*. McGraw-Hill, New York.

Svanberg, K., 1987. The method of moving asymptotes – A new method for structural optimization. *Int. J. Num. Meth. Eng.*, Vol. 24, pp. 359–373.

Taylor, C. F., 1985. *The Internal Combustion Engine in Theory and Practice*, 2nd. ed. MIT Press, Cambridge.

Torczon, V. J., 1997. On the convergence of pattern search algorithms. *SIAM J. of Optimization*, Vol. 7, pp. 1–25.

Torczon, V. J., and M. W. Trosset, 1998. From evolutionary operation to parallel direct search: Pattern search algorithms for numerical optimization. *Computing Science and Statistics*, Vol. 29, pp. 396–401.

Ullman, D. G., 1992. *The Mechanical Design Process*. McGraw-Hill, New York.

Unklesbay, K., Staats, G. E., and D. L. Creighton, 1972. Optimal design of pressure vessels. *ASME Paper No. 72-PVP-2*.

Vanderplaats, G. N., 1984. *Numerical Optimization Techniques for Engineering Design*. McGraw-Hill, New York. Third edition (with software), Vanderplaats Research and Development, Colorado Springs, 1999.

Vincent, T. L., 1983. Game theory as a design tool. *J. of Mechanisms, Transmissions, and Automation in Design*, Vol. 105, pp. 165–170.

Vincent, T. L., and W. J. Grantham, 1981. *Optimality in Parametric Systems*. Wiley Interscience, New York.

von Neumann, J., and O. Morgenstern, 1947. *Theory of Games and Economic Behavior*, 2nd ed. Princeton University Press, Princeton, New Jersey.

Wagner, T. C., 1993. *A General Decomposition Methodology for Optimal System Design*. Doctoral dissertation, University of Michigan, Ann Arbor.

Wainwright, S. A., Biggs, W. D., Currey, J. D., and J. M. Gosline, 1982. *Mathematical Design in Organisms*. Princeton University Press, Princeton, New Jersey.

Walton, J. W., 1991. *Engineering Design: From Art to Practice*. West Publishing, St. Paul, Minnesota.

Wellstead, P. E., 1979. *Introduction to Physical System Modeling*. Academic Press, London.

Wilde, D. J., 1975. Monotonicity and dominance in optimal hydraulic cylinder design. *Trans. ASME J. of Eng. for Industry*, Vol. 94, No. 4, pp. 1390–1394.

Wilde, D. J., 1978. *Globally Optimal Design*. Wiley-Interscience, New York.

Wilde, D. J., and C. S. Beightler, 1967. *Foundations of Optimization*. Prentice-Hall, Englewood Cliffs, New Jersey. Second Edition: Beightler, C. S., Phillips, D. T., and D. J. Wilde, 1979.

Williams, H. P., 1978. *Model Building in Mathematical Programming*. Wiley-Interscience, Chichester, England.

Wismer, D. A. (ed.), 1971. *Optimization Methods for Large-Scale Systems – With Applications*. McGraw-Hill, New York.

Wolfe, P., 1963. Methods of nonlinear programming. In *Recent Advances in Mathematical Programming* (Graves, R. L., and P. Wolfe, eds.). McGraw-Hill, New York.

Yuan, Y., 1995. On the convergence of a new trust region algorithm. *Numerische Mathematik*, Vol. 70, pp. 515–539.

Zangwill, W. I., 1969. *Nonlinear Programming, A Unified Approach.* Prentice-Hall, Englewood Cliffs, New Jersey.

Zener, C., 1971. *Engineering Design by Geometric Programming.* Wiley-Interscience, New York.

Zhou, J. L., and A. L. Tits, 1996. An SQP algorithm for finely discretized continuous minimax problems and other minimax problems with many objective functions. *SIAM J. on Optimization*, Vol. 6, No. 2, pp. 461–487.

Zoutendijk, G., 1960. *Methods of Feasible Directions.* Elsevier, Amsterdam.

Author Index

Subject Index

Made in the USA
Lexington, KY
05 February 2012